Supported by Key Project for the Development of State Facilities and Information Infrastructure for Science and Technology: National Specimen Information Infrastructure (2005DKA21401)
国家科技基础条件平台：标本资源共享平台建设项目（2005DKA21401）资助

Type Specimens in China National Herbarium (PE)

Volume 4

ANGIOSPERMAE (1)

中国国家植物标本馆（PE）模式标本集

第4卷

被子植物门（1）

Institute of Botany, Chinese Academy of Sciences Edit
The editor of this volume LIN Qi

中国科学院植物研究所 编
本卷主编 林 祁

河南科学技术出版社
Henan Science and Technology Press

图书在版编目(CIP)数据

中国国家植物标本馆模式标本集. 第4卷:中国科学院植物研究所编;林祁分册主编. —郑州:河南科学技术出版社,2015.10
ISBN 978-7-5349-7502-8

Ⅰ.①中… Ⅱ.①中…②林… Ⅲ.①植物—标本—中国—图集 Ⅳ.①Q94-34

中国版本图书馆CIP数据核字(2014)第297920号

The editor of this volume LIN Qi

Edited by BAN Qin, CHEN Shurong, DU Yufen, LI Qianyun, LIN Qi, LIN Yun, MA Xintang, SHI Qingchun, SUN Qian, TIAN Xiya, WANG Zhongtao, WU Tingting, YANG Zhirong

(Names are arranged in alphabetical order)

主　编　林　祁

编著者(以姓氏拼音为序)

班　勤　陈淑荣　杜玉芬　李倩云　林　祁　林　云
马欣堂　石青春　孙　茜　田希娅　王忠涛　吴婷婷
杨志荣

生物中国总策划:周本庆

出版发行:河南科学技术出版社
地址:郑州市经五路66号　邮编:450002
电话:(0371)65737028　65788613
网址:www.hnstp.cn
策划编辑:周本庆　杨秀芳
责任编辑:陈　艳
责任校对:柯　姣
整体设计:张　伟
责任印制:张艳芳
印　　刷:北京盛通印刷股份有限公司
经　　销:全国新华书店
幅面尺寸:240mm×345mm　印张:66　字数:180千字
版　　次:2015年10月第1版　2015年10月第1次印刷
定　　价:1200.00元

Introduction

Predecessors of China National Herbarium (abbreviated code PE), Institute of Botany, Chinese Academy of Sciences are the Herbarium, Department of Botany, the Fan Memorial Institute of Biology, Peiping (1928), as well as the Herbarium, Institute of Botany, the National Academy of Peiping (1929). Now PE has developed into the largest herbarium in Asia. The current collections contain more than 2,650,000 specimens in the herbarium, including mosses, ferns, seed plants, seed collections and plant fossil samples. Among them, there are about 20,000 type specimens (holotype, isotype, lectotype, isolectotype, neotype, isoneotype, epitype, isoepitype, paratype, isoparatype, syntype, isosyntype).

Type specimens in this book were produced by selecting the most important type specimen deposited at PE under the same scientific name (species, subspecies, variety and form), and then they were also reviewed and scanned. After compilation, ***Type Specimens in China National Herbarium (PE)*** which consists of 12 volumes is completed. The taxa are arranged by family according to the system of ***Flora Bryophytorum Sinicorum*** and ***Flora Reipublicae Popularis Sinicae***. Infra-family taxa are alphabetized by genera, species, subspecies, varieties and forms. The explanation of each taxon is listed in the figure caption with Chinese name, scientific name, original publication, nature of specimen (holotype/ isotype/ lectotype/ isolectotype/ neotype/ isoneotype/ epitype/ isoepitype/ paratype/ isoparatype/ syntype/ isosyntype), type locality (country/ province/ county/ mountain if present), altitude, collection date, collector and collection number. The collector and type locality in this book follow ***Index Herbariorum Sinicorum*** (L. K. Fu, 1993) and ***Gazetteer of China—An Index to the Atlas of the People's Republic of China*** (Chinese Academy of Surveying & Mapping, 1997) respectively.

This book is a very important work for researching and identifying Chinese plants. It could also be used as a reference by plant taxonomists and people from botanic research institutions, educational institutions and production departments at home and abroad.

Volume 4 of ***Type Specimens in China National Herbarium (PE)*** includes 489 type specimens from Sparganiaceae to Liliaceae (1) (Angiospermae, Monocotyledoneae), comprising 267 holotypes, 185 isotypes, 9 lectotypes, 1 syntype, 1 isosyntype, 14 paratypes, 12 isoparatypes, and belonging to 13 families, 94 genera, 411 species, 4 subspecies, 72 varieties and 2 forms.

Greatest thanks to Prof. Li Liangqian and Prof. Zhang Xianchun, curators of China National Herbarium (PE), for their support and help throughtout the publication of the book.

前　言

中国科学院植物研究所国家植物标本馆（缩写代号 PE）的前身是 1928 年成立的北平静生生物调查所植物标本室和 1929 年成立的北平研究院植物研究所标本室，PE 现已发展为亚洲最大的植物标本馆，目前馆藏植物标本 265 万余份，包括苔藓植物标本、蕨类植物标本、种子植物标本、种子标本和植物化石标本，其中模式标本近 2 万份（含主模式、等模式、后选模式、等后选模式、新模式、等新模式、附加模式、等附加模式、副模式、等副模式、合模式、等合模式）。

书中所收录的模式标本是在同一学名下（种、亚种、变种、变型）遴选出一份最重要的馆藏模式标本，经整理并扫描后编撰而成《中国国家植物标本馆（PE）模式标本集》（共 12 卷）。全书各科依据《中国苔藓志》及《中国植物志》系统排列，属、种、亚种、变种、变型的名称按字母顺序排列。每张扫描模式标本相片的图注解释均标注中名、学名、原始文献、模式类型（主模式、等模式、后选模式、等后选模式、新模式、等新模式、附加模式、等附加模式、副模式、等副模式、合模式、等合模式）、采集地点（国名、省名、县名、山名）、海拔、采集时间（年月日）、采集人和采集号。本书中的采集人根据《中国植物标本馆索引》(傅立国，1993) 书写，采集地根据《中国地名录——中华人民共和国地图集地名索引》（国家测绘局地名研究所，1997）书写。

本书是一部研究与鉴定中国植物的重要著作，可供国内外植物分类学者及有关植物学科研、教学和生产部门人员参考。

第 4 卷包括被子植物门单子叶植物纲黑三棱科至百合科（1）的模式标本，共 489 份，含 267 份主模式、185 份等模式、9 份后选模式、1 份合模式、1 份等合模式、14 份副模式、12 份等副模式，隶属于 13 科、94 属、411 种、4 亚种、72 变种和 2 变型。

感谢中国国家植物标本馆（PE）馆长李良千研究员和张宪春研究员在本书编撰过程中给予的支持和帮助。

林　祁

2014 年 2 月

Contents
目　录

Sparganiaceae

黑三棱科

heisanlengke

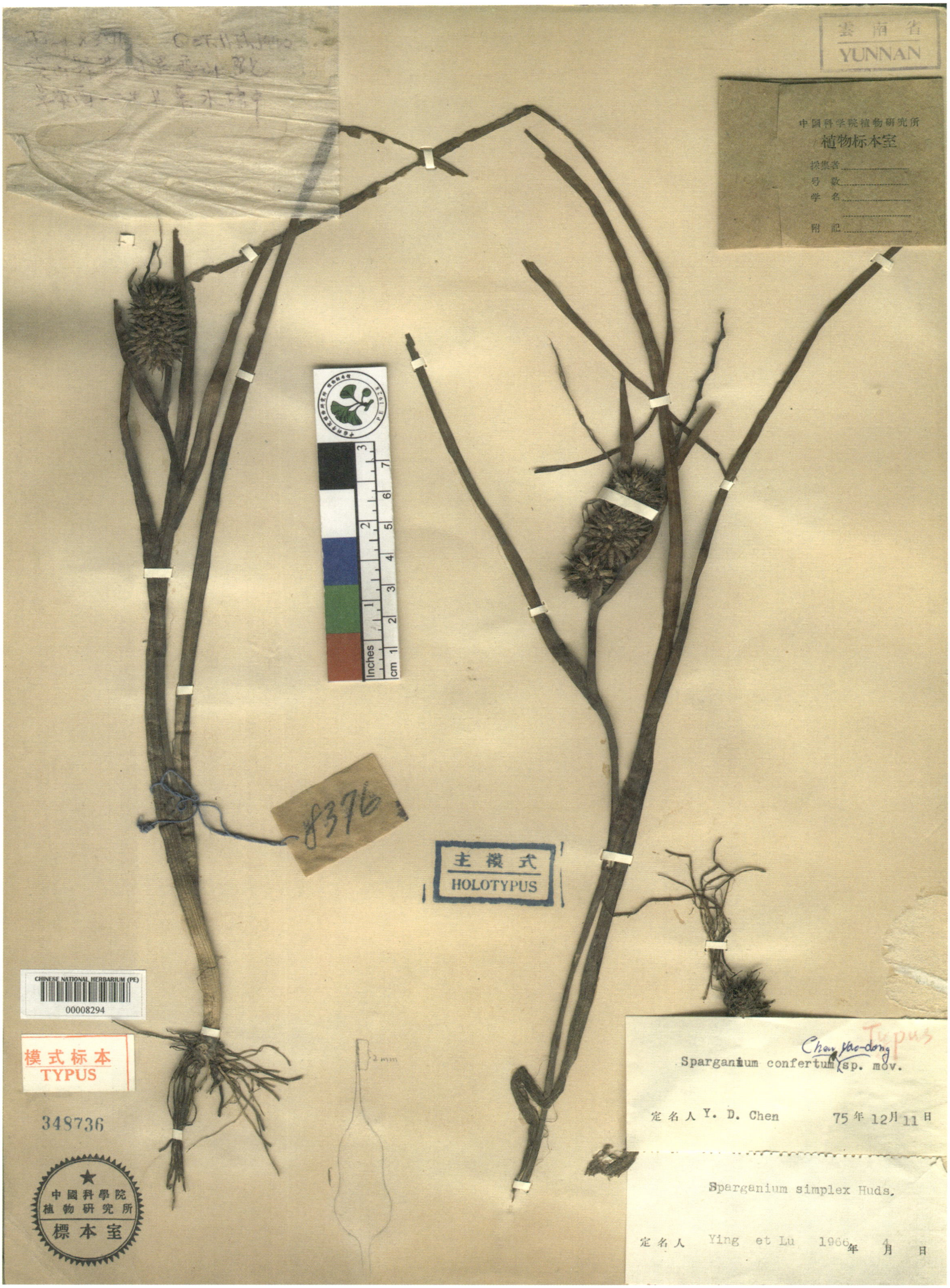

穗状黑三棱 ***Sparganium confertum*** Y. D. Chen in Acta Phytotax. Sin. 19(1): 53, pl. 3: 2. 1981. **Holotype:** China. Yunnan: Gongshan, alt. 3100 m, 1940-10-11, K. M. Feng 8376.

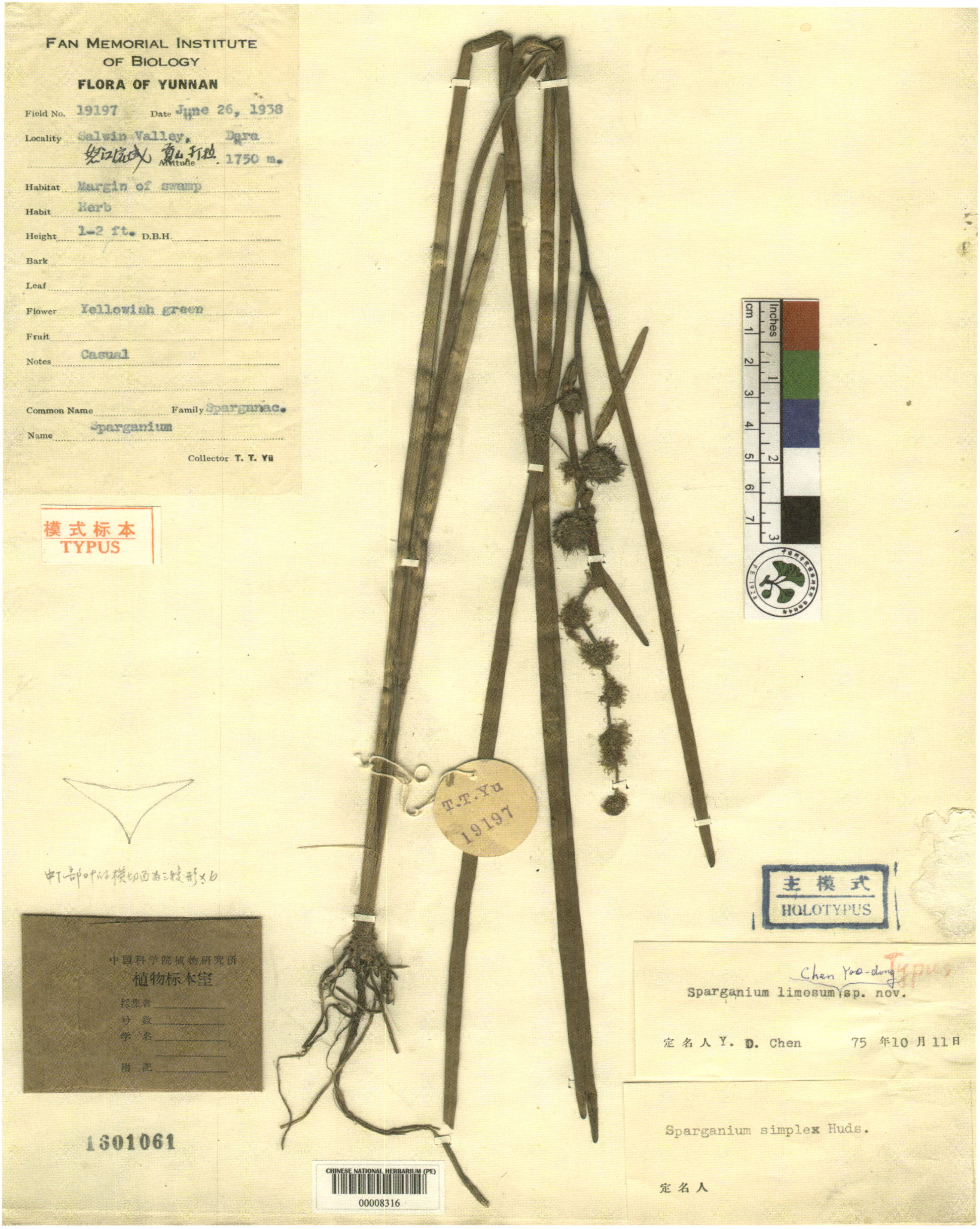

沼生黑三棱 ***Sparganium limosum*** Y. D. Chen in Acta Phytotax. Sin. 19(1): 52, pl. 3: 1. 1981. **Holotype:** China. Yunnan: Gongshan, alt. 1750 m, 1938-06-26, T. T. Yu 19197.

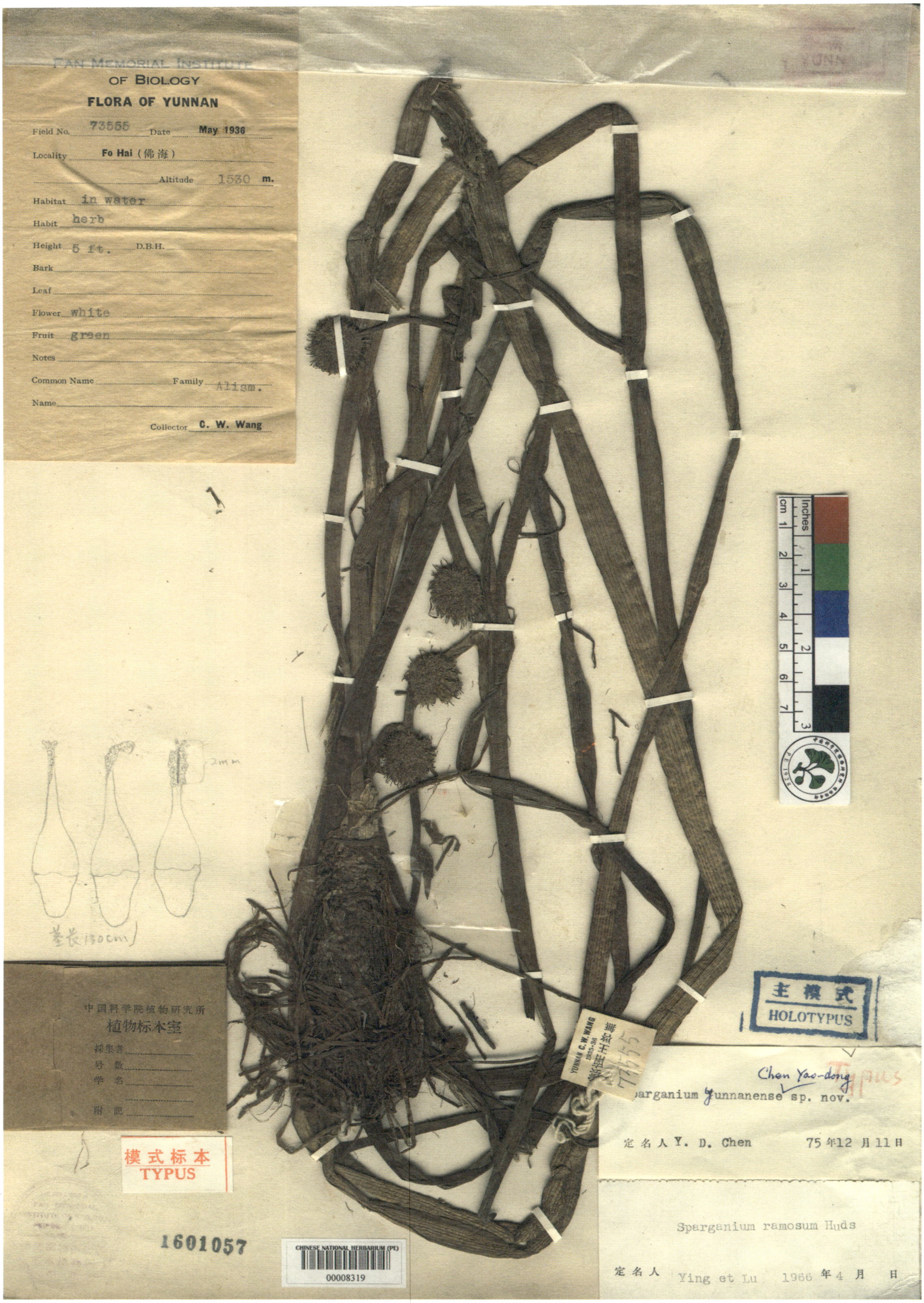

云南黑三棱 ***Sparganium yunnanense*** Y. D. Chen in Acta Phytotax. Sin. 19(1): 54, pl. 3: 3. 1981.**Holotype:** China. Yunnan: Fohai (= Menghai), alt. 1530 m, 1936-05-??, C. W. Wang 73555..

Potamogetonaceae

眼子菜科

yanzicaike

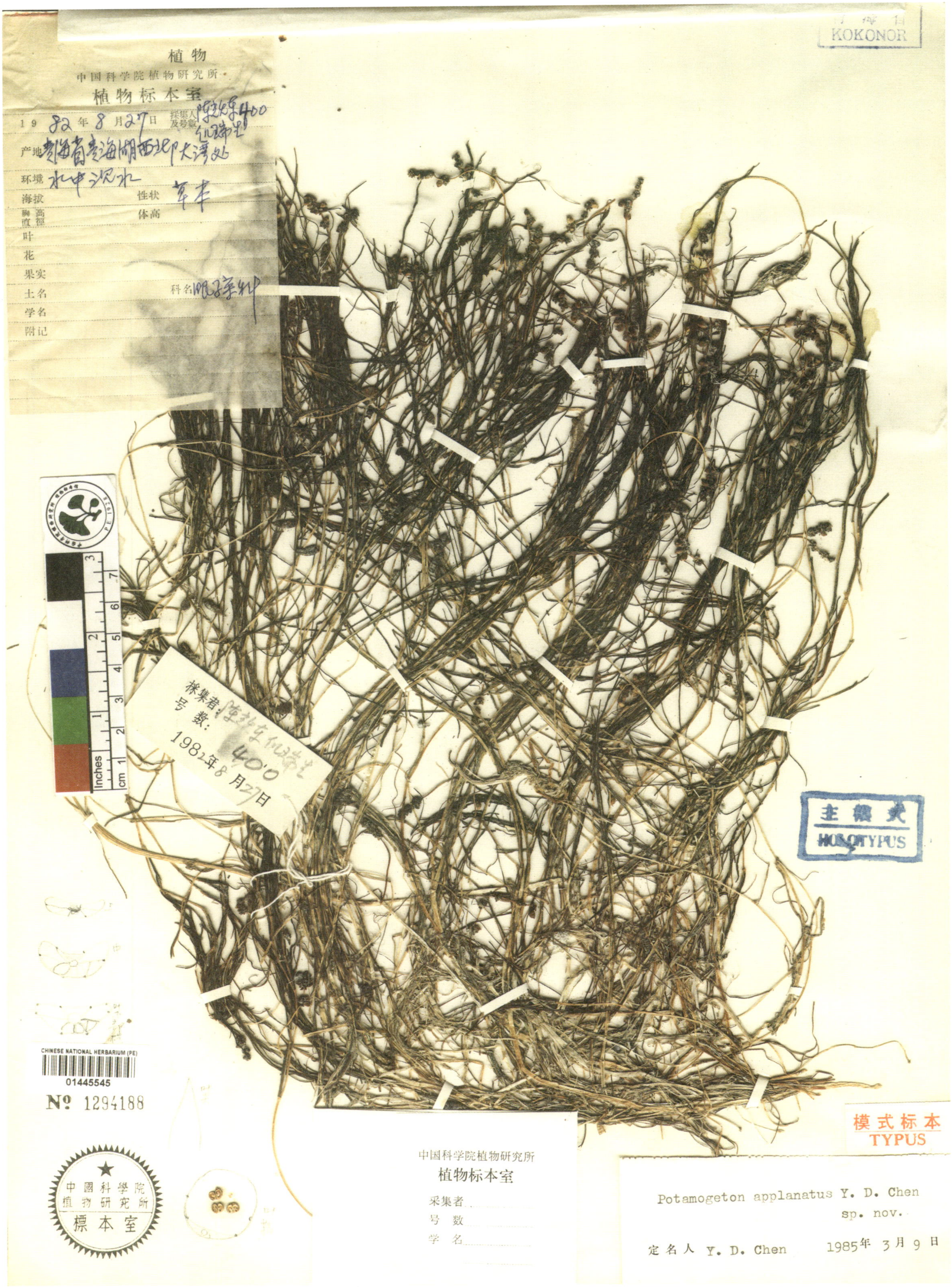

扁茎眼子菜 ***Potamogeton applanatus*** Y. D. Chen in Acta Hydrobiol. Sin. 11(3): 234, f. 1: 3. 1987. **Holotype:** China. Qinghai: Qinghai Lake, alt. 3300 m, 1982-08-27, Y. D. Chen & R. S. Ni 400.

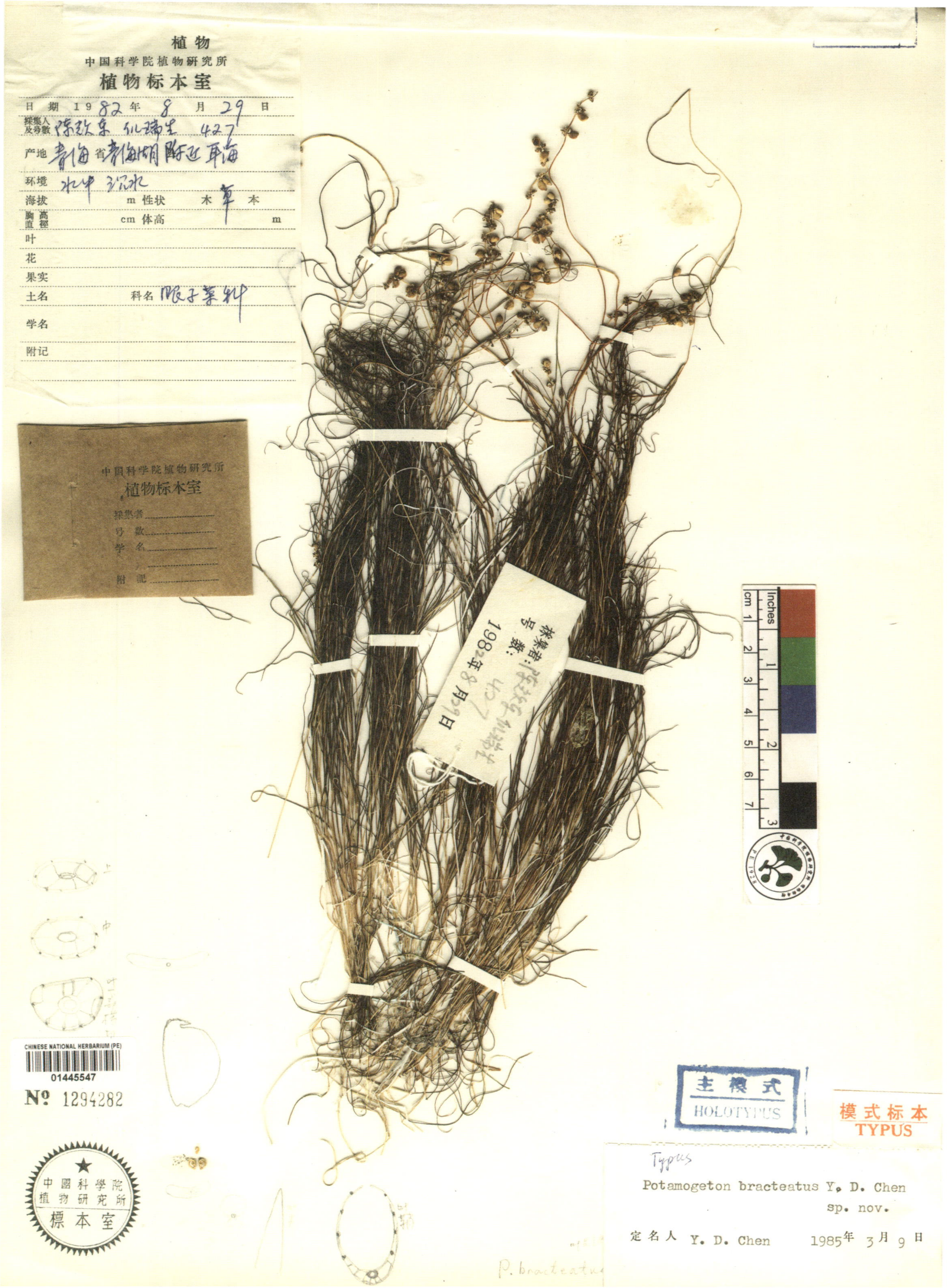

具苞眼子菜 ***Potamogeton bracteatus*** Y. D. Chen in Acta Hydrobiol. Sin. 11(3): 234, f. 1: 4. 1987. **Holotype:** China. Qinghai: Qinghai Lake, alt. 3300 m, 1982-08-29, Y. D. Chen & R. S. Ni 427.

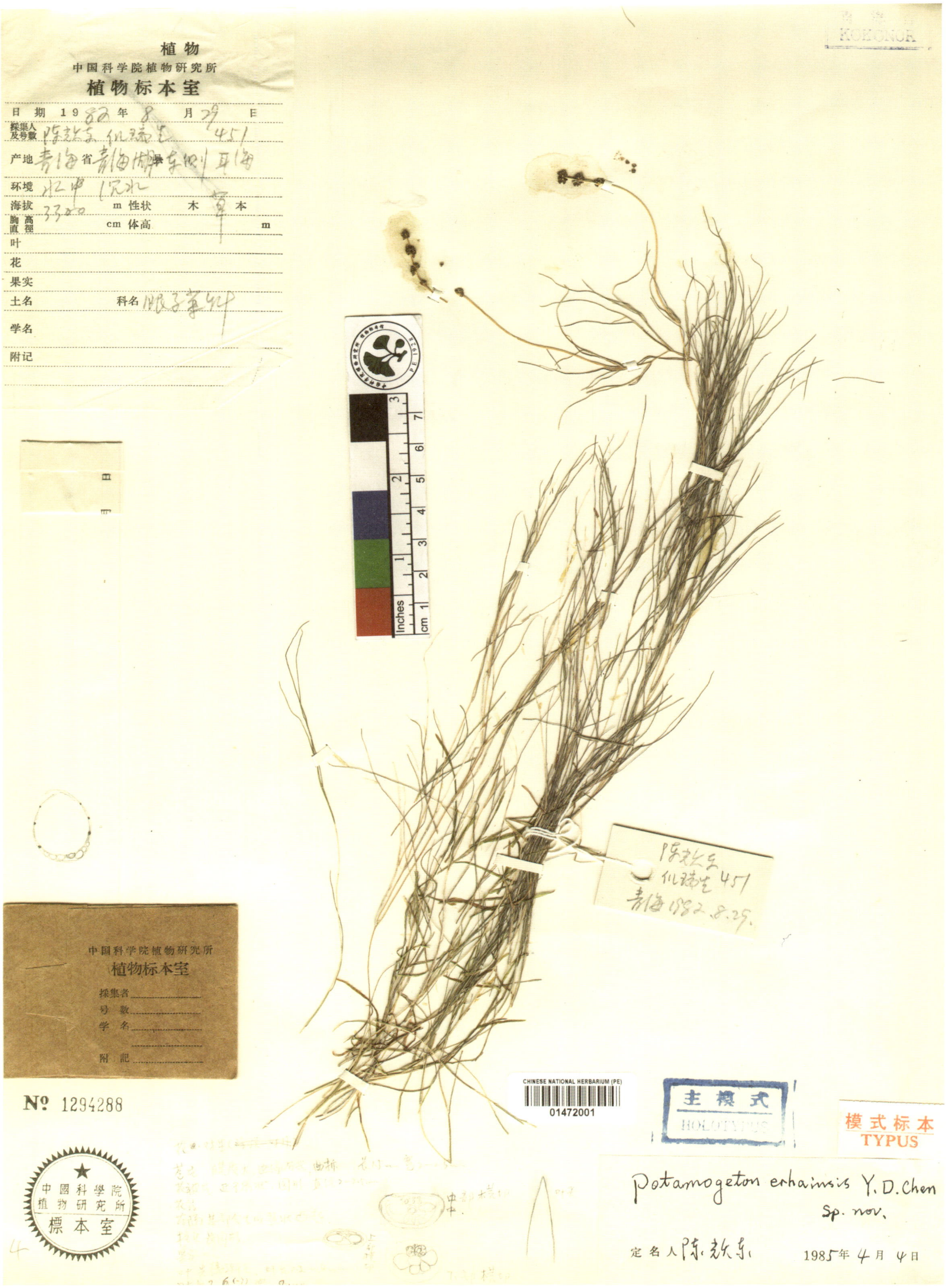

耳海眼子菜 ***Potamogeton erhaiensis*** Y. D. Chen in Acta Hydrobiol. Sin. 11(3): 234, f. 1: 2. 1987. **Holotype:** China. Qinghai: Erhai Lake, alt. 3300 m, 1982-08-29, Y. D. Chen & R. S. Ni 451.

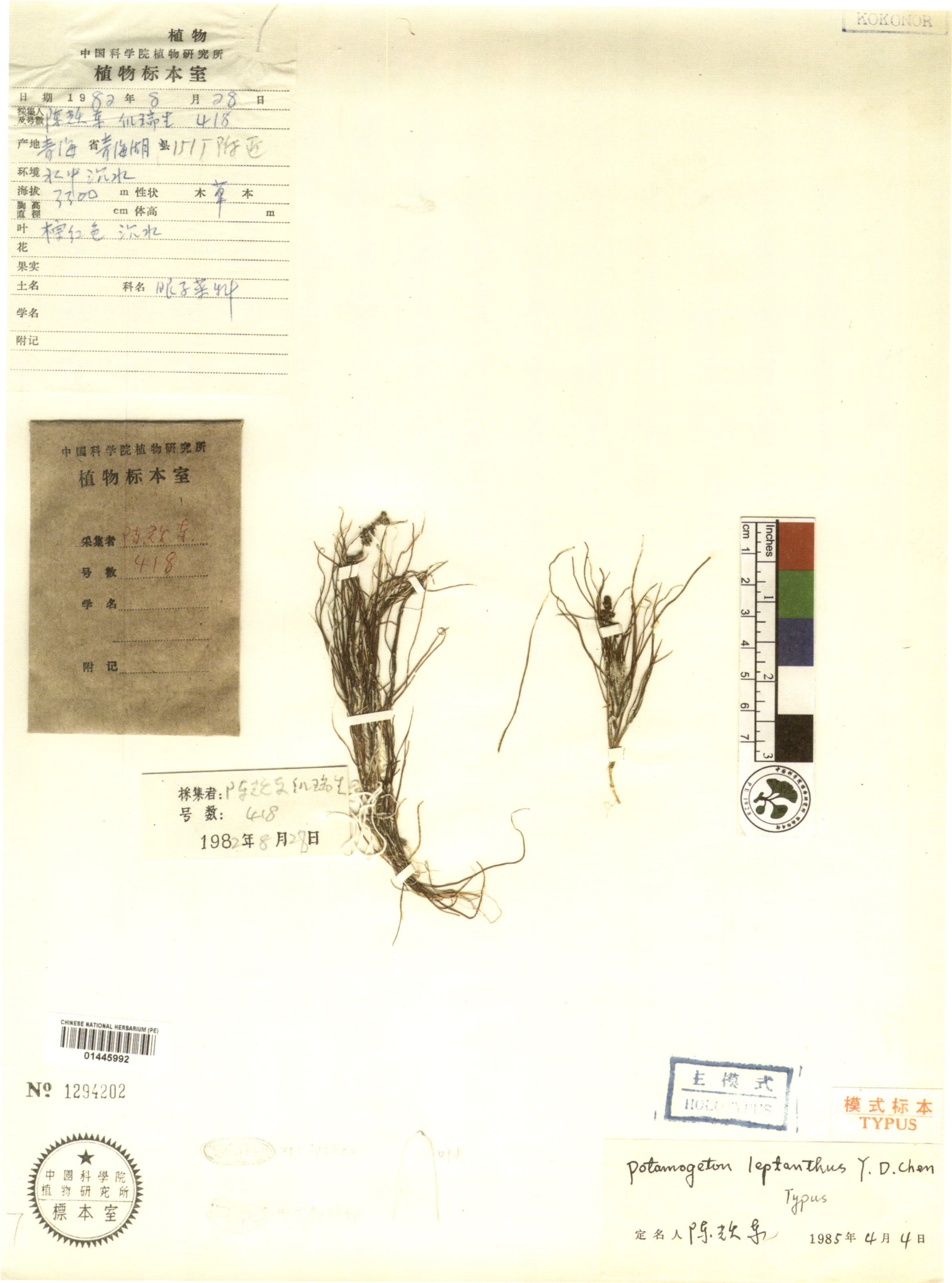

柔花眼子菜 ***Potamogeton leptanthus*** Y. D. Chen in Acta Hydrobiol. Sin. 11(3): 235, f. 2: 1. 1987. **Holotype:** China. Qinghai: Qinghai Lake, alt. 3300 m, 1982-08-28, Y. D. Chen & R. S. Ni 418.

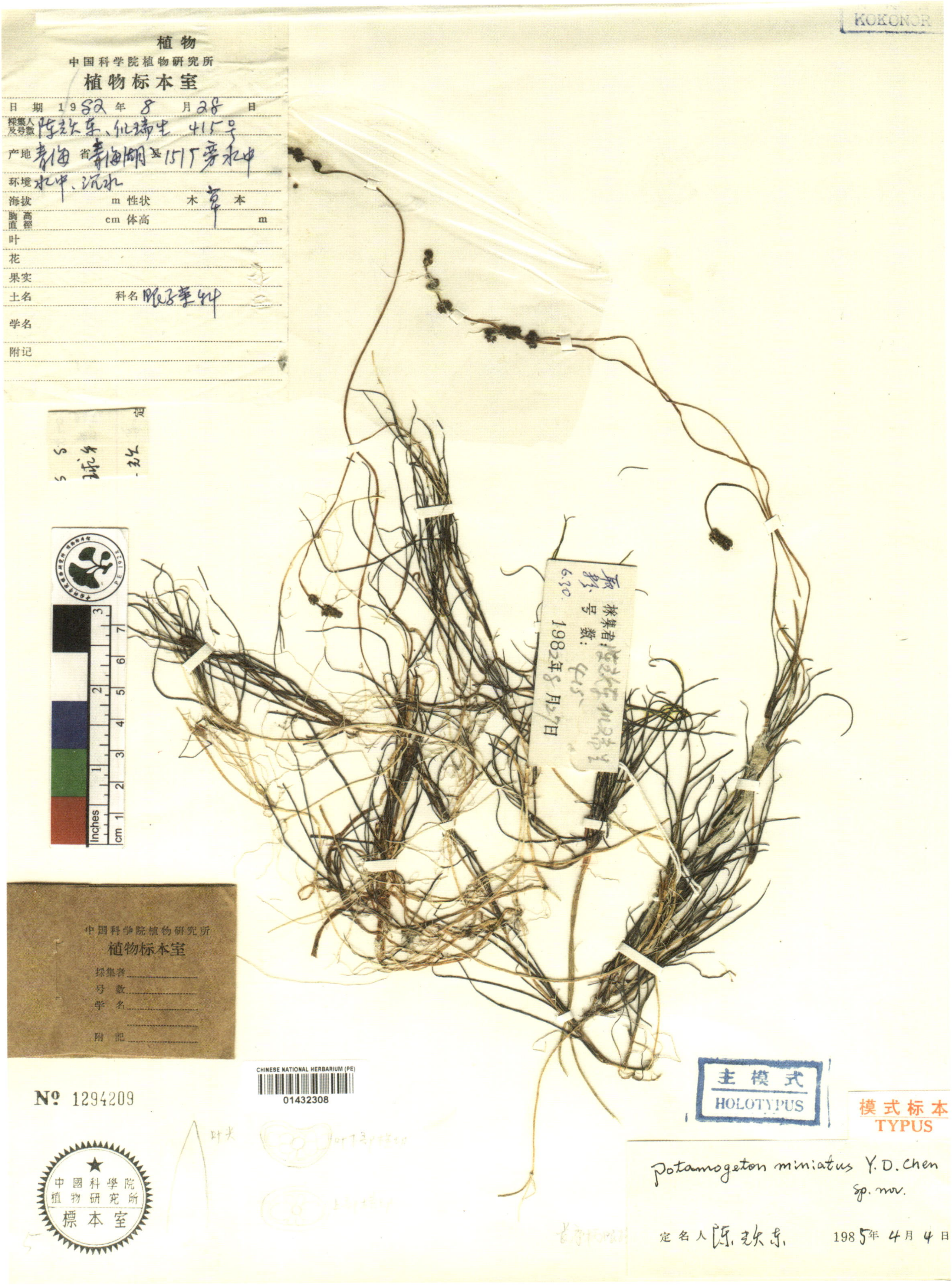

红梗眼子菜 ***Potamogeton miniatus*** Y. D. Chen in Acta Hydrobiol. Sin. 11(3): 234, f. 1: 1. 1987. **Holotype:** China. Qinghai: Qinghai Lake, alt. 3300 m, 1982-08-28, Y. D. Chen & R. S. Ni 415.

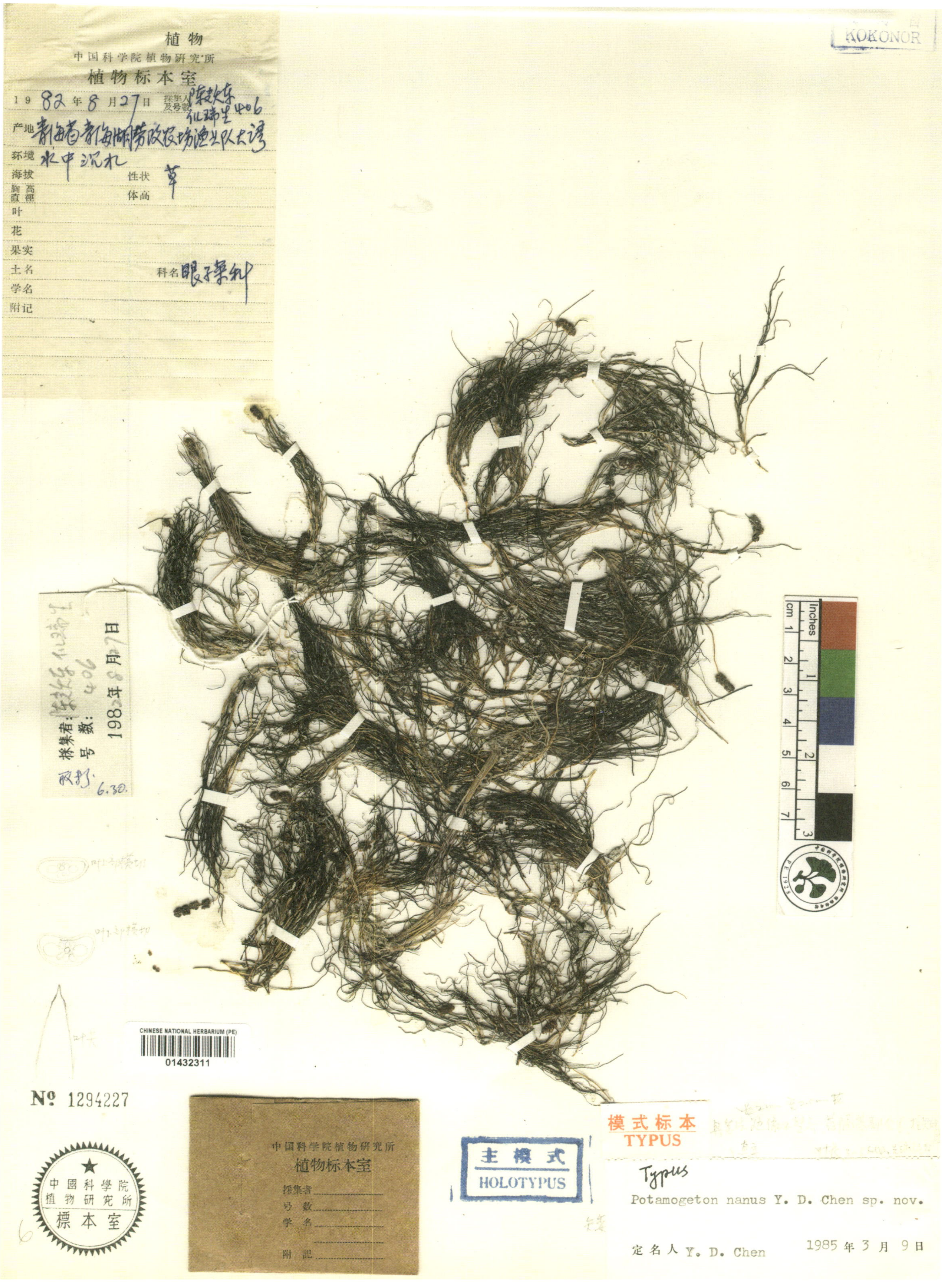

矮眼子菜 ***Potamogeton nanus*** Y. D. Chen in Acta Hydrobiol. Sin. 11(3): 235, f. 2: 2. 1987. **Holotype:** China. Qinghai: Qinghai Lake, alt. 3300 m, 1982-08-27, Y. D. Chen & R. S. Ni 406.

青海角果藻 ***Zannichellia qinghaiensis*** Y. D. Chen in Acta Hydrobiol. Sin. 11(3): 235, f. 2: 3. 1987. **Holotype:** China. Qinghai: Qinghai Lake, alt. 3245 m, 1982-09-03, Y. D. Chen & R. S. Ni 442.

Alismataceae

泽泻科

zexieke

小慈姑 ***Sagittaria potamogetifolia*** Merr. in Sunyatsenia 1(4): 189. 1934. **Isotype:** China. Guangdong: Sunyi (=Xinyi), 1929-06-15, Y. Tsiang 2726.

Hydrocharitaceae

水鳖科

shuibieke

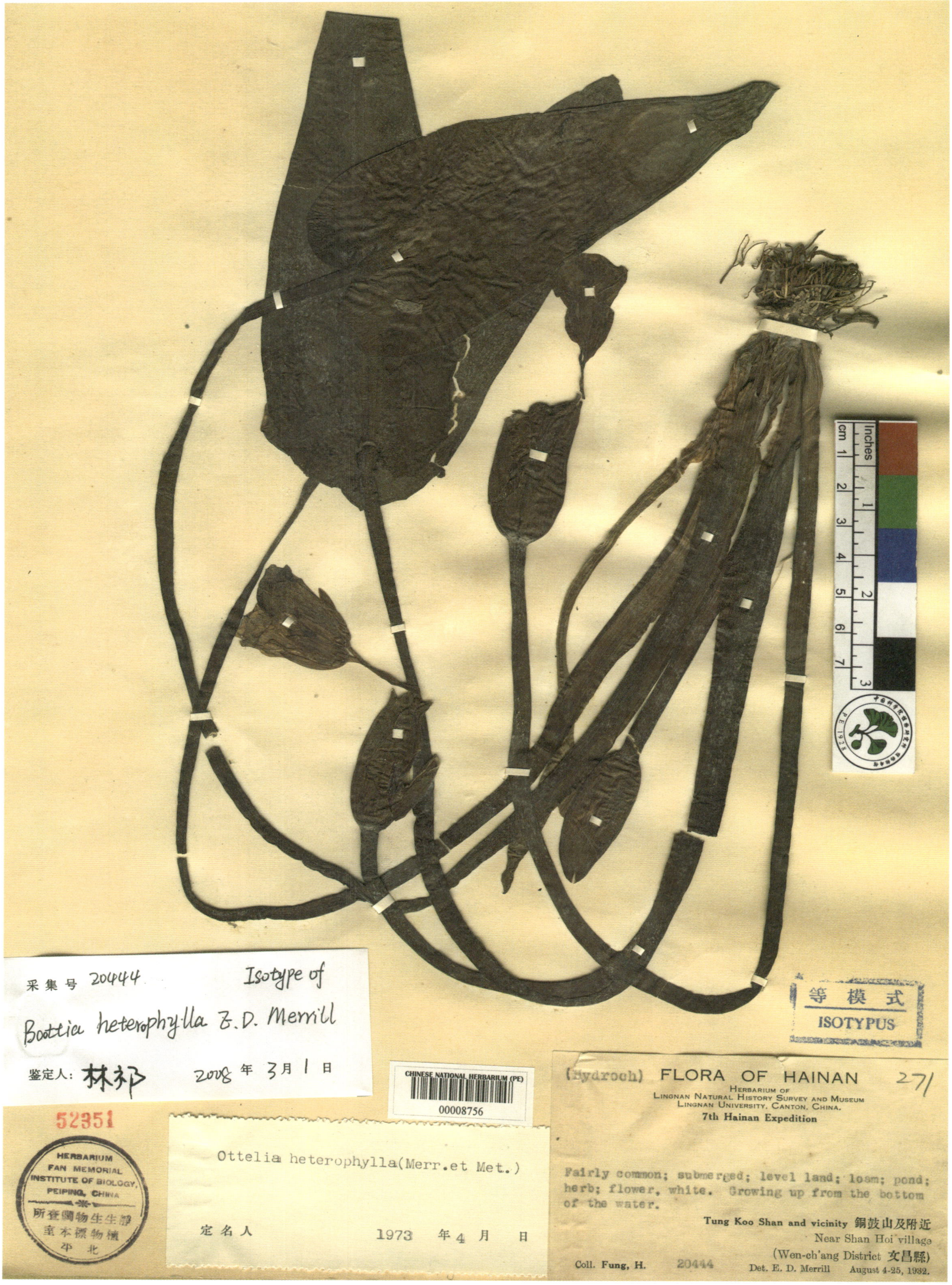

水菜花 ***Boottia heterophylla***. Merr. in Lingnan Sci. Journ. 17: 568, pl. 24, 25. 1938. **Isotype:** China. Hainan: Wenchang, 1932-08-(04-25), H. Fung 20444.

Gramineae

禾本科

hebenke

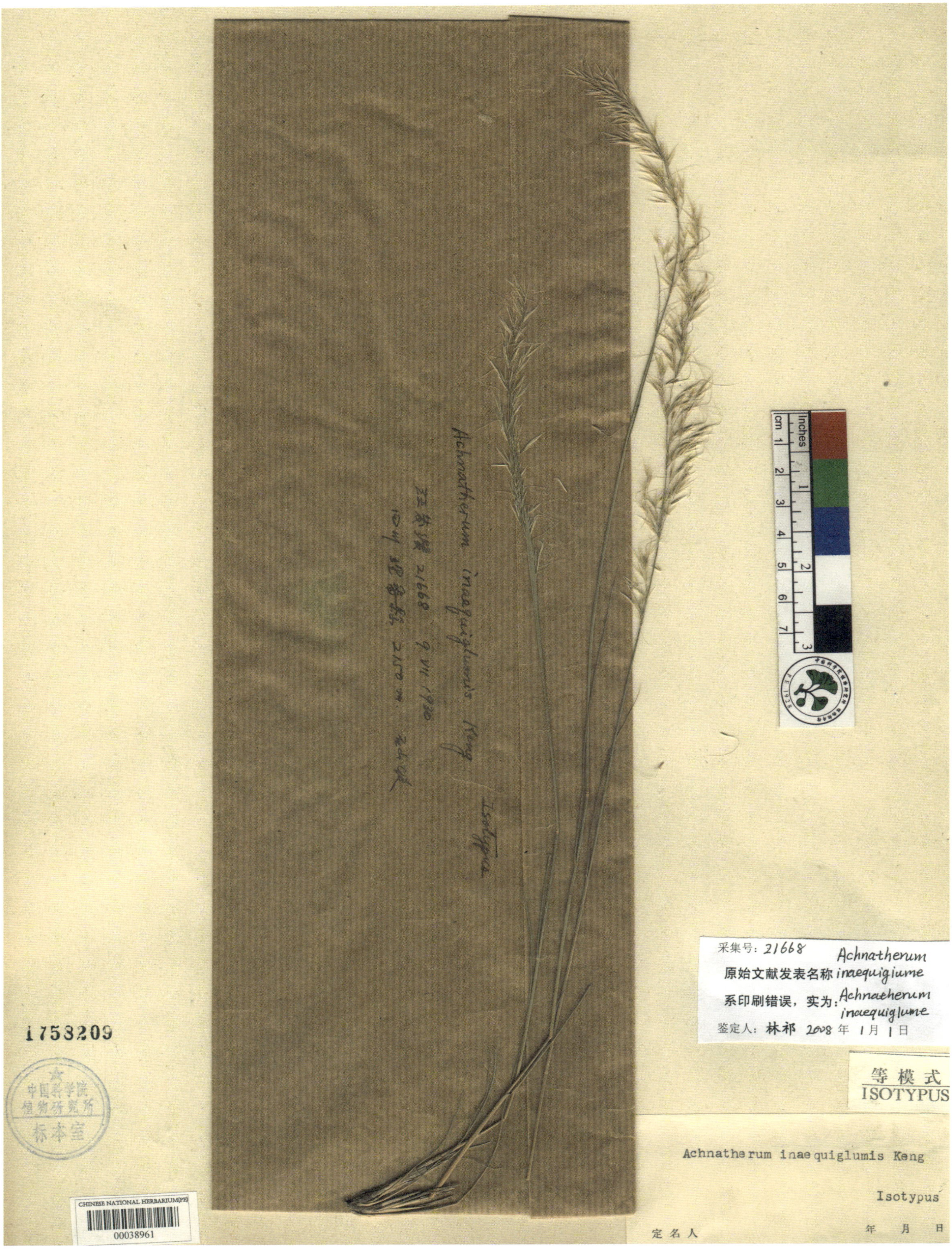

异颖芨芨草 ***Achnatherum inaequiglume***. Keng, Fl. Tsinling. 1(1): 443. 1976. **Isotype:** China. Sichuan: Lihsien, alt. 2150 m, 1930-07-09, F. T. Wang 21668.

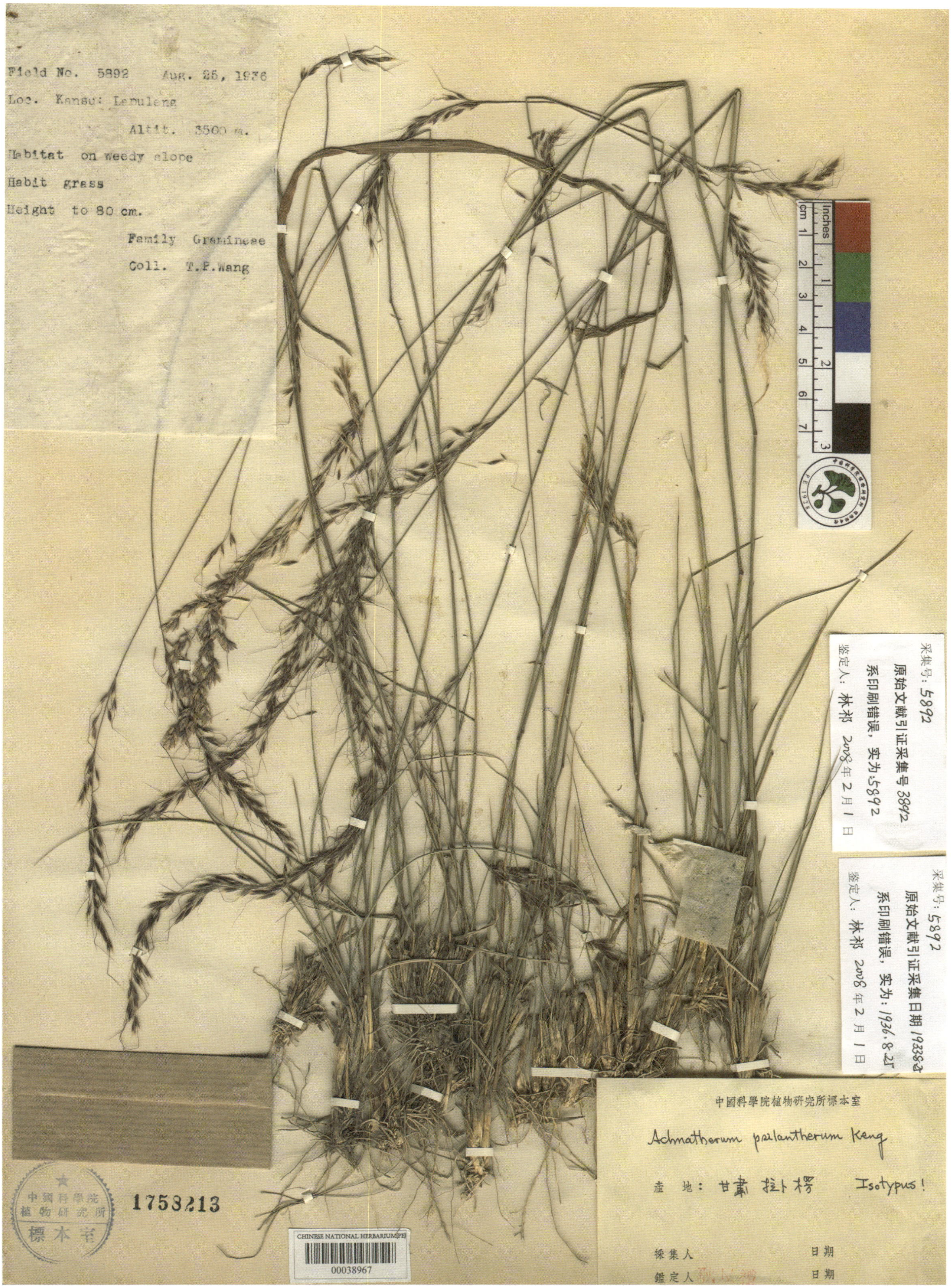

光药芨芨草 ***Achnatherum psilantherum*** Keng ex P. C. Keng in Bull. Bot. Res., Harbin 4(3): 198. 1984. **Isotype:** China. Gansu: Xiahe, Labrang, alt. 2550 m, 1936-08-25, T. P. Wang 5892.

黑药鹅观草 ***Agropyron melantherum*** Keng in Sunyatsenis 6(1): 62. 1941. **Isotype:** China. Qinghai: Baian Khara Mts, 1935-07-12, C. W. Yao 832.

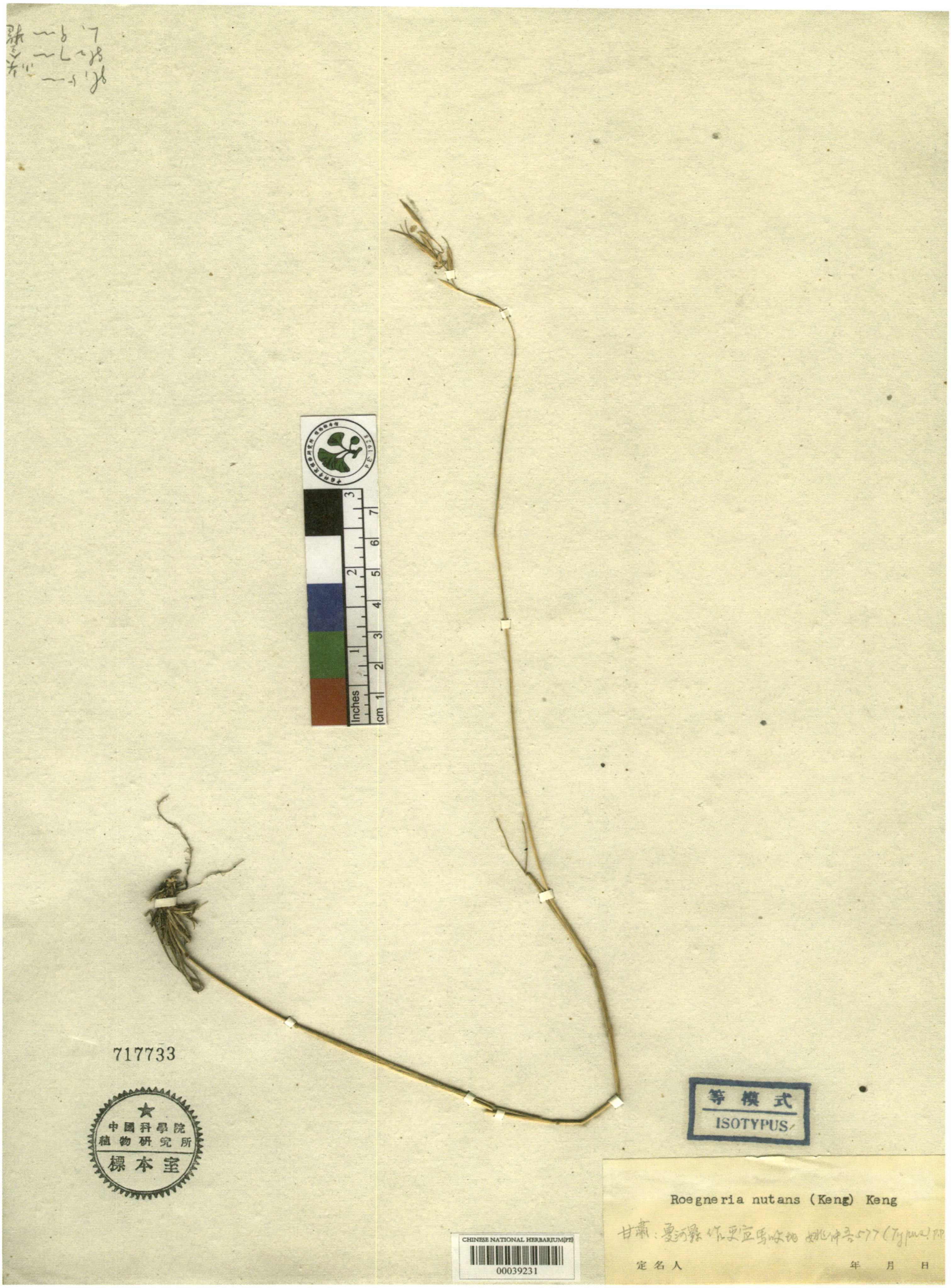

垂穗鹅观草 ***Agropyron nutans*** Keng in Sunyatsenia 6(1): 63. 1941. **Isotype**: China. Gansu: Xiahe, 1934-10-07, C. W. Yao 503.

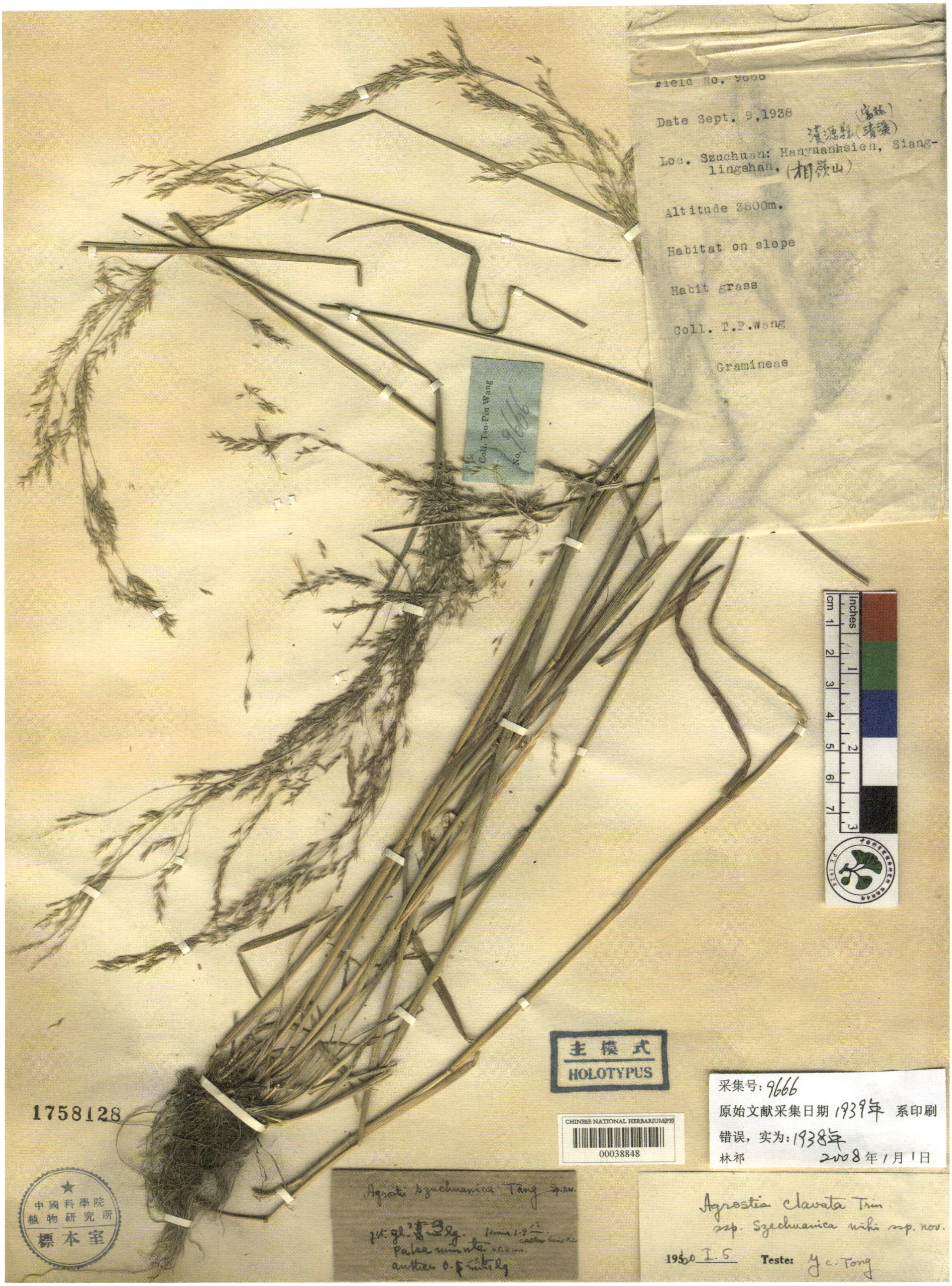

四川剪股颖 ***Agrostis clavata*** Trin. var. ***szechuanica*** Y. C. Tong ex Y. C. Yang in Bull. Bot. Res., Harbin 4(4): 99. 1984. **Isotype:** China. Sichuan: Hanyuan, alt. 3800 m, 1938-09-09, T. P. Wang 9666.

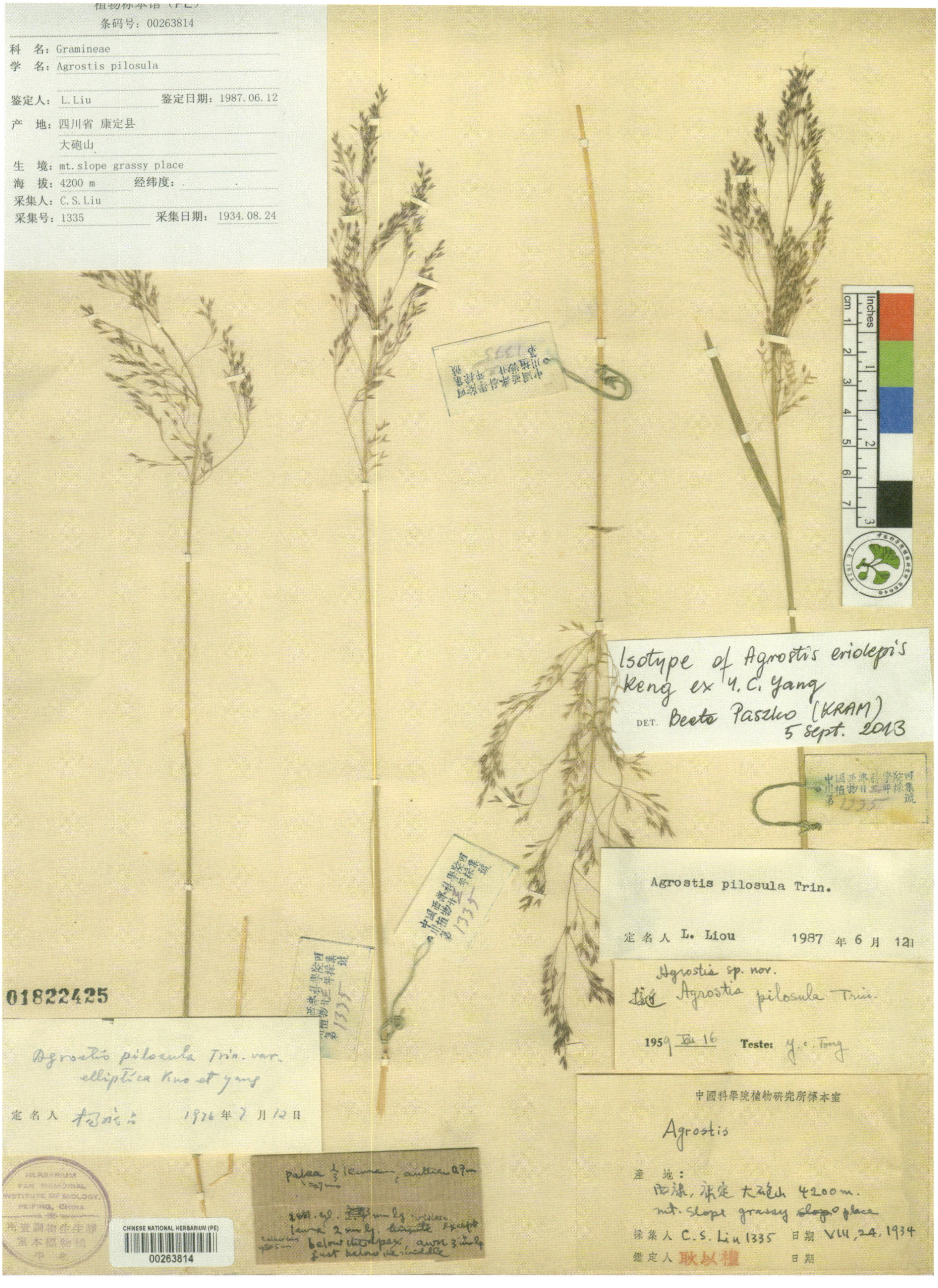

柔毛剪股颖 ***Agrostis eriolepis*** Keng ex Y. C. Yang in Bull. Bot. Res., Harbin 4(4): 97, f. 1. 1984. **Isotype:** China. Sichuan: Kangding, alt. 4200 m, 1934-08-24, C. S. Liu 1335.

线序剪股颖 ***Agrostis sinkiangensis*** Y. C. Yang in Bull. Bot. Res., Harbin 4(4): 98, f. 2. 1984. **Holotype:** China. Xinjiang: Wulistai district, 1957-08-30, K. C. Kuan 3968.

耳形巴西竹 ***Alvimia auriculata*** Londono & Soderstrom in Amer. Journ. Bot. 75(6): 834, f. 3-7, 12. 1988. **Isotype:** Brazil. Bahia: Mun. de Ilheus, 1976-05-21, C. E. Calderon, T. S. dos Santos & L. B. Oliveira 2444.

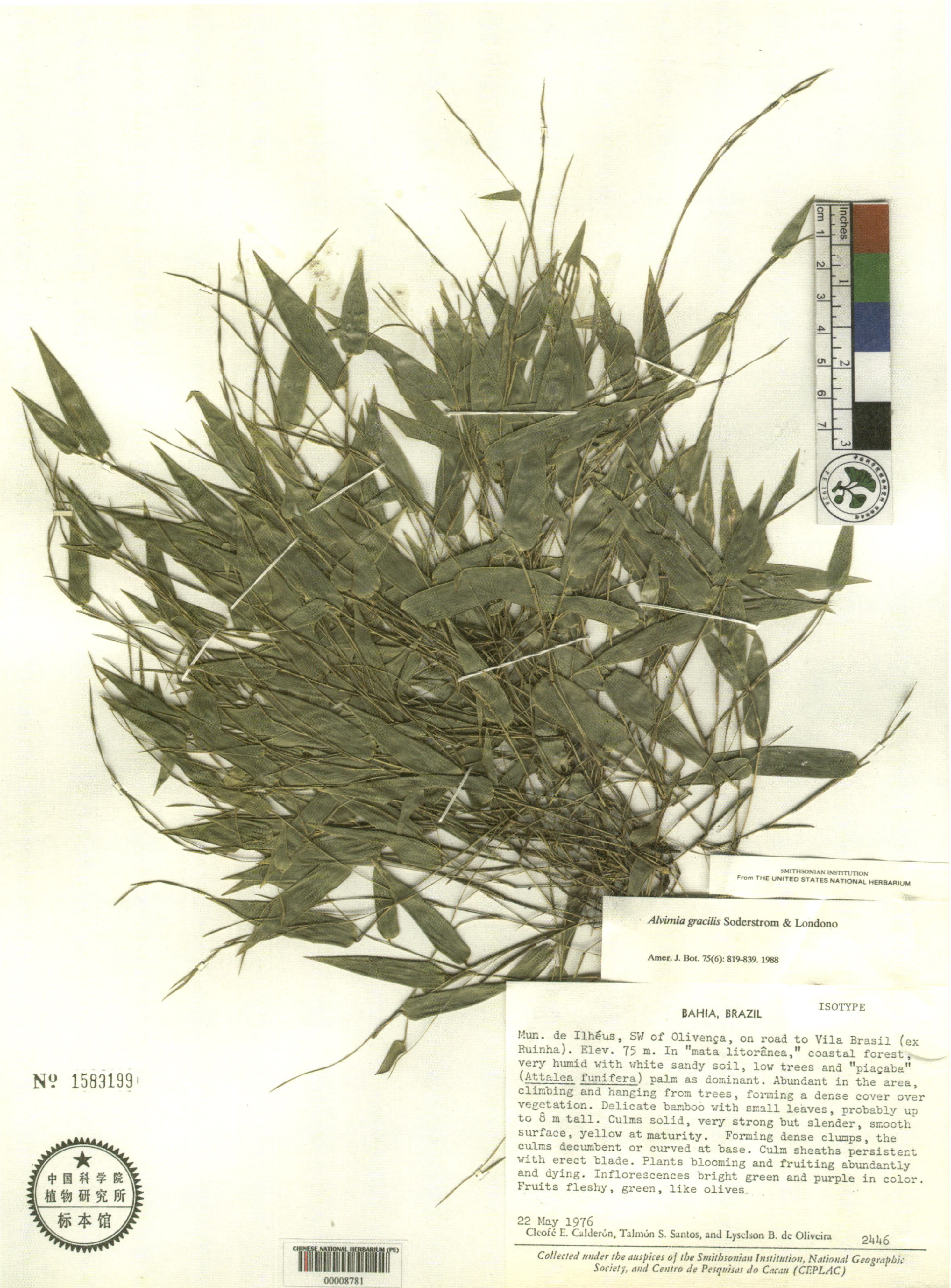

纤细巴西竹 ***Alvimia gracilis*** Soderstrom & Londono in Amer. Journ. Bot. 75(6): 835, f. 8, 11, 13-16. 1988. **Isotype:** Brazil. Bahia: Mun. de Ilheus, 1976-05-22, C. E. Calderon, T. S. dos Santos & L. B. Oliveira 2446.

披针叶巴西竹 ***Alvimia lancifolia*** Soderstrom & Londono in Amer. Journ. Bot. 75(6): 837, f. 1-2, 9, 17. 1988. **Isotype:** Brazil. Bahia: Mun. de Ilheus, 1977-04-08, C. E. Calderon 2456 .

高原三芒草 ***Aristida alpina*** L. Liou, Fl. Xizang. 5: 82. 1987. **Holotype:** China. Xizang: Lhasa, alt. 4500 m, 1972-10-06, Exped. Medic. Pl. Xizang 2058 .

短芒草 ***Aristida brevissima*** L. Liou, Fl. Xizang. 5: 81. 1987. **Holotype:** China. Xizang: Chagyab, alt. 3000 m, 1976-08-10, Qhinghai-Xizang Exped. 9512 .

糙三芒草 ***Aristida scabrescens*** L. Liou, Fl. Xizang. 5: 84. 1987. **Holotype:** China. Xizang: Qamdo, alt. 4100 m, 1974-06-24, Qhinghai-Xizang Exped. 97.

藏布三芒草 ***Aristida tsangpoensis*** L. Liou, Fl. Xizang. 5: 82. 1987. **Holotype:** China. Xizang: Nangxian, alt. 3000 m, 1975-09-19, Qinghai-Xizang Exped. 7896.

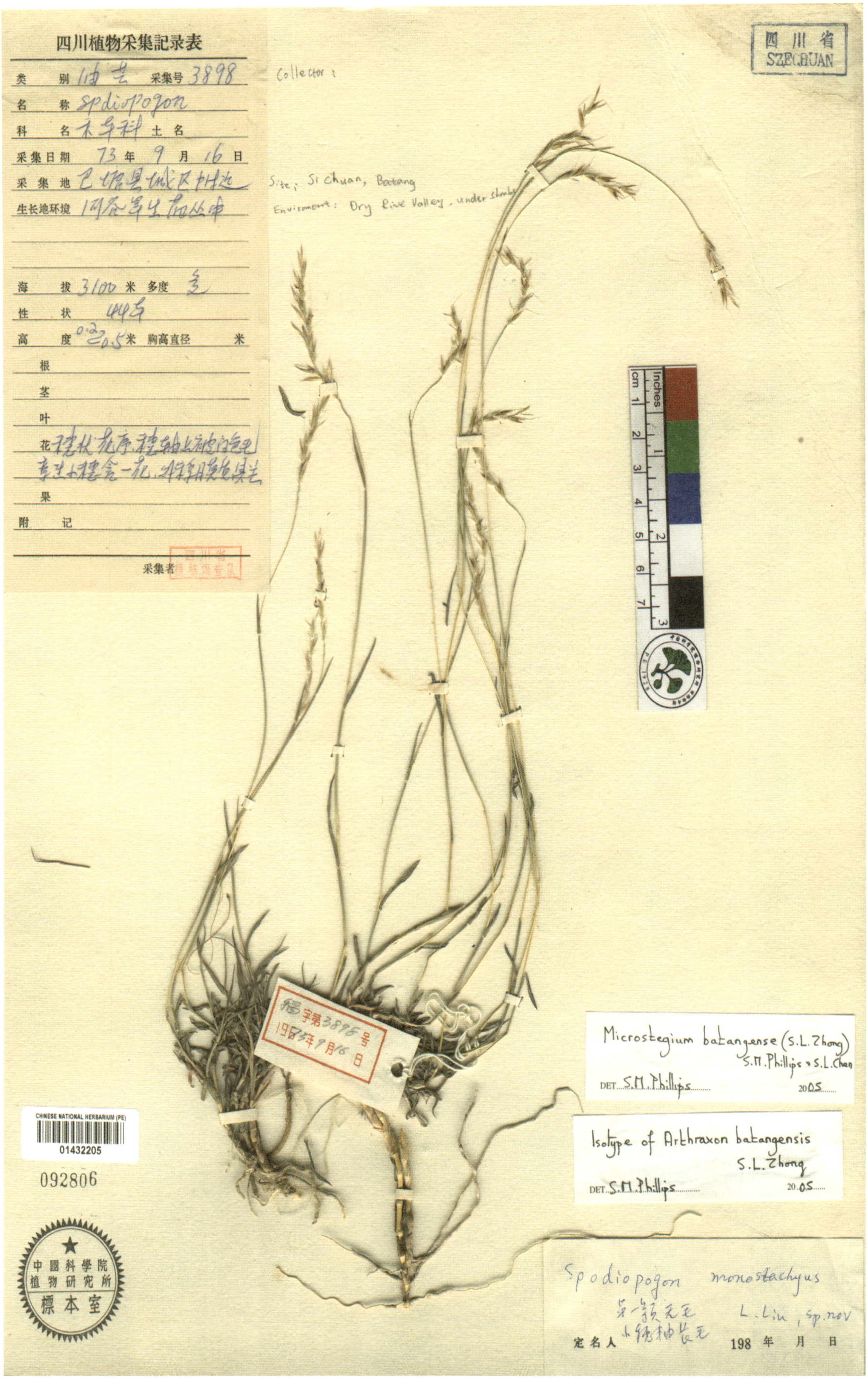

巴塘荩草 ***Arthraxon batangensis*** S. L. Zhong in Journ. SW Agric. Univ. 1982(4): 79, f. 4. 1982. **Isotype:** China. Sichuan: Batang, alt. 3100 m, 1973-9-16, Sichuan Vegetatio Exped. 3898.

文培箭竹 ***Arundinaria fangiana*** Camus in Journ. Arn. Arb. 11(3): 192. 1930. **Isotype:** China. Sichuan: Emei, Emeishan, alt. 3000~3300 m, 1928-08-15. W. P. Fang 3002.

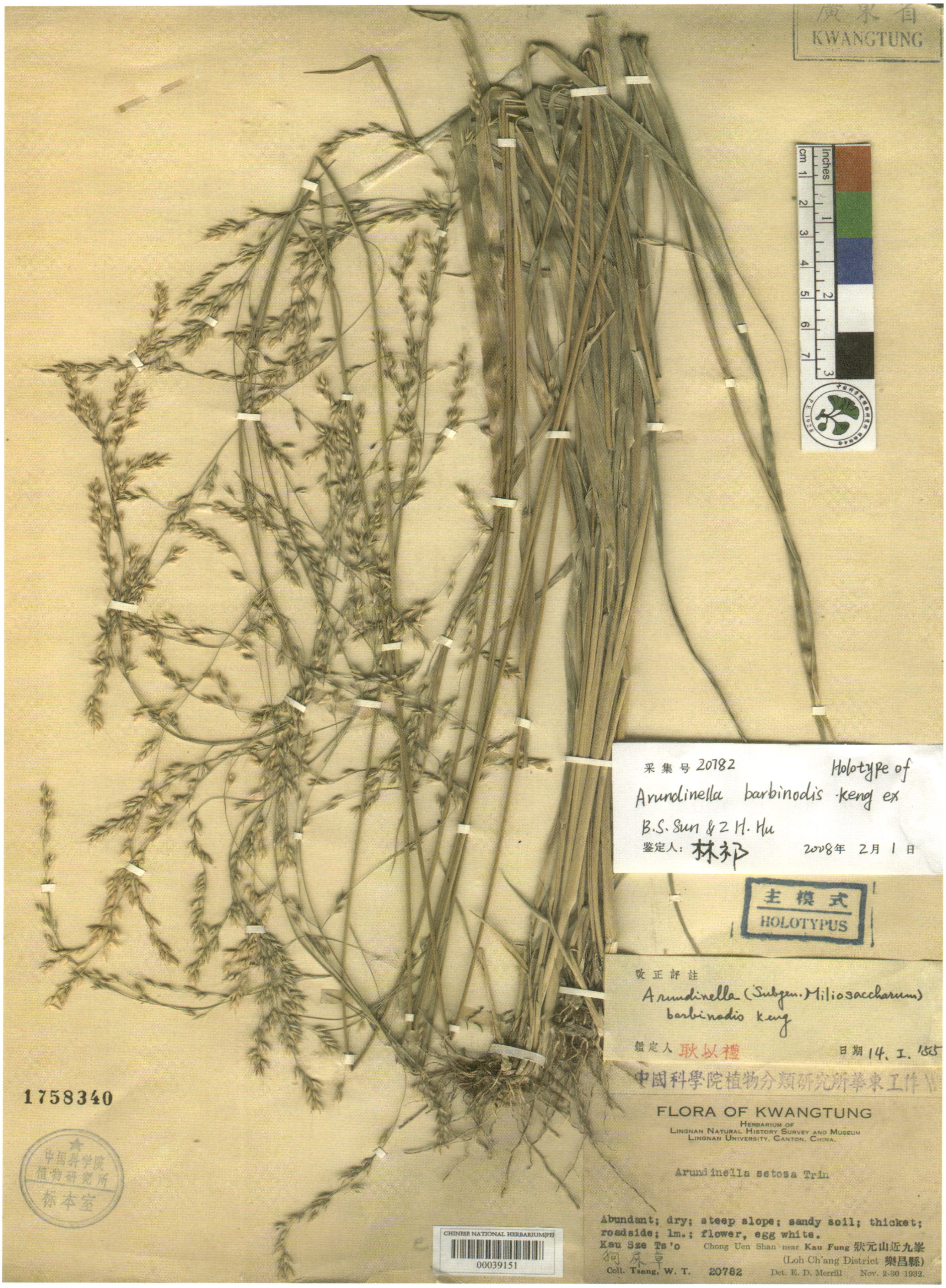

毛节野古草 ***Arundinella barbinodis*** Keng ex B. S. Sun & Z. H. Hu in Acta Bot. Yunnan. 2(3): 329. 1980. **Holotype:** China. Guangdong: Lechang, 1932-11-(02-30), W. T. Tsang 20782.

多节野古草 ***Arundinella nodosa*** B. S. Sun & Z. H. Hu in Acta Bot. Yunnan. 2(3): 327. 1980. **Isoparatype:** China. Yunnan: Jinghong, alt. 720 m, 1936-10-??, C. W. Wang 79918.

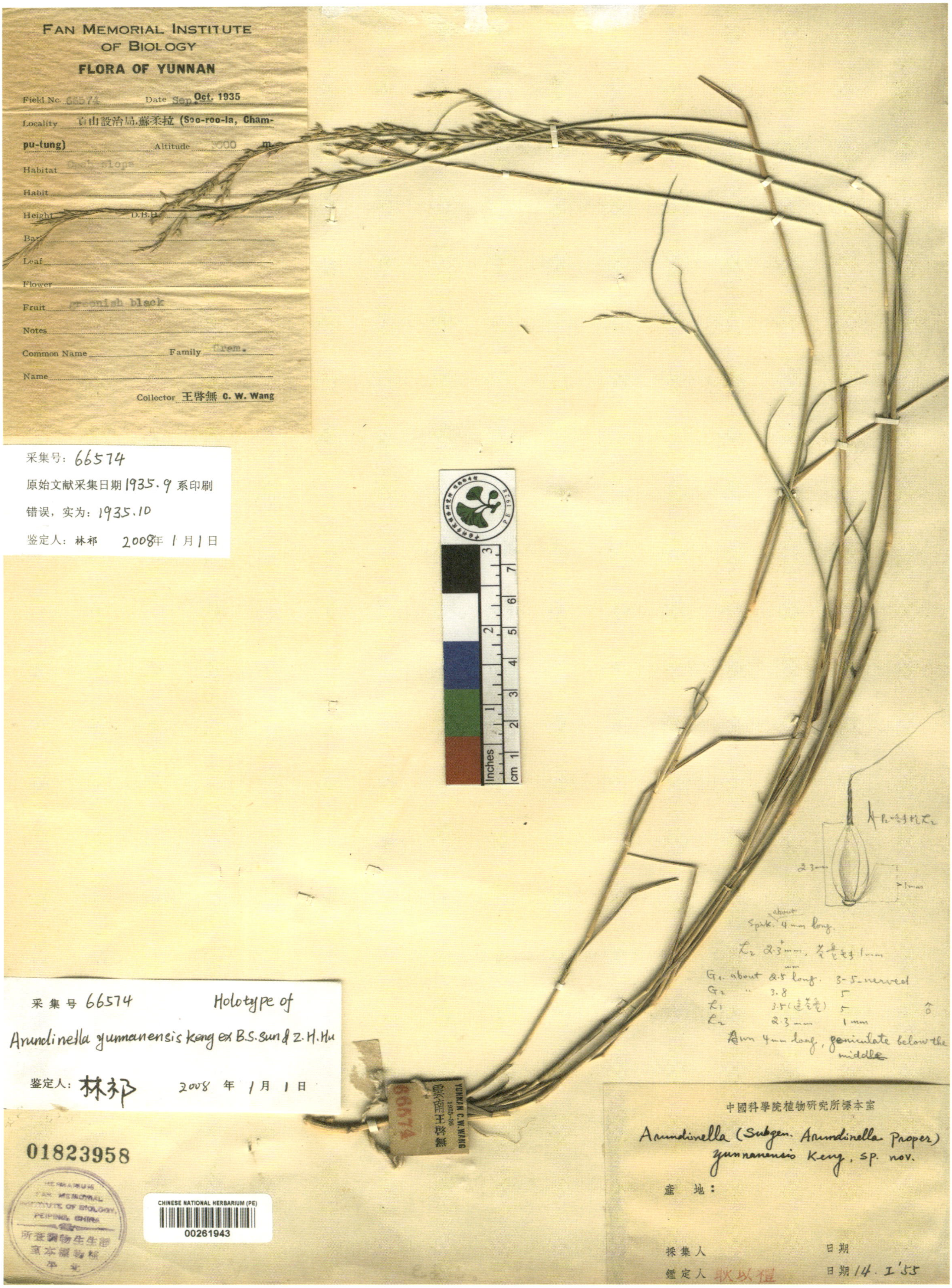

云南野古草 ***Arundinella yunnanensis*** Keng ex B. S. Sun & Z. H. Hu in Acta Bot. Yunnan. 2(3): 326. 1980. **Holotype:** China. Yunnan: Gongshan, alt. 3000 m, 1935-09-??, C. W. Wang 66574.

冷杉异燕麦 ***Avena abietorum*** Ohwi in Acta Phytotax. Geobot. 2(3): 162. 1933. **Isotype:** China. Taiwan: Mt. Niitaka, J. Ohwi 3675.

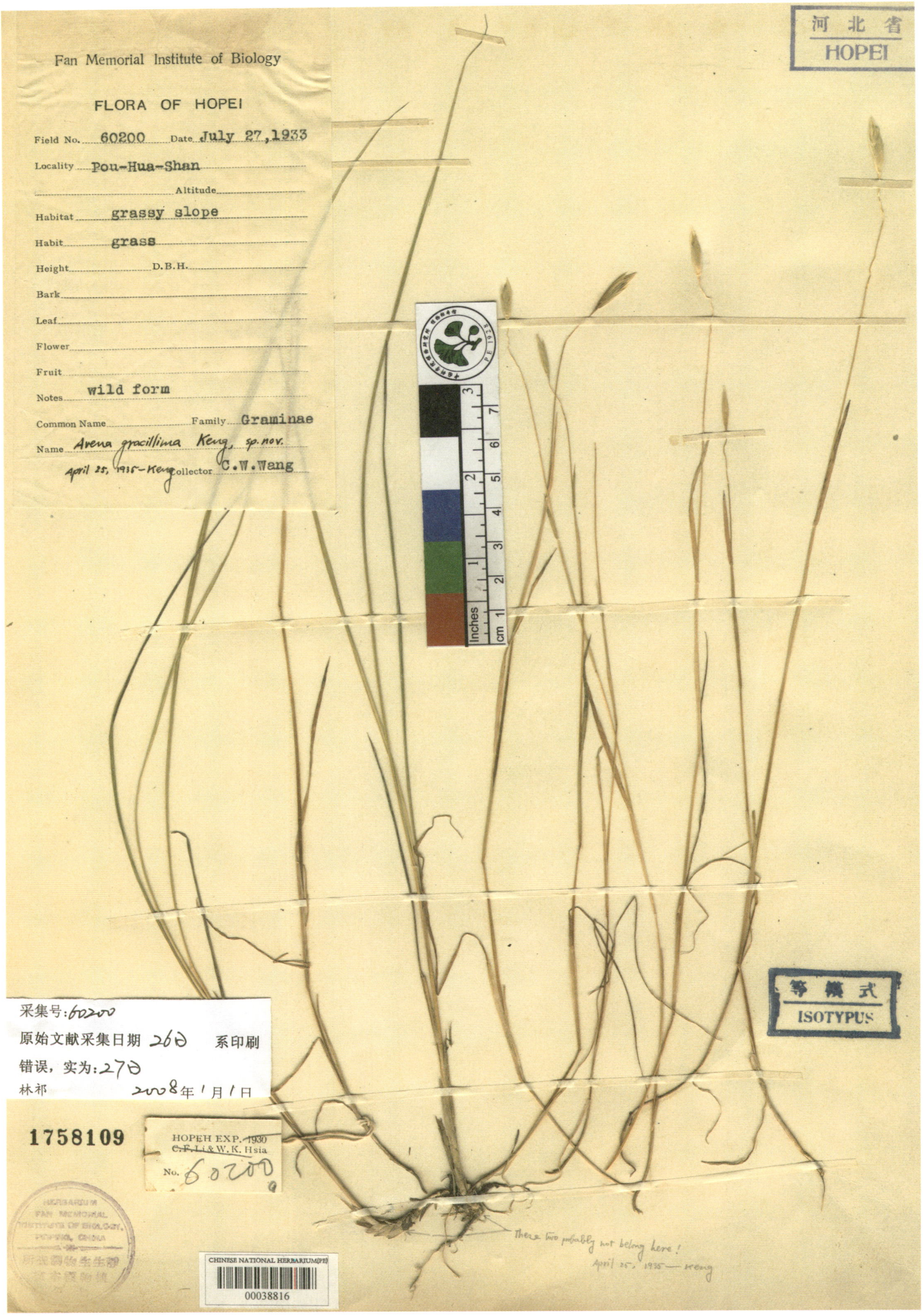

纤细叶燕麦 ***Avena gracillima*** Keng in Bull. Fan Mem. Inst. Biol., Bot. Ser. 7(1): 36. 1936. **Isotype:** China. Hebei: Pouhuashan, 1933-07-27, C. W. Wang 60200a.

光花异燕麦 ***Avena leiantha*** Keng in Bull. Fan Mem. Inst. Biol., Bot. Ser. 7(1): 35. 1936. **Isotype:** China. Shanxi: Lishih, Mienshan, alt. 1500 m, 1929-05-30, T. Tang 895.

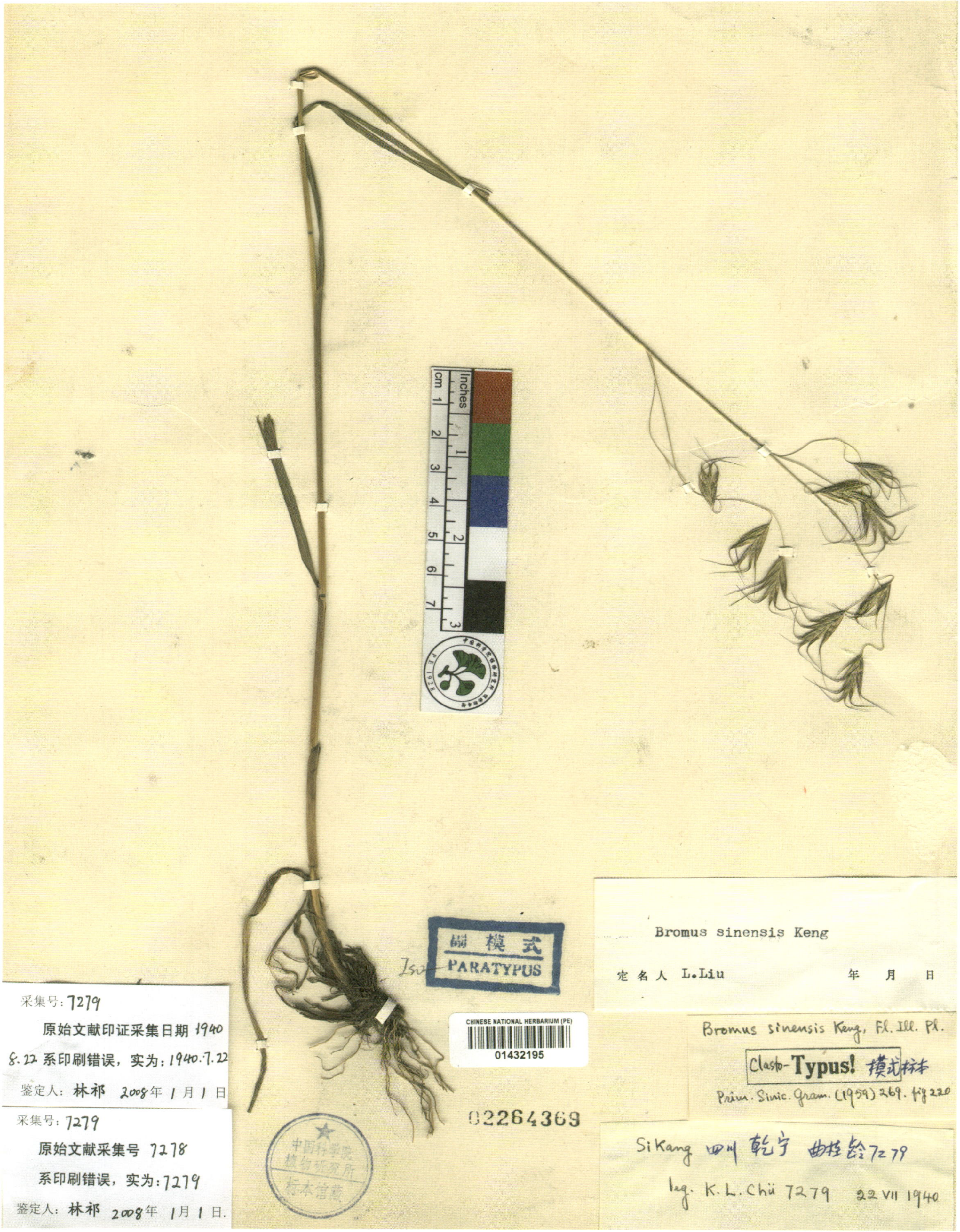

华雀麦 ***Bromus sinensis*** Keng ex P. C. Keng in Acta Bot. Yunnan. 4(4): 349. 1982. **Isoparatype:** China. Sichuan: Qianning, 1940-07-22, K. L. Chu 7279.

细弱隐子草 ***Cleistogenes gracilis*** Keng ex P. C. Keng in Acta Bot. Sin. 9(1): 69. 1960. **Isotype:** China. Shaanxi: Wukung, 1940-??-??, K. S. Tian s. n..

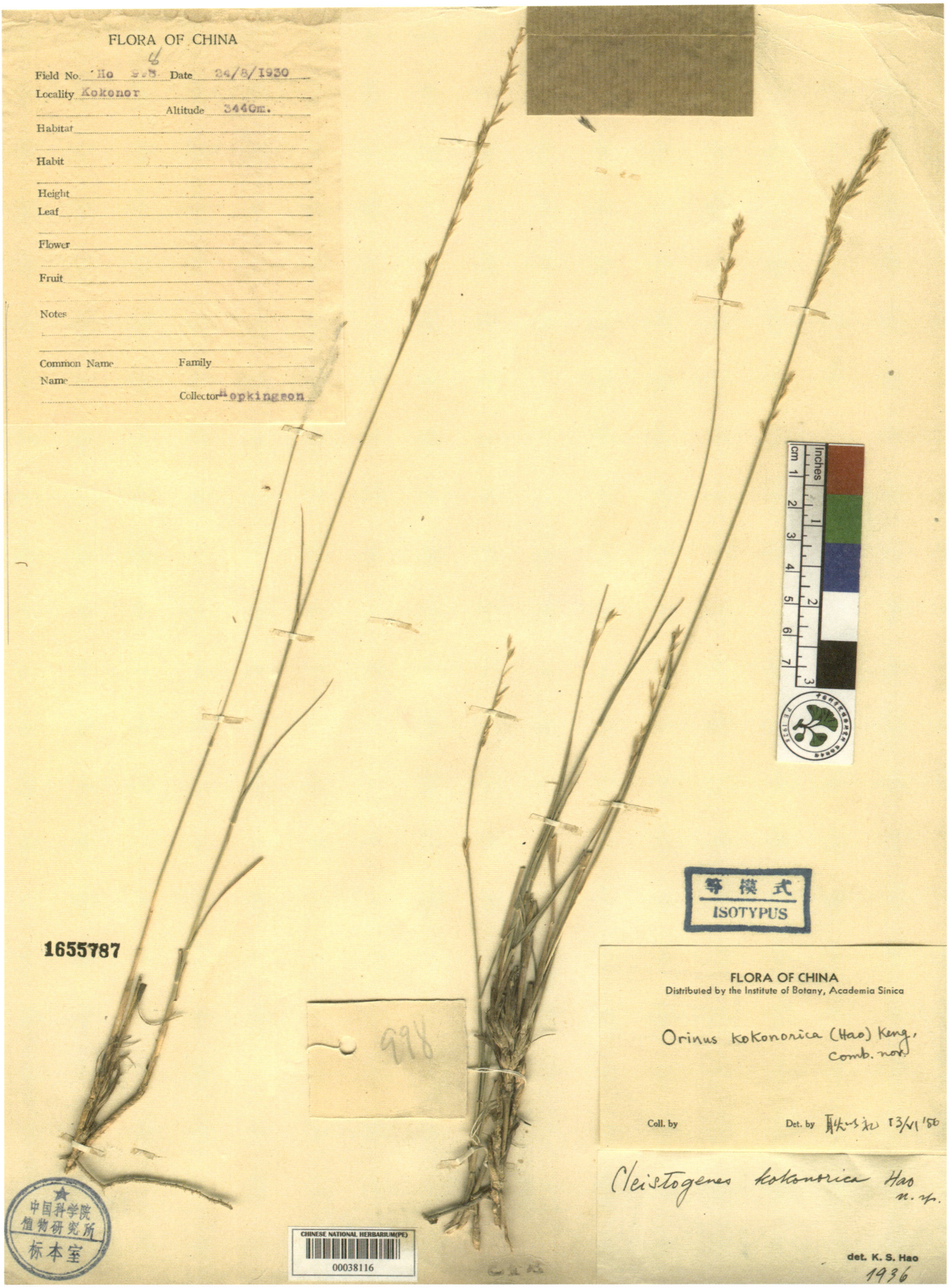

青海固沙草 ***Cleistogenes kokonorica*** K. S. Hao in Bot. Jahrb. Syst. 68(5): 582. 1938. **Isotype:** China. Qinghai: Tsi-gi-gan-ba-Gebiet, alt. 3340 m, 1930-08-24, Hopkingson (= K. S. Hao) 998.

长花隐子草 ***Cleistogenes longiflora*** Keng ex P. C. Keng in Acta Bot. Sin. 9(1): 69. 1960. **Isotype:** China. Hebei: Lanyuan, alt. 810 m, 1934-07-13, K. M. Liou 2853.

小尖隐子草 ***Cleistogenes mucronata*** Keng ex P. C. Keng in Acta Bot. Sin. 9(1): 70. 1960. **Isotype** China. Gansu: Precise locality not known, T. P. Wang s. n.

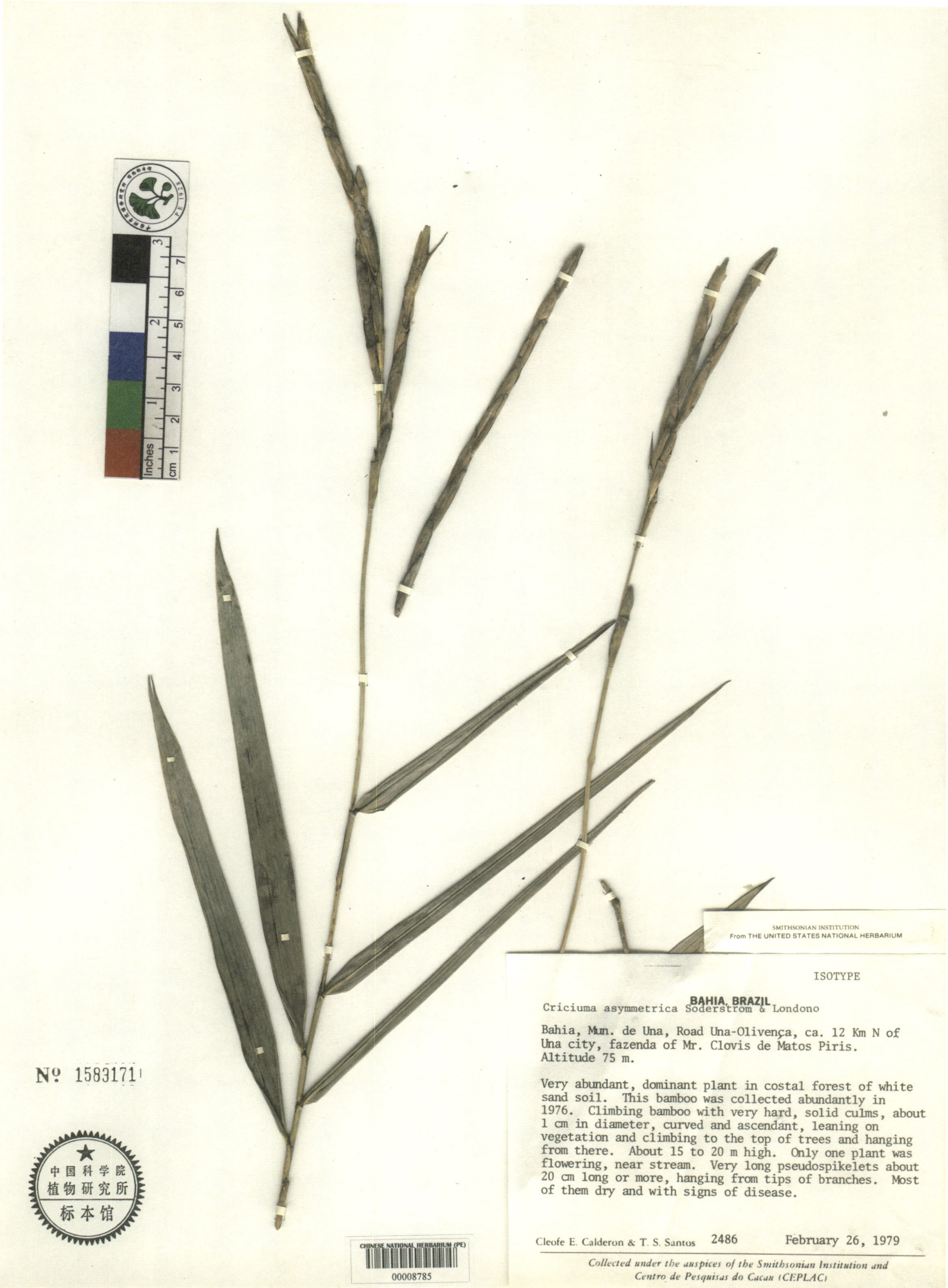

歪竹 ***Criciuma asymmetrica*** Soderstrom & Londono in Amer. Journ. Bot. 74(1): 35, f. 1, 2, 5a, b. 1987. **Isotype:** Brazil. Bahia: Mun. de Ilheus, alt. 75 m, 1979-02-26, C. E. Calderon & T. S. Santos 2486.

热带美洲草 ***Cryptochloa decumbense*** Soderstrom & Zuloaga in Brittonia. 37(1): 29, f. 5. 1985. **Isotype:** Panama. Cerro Gefe, 1968-03-03, C. E. Calderon 2074.

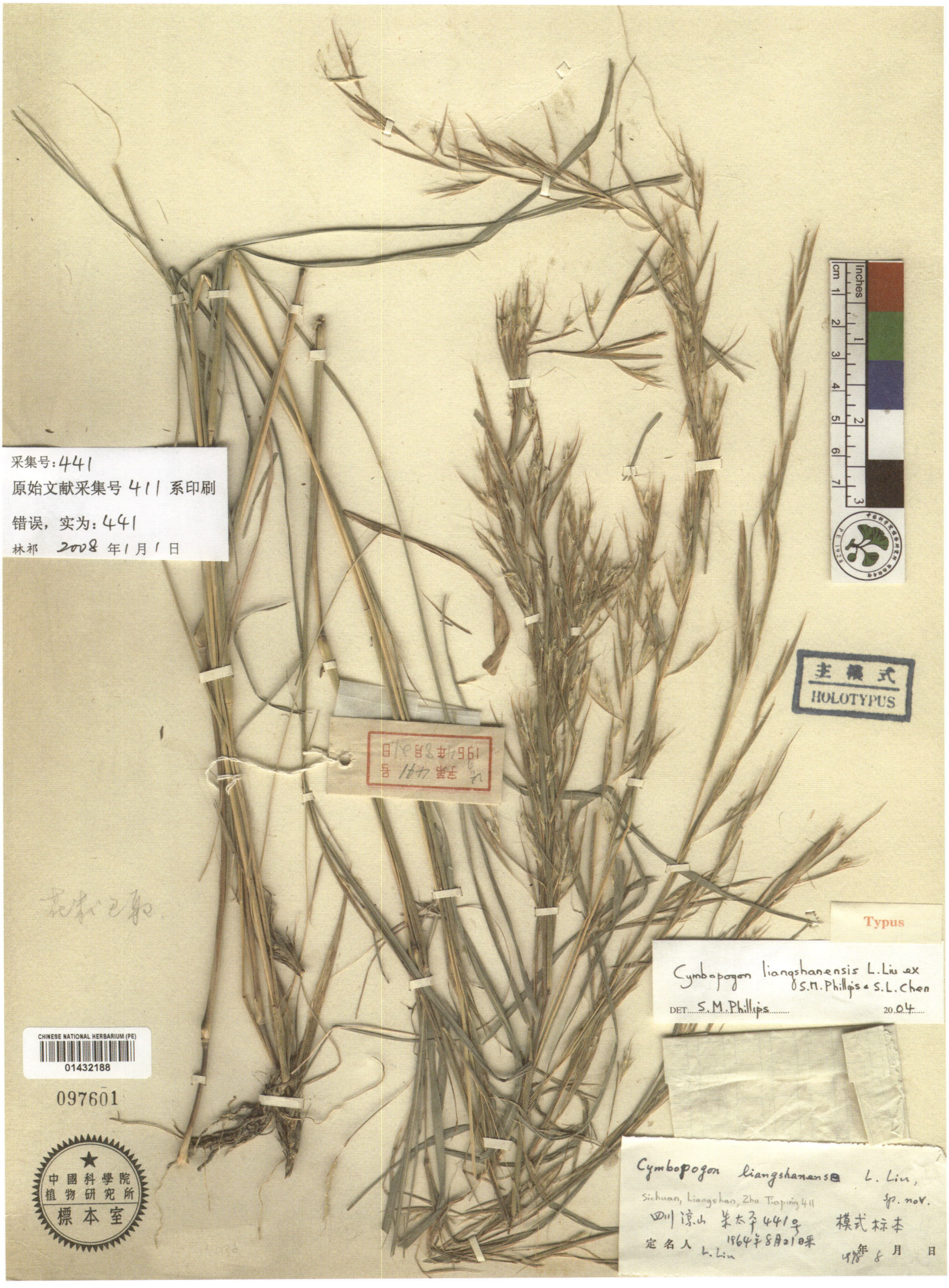

凉山香茅 ***Cymbopogon liangshanensis*** L. Liou ex S. M. Phillips & H. Peng in Novon. 15(3): 471. 2005. **Holotype:** China. Sichuan: Liangshan, 1964-08-21, T. P. Zhu 441.

短枝发草 ***Deschampsia ivanovae*** Tzvelev in Bot. Mat. Gerb. Bot. Inst. Kom. Akad. Nauk SSSR 21: 49. 1961.**Isotype:** China. Gansu: Jetung, 1872-08-(06-18), N. M. Przewalski 444.

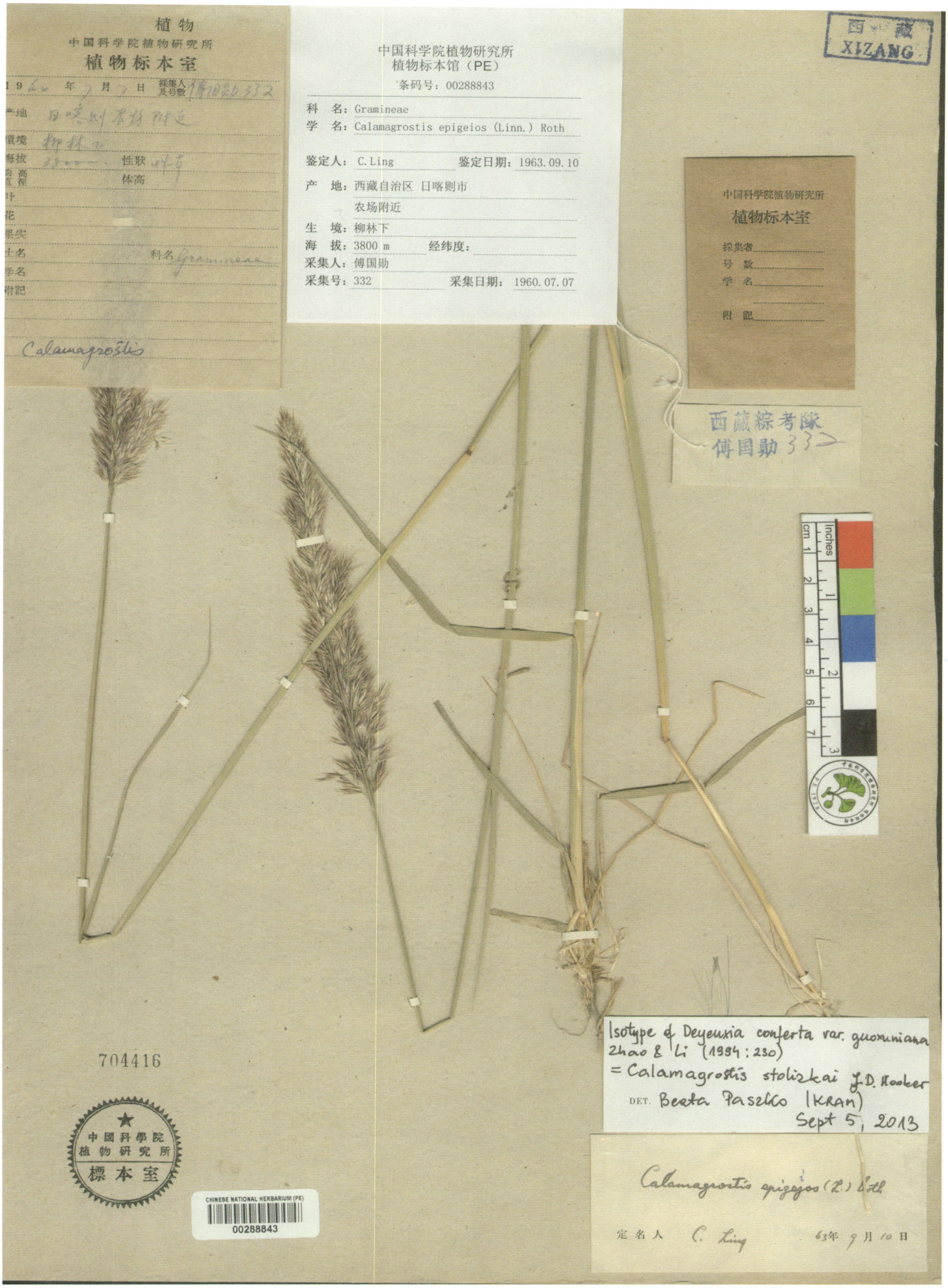

国勋野青茅 ***Deyeuxia conferta*** Keng var. ***guoxuniana*** N. X. Zhao & M. F. Li in Acta Bot. Yunnan. 16(3): 230. 1994.
Isotype: China. Xizang: Xigazê, alt. 3800 m, 1960-07-07, G. X. Fu & J. W. Zhang 332.

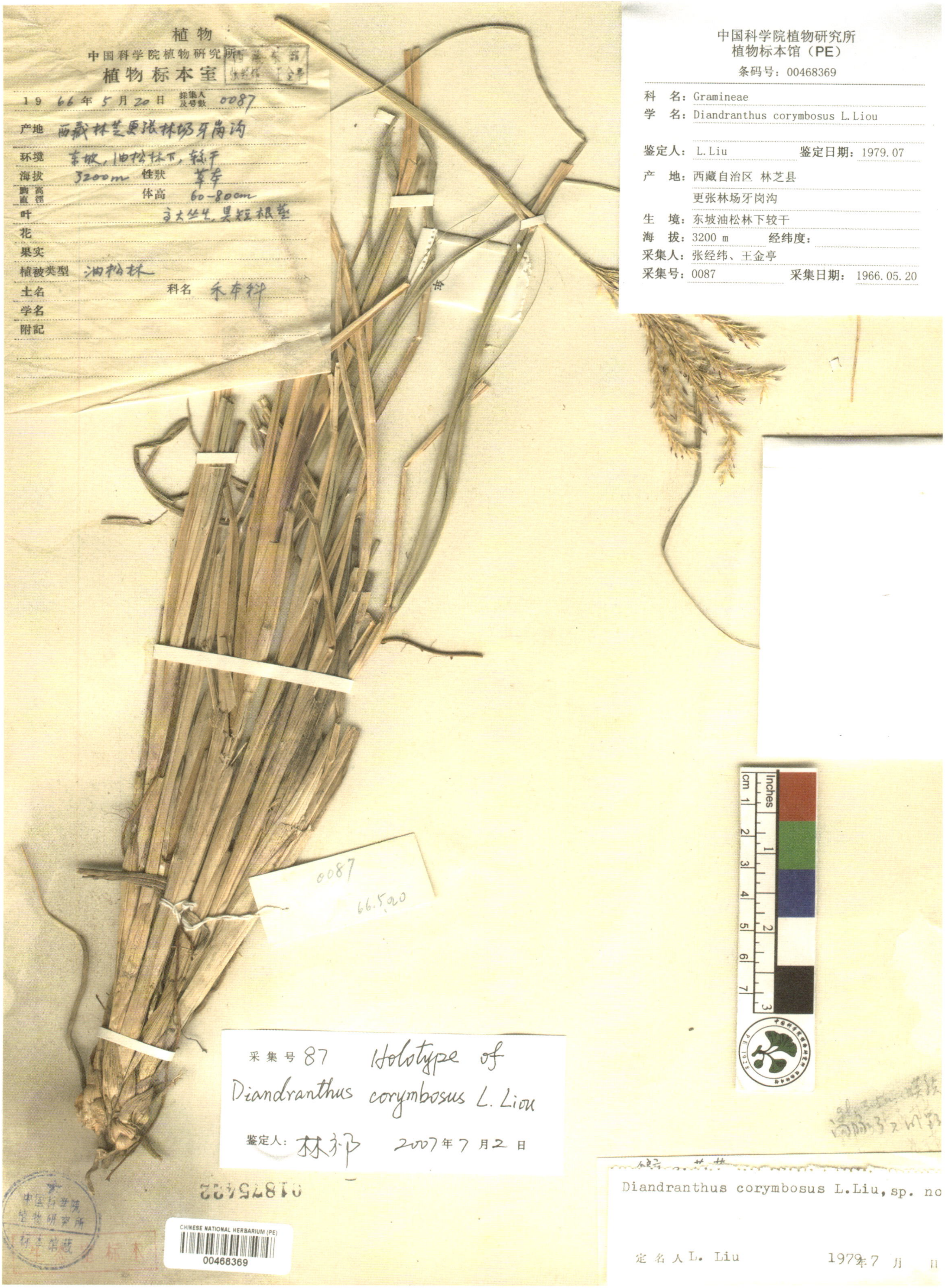

伞房双药芒 ***Diandranthus corymbosus*** L. Liou, Fl. Xizang. 5: 312. 1987. **Holotype:** China. Xizang: Nyingchi, alt. 3200 m, 1966-05-20, J. W. Zhang & J. T. Wang 87.

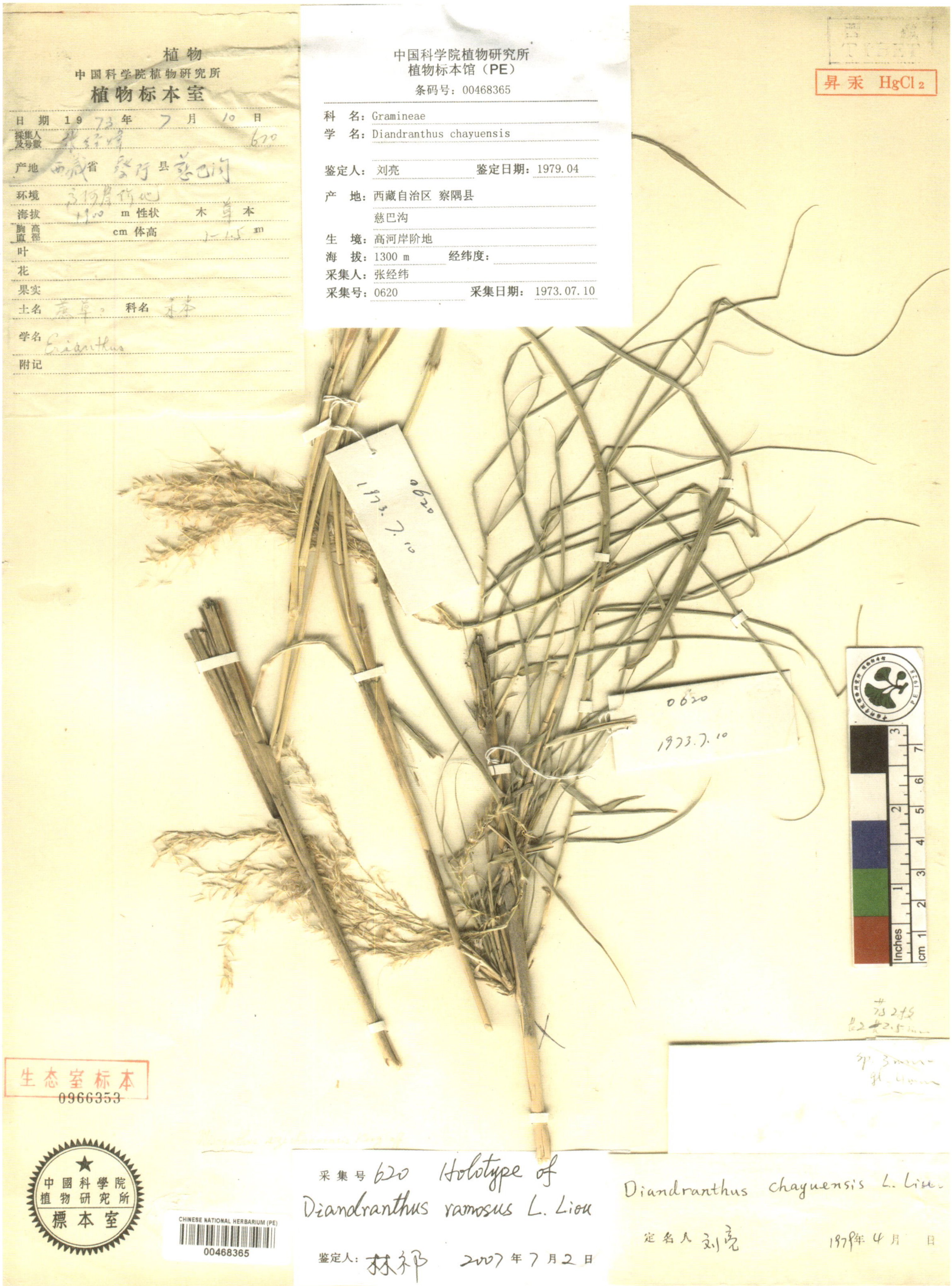

分枝双药芒 ***Diandranthus ramosus*** L. Liou, Fl. Xizang. 5: 310. 1987. **Holotype:** China. Xizang: Zayü, alt.1300 m, 1973-07-10, J. W. Zhang 620.

西藏双药芒 ***Diandranthus tibeticus*** L. Liou, Fl. Xizang. 5: 307. 1987. **Lectotype** (designated by Q. Sun & al. in Bot. Journ. Linn. Soc. 164: 208. 2010.): China. Xizang: Bomi, alt. 2080 m, 1965-07-23, T. S. Ying & D. Y. Hong 650799.

横断山马唐 ***Digitaria hengduanensis*** L. Liou in Bot. Res. Acad. Sin. 4(4): 37. 1989. **Holotype:** China. Yunnan: Yuanmou, alt. 1200 m, 1963-09-13, Jinshajiang Exped. 63-6847.

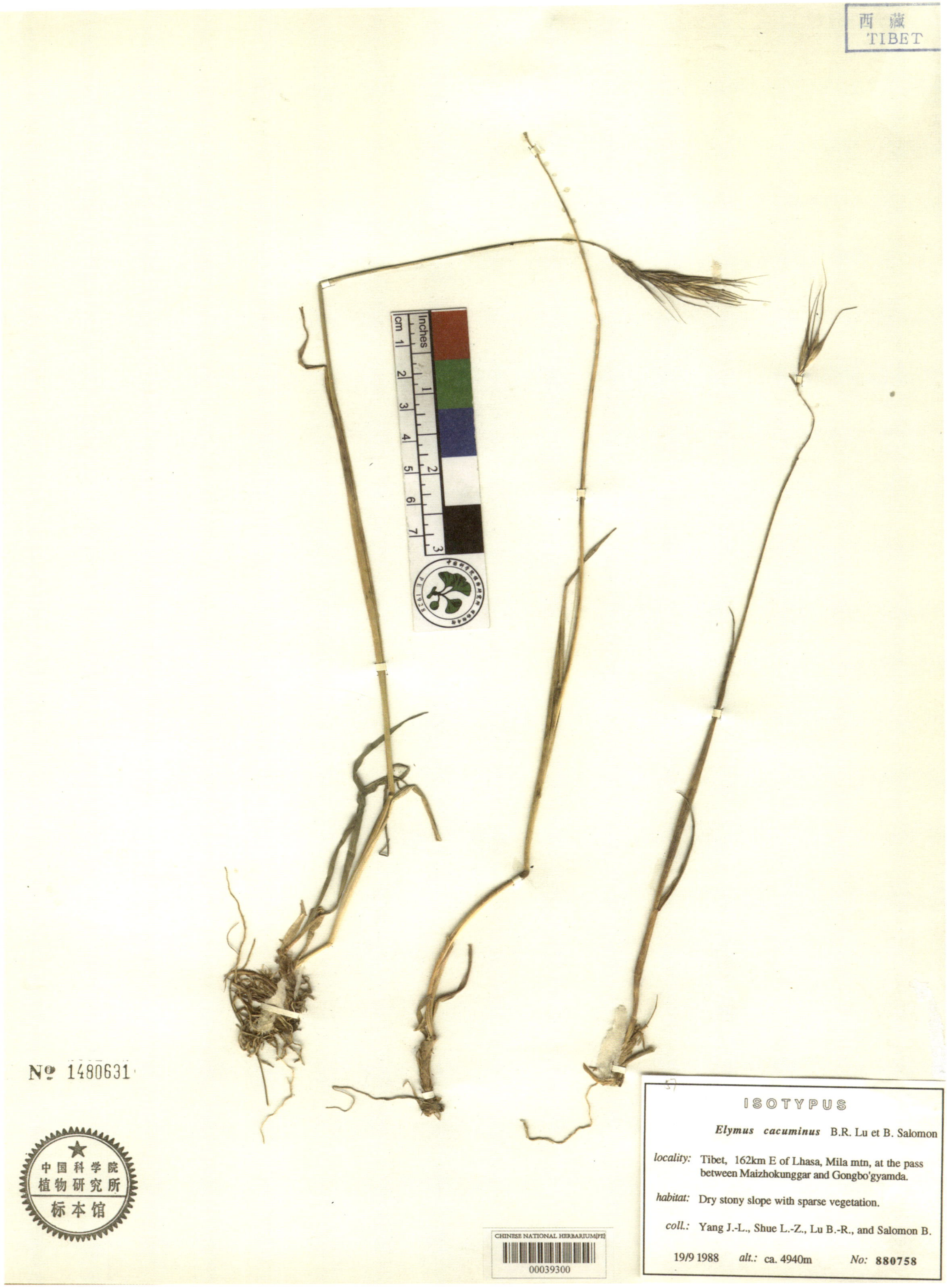

峰峦披碱草 ***Elymus cacuminis*** B. R. Lu & B. Salomon in Nord. Journ. Bot. 13(4): 355, f. 1. 1993 “cacuminus”. **Isotype:** China. Xizang: Lhasa, alt. 4940 m, 1988-09-19, J. L. Yang & al. 880758.

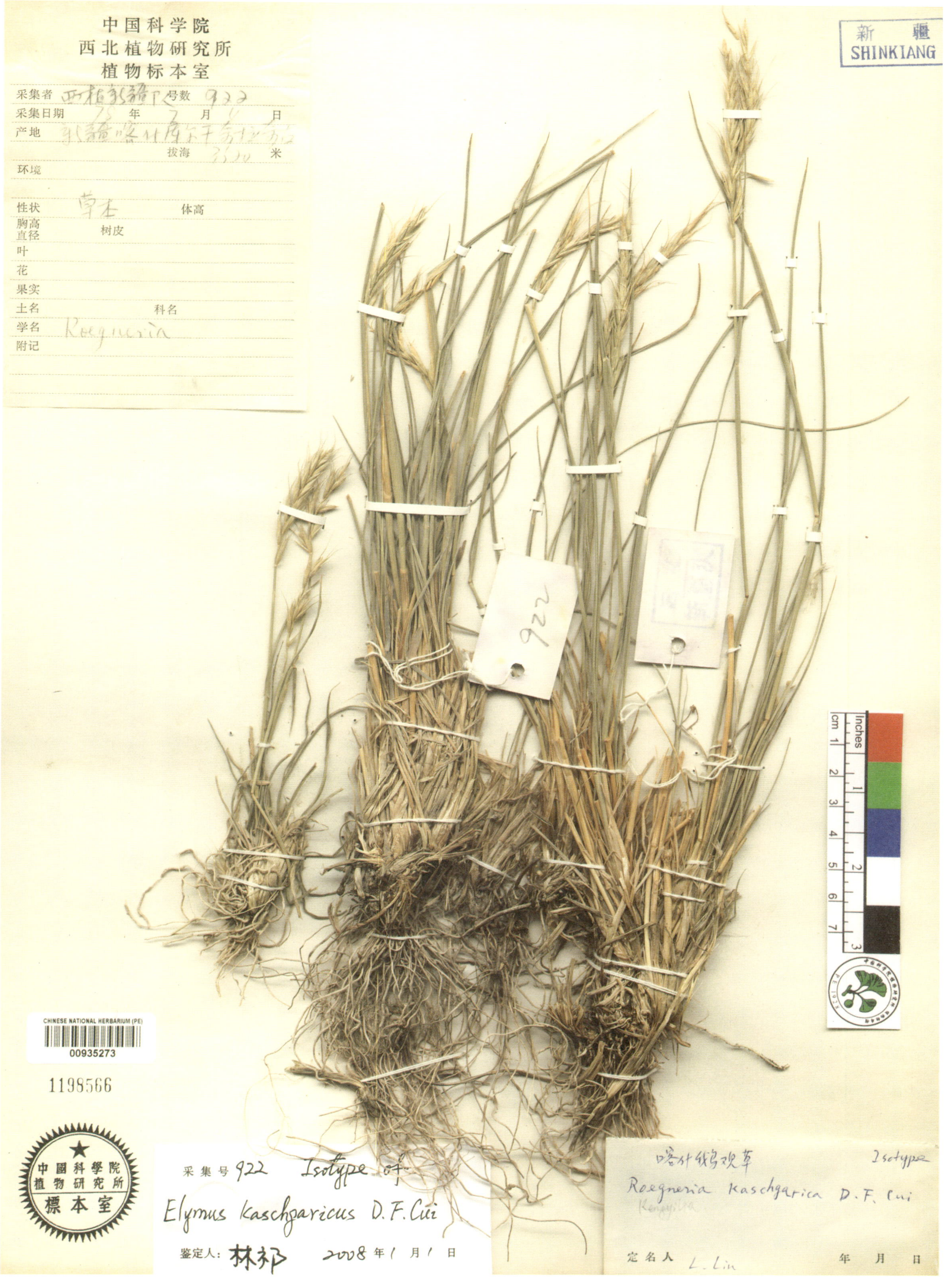

喀什披碱草 ***Elymus kaschgaricus*** D. F. Cui in Bull. Bot. Res., Harbin 10(3): 27, f. 3. 1990. **Isotype:** China. Xinjiang: Kashi, alt. 3520 m, 1978-07-04, Xinjiang Exped. 922.

反折披碱草 ***Elymus retroflexus*** B. R. Lu & B. Salomon in Nord. Journ. Bot. 13(4): 355, f. 2. 1993. **Isotype:** China. Xizang: Gongboyamda, alt. 4100 m, 1988-09-21, J. L. Yang & al. 880786.

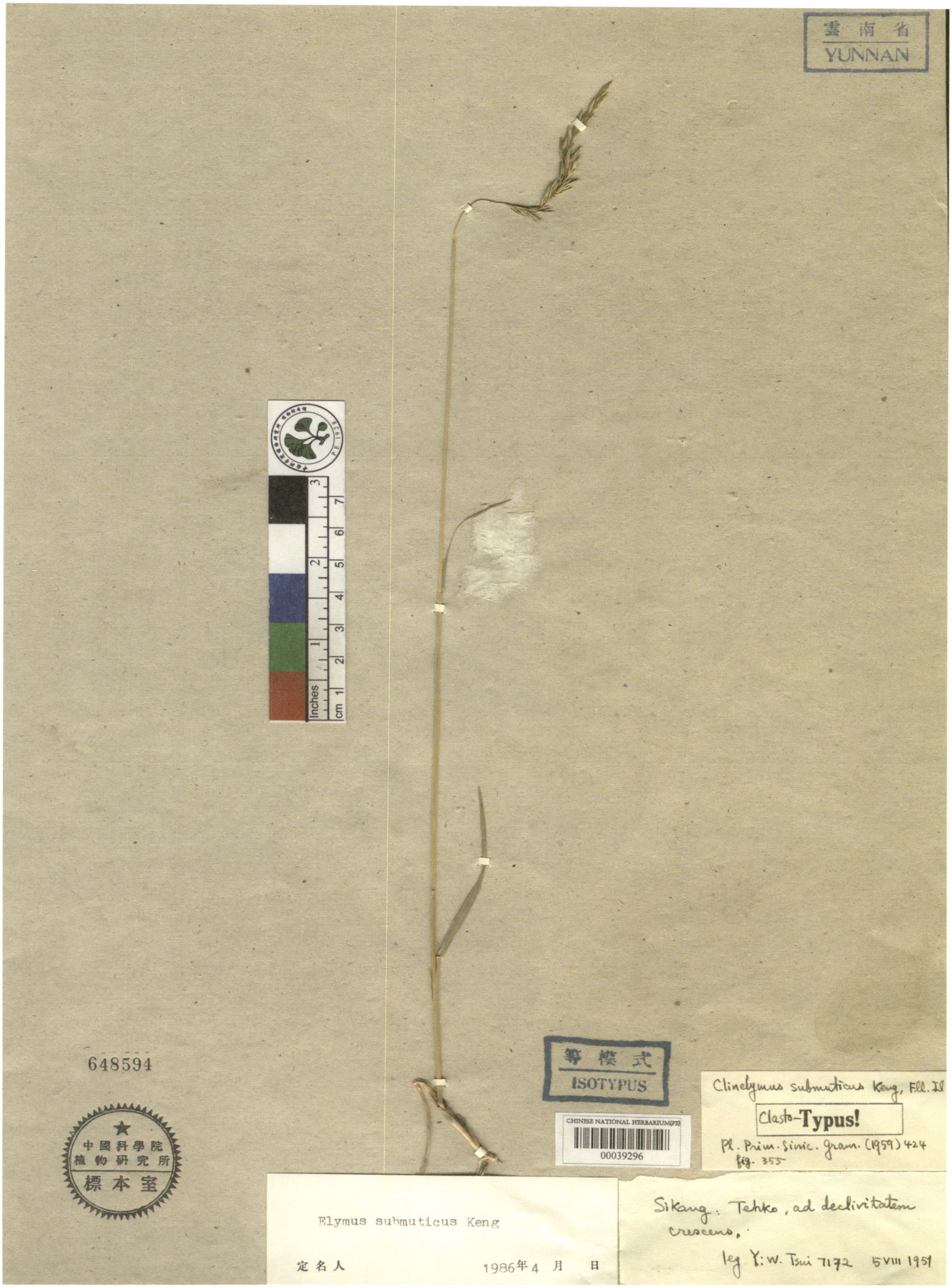

无芒披碱草 ***Elymus submuticus*** Keng ex P. C. Keng in Bull. Bot. Res., Harbin 4(3): 192. 1984. **Isotype:** China. Sichuan: Tehkeh (=Dege), 1951-08-05, Y. W. Cui 7172.

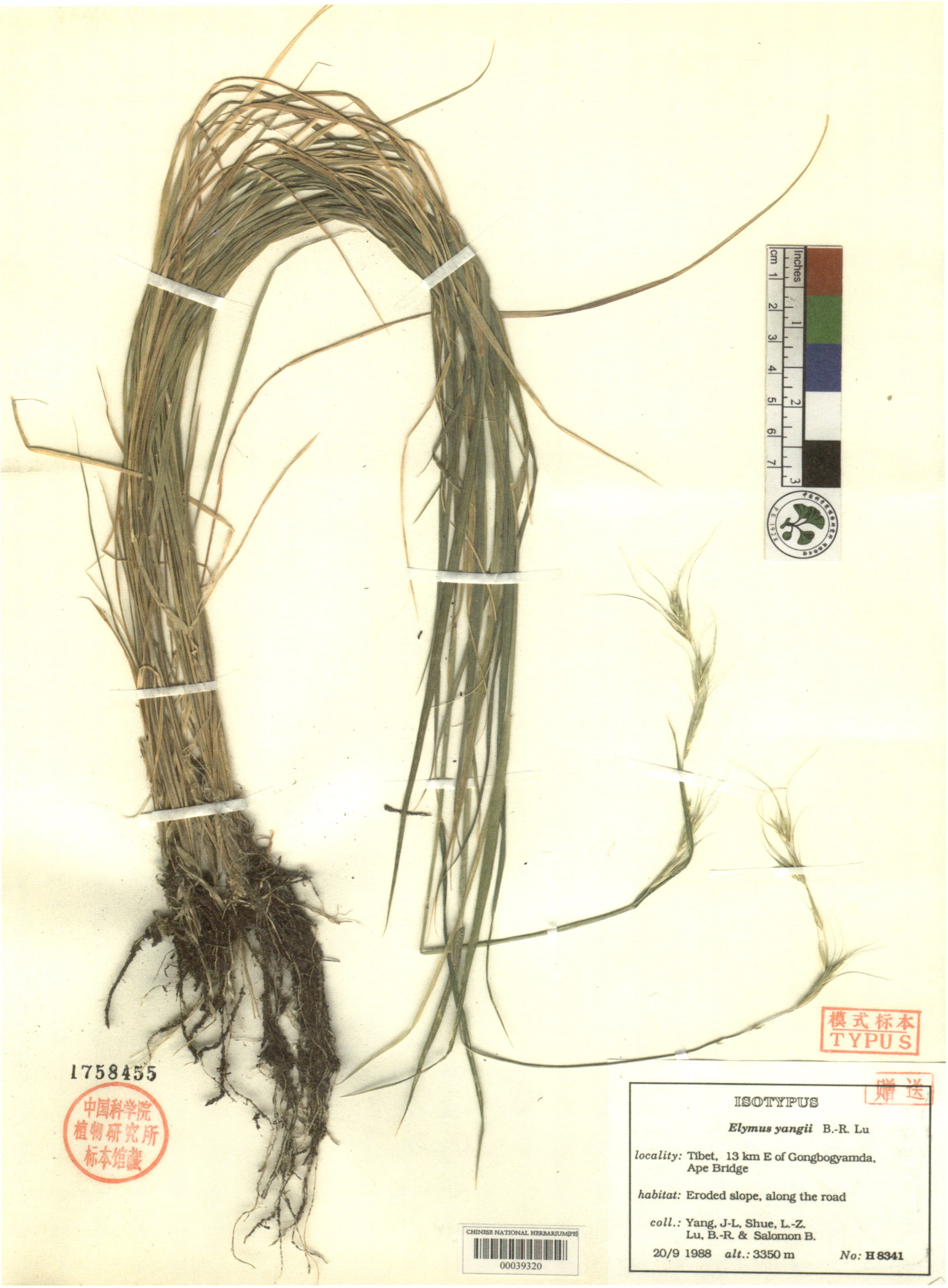

杨氏披碱草 ***Elymus yangii*** B. R. Lu in Willdenowia. 22(1-2): 129, f. 1. 1992. **Isotype:** China. Xizang: Gongbogyamda, alt. 3350 m, 1988-09-20, J. L. Yang & al. 8341.

粗糙九顶草 ***Enneapogon asperatus*** C. E. Hubb. in Hooker's Icon. Pl. 34: t. 3337. 1937. **Isotype:** Australia. Queensland: Gregory North District, Duchess, alt. 350 m, 1931-02-09, C. E. Hubbard. 7348.

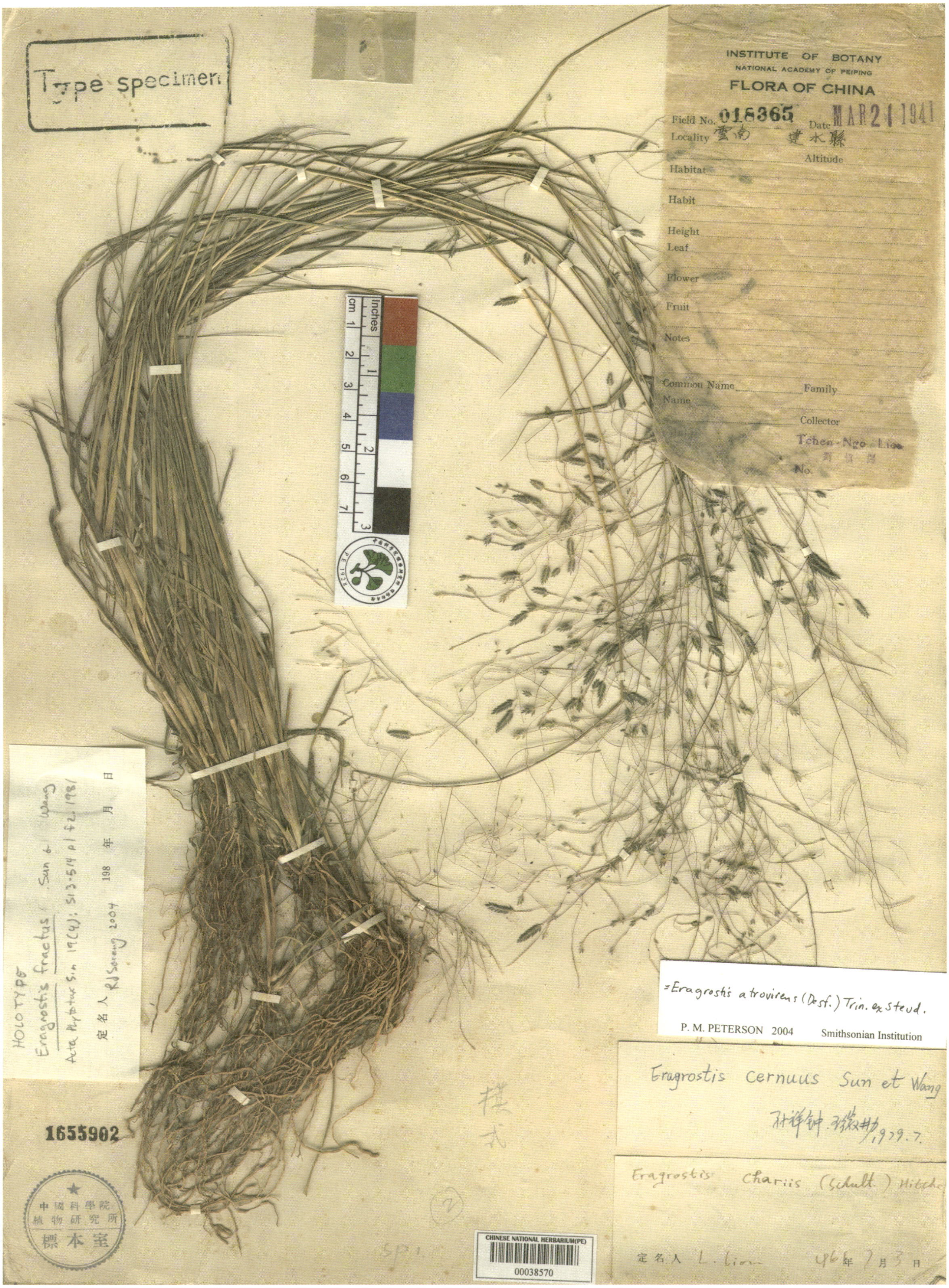

垂穗画眉草 ***Eragrostis fractus*** S. C. Sun & H. Q. Wang in Acta Phytotax. Sin. 19(4): 513, f. 1:3. 1981. **Holotype:** China. Yunnan: Jianshui, 1941-03-21, T. N. Liou 18365.

广 西 画 眉 草 ***Eragrostis guangxiensis*** S. C. Sun & H. Q. Wang in Acta Phytotax. Sin. 19(4): 514, f. 1: 4. 1981.
Holotype: China. Guangxi: Maoqiao, 1955-05-21, Guangxi Exped. 199.

美丽画眉草 ***Eragrostis pulchra*** S. C. Sun & H. Q. Wang in Acta Phytotax. Sin. 19(4): 513, f. 1: 1. 1981. **Holotype:** China. Hainan: Precise locality not known, 1955-??-??, Hainan Exped. 540.

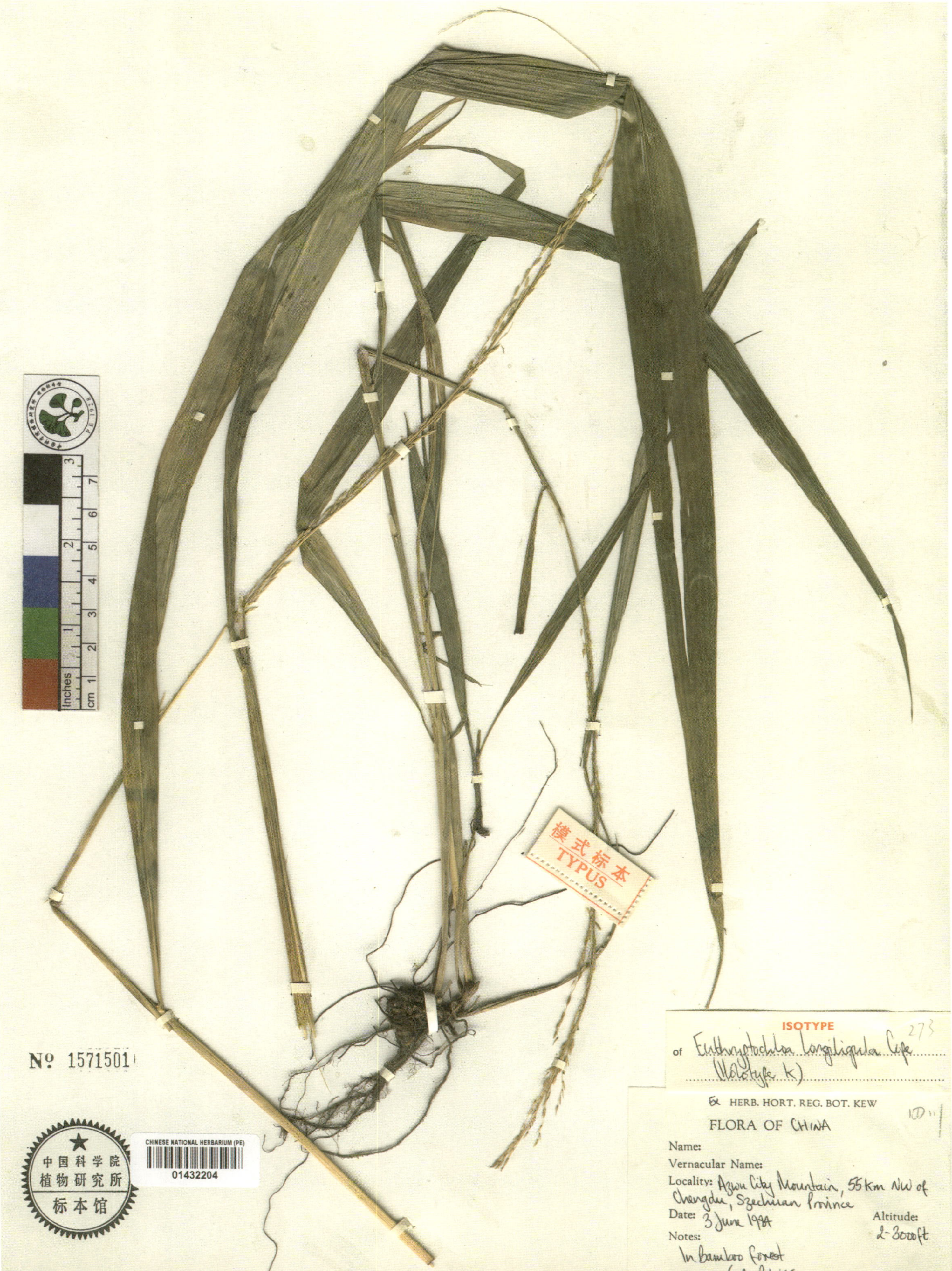

长叶舌显子草 ***Euthryptochloa longiligula*** Cope in Kew Bull. 42(3): 707. 1987. **Isotype:** China. Sichuan: Chengdu, alt. 670~1000 m, 1984-06-03, T. B. Ryves s. n.

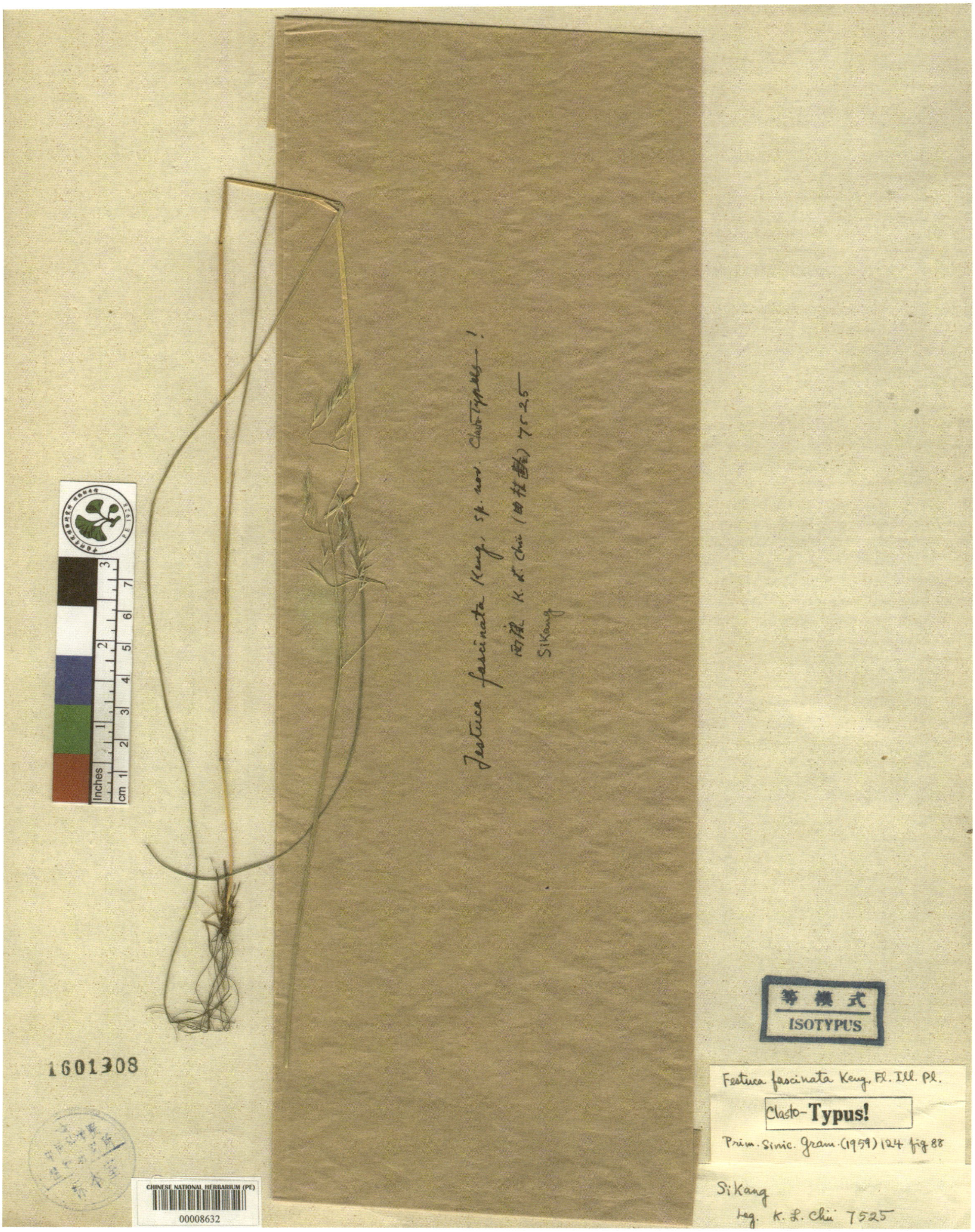

蛊羊茅 ***Festuca fascinata*** Keng ex S. L. Lu in Acta Phytotax. Sin. 30(6): 533. 1992. **Isotype:** China. Sichuan: Precise locality not known, K. L. Chu 7525.

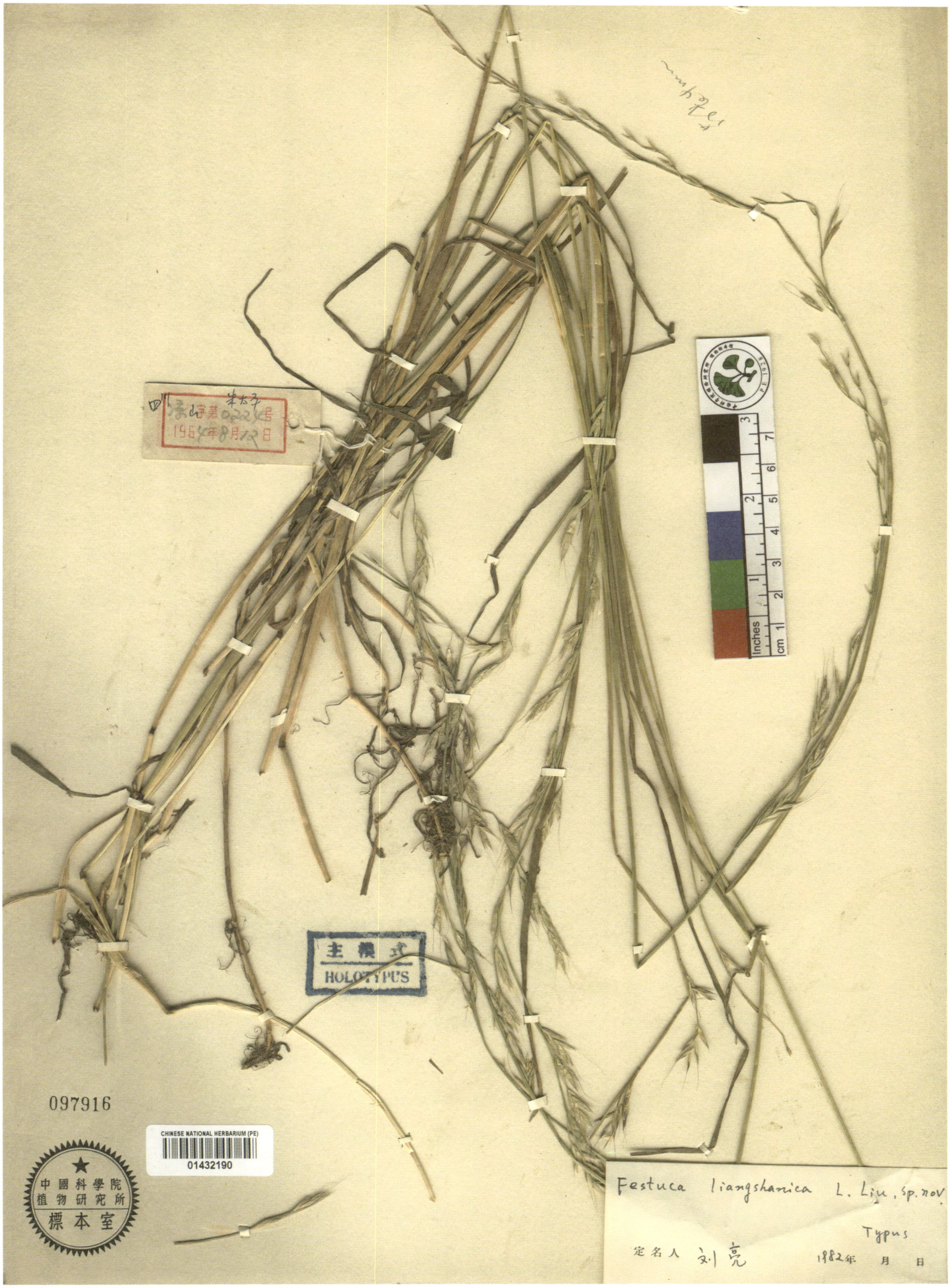

凉山羊茅 ***Festuca liangshanica*** L. Liu, Fl. Reip. Pop. Sin. 9(2): 387. 2002. **Holotype:** China. Sichuan: Xichang, alt. 1200 m, 1964-08-12, T. P. Zhu 224.

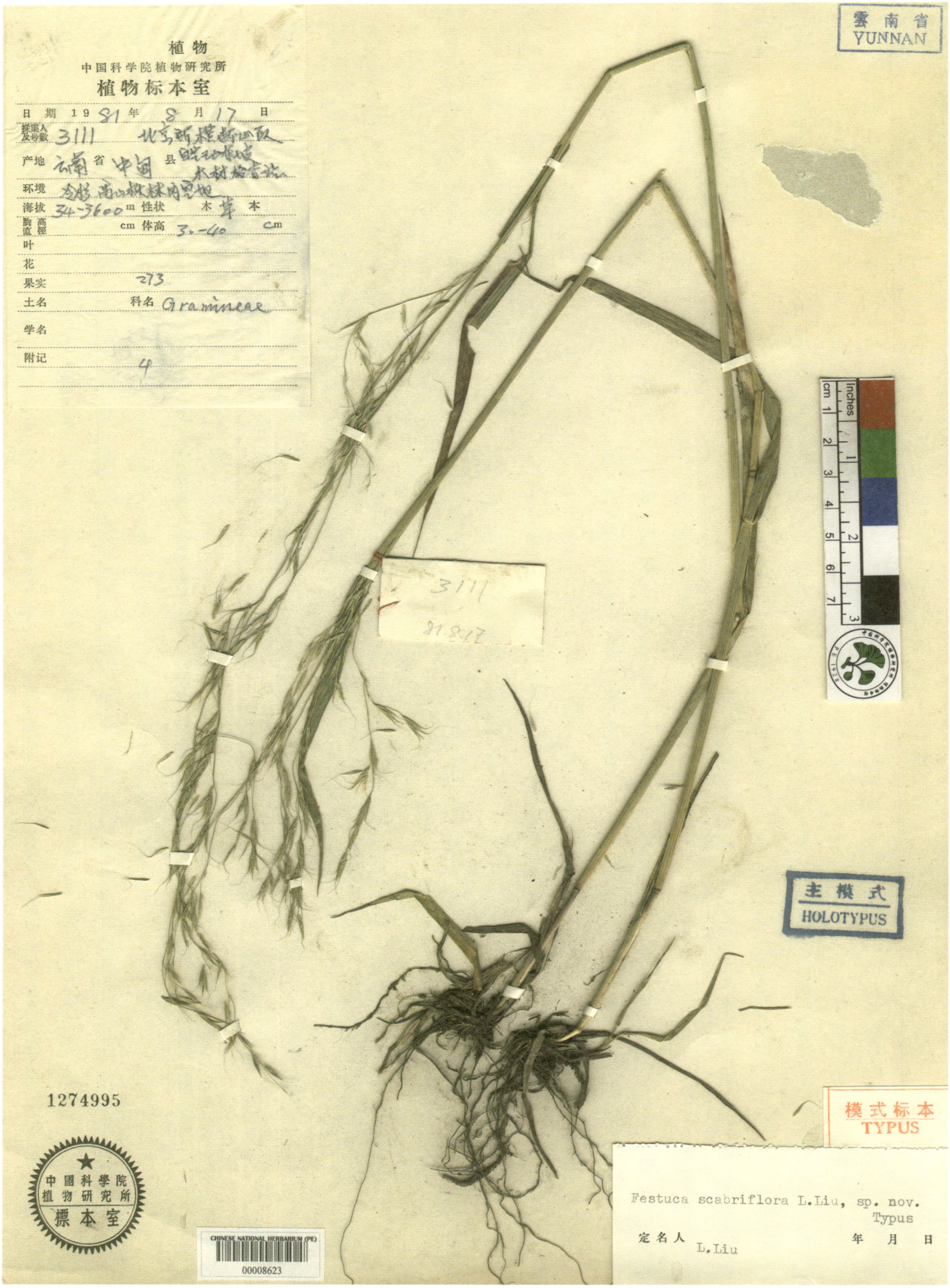

糙花羊茅 ***Festuca scabriflora*** L. Liu, Fl. Reip. Pop. Sin. 9(2): 387. 2002. **Holotype:** China. Yunnan: Zhongdian, alt. 3400~3600 m, 1981-08-17, Hengduanshan Exped. 3111.

中华羊茅 ***Festuca sinensis*** Keng ex S. L. Lu in Acta Phytotax. Sin. 30(6): 536. 1992. **Isotype:** China. Sichuan: Precise locality not known, 1941-07-15, K. L. Chu 7443.

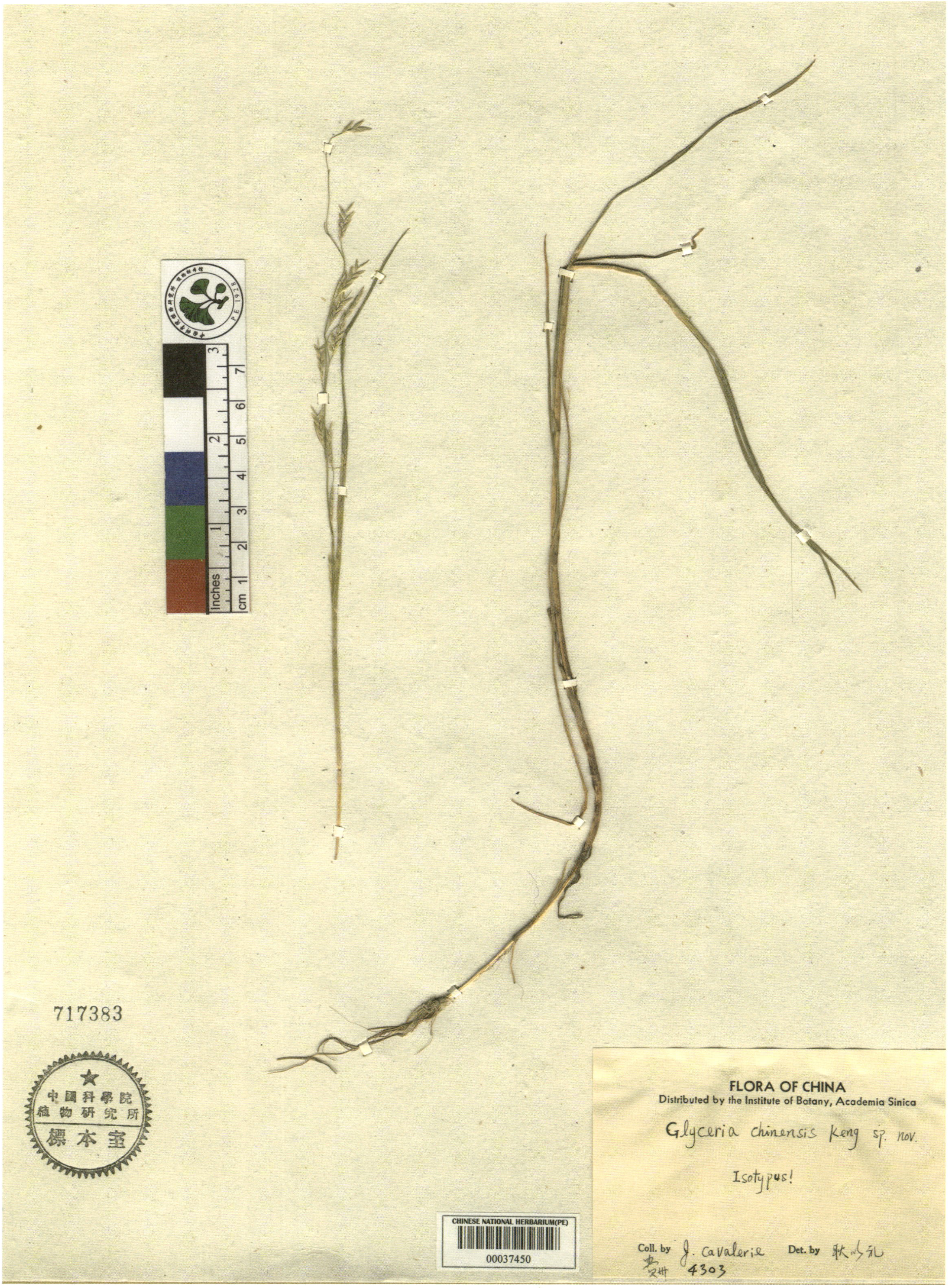

中华甜茅 ***Glyceria chinensis*** Keng ex Z. L. Wu in Acta Phytotax. Sin. 30(2): 174. 1992. **Isotype:** China. Guizhou: Xingyi, 1917-??-??, J. Cavalerie 4303.

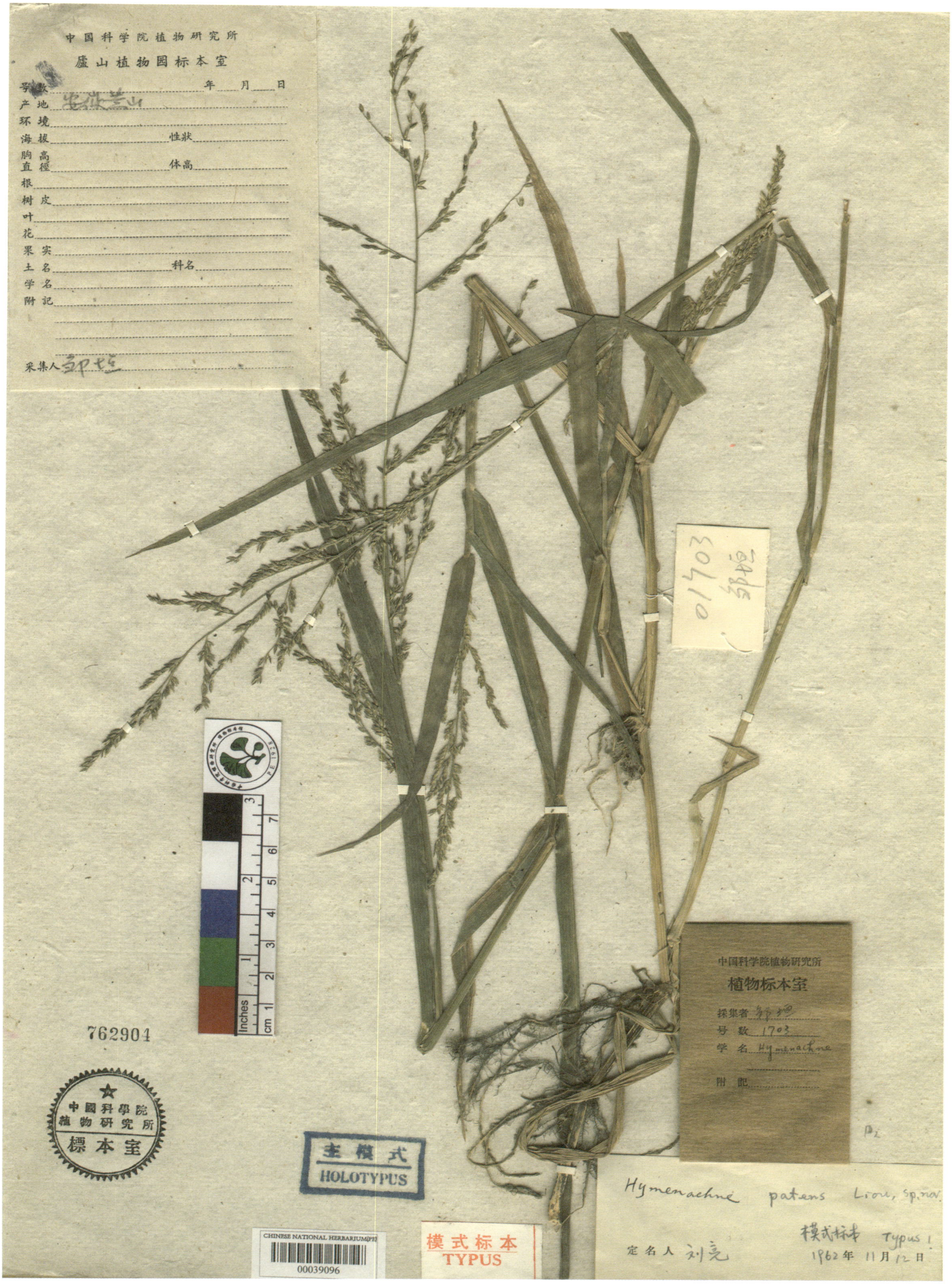

展穗膜稃草 ***Hymenachne patens*** L. Liou in Bot. Res. Acad. Sin. 4: 35, f. 1. 1989. **Holotype:** China. Anhui: Huangshan, Y. Zou 1703.

满山跑 ***Indosasa abbo-hispidula*** Q. H. Dai & C. F. Huang in Journ. Bamboo Res. 3(1): 47, f. 1. 1984. **Isotype:** China. Guangxi: Nanning, 1984-04-10, Q. H. Dai & C. F. Huang 8305.

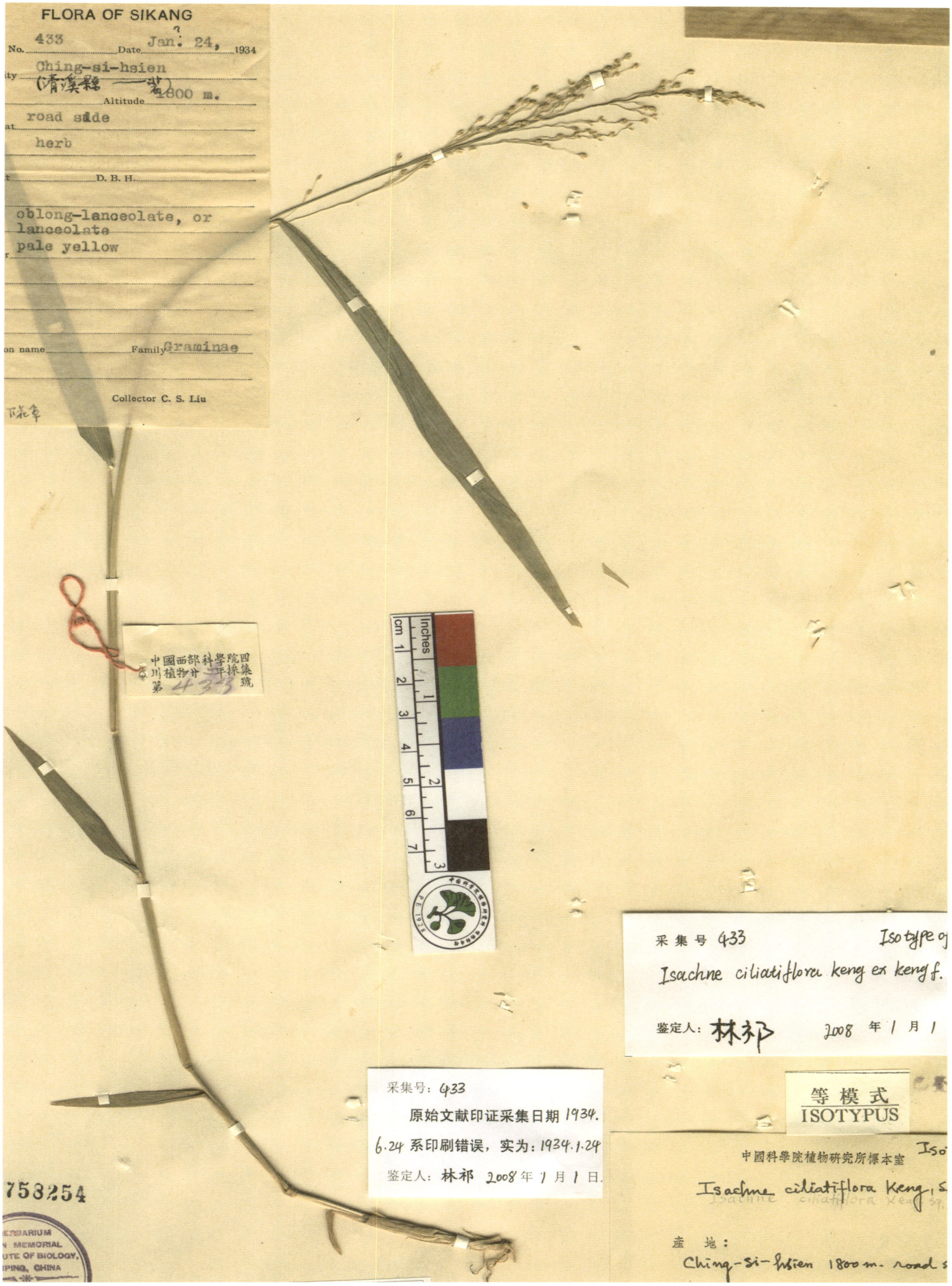

纤毛柳叶箬 ***Isachne ciliatiflora*** Keng ex P. C. Keng in Acta Phytotax. Sin. 10(1): 13. 1965. **Isotype:** China. Sichuan: Qingxi, 1934-01-24, C. S. Liu 433.

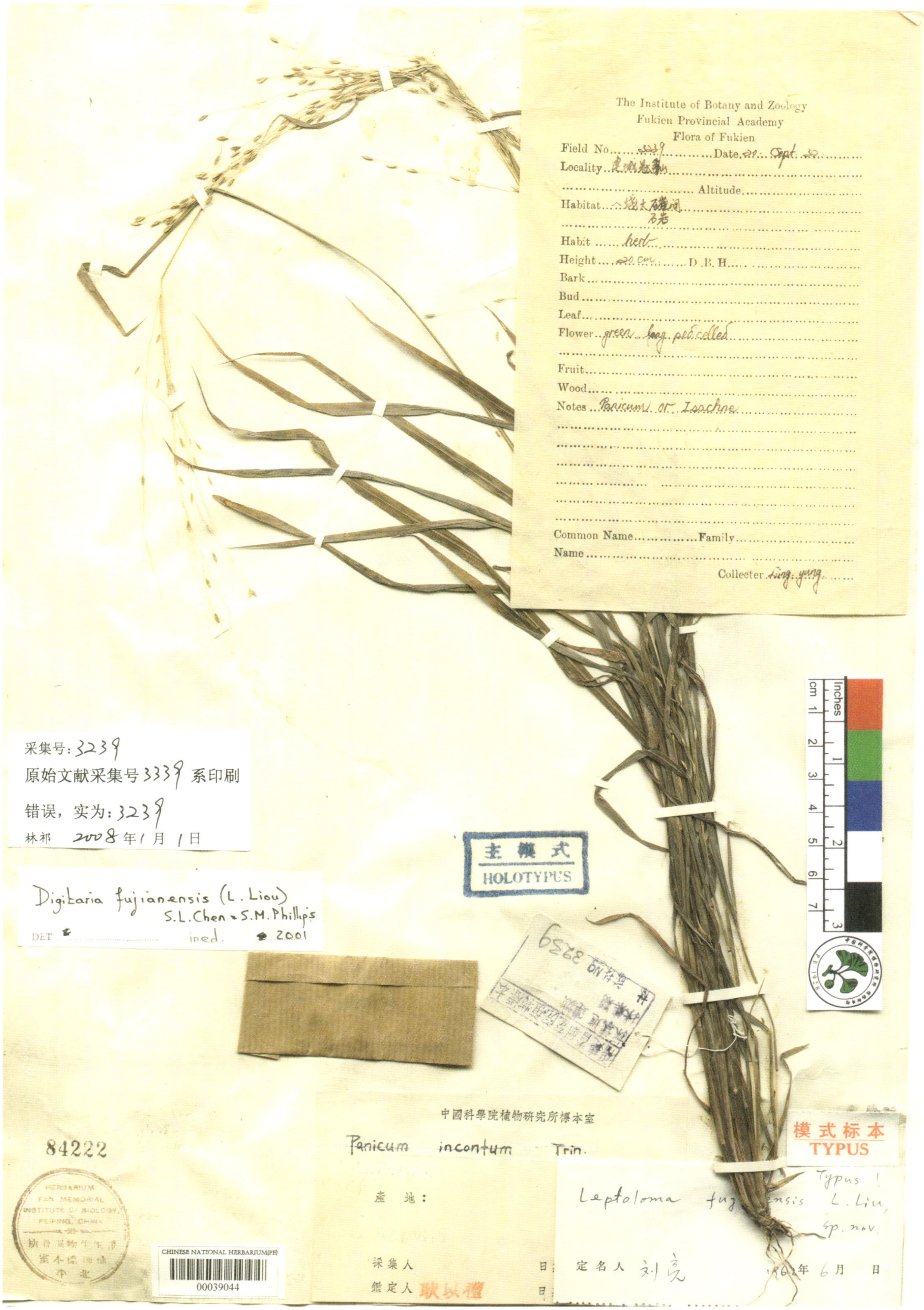

福建薄稃草 ***Leptoloma fujianensis*** L. Liou in Bot. Res. Acad. Sin. 1: 41, f. 1. 1983. **Holotype:** China. Fujian: Liancheng, 1932-09-20, Y. Ling 3239.

扇穗茅 ***Littledalea racemosa*** Keng in Contr. Biol. Lab. Sci. Soc. China, Bot. Ser. 9(2): 136, f. 15. 1934. **Isotype:** China. Sichuan: Sungpan, 1928-08-09, W. P. Fang 4096.

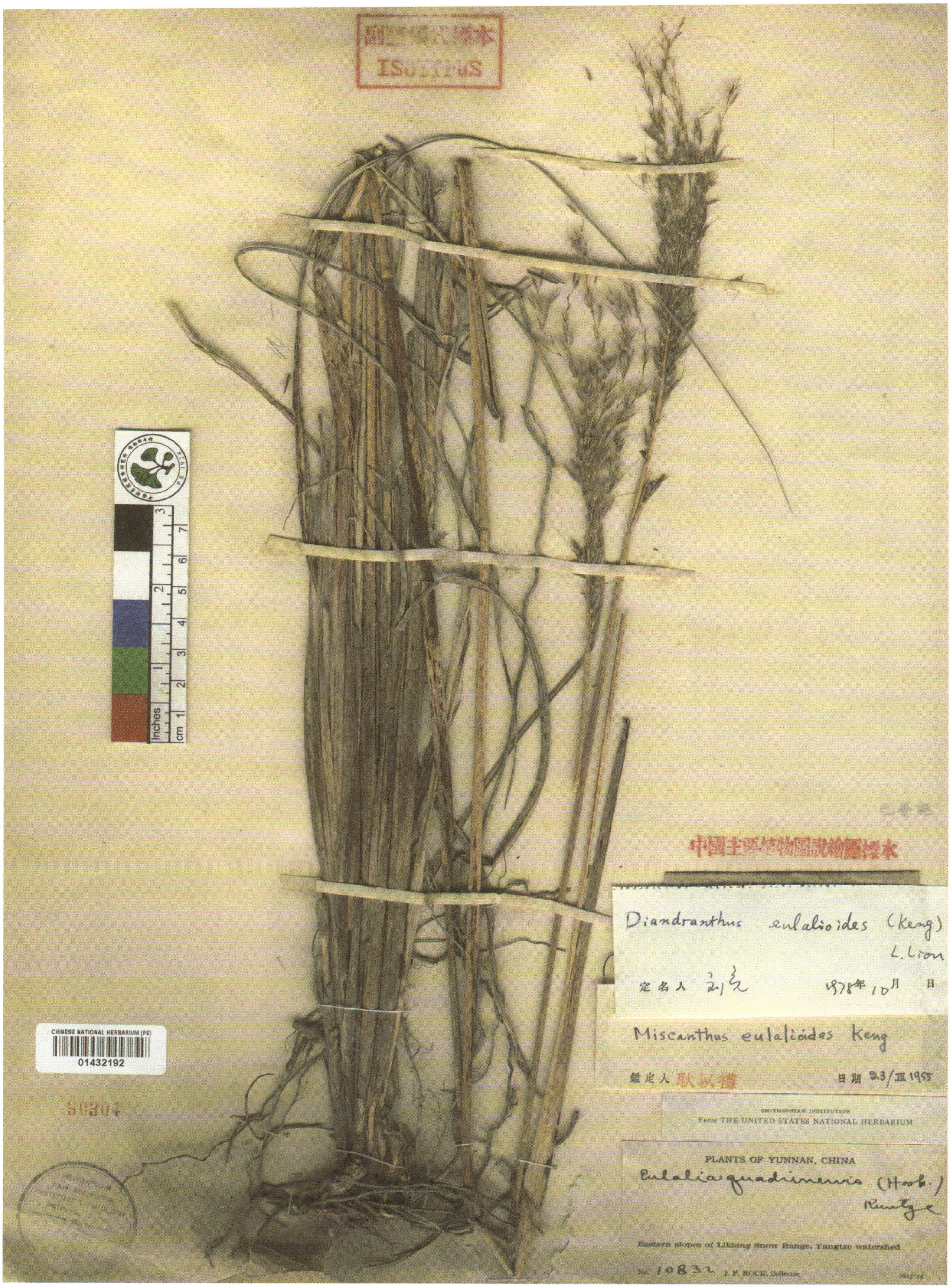

丽江芒 ***Miscanthus eulalioides*** Keng in Sinensia 10(1-6): 288. 1939. **Isotype:** China. Yunnan, Lijiang, 1923-??-??, J. F. Rock 10832.

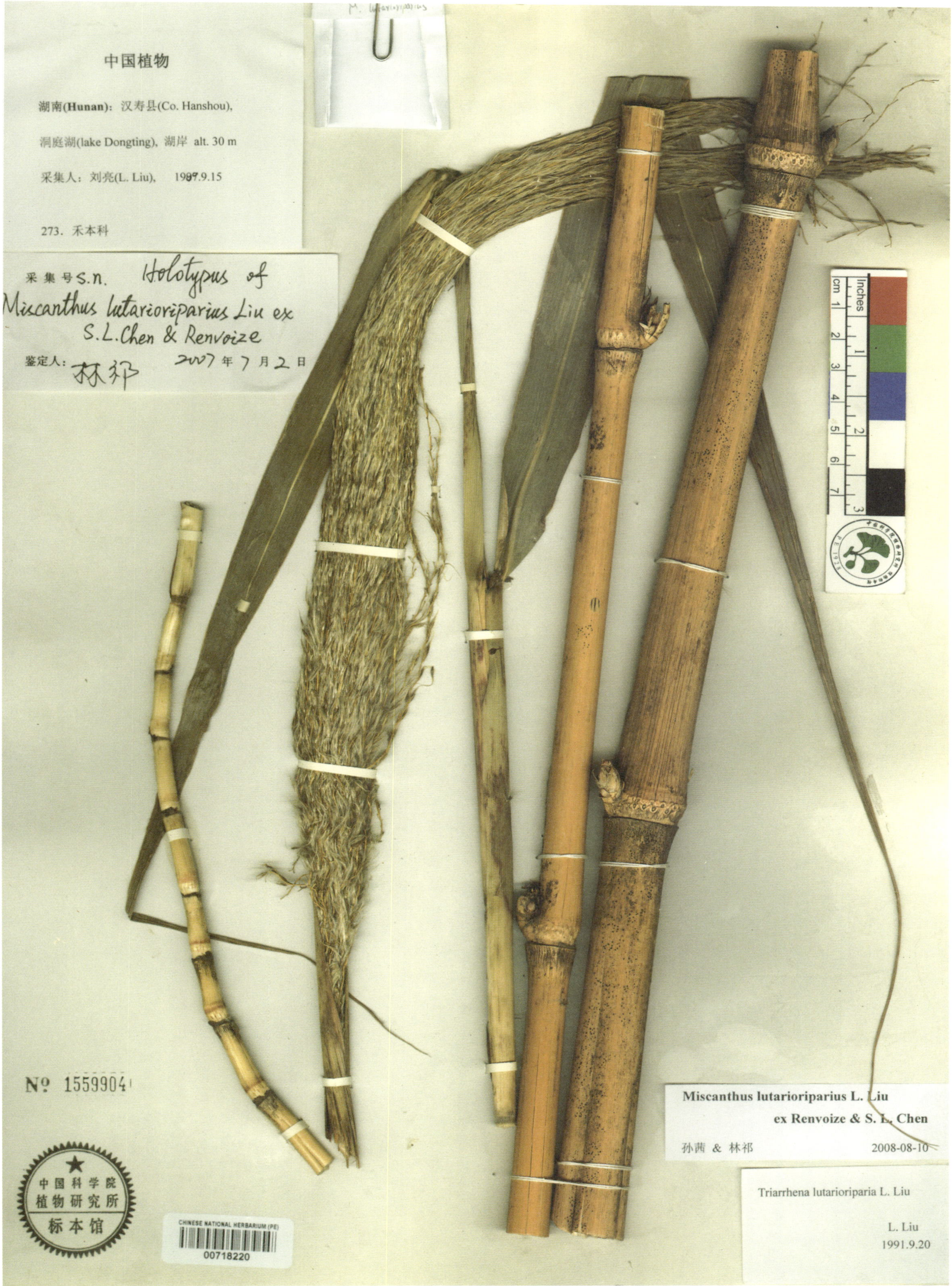

南荻 ***Miscanthus lutarioriparius*** L. Liu ex S. L. Chen & Renvoize in Kew Bull. 60: 605, f. 1. 2005. **Holotype:** China. Hunan: Hanshou, Lake Dongting, alt. 30 m, 1989-09-15, L. Liu s. n.

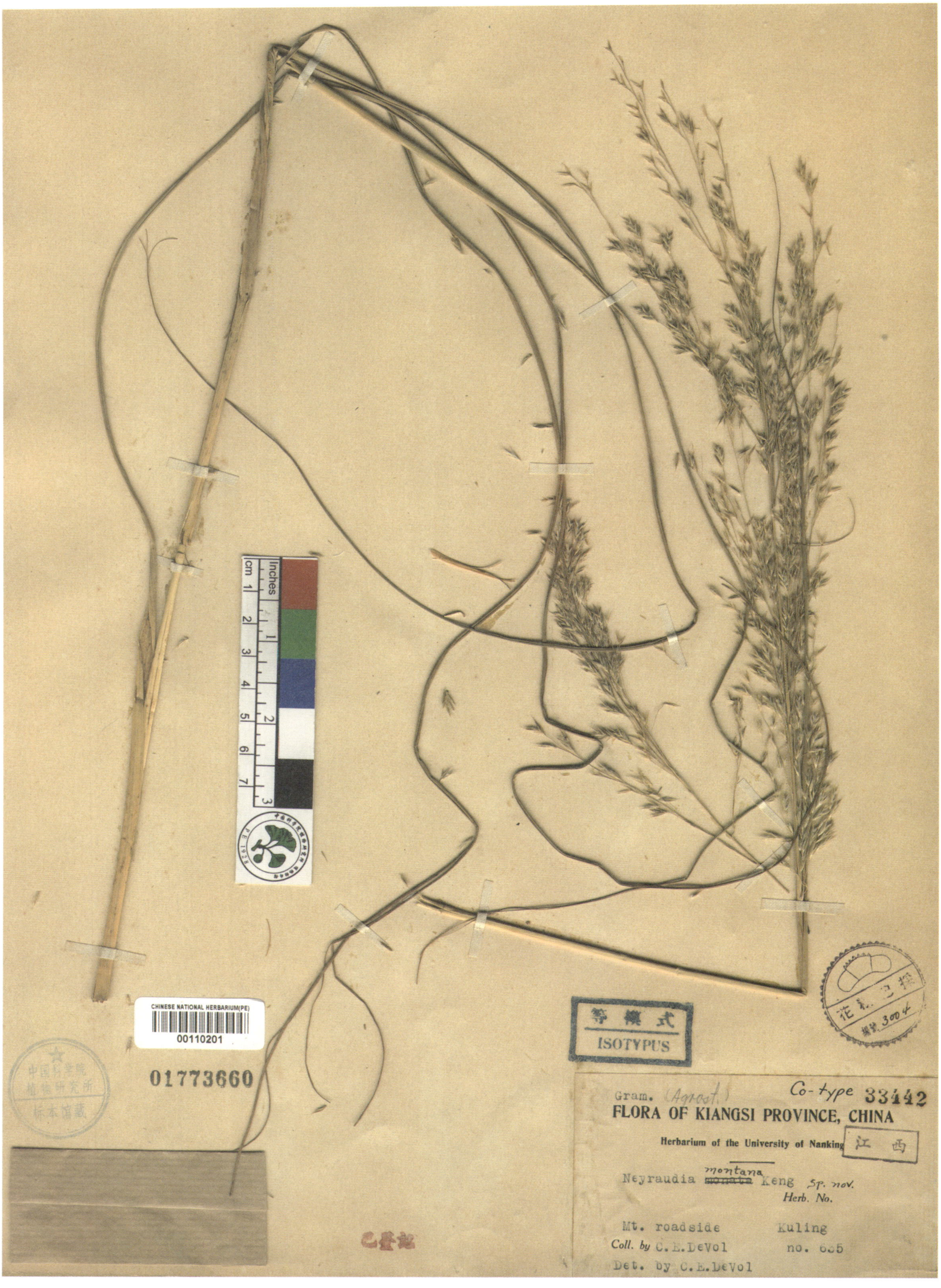

山类芦 ***Neyraudia montana*** Keng in Sinensia 6(1-6): 151. 1935. **Isotype:** China. Jiangxi: Lushan, 1934-08-28, C. E. Devol 635.

高秆固沙草 ***Orinus alticulmus*** L. B. Cai & Tong Lin Zhang in Novon 18: 275, f. 1. 2008. **Isotype:** China. Qinghai: Xining, Yangjiazhai, alt. 2450 m, 1985-08-14, Y. H. Wu 1913.

鸡爪草 ***Orinus anomala*** Keng ex P. C. Keng in Acta Bot. Sin. 9(1): 68. 1960. **Isotype:** China. Sichuan: Precise locality not known, 1940-07-25, K. L. Chu 7469.

中华芨芨草 ***Oryzopsis chinensis*** A. S. Hitchcock in Proc. Biol. Soc. Washington 43: 92.1930. **Isotype:** China. Shansi: Mt. Path, alt. 1200 m, 1929-05-18, T. Tang 788.

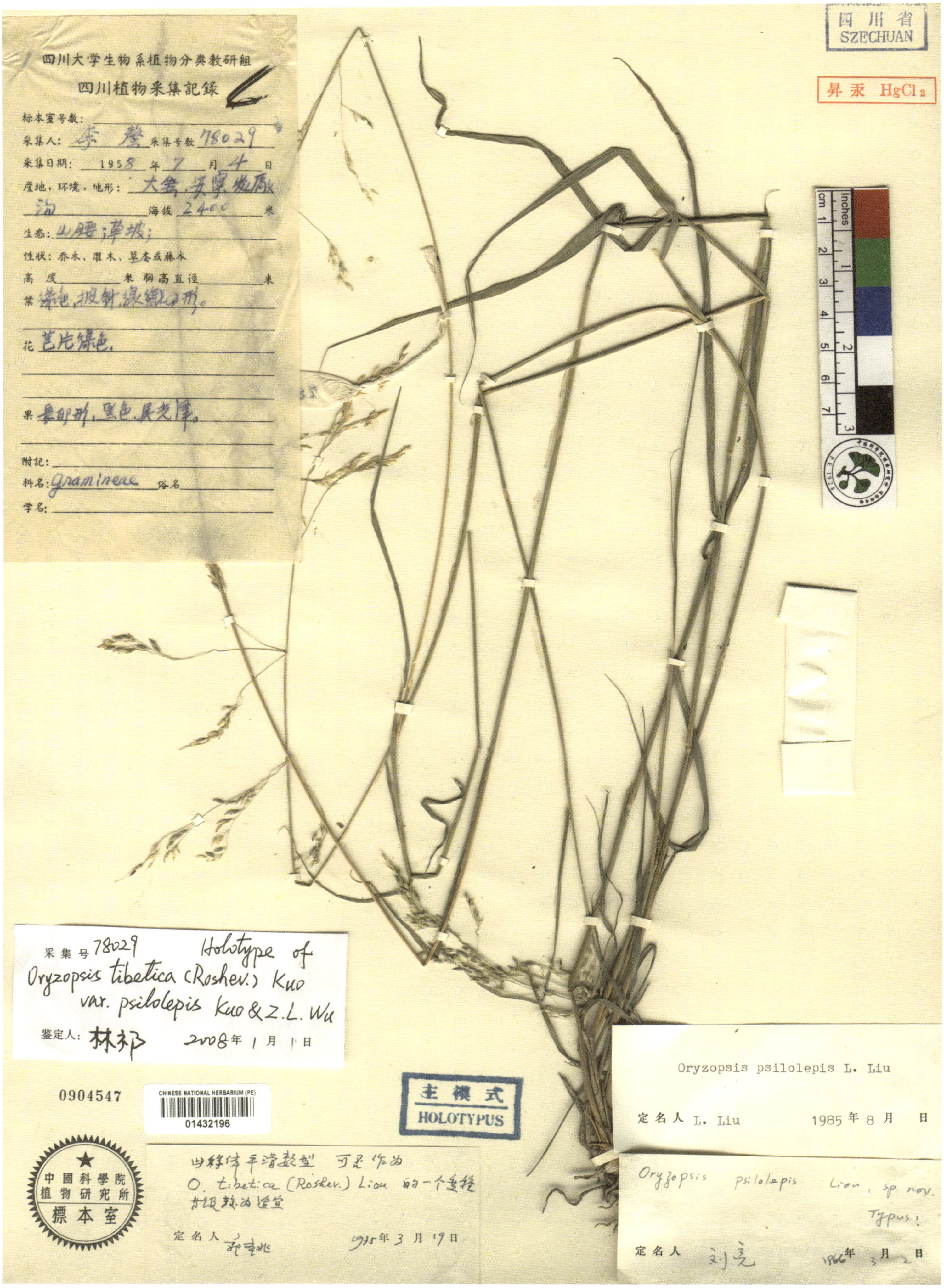

光稃落芒草 ***Oryzopsis tibetica*** (Roshev.) P. C. Kuo var. ***psilolepis*** P. C. Kuo & Z. L. Wu in Acta Phytotax. Sin. 19(4): 435. 1981. **Holotype:** China. Sichuan: Dajiang, Anning, alt. 2400 m, 1958-07-04, X. Li 78029.

南峤滇竹 ***Oxytenanthera parviflora*** P. C. Keng in Acta Phytotax. Sin. 6(4): 358, pl. 57. 1957. **Isotype:** China. Yunnan: Nanchiao (=Wufu), alt. 1460 m, 1936-06-??, C. W. Wang 75259.

无稃细柄黍 ***Panicum psilopodium*** Trin. var. ***epaleatum*** Keng ex S. L. Chen, T. D. Zhuang & X. L. Yang in Bull. Bot. Res., Harbin 4(3): 198. 1984. **Isotype:** China. Guizhou: Dushan, alt. 500 m, 1930-09-07, Y. Tsiang 6912.

长序狼尾草 ***Pennisetum longissimum*** S. L. Chen & Y. X. Jin in Bull. Bot. Res., Harbin 4(1): 65, pl. 2. 1984. **Paratype:** China. Shaanxi: Foping, alt. 1050 m, 1952-07-16, K. T. Fu 5048.

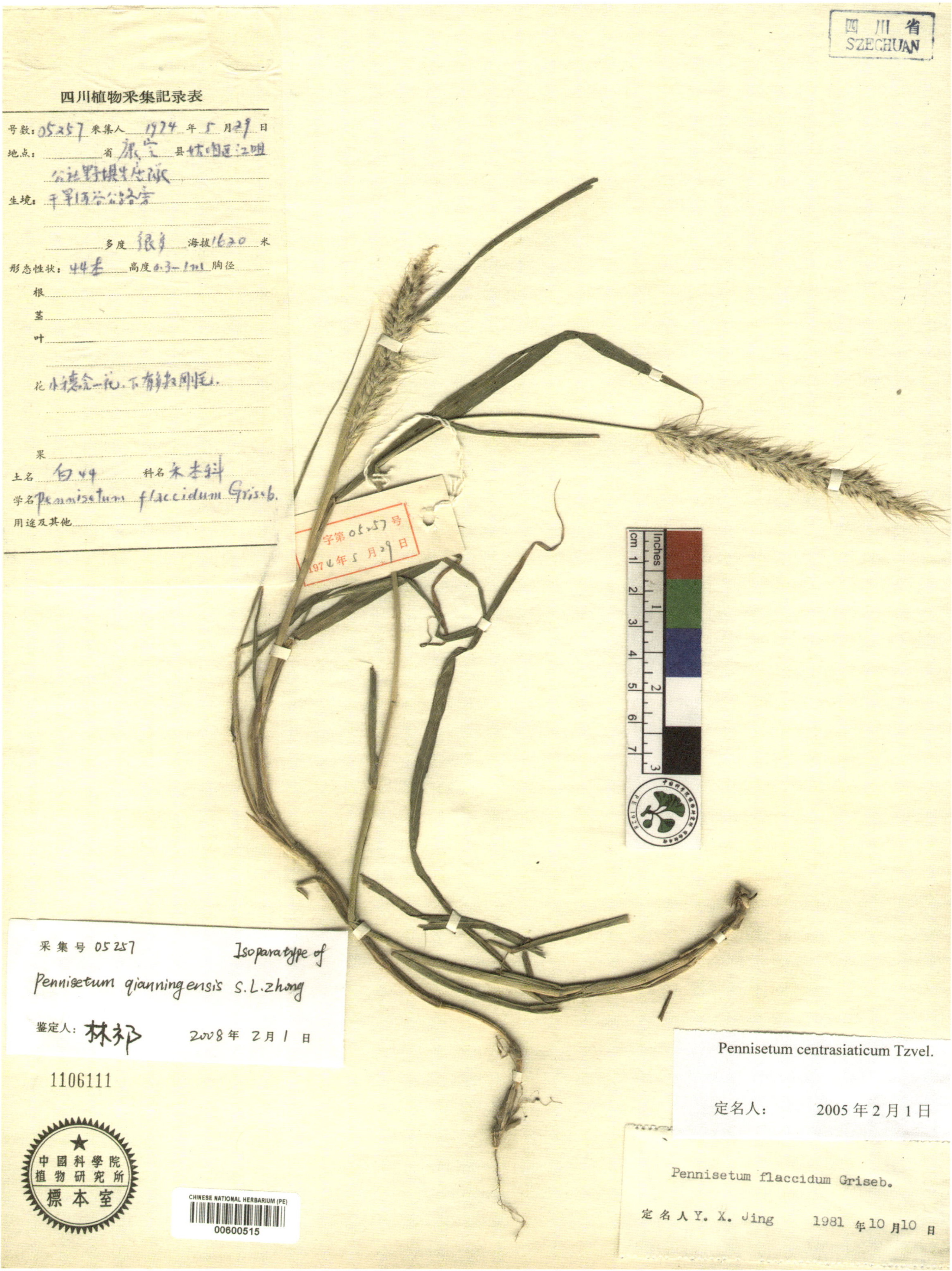

乾宁狼尾草 ***Pennisetum qianningensis*** S. L. Zhong in Journ. SW Agri. Coll. 1982(4): 75, f. 1. 1982. **Isoparatype:** China. Sichuan, Kangding, alt. 1620 m, 1974-05-29, W Sichuan Exped. 5257.

花毛竹 ***Phyllostachys pubescens*** Mazel. f. ***huamozhu*** T. H. Wen in Acta Phytotax. Sin. 16(4): 99. 1978. **Holotype:** China. Zhejiang: Moganshan, 1958-07-30, T. H. Wen 58701.

红后竹 ***Phyllostachys rubicunda*** T. H. Wen in Acta Phytotax. Sin. 16(4): 98. 1978. **Holotype:** China. Zhejiang: Anji, 1961-05-23, T. H. Wen 61528.

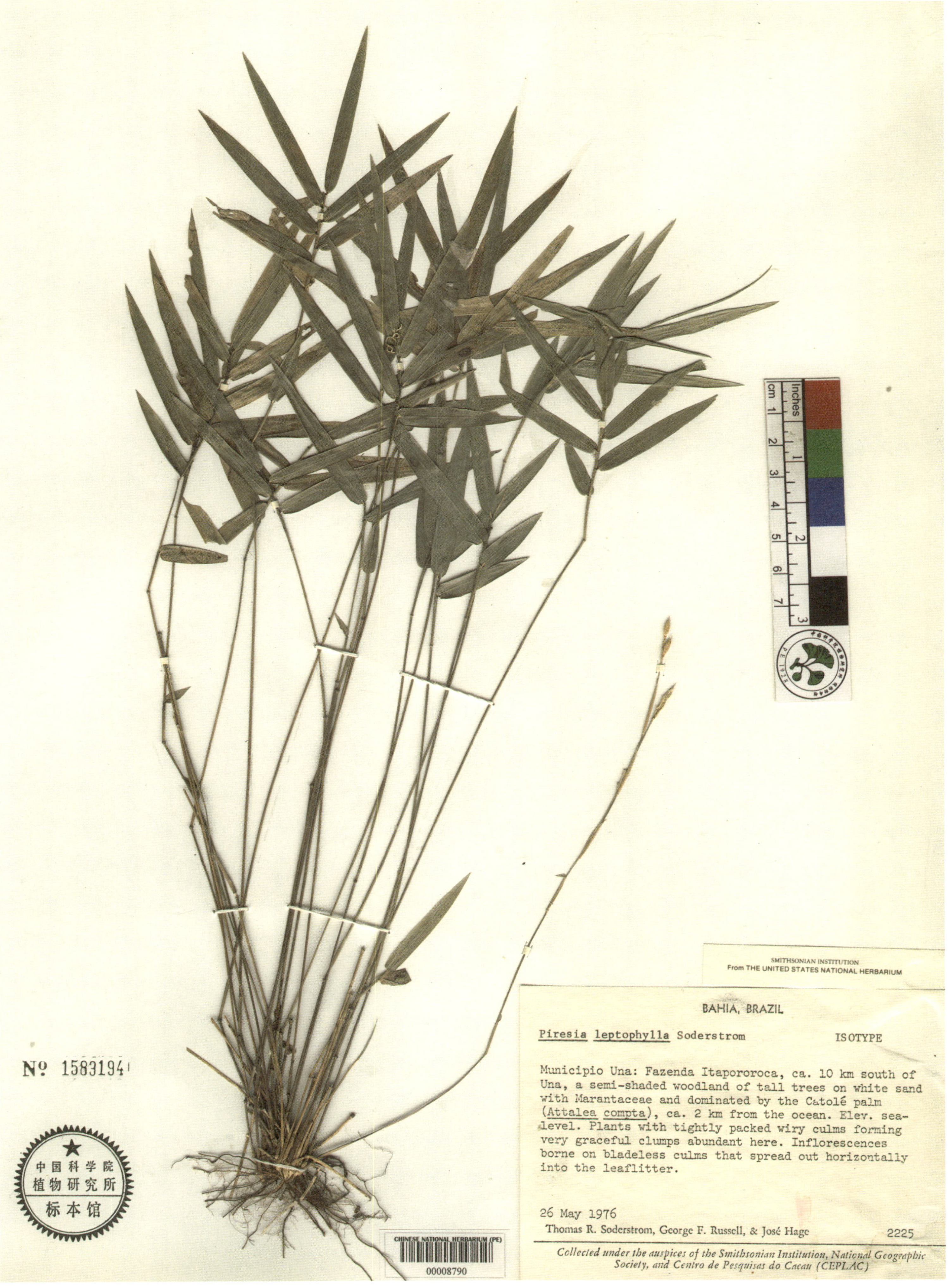

狭长叶矮竹 ***Piresia leptophylla*** Soderstrom in Brittonia. 34(2): 206, f. 4. 1982. **Isotype:** Brazil. Bahia: Municipio Una, 1976-05-26, T. R. Soderstrom, G. Russel & J. Hage 2225.

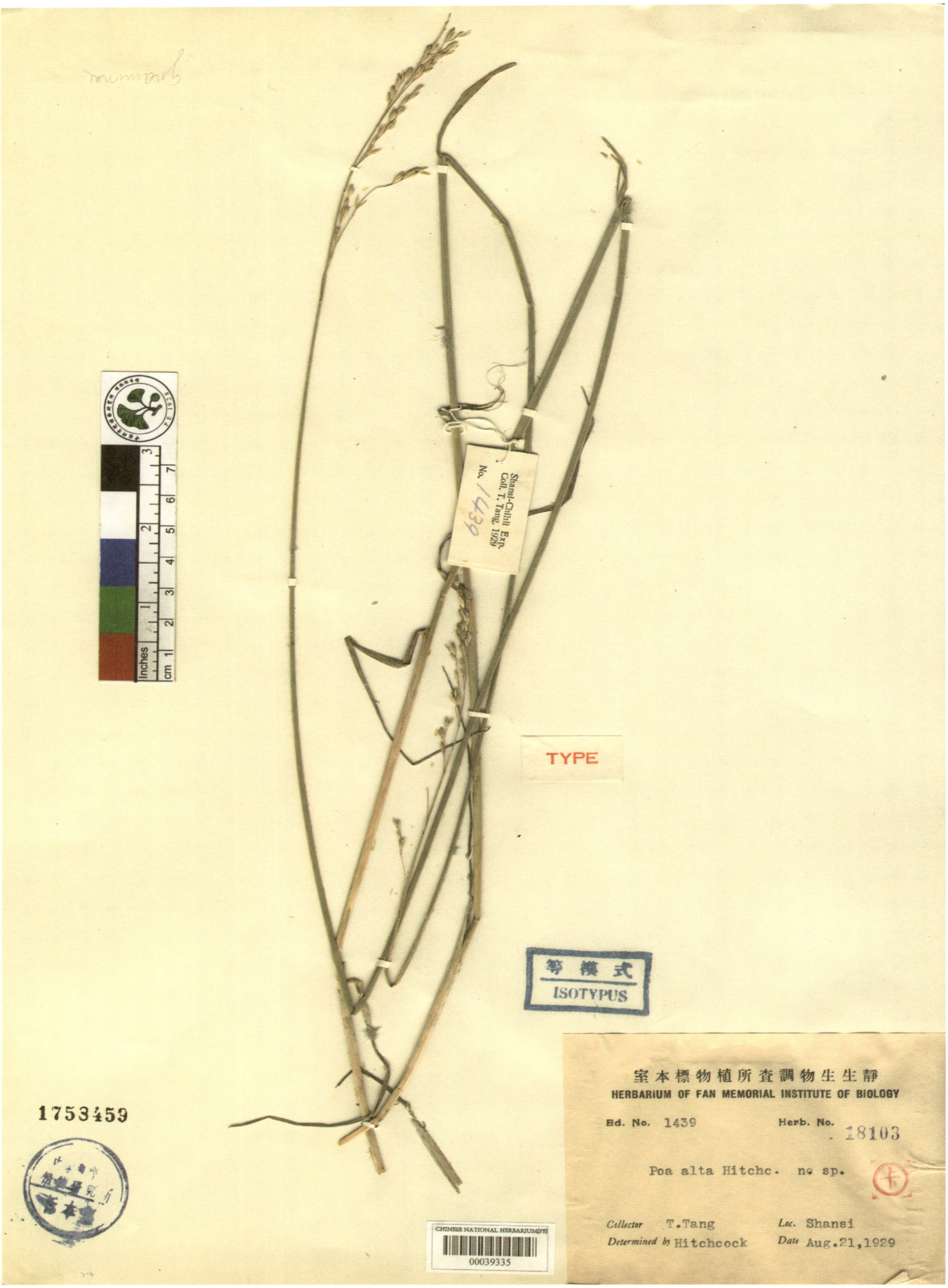

高株早熟禾 ***Poa alta*** A. S. Hitchcock in Proc. Biol. Soc. Washington 43: 93. 1930. **Isotype:** China. Shanxi: Ningwu, 1929-08-21, T. Tang 1439.

波密早熟禾 ***Poa bomiensis*** C. Ling in Acta Phytotax. Sin. 17(1): 101, pl.1: 1-3. 1979. **Holotype:** China. Xizang: Bomi, alt. 4200 m, 1965-08-03, T. S. Ying. & D. Y. Hong 650863.

藏北早熟禾 ***Poa boreali-tibetica*** C. Ling in Acta Phytotax. Sin. 17(1): 102, f. 3. 1979. **Holotype:** China. Xizang: Baingoin, alt. 4900 m, 1976-08-03, Qinghai-Xizang Exped. 12143.

疏花早熟禾 ***Poa chalarantha*** Keng ex L. Liu, Fl. Reip. Pop. Sin. 9(2): 390. 2002. **Isotype:** China. Sichuan: Yulingong, 1940-06-01, K. L. Chu 7012.

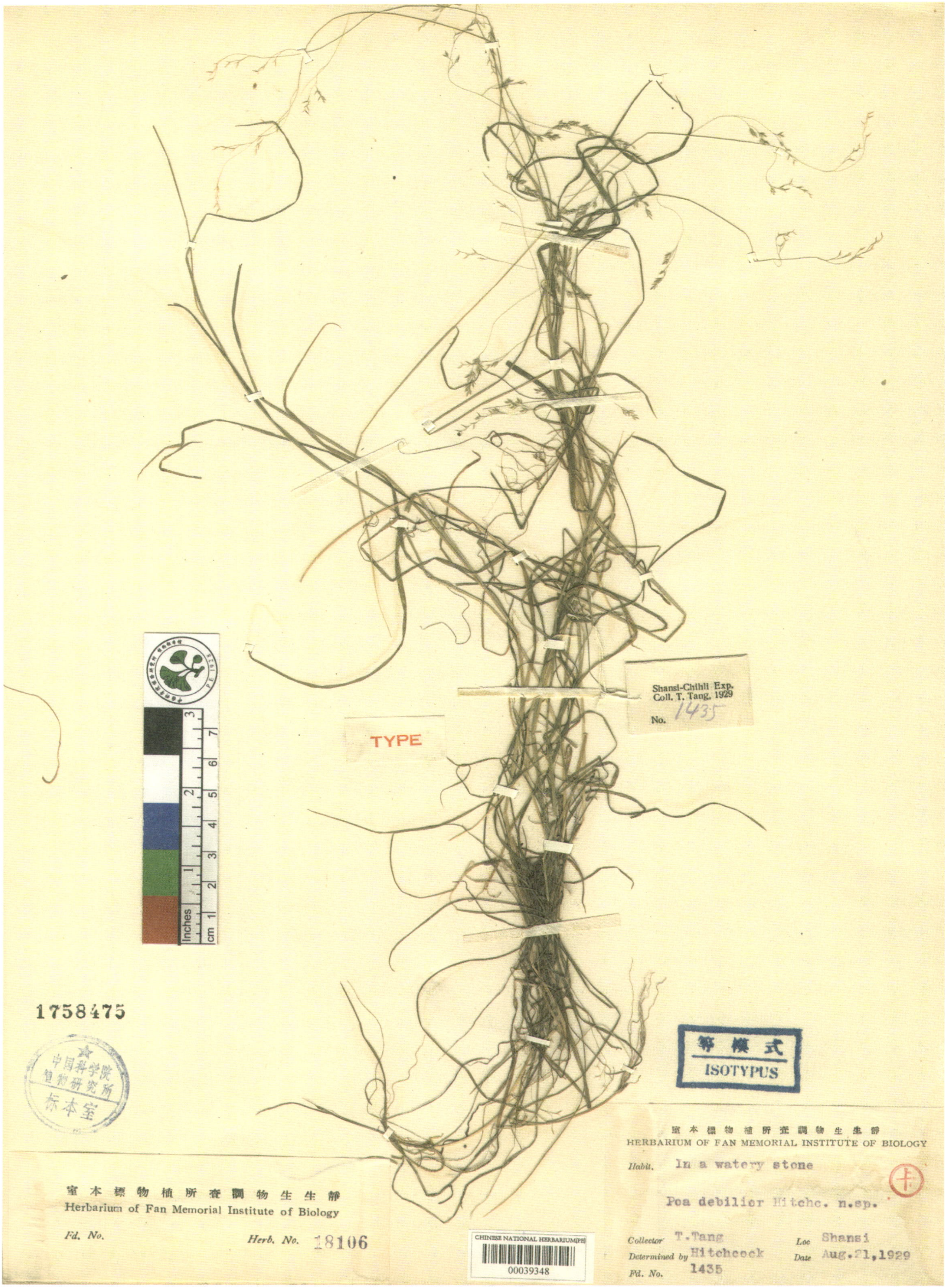

垂枝早熟禾 ***Poa debilior*** A. S. Hitchcock in Proc. Biol. Soc. Washington 43: 93. 1930. **Isotype:** China. Shanxi: Ningwu, alt. 2000 m, 1929-08-21, T. Tang 1435.

蛊早熟禾 ***Poa fascinata*** Keng ex L. Liu, Fl. Reip. Pop. Sin. 9(2): 392. 2002. **Isotype:** China. Gansu: Minxian, alt. 3700 m, 1937-10-31, T. P. Wang 7661.

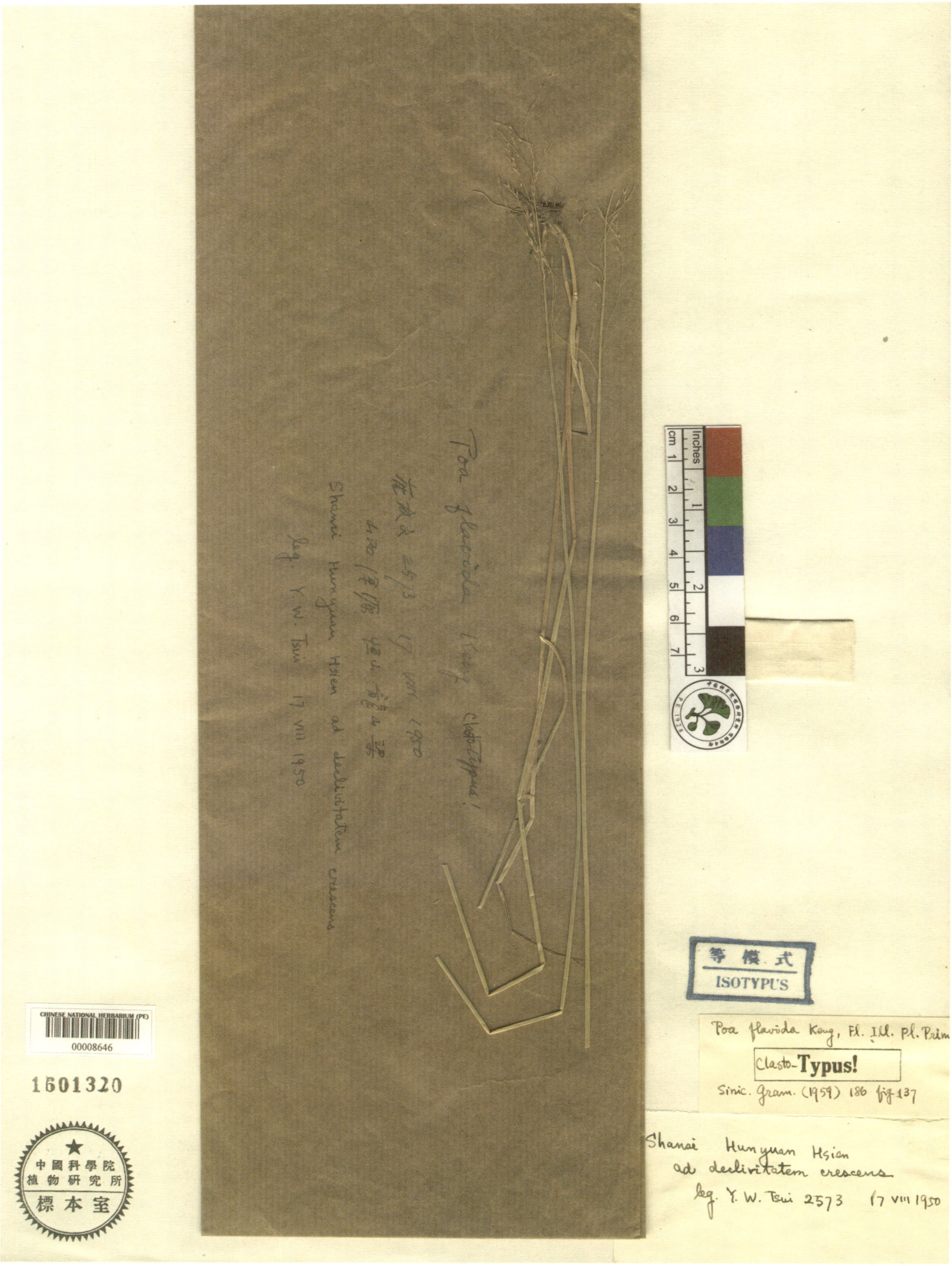

黄色早熟禾 ***Poa flavida*** Keng ex L. Liu, Fl. Reip. Pop. Sin. 9(2): 393. 2002. **Isotype:** China. Shanxi: Hunyuan, 1950-08-17, Y. Y. Tsui 2573.

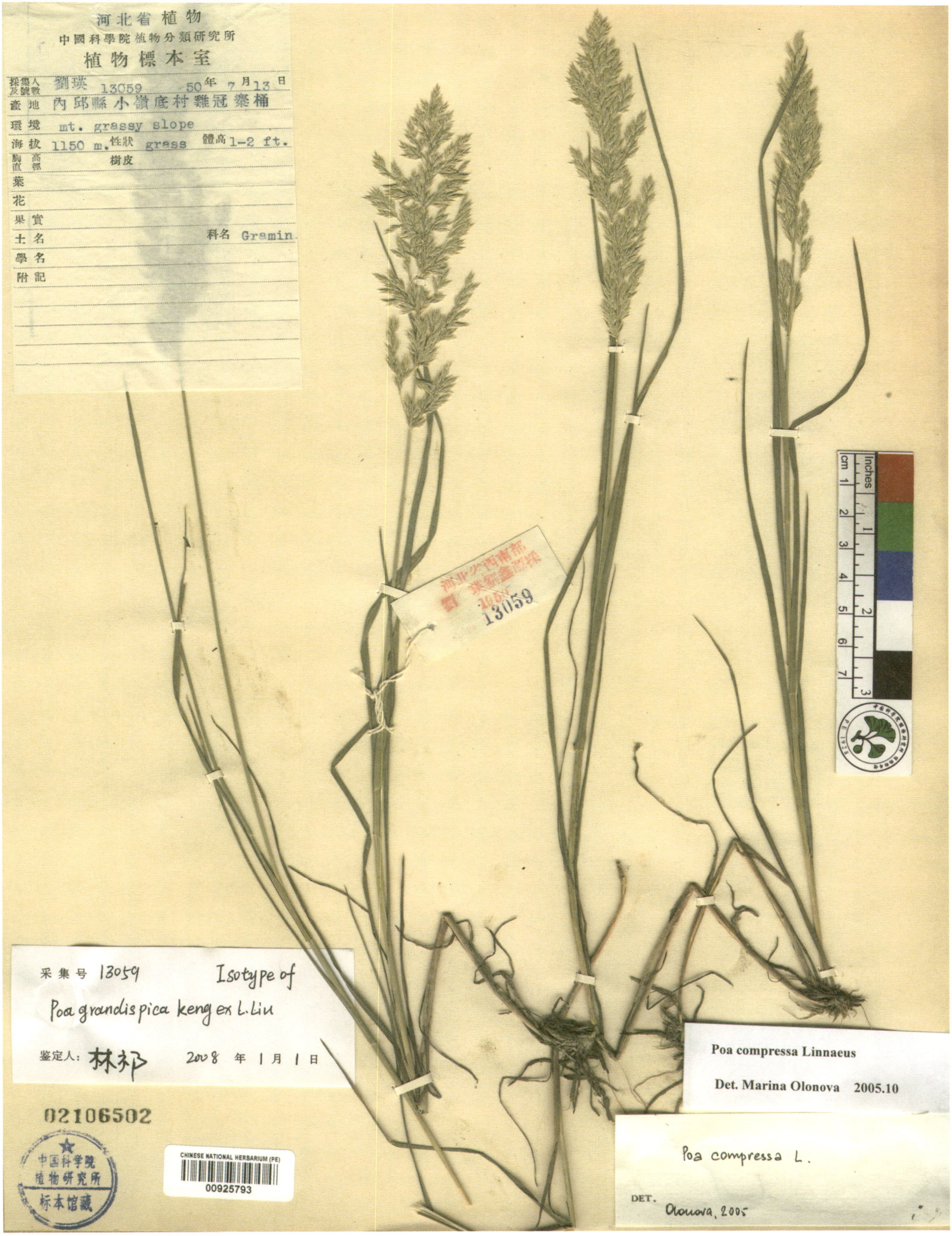

大穗早熟禾 ***Poa grandispica*** Keng ex L. Liu, Fl. Reip. Pop. Sin. 9(2): 394. 2002. **Isotype:** China. Hebei: Neiqiu, alt. 1150 m, 1950-07-13, Y. Liu 13059.

堇色早熟禾 ***Poa ianthina*** Keng ex S. Chen, Fl. Intramongol. 260, pl. 36, f. 1-3. 1983. **Isotype:** China. Shanxi, Wutai, Wutaishan, 1929-07-18, T. Tang 1136.

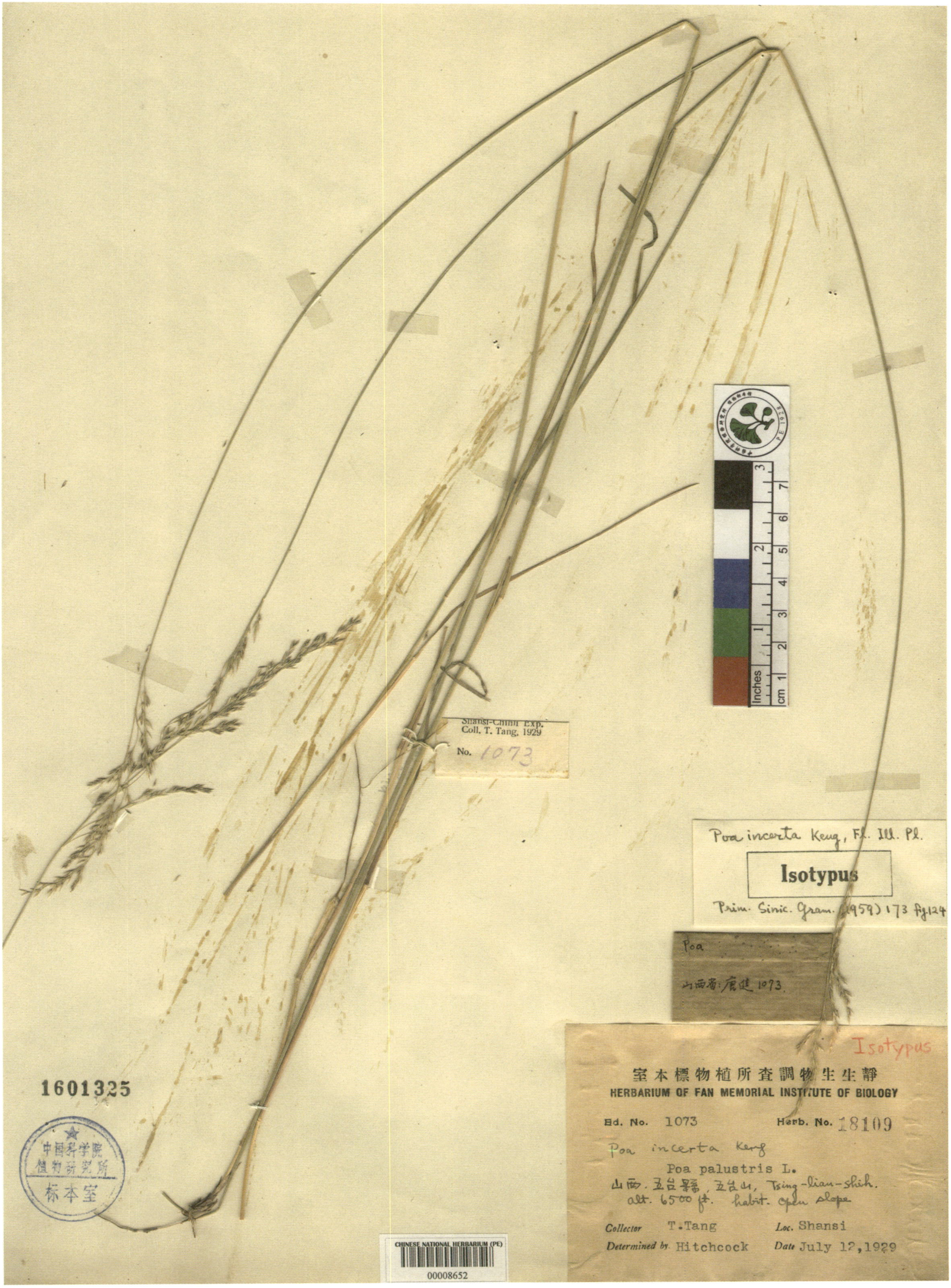

疑早熟禾 ***Poa incerta*** Keng ex L. Liu, Fl. Reip. Pop. Sin. 9(2): 395. 2002. **Isotype:** China. Shanxi: Wutai, Wutaishan, alt. 2170 m, 1929-07-12, T. Tang 1073.

印度早熟禾 ***Poa indattenuata*** Keng ex L. Liu, Fl. Reip. Pop. Sin. 9(2): 396. 2002. **Isotype:** China. Qinghai: Jishishan, 1935-07-16, C. W. Yao 855.

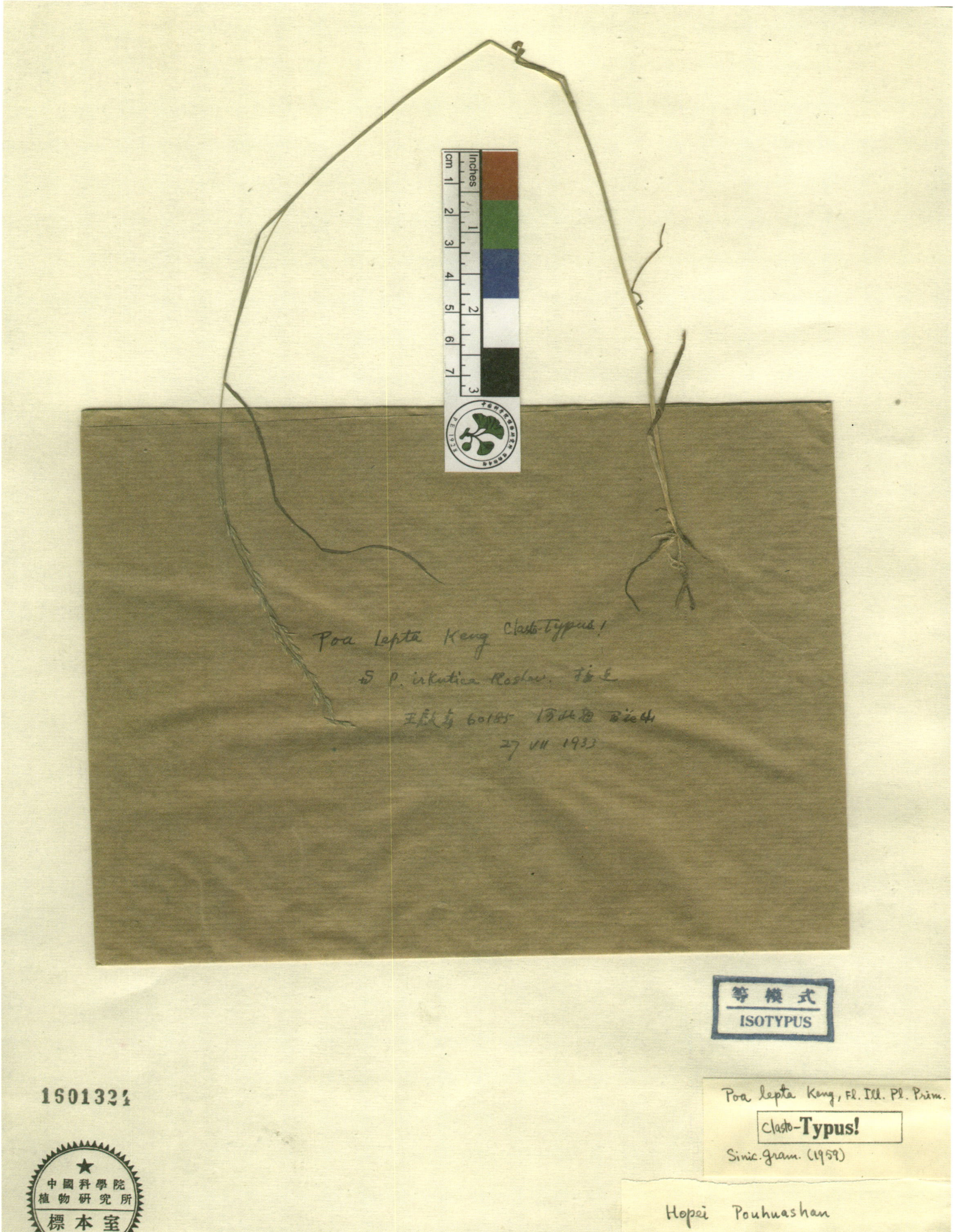

柔软早熟禾 ***Poa lepta*** Keng ex L. Liu, Fl. Reip. Pop. Sin. 9(2): 396. 2002. **Isotype:** China. Beijing: Baihuashan, 1933-07-27, C. W. Wang 60185.

石生早熟禾 ***Poa lithophila*** Keng ex L. Liu, Fl. Reip. Pop. Sin. 9(2): 397, pl. 19: 5. 2002. **Isotype:** China. Sichuan: Dege, alt. 2800-4300 m, 1951-07-15, Y. W. Tsui 4413.

大颖早熟禾 ***Poa macrolepis*** Keng ex L. Liu, Fl. Reip. Pop. Sin. 9(2): 398, pl. 19: 10. 2002. **Isotype:** China. Sichuan: Precise locality not known, 1951-08-23, Y. W. Tsui 5511.

小药早熟禾 ***Poa micrandra*** Keng, Fl. Tsinling. 1(1): 435. 1976. **Isotype:** China. Sichuan: Tianquan, 1939-06-19, K. L. Chu 2346.

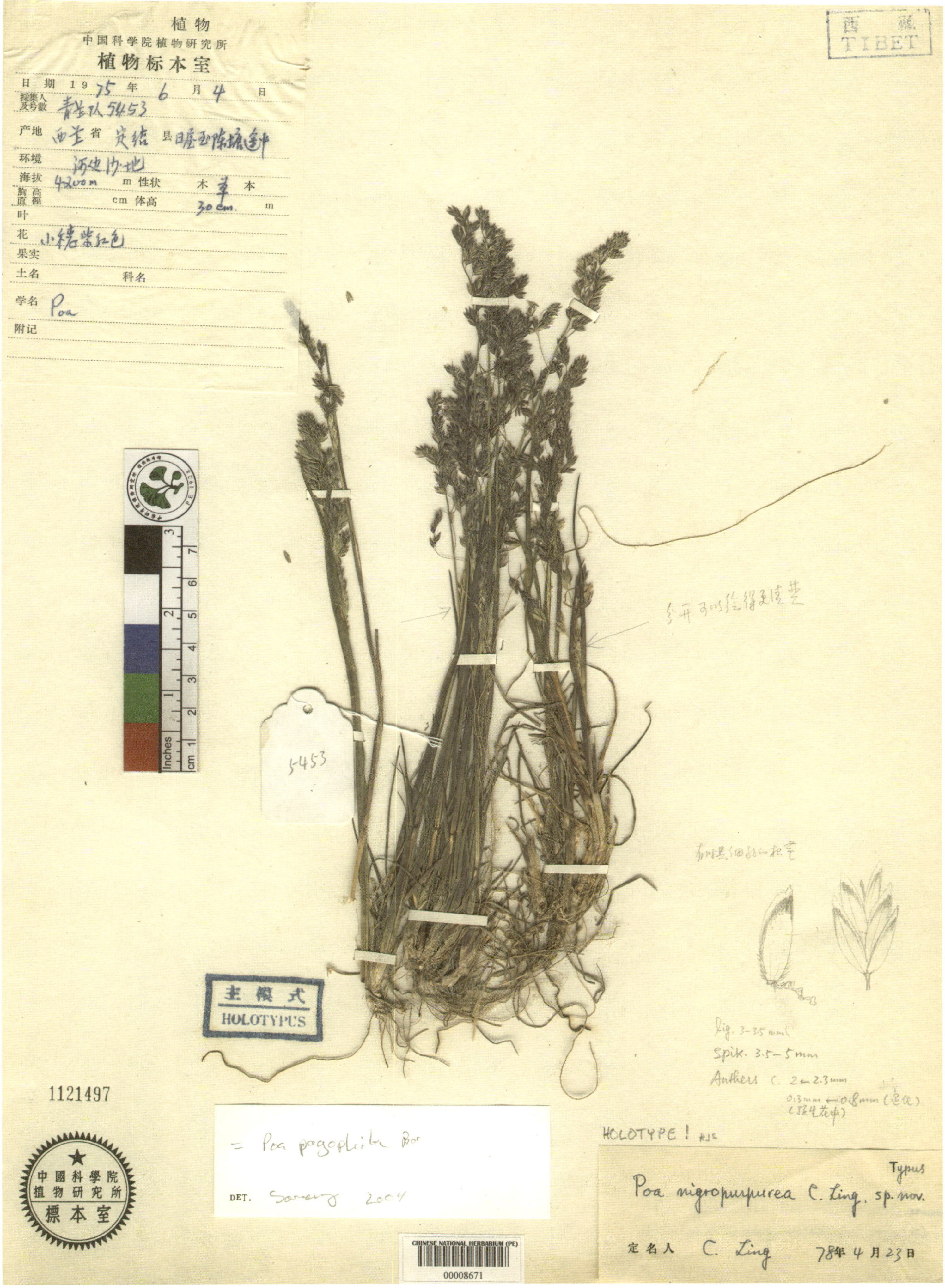

紫黑早熟禾 ***Poa nigropurpurea*** C. Ling in Acta Phytotax. Sin. 17(1): 101, f. 2: 4-6. 1979. **Holotype:** China. Xizang: Dingjie, alt. 4200 m, 1975-06-04, Qinghai-Xizang Exped. 5453.

尼木早熟禾 ***Poa nimuana*** C. Ling in Acta Phytotax. Sin. 17(1): 103, f. 1: 4-6. 1979. **Holotype:** China. Xizang: Nimu, alt. 5300 m, 1960-08-10, K. H. Fu 520.

山地早熟禾 ***Poa orinosa*** Keng, Fl. Tsinling. 1(1): 439. 1976. **Isotype:** China. Shaanxi: Taibaishan, alt. 2600 m, 1933-08-05, T. P.Wang 1859.

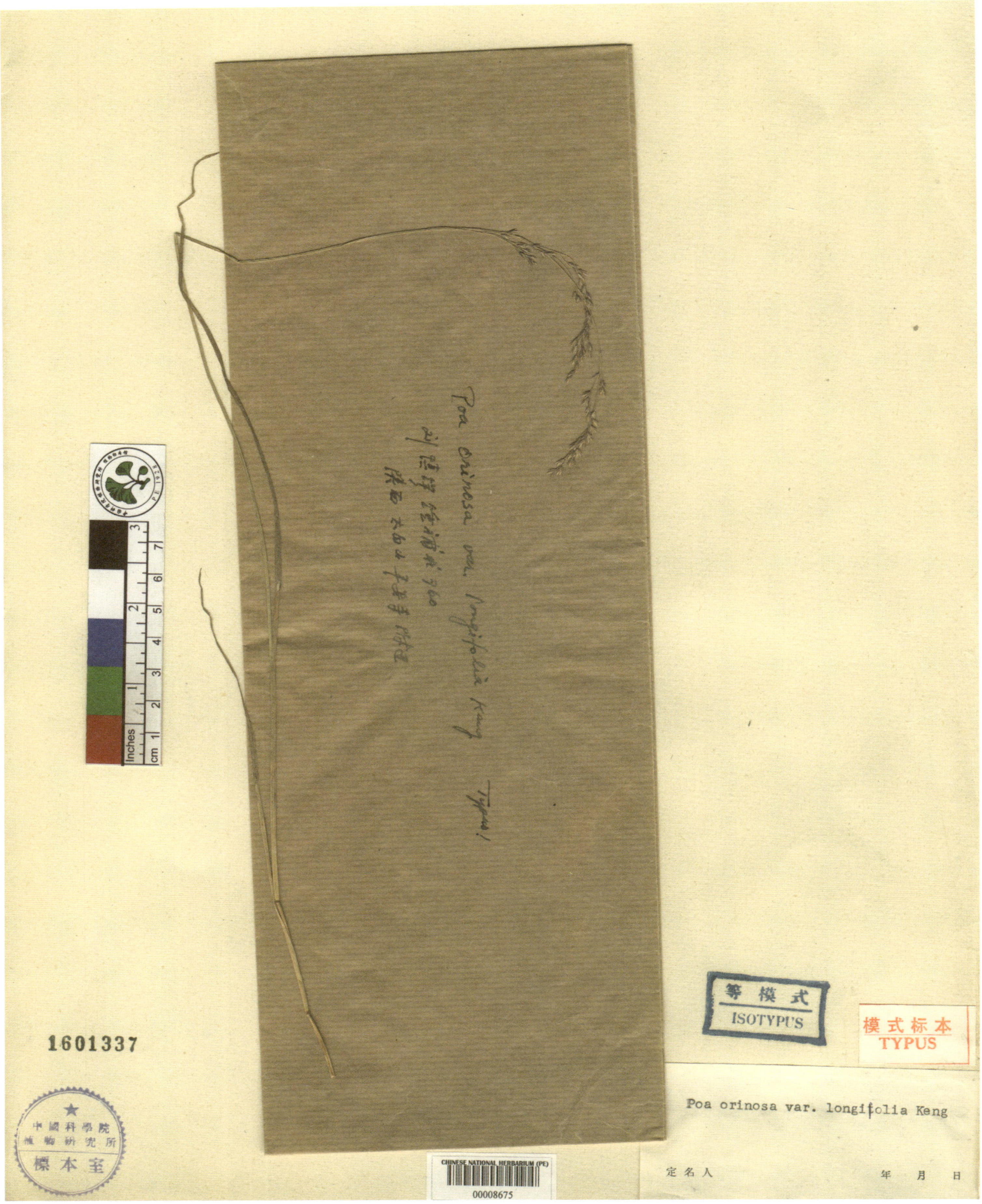

长叶早熟禾 ***Poa orinosa*** Keng var. ***longifolia*** Keng, Fl. Tsinling. 1(1): 439. 1976. **Isotype:** China. Shaanxi: Taibaishan, 1937-10-09, T. N. Liou & P. C. Tsoong 760.

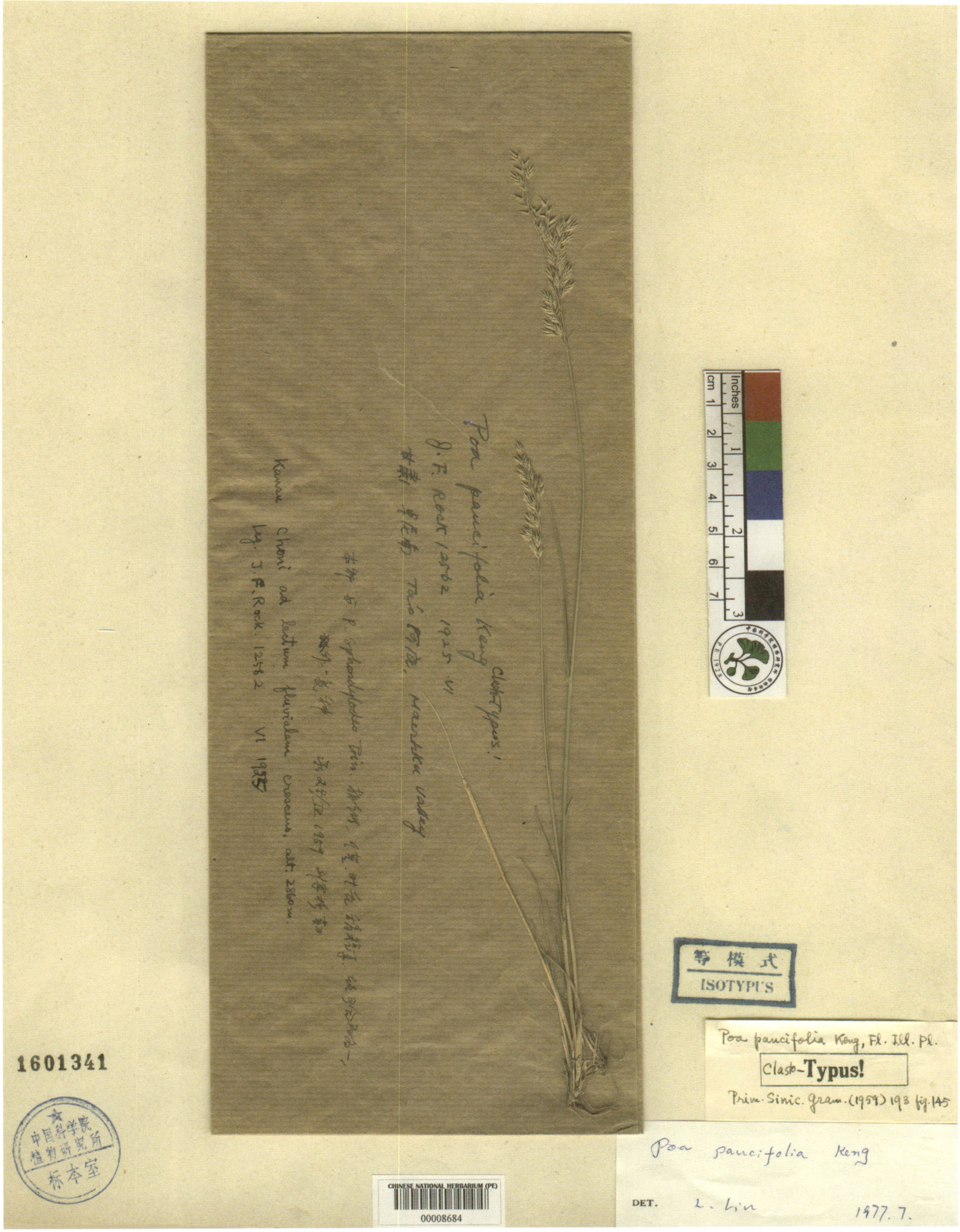

少叶早熟禾 ***Poa paucifolia*** Keng ex S. Chen, Fl. Intramongol. 261, pl. 36, f. 10-11. 1983. **Isotype:** China. Gansu: Jone, 1925-06-??, J. F. Rock 12562.

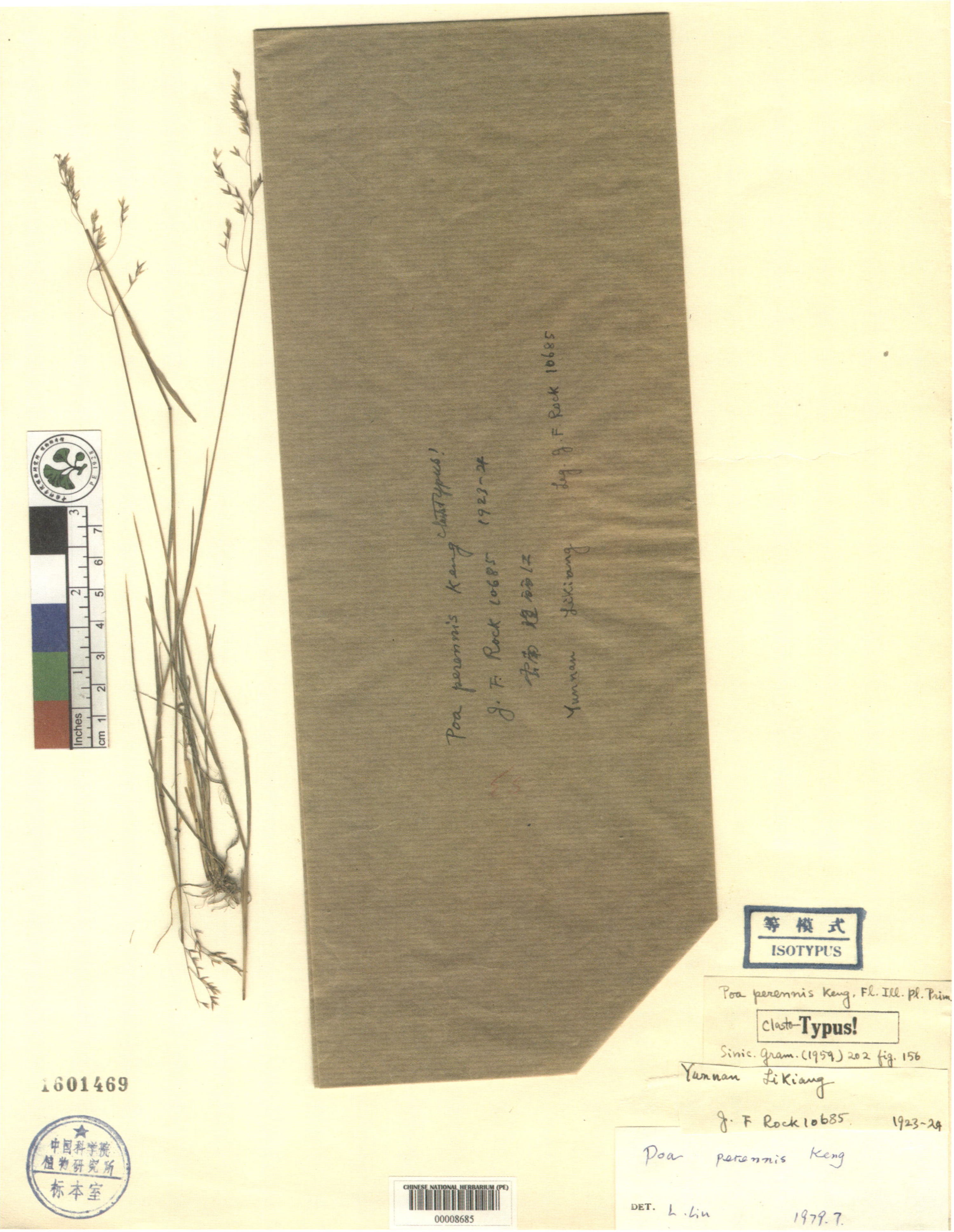

宿生早熟禾 ***Poa perennisa*** Keng ex P. C. Keng in Acta Bot. Yunnan. 4(3): 276. 1982. **Isotype:** China. Yunnan: Lijiang, (1923-1924)-??-??, J. F. Rock 10685.

多叶早熟禾 ***Poa plurifolia*** Keng, Fl. Tsinling. 1(1): 438. 1976. **Isotype:** China. Beijing: Xishan, 1930-06-30, K. S. Hao 4553.

多节早熟禾 ***Poa plurinodis*** Keng in Acta Bot. Yunnan. 4(3): 275. 1982. **Paratype:** China. Yunnan: Bijiang, alt. 4000 m, 1934-08-23, H. T. Tsai 58125.

罗氏早熟禾 ***Poa rossbergiana*** K. S. Hao in Bot. Jahrb. Syst. 68(5): 581. 1938. **Isotype:** China. Gansu: Tasiouszu (=Ngolashan), alt. 4000 m, 1930-09-09, Hopkingson (=K. S. Hao) 1197.

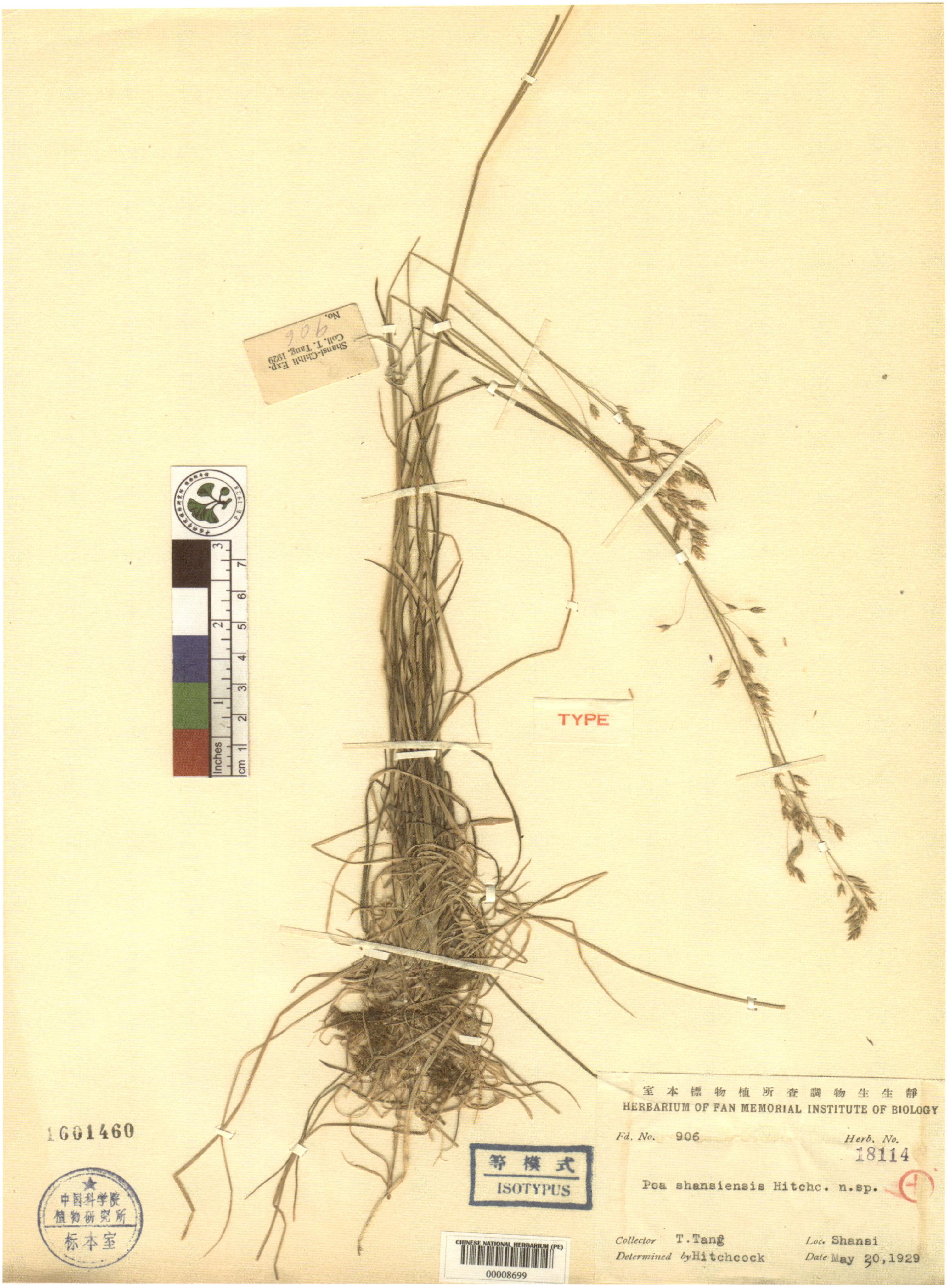

山西早熟禾 ***Poa shansiensis*** A. S. Hitchcock in Proc. Biol. Washington 43: 93. 1930. **Isotype:** China. Shanxi: Linshih, alt. 1600~1700 m, 1929-05-30, T. Tang 906.

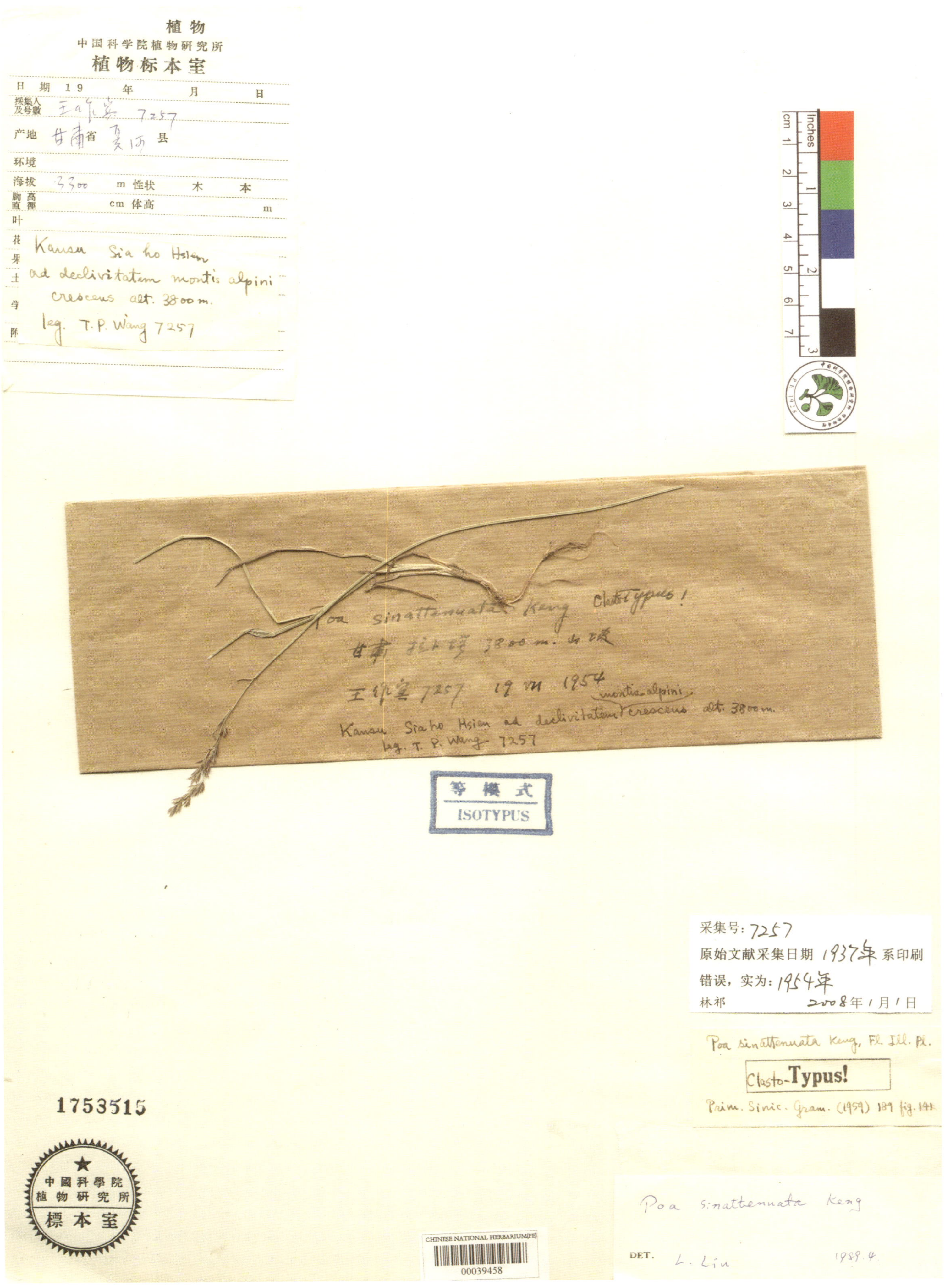

中华早熟禾 *Poa sinattenuata* Keng, Fl. Tsinling. 1(1): 437. 1976. **Isotype:** China. Gansu: Xiahe, alt. 3800 m, 1954-07-19, T. P. Wang 7257.

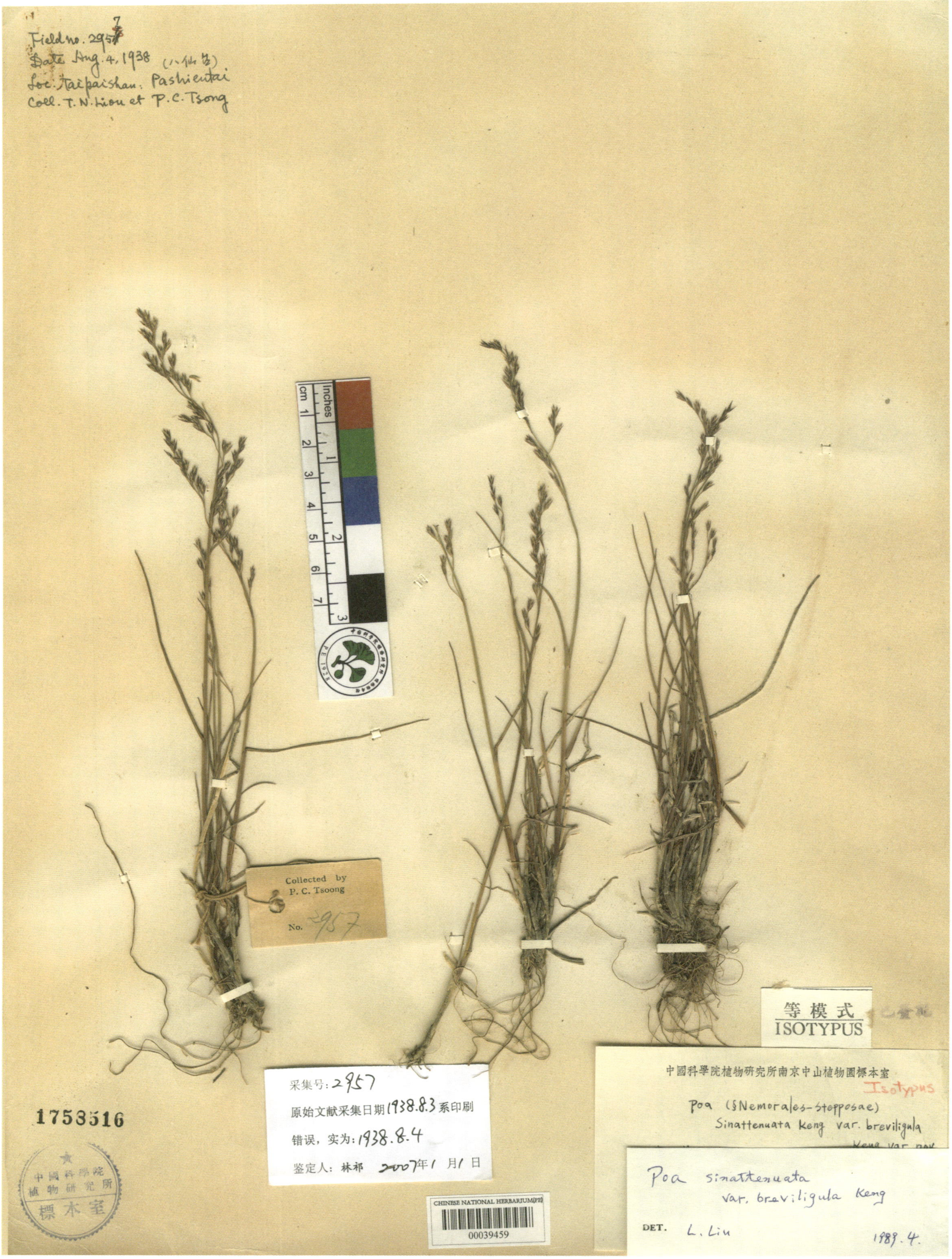

短舌早熟禾 ***Poa sinattenuata*** Keng var. ***breviligula*** Keng, Fl. Tsinling. 1(1): 437. 1976. **Isotype:** China. Shaanxi: Taibaishan, 1938-08-04, T. N. Liou & P. C. Tsoong 2957.

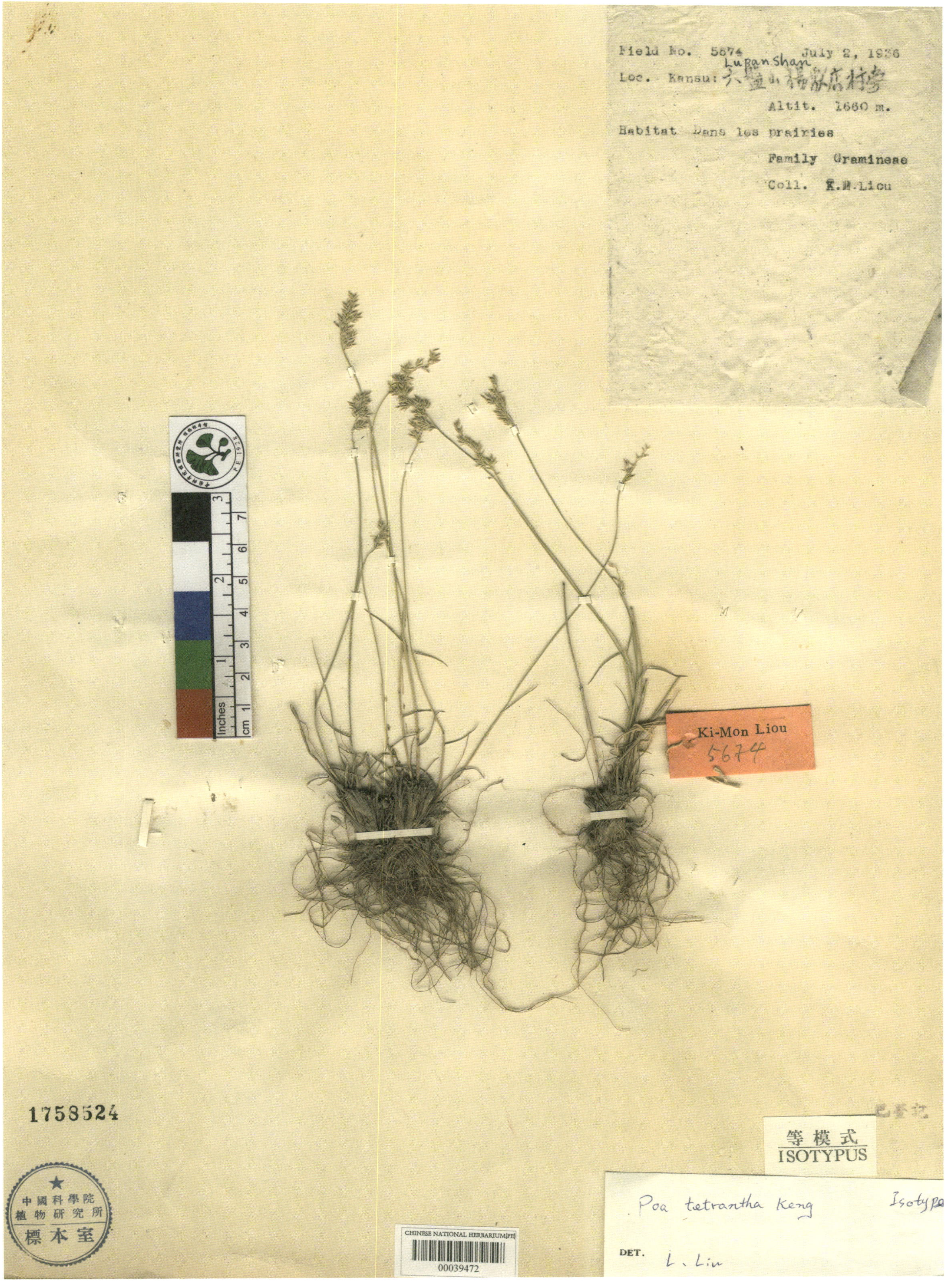

四花早熟禾 ***Poa tetrantha*** Keng ex L. Liu, Fl. Reip. Pop. Sin. 9(2): 403. 2002. **Isotype:** China. Gansu: Liupanshan, alt. 1660 m, 1936-07-02, K. M. Liou 5674.

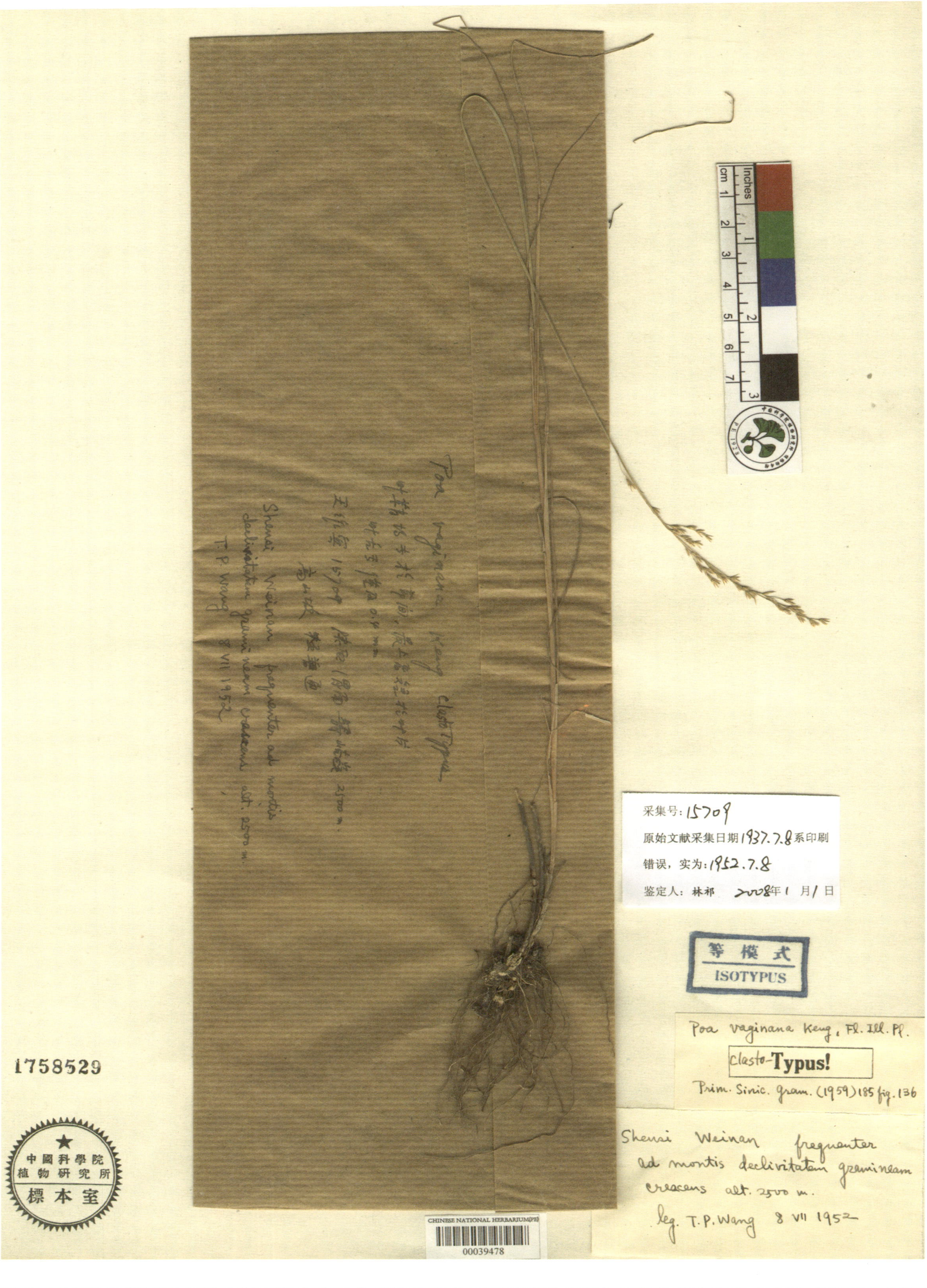

长鞘早熟禾 ***Poa vaginans*** Keng, Fl. Tsinling. 1(1): 435. 1976. **Isotype:** China. Shaanxi: Weinan, alt. 2500 m, 1952-07-08, T. P. Wang 15709.

多变早熟禾 ***Poa varia*** Keng ex L. Liu, Fl. Reip. Pop. Sin. 9(2): 404. 2002. **Isotype:** China. Gansu: Lintan, alt. 2800 m, 1936-07-18, T. P. Wang 5066.

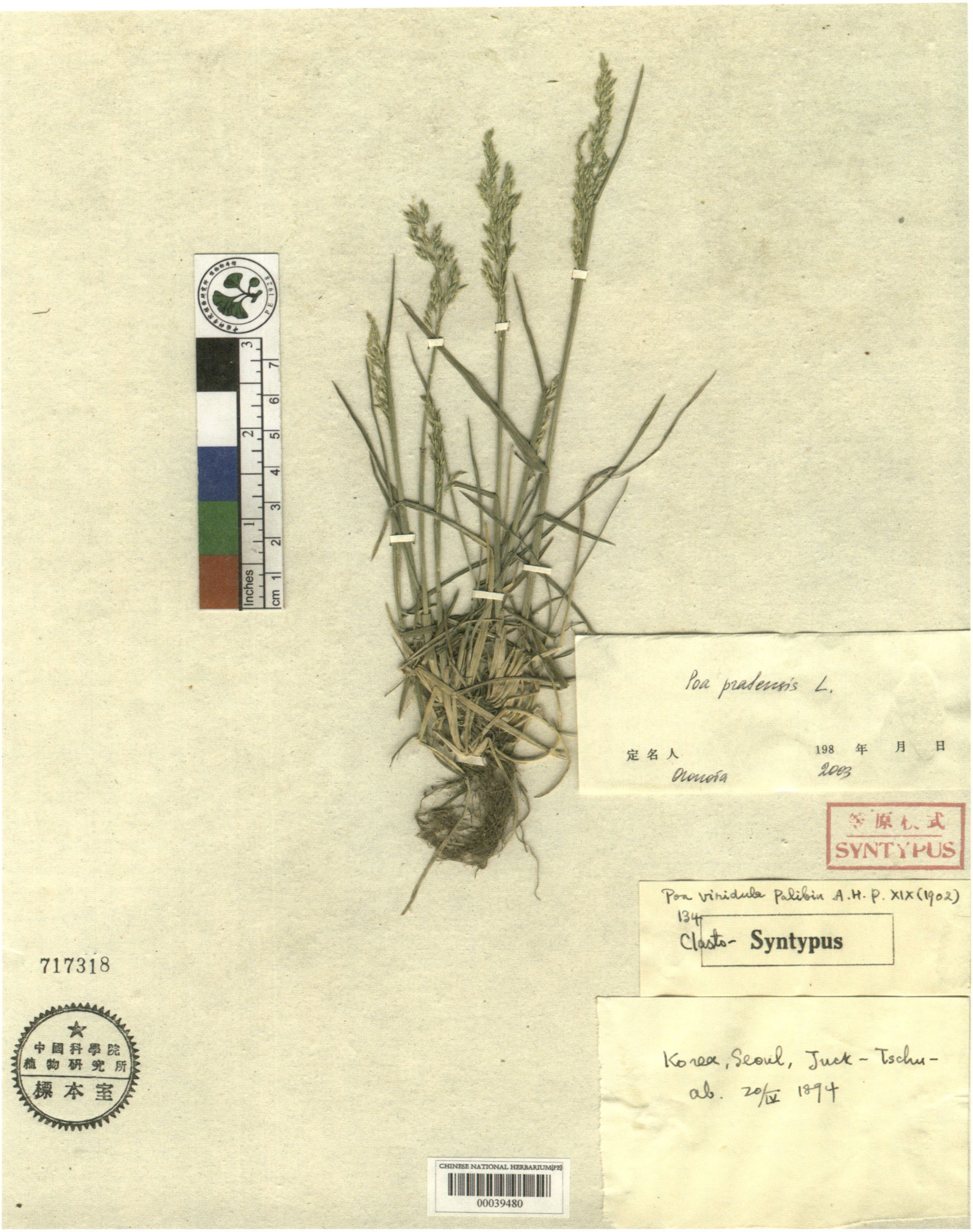

绿早熟禾 ***Poa viridula*** I. V. Palibin in Trudy Imp. St.-Peterb. Bot. Sada 19(2): 134. 1901. **Syntype:** P. O. Korea. Seoul: Mt. Yisan, 1894-05-28, A. Sontag s. n.

雅江早熟禾 ***Poa yakiangensis*** L. Liu, Fl. Reip. Pop. Sin. 9(2): 405, pl. 24: 2. 2002. **Holotype:** China. Sichuan: Yajiang, alt, 3700 m, 1959-08-01, S. Jiang 2630.

德格碱茅 ***Puccinellia degeensis*** L. Liu, Fl. Reip. Pop. Sin. 9(2): 405, pl. 29: 8. 2002. **Holotype:** China. Sichuan: Dege, alt. 3600 m, 1974-06-22, Sichuan Exped. 7076.

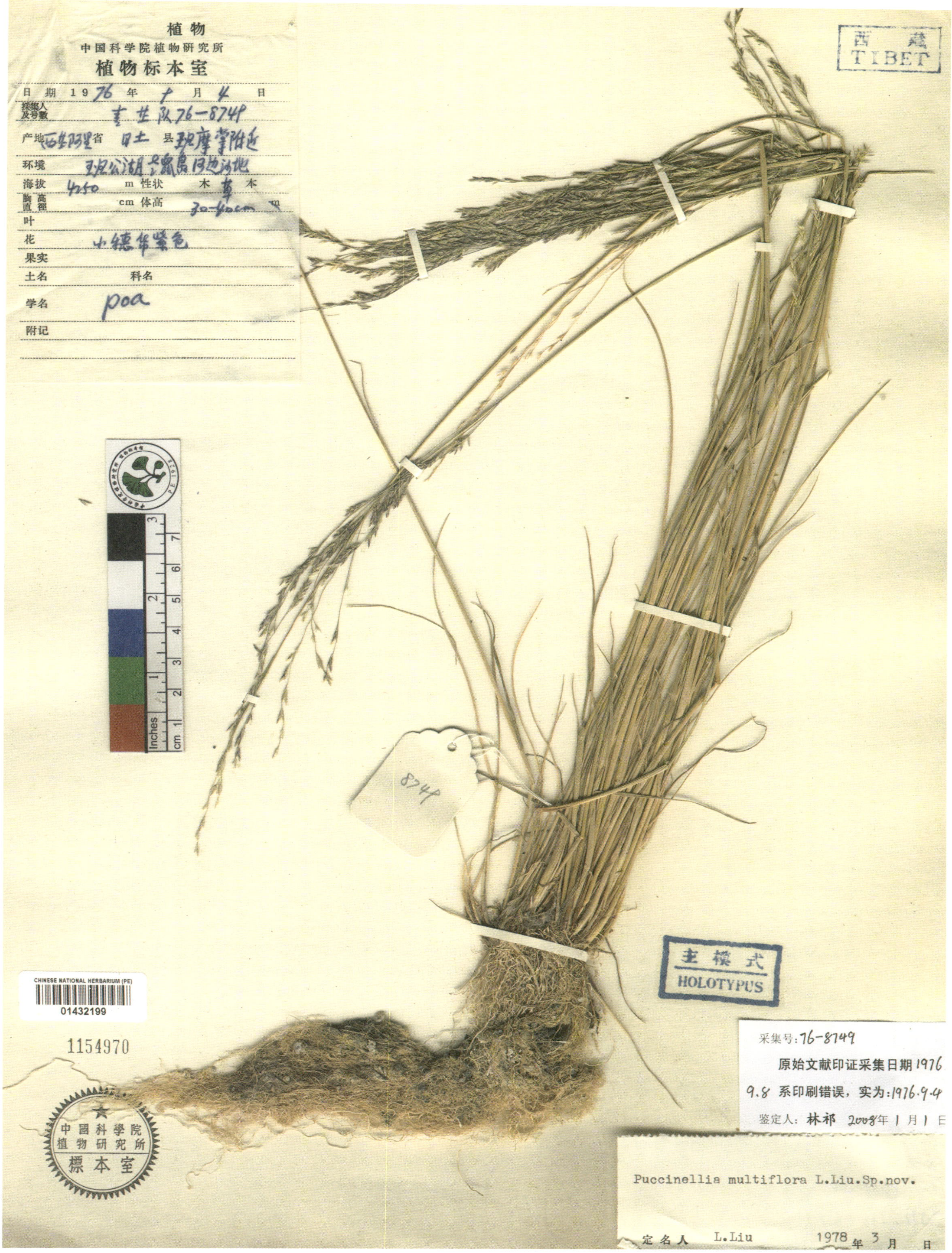

多花碱茅 ***Puccinellia multiflora*** L. Liou, Fl. Xizang. 5: 123. 1987. **Holotype:** China. Xizang: Rutog, alt. 4250 m, 1976-09-04, Qinghai-Xizang Exped. 76-8749.

坚碱茅 ***Puccinellia strictura*** L. Liu, Fl. Reip. Pop. Sin. 9(2): 405, pl. 30: 16. 2002. **Holotype:** China. Xizang: Shuanghu, 1976-08-02, Qinghai-Xizang Exped. 12118.

高株鹅观草 ***Roegneria altissima*** Keng in Journ. Nanjing Univ. (Biol.) 1963(1): 53. 1963. **Holotype:** China. Sichuan: Taswarung (=Chawalong), alt. 3400 m, 1935-09-??, C. W. Wang 66409.

假花鳞草 ***Roegneria anthosachnoides*** Keng in Journ. Nanjing Univ. (Biol.) 1963(1): 65. 1963. **Isotype:** China. Yunnan: Bijiang, alt. 4000 m, 1934-08-26, H. T. Tsai 58184.

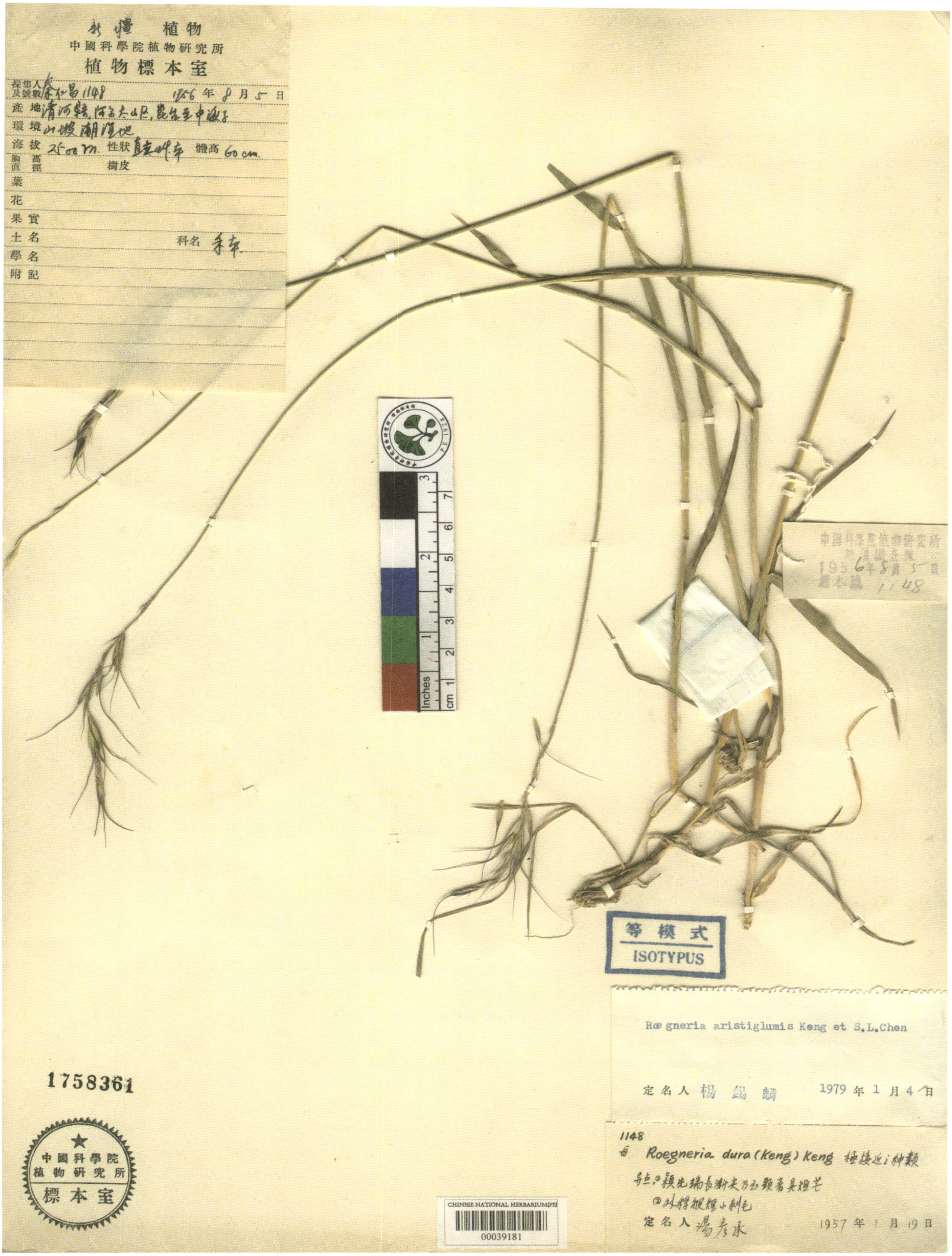

芒颖鹅观草 ***Roegneria aristiglumis*** Keng & S. L. Chen in Journ. Nanjing Univ. (Biol.) 1963(1): 55, pl. 4. 1963. **Isotype:** China. Xinjiang: Qinghe, alt. 2500 m, 1956-08-05, R. C. Ching 1148.

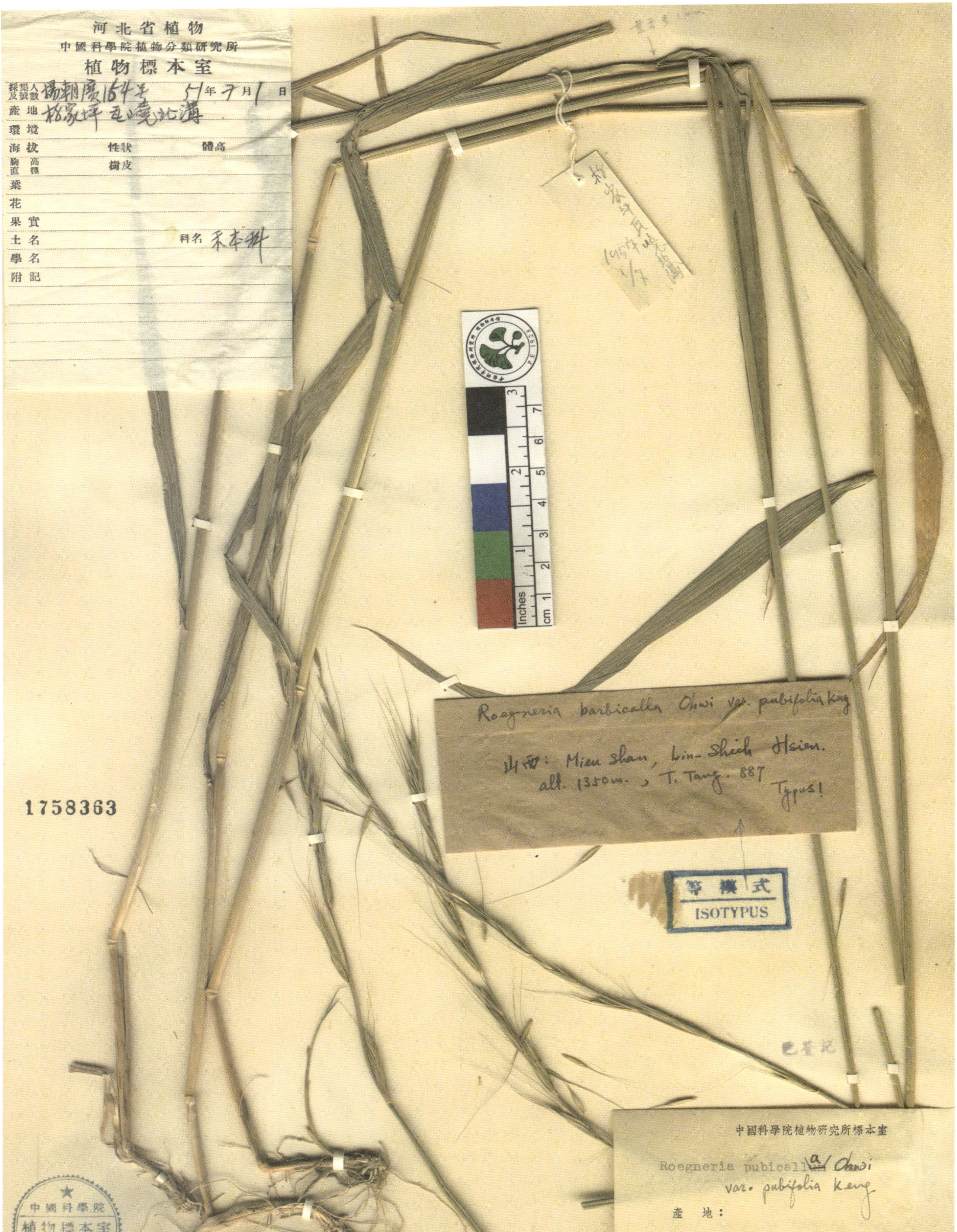

毛叶毛盘草 ***Roegneria barbicalla*** Ohwi var. ***pubifolia*** Keng in Journ. Nanjing Univ. (Biol.) 1963(1): 25. 1963. **Isotype:** China. Shaanxi: Linshi, alt. 1350 m, 1929-05-29, T. Tang 887.

毛节毛盘草 ***Roegneria barbicalla*** Ohwi var. ***pubinodis*** Keng in Journ. Nanjing Univ. (Biol.) 1963(1): 24. 1963. **Isotype:** China. Hebei: Neiqiu, 1951-06-17, H. Y. Liou 441.

短柄鹅观草 ***Roegneria brevipes*** Keng in Journ. Nanjing Univ. (Biol.) 1963(1): 49. 1963. **Isotype:** China. Qinghai: Huangyuan, 1944-08-08, Y. L. Keng & P. C. Keng 5246.

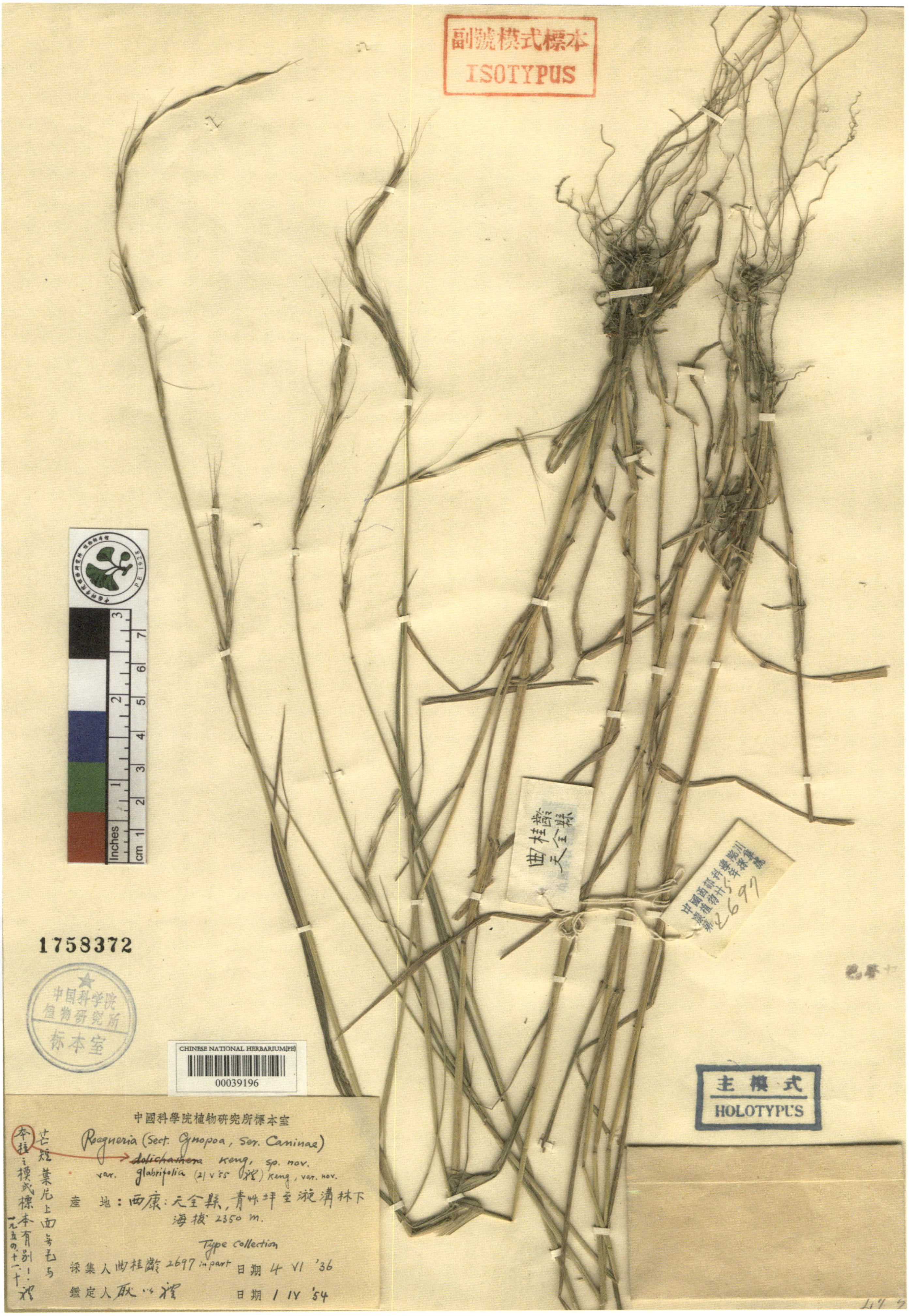

光叶鹅观草 ***Roegneria dolichathera*** Keng var. ***glabrifolia*** Keng in Journ. Nanjing Univ. (Biol.) 1963(1): 19. 1963.

Holotype: China. Sichuan: Tianquan, alt. 2350 m, 1936-06-04, K. L. Chu 2697.

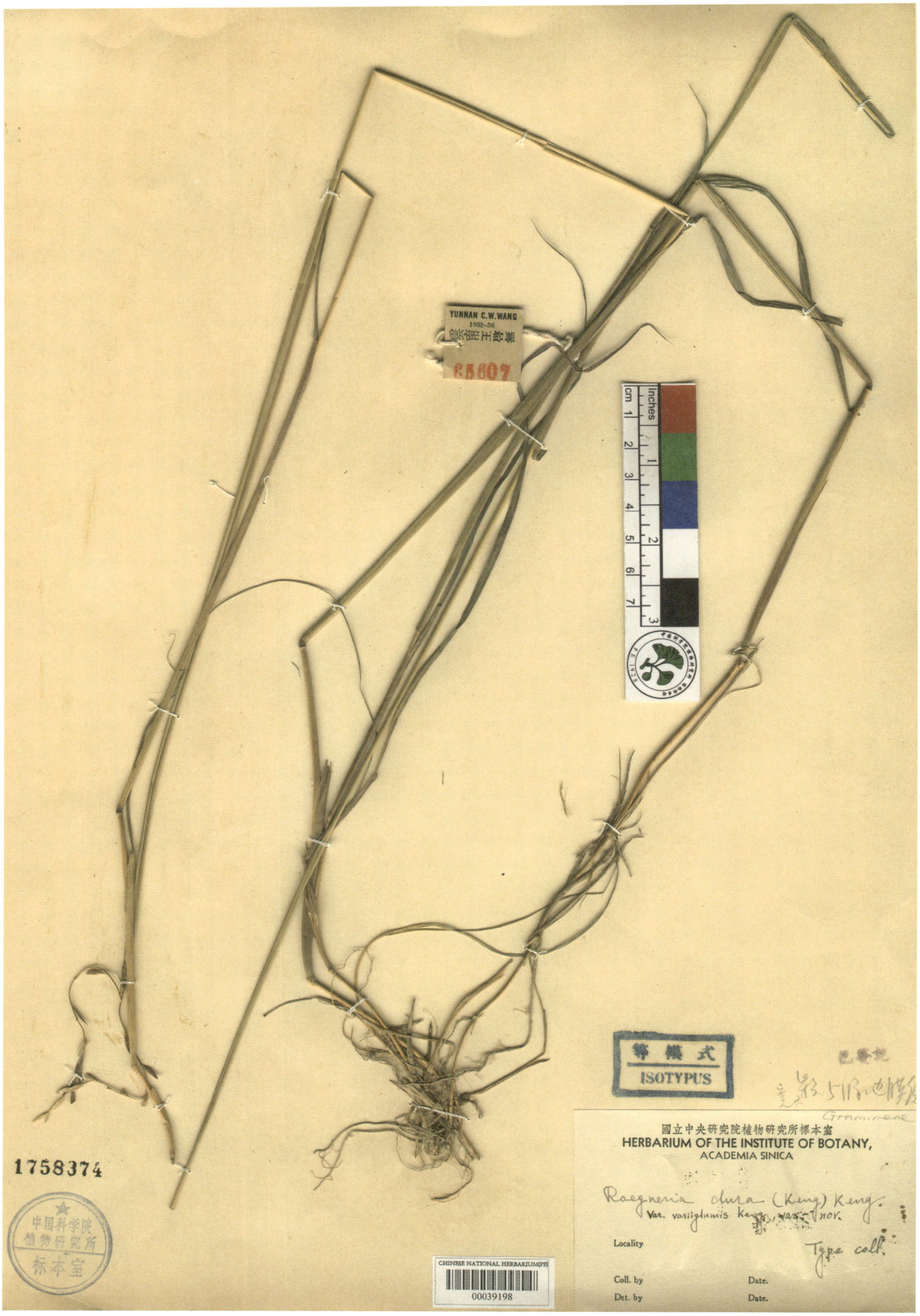

变颖鹅观草 ***Roegneria dura*** (Keng) Keng var. ***variiglumis*** Keng in Journ. Nanjing Univ. (Biol.) 1963(1): 54. 1963.
Isotype: China. Sichuan: Tsawarung (=Chawalong), alt. 3700 m, 1935-08-??, C. W. Wang 65607.

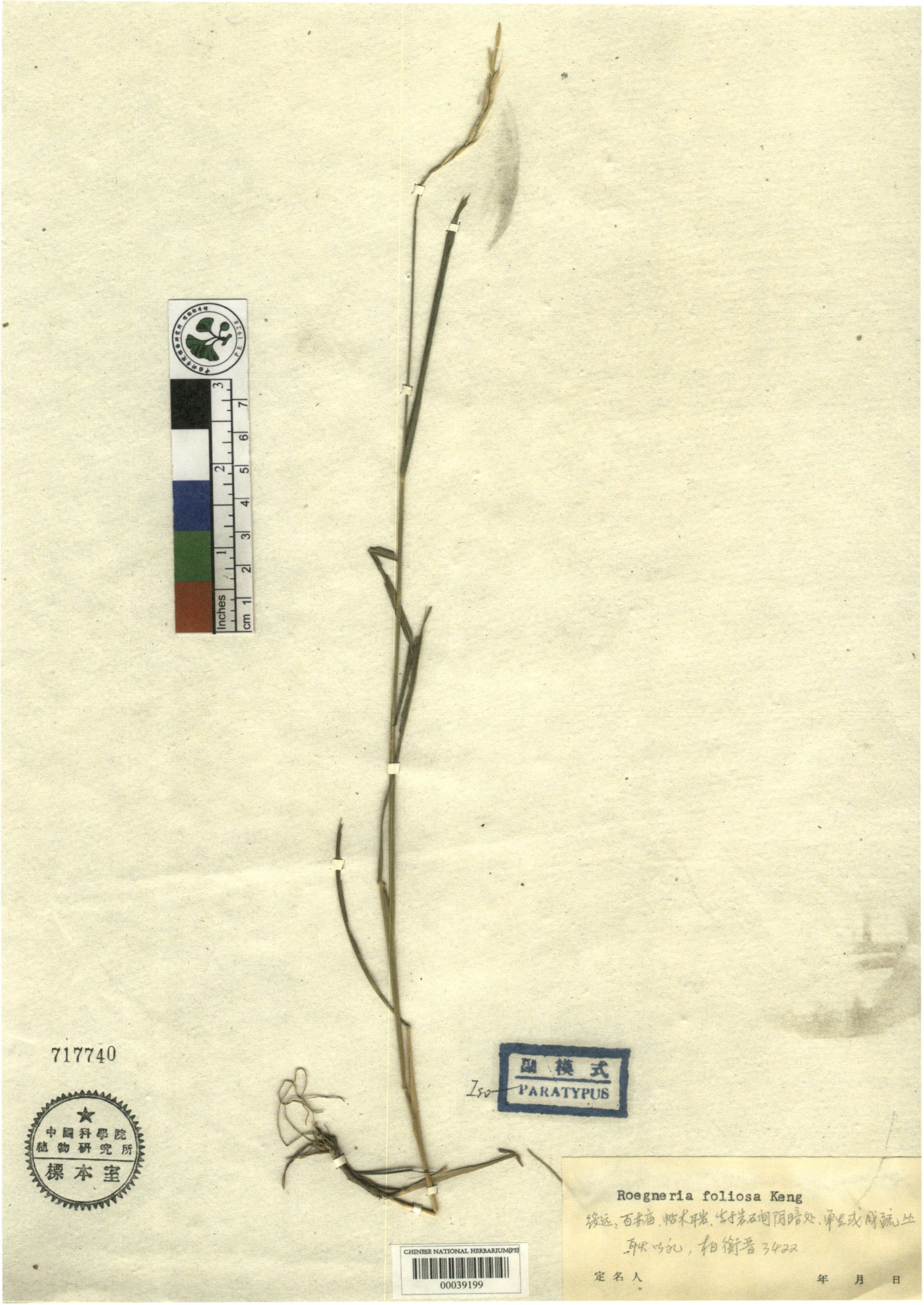

多叶鹅观草 ***Roegneria foliosa*** Keng in Journ. Nanjing Univ. (Biol.) 1963(1): 32. 1963. **Isoparatype:** China. Nei Mongol: Bailinsi, alt. 3700 m, Y. L. Keng & Y. C. Yang 3422.

毛鞘台湾鹅观草 ***Roegneria formosana*** (Honda) Ohwi var. ***pubigera*** Keng in Journ. Nanjing Univ. (Biol.) 1963(1): 60. 1963. **Isotype:** China. Taiwan: Fan. Mem, Inst. Biol. No. 82369.

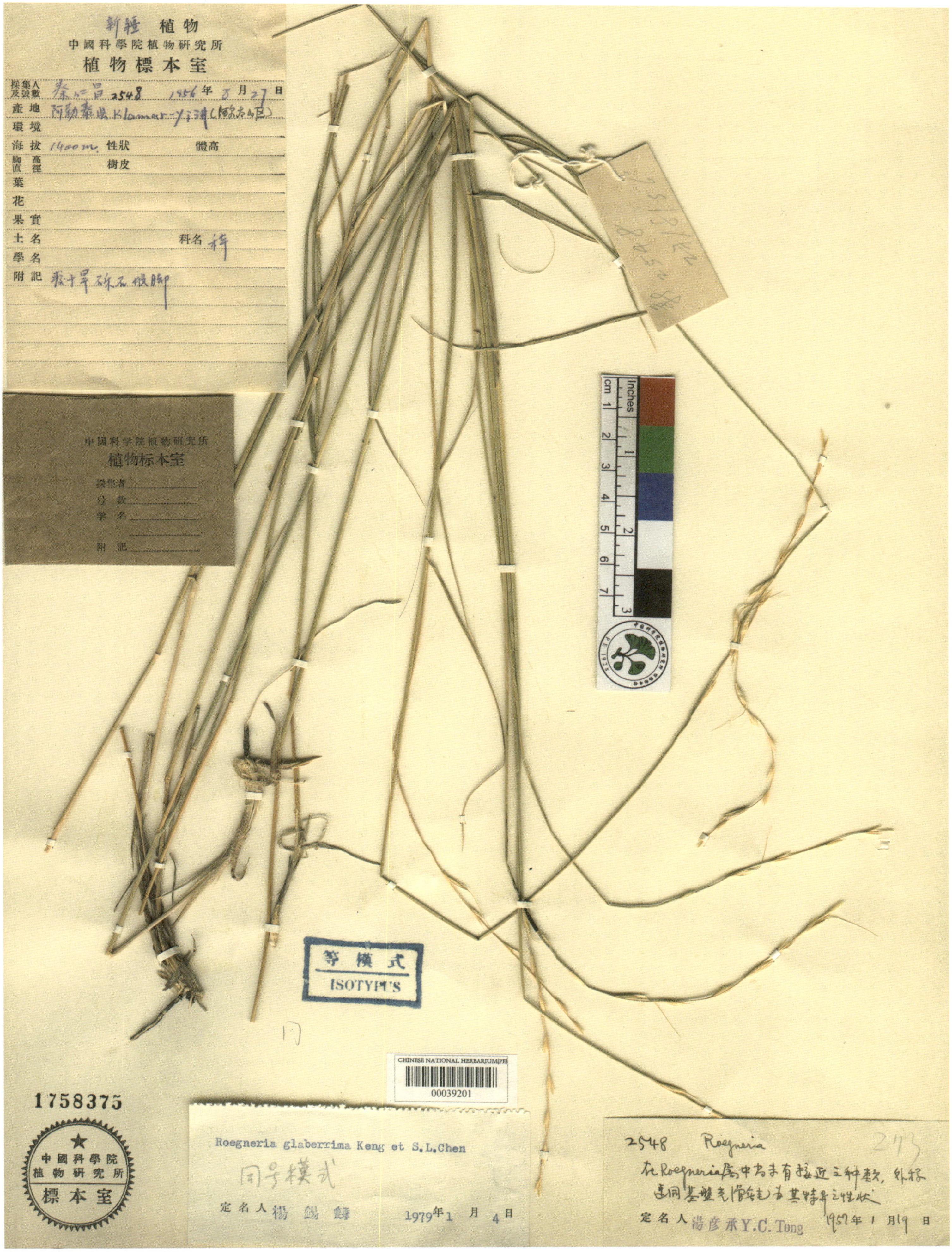

光穗鹅观草 ***Roegneria glaberrima*** Keng & S. L. Chen in Journ. Nanjing Univ. (Biol.) 1963(1): 72. 1963. **Isotype:** China. Xinjiang: Altay, alt. 1400 m, 1956-08-27, R. C. Ching 2548.

马格草 ***Roegneria glaucifolia*** Keng in Journ. Nanjing Univ. (Biol.) 1963(1): 57. 1963. **Isotype:** China. Xizang: Mage, 1952-09-09, P. C. Tsoong 5365.

大鹅观草 ***Roegneria grandis*** Keng in Journ. Nanjing Univ. (Biol.) 1963(1): 45. 1963. **Holotype:** China. Shaanxi: Huh Hsien (=Hu Xian), 1951-07-07, B. C. Kuo 149.

疏花鹅观草 ***Roegneria laxiflora*** Keng in Journ. Nanjing Univ. (Biol.) 1963(1): 75. 1963. **Isotype:** China. Sichuan: Kantse (=Ganzi), 1951-07-11, Y. W. Tsui 4338.

光脊鹅观草 ***Roegneria leiotropis*** Keng in Journ. Nanjing Univ. (Biol.) 1963(1): 58. 1963. **Isotype:** China. Yunnan: Lijiang, (1923-1924)-??-??, J. F. Rock 10646.

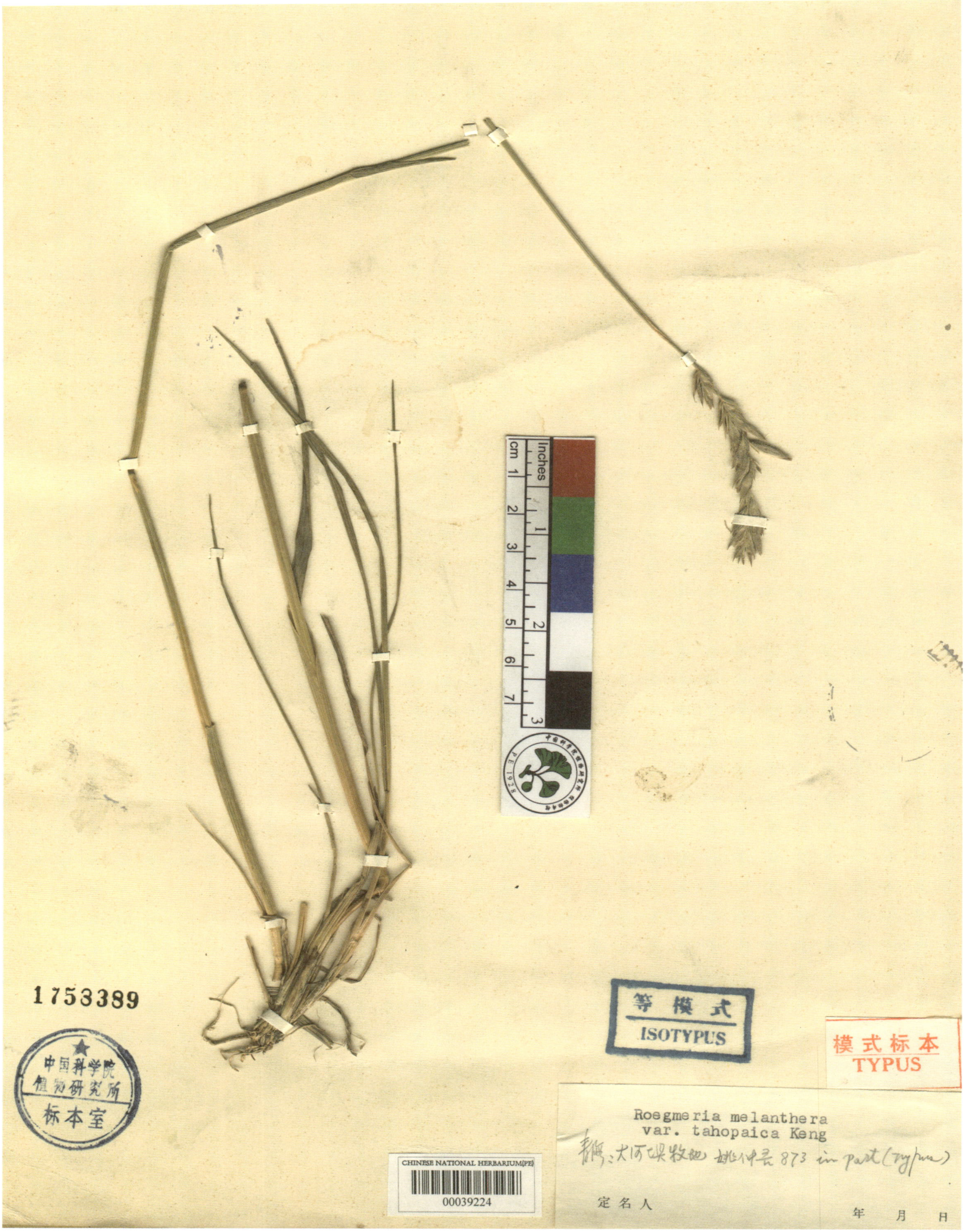

大河坝黑药草 ***Roegneria melanthera*** (Keng) Keng var. ***tahopaica*** Keng in Journ. Nanjing Univ. (Biol.) 1963(1): 78. 1963. **Isotype:** China. Qinghai: Tahopai (=Daheba), 1935-07-19, C. W. Yao 873.

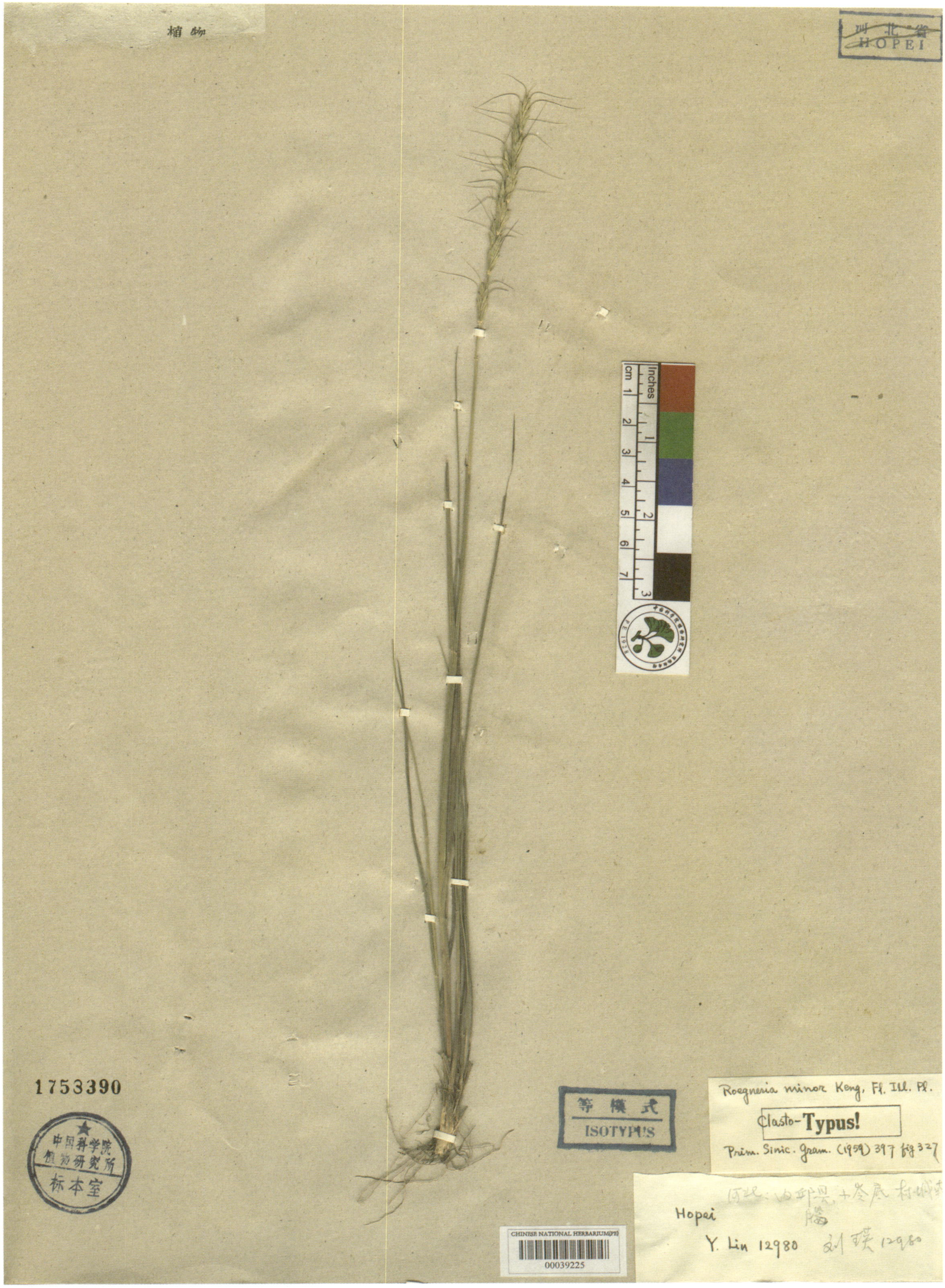

小株鹅观草 ***Roegneria minor*** Keng in Journ. Nanjing Univ. (Biol.) 1963(1): 71. 1963. **Isotype:** China. Hebei: Neiqiu, alt. 1600 m, 1950-07-03, Y. Liou 12980.

小颖鹅观草 ***Roegneria parvigluma*** Keng in Journ. Nanjing Univ. (Biol.) 1963(1): 47. 1963. **Isotype:** China. Sichuan: Kangding, 1951-06-30, Y. W. Tsuei 4162.

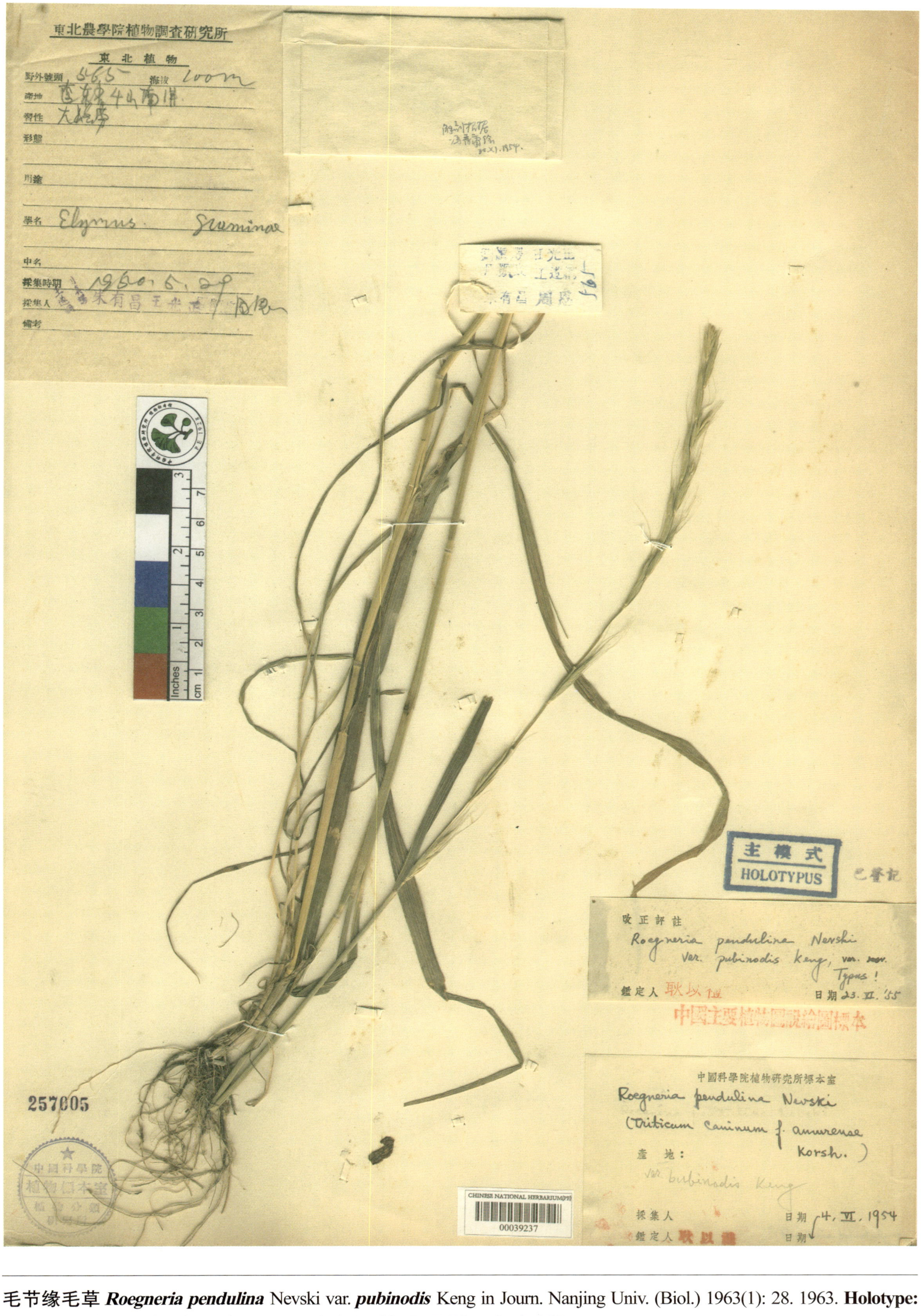

毛节缘毛草 ***Roegneria pendulina*** Nevski var. ***pubinodis*** Keng in Journ. Nanjing Univ. (Biol.) 1963(1): 28. 1963. **Holotype:** China. Liaoning: Qianshan, alt. 100 m, 1950-05-29, T. N. Liou & al. 565.

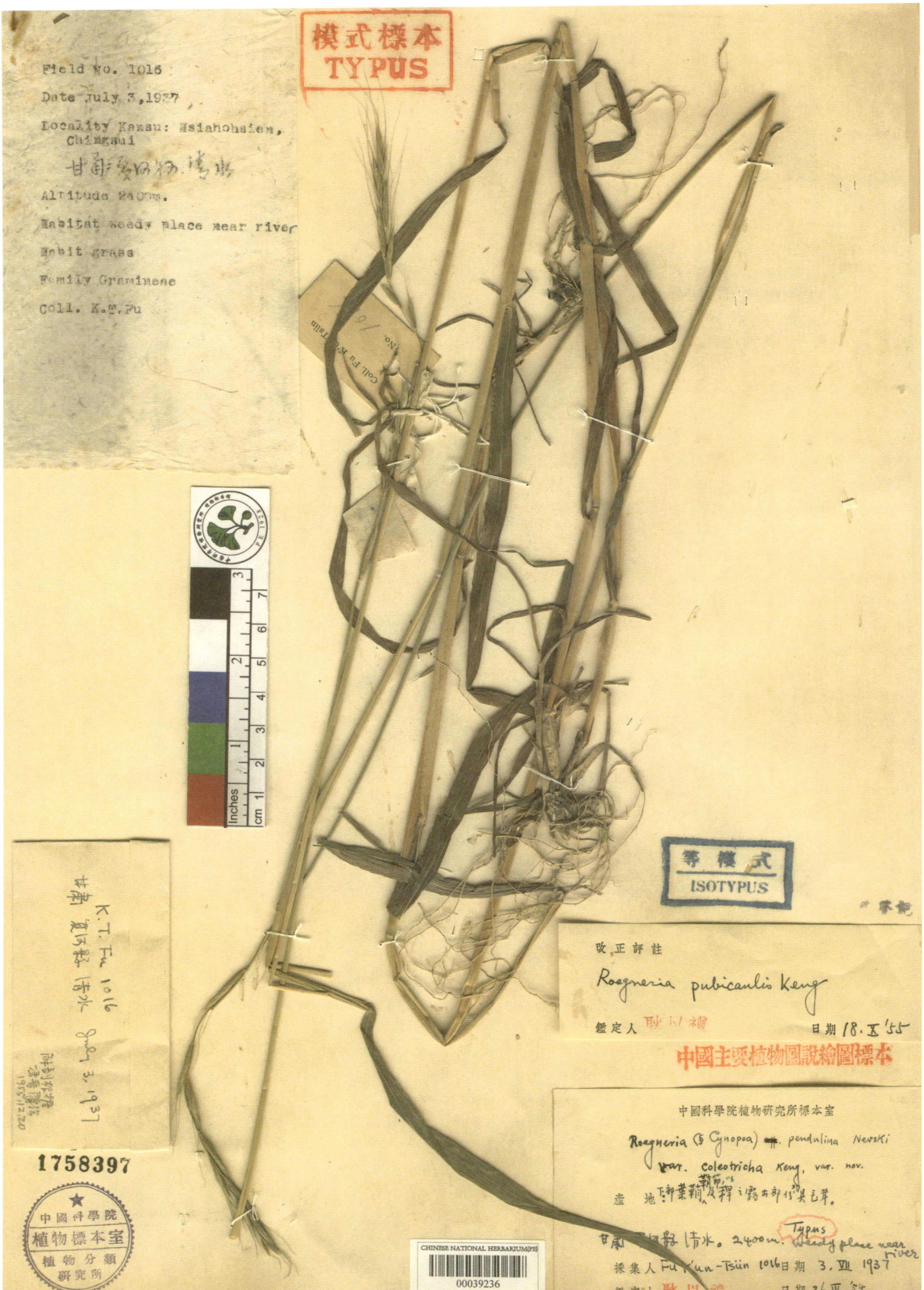

毛秆鹅观草 ***Roegneria pubicaulis*** Keng in Journ. Nanjing Univ. (Biol.) 1963(1): 30. 1963. **Isotype:** China. Gansu: Xiahe, alt. 2400 m, 1937-07-03, K. T. Fu 1016.

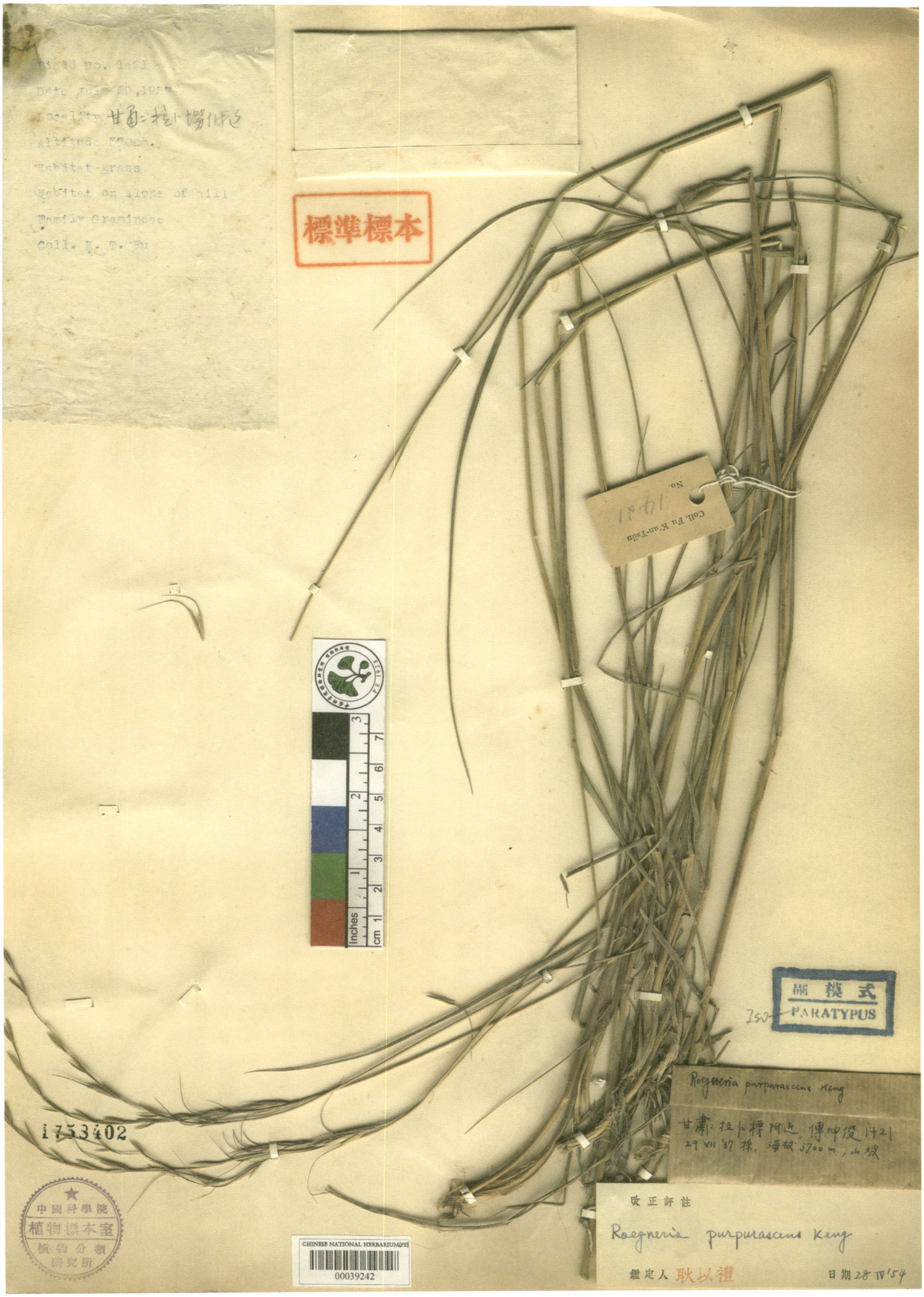

紫穗鹅观草 ***Roegneria purpurascens*** Keng in Journ. Nanjing Univ. (Biol.) 1963(1): 56. 1963. **Isoparatype:** China. Gansu: Longde, alt. 3700 m, 1937-07-29, K. T. Fu 1421.

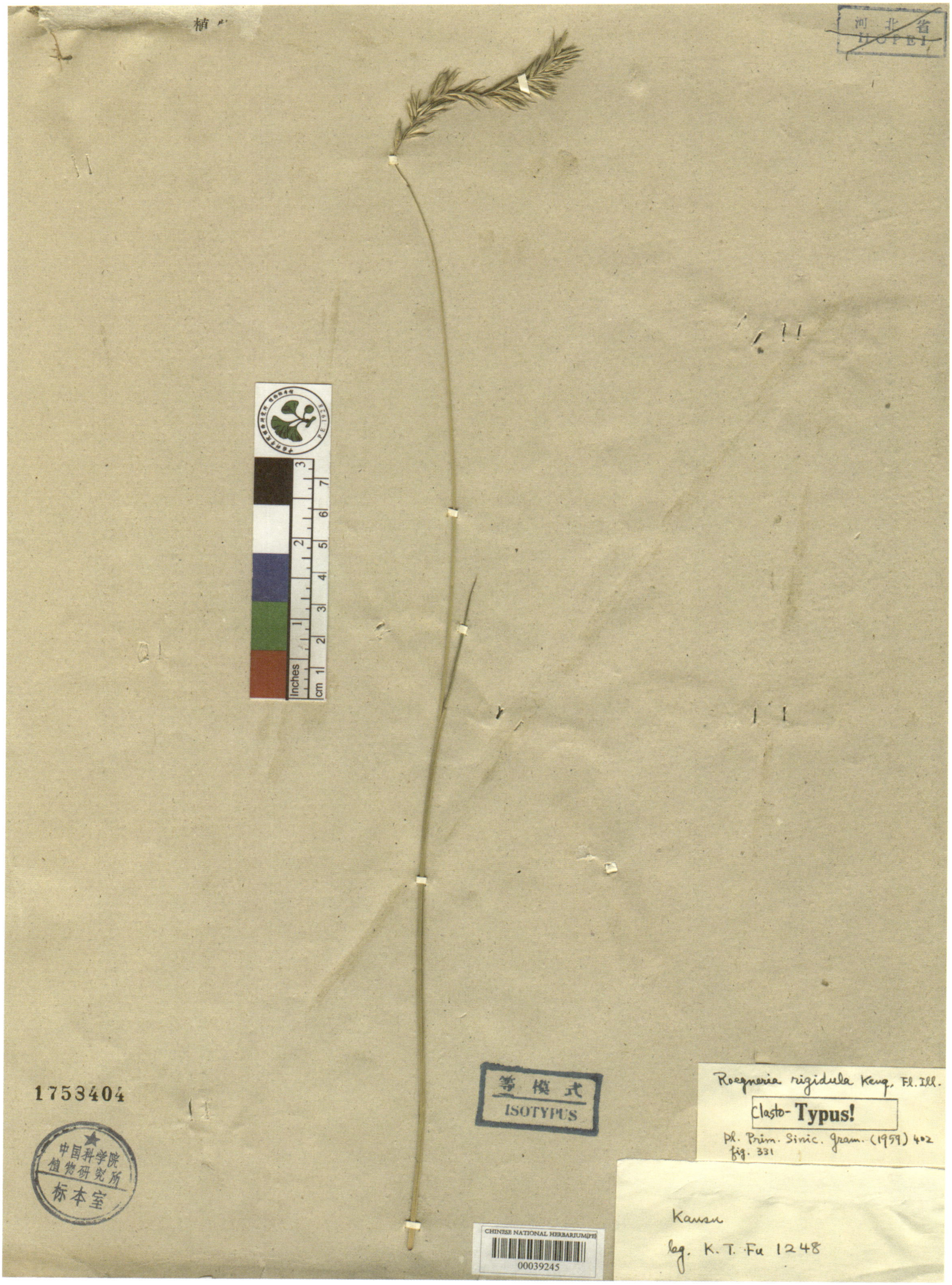

硬秆鹅观草 ***Roegneria rigidula*** Keng in Journ. Nanjing Univ. (Biol.) 1963(1): 77. 1963. **Isotype:** China. Gansu: Xiahe, alt. 3150 m, 1937-07-31, K. T. Fu 1248.

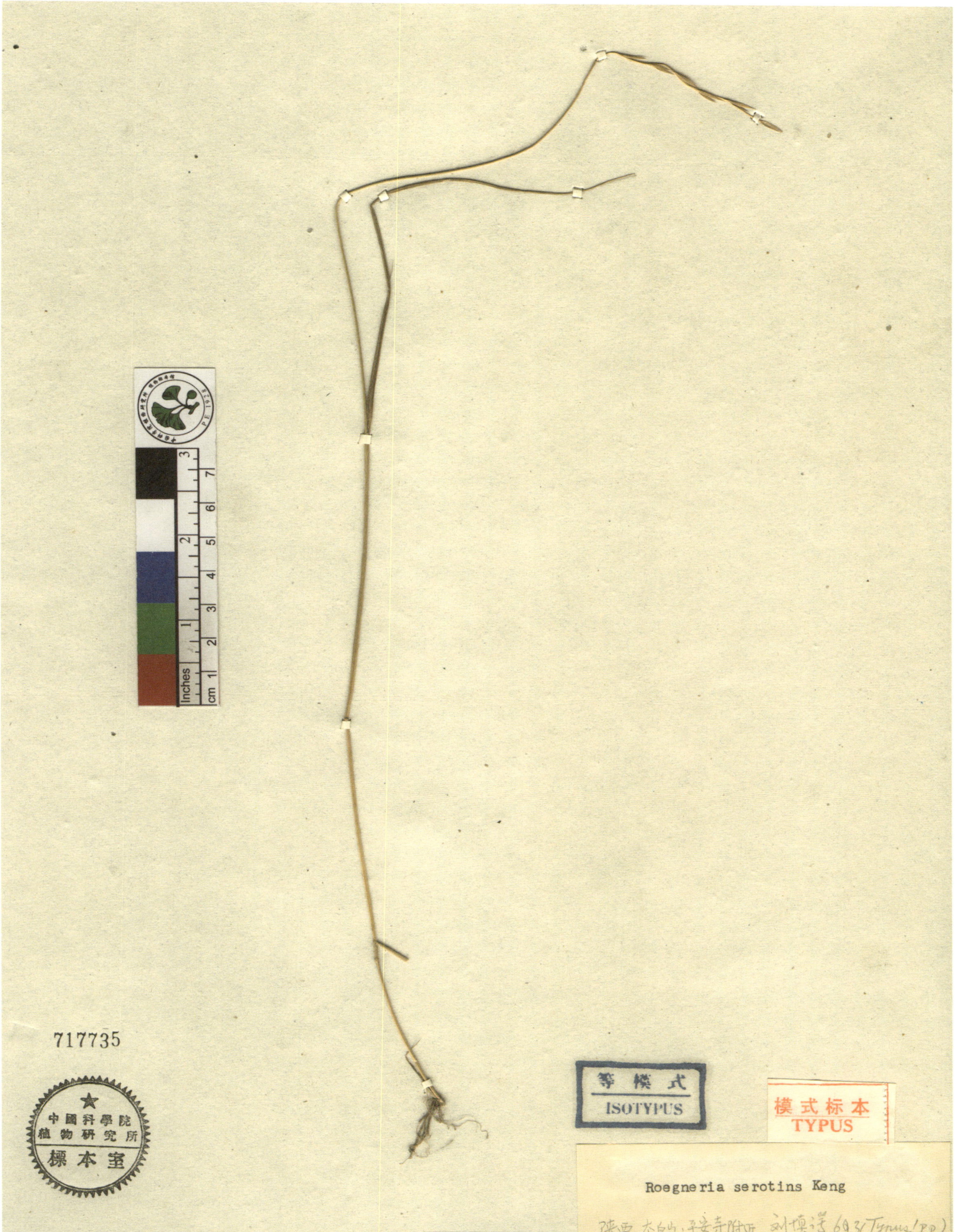

秋鹅观草 ***Roegneria serotins*** Keng in Journ. Nanjing Univ. (Biol.) 1963(1): 50. 1963. **Isotype:** China. Shaanxi: Taibaishan, 1937-09-07, T. N. Liou & P. C. Tsoong 693.

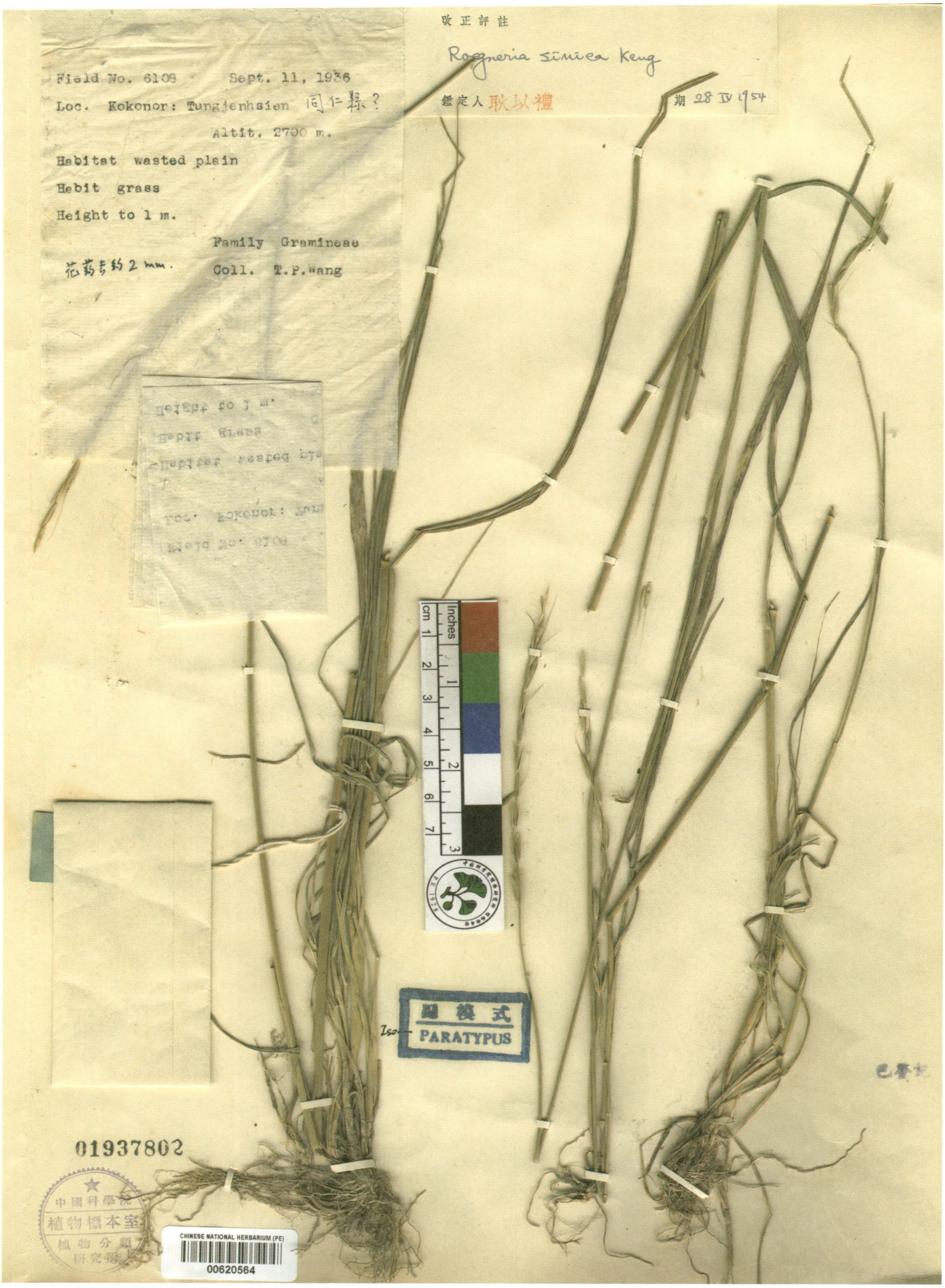

中华鹅观草 ***Roegneria sinica*** Keng in Journ. Nanjing Univ. (Biol.) 1963(1): 33. 1963. **Isoparatype:** China. Qinghai: Tongren, alt. 2700 m, 1936-09-11, T. P. Wang 6108.

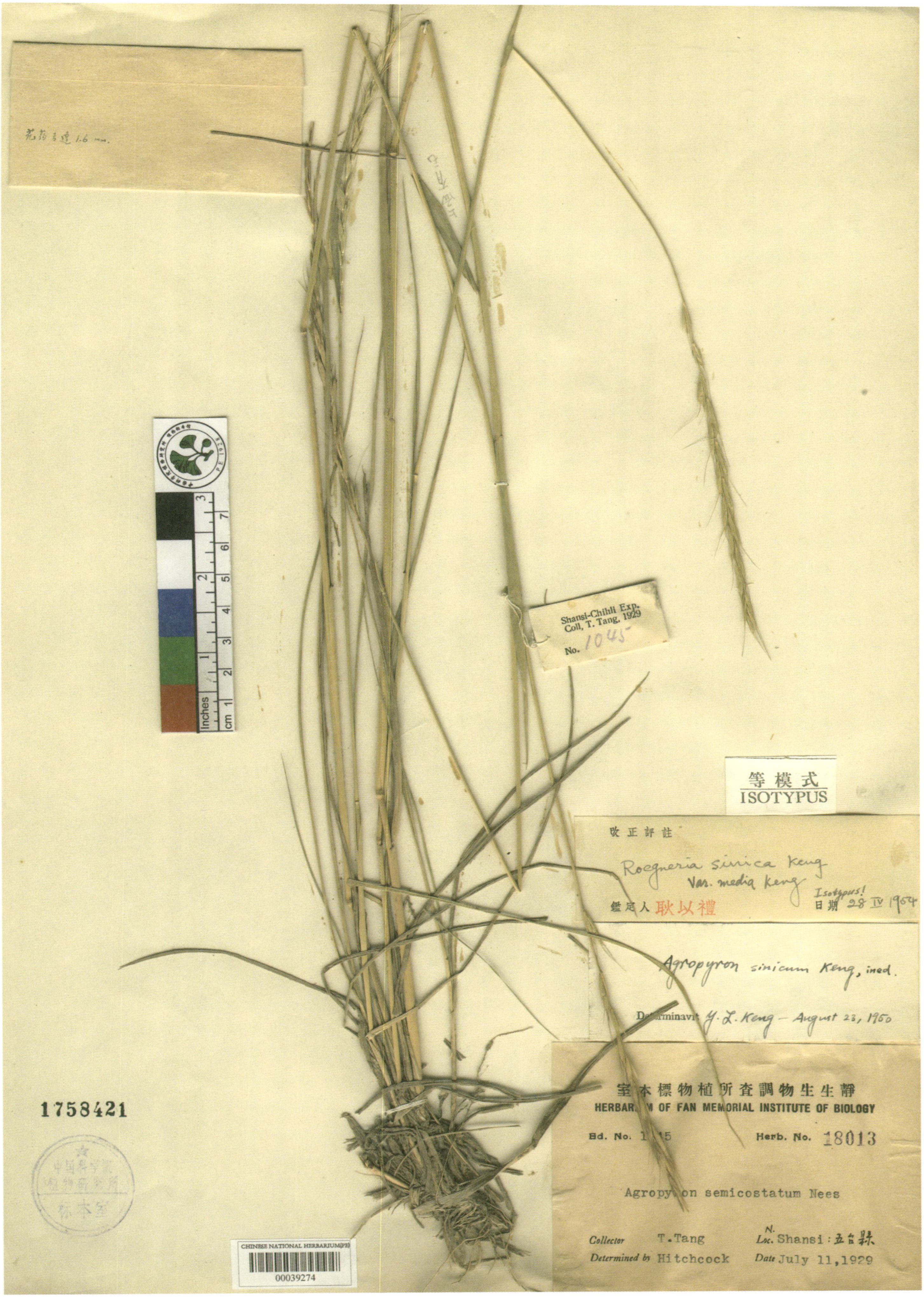

中间鹅观草 ***Roegneria sinica*** Keng var. ***media*** Keng in Journ. Nanjing Univ. (Biol.) 1963(1): 35. 1963. **Isotype:** China. Shanxi:Wutai, alt. 3500~4000 m, 1929-07-11, T. Tang 1045.

多变鹅观草 ***Roegneria varia*** Keng in Journ. Nanjing Univ. (Biol.) 1963(1): 70. 1963. **Isoparatype:** China. Gansu: Tongwei, Y. L. Keng & P. C. Keng 5951.

绿穗鹅观草 ***Roegneria viridula*** Keng & S. L. Chen in Journ. Nanjing Univ. (Biol.) 1963(1): 39, f. 2. 1963. **Isotype:** China. Xinjiang: Urumqi, 1956-07-14, R. C. Ching 394.

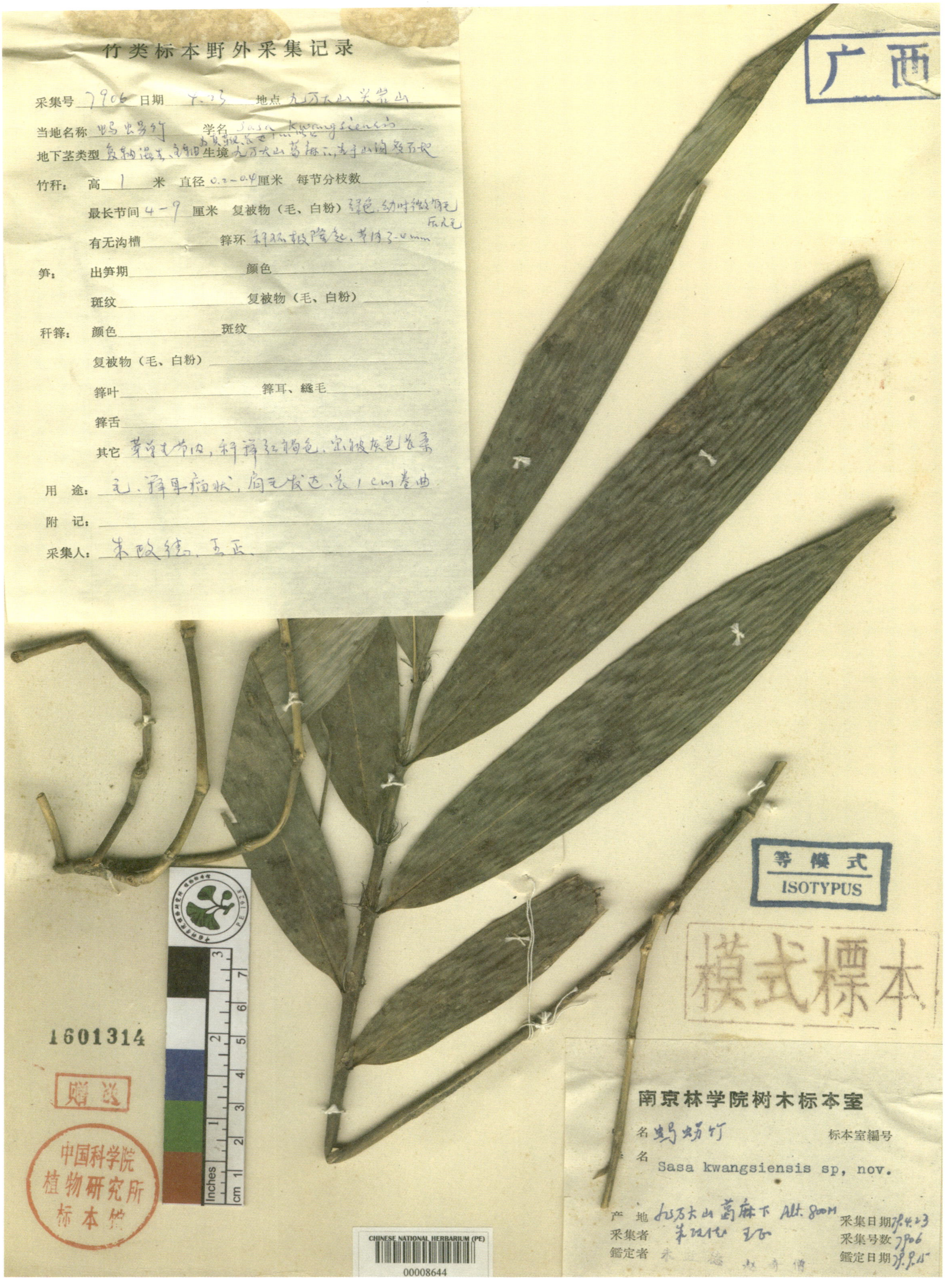

广西赤竹 ***Sasa guangxiensis*** C. D. Chu & C. S. Chao in Journ. Nanjing Technolog. Coll. For. Prod. 1981(3): 34, f. 3. 1981. **Isotype:** China. Guangxi: Rongshui, alt. 800 m, 1979-04-23, C. D. Chu & Z. Wang 7906.

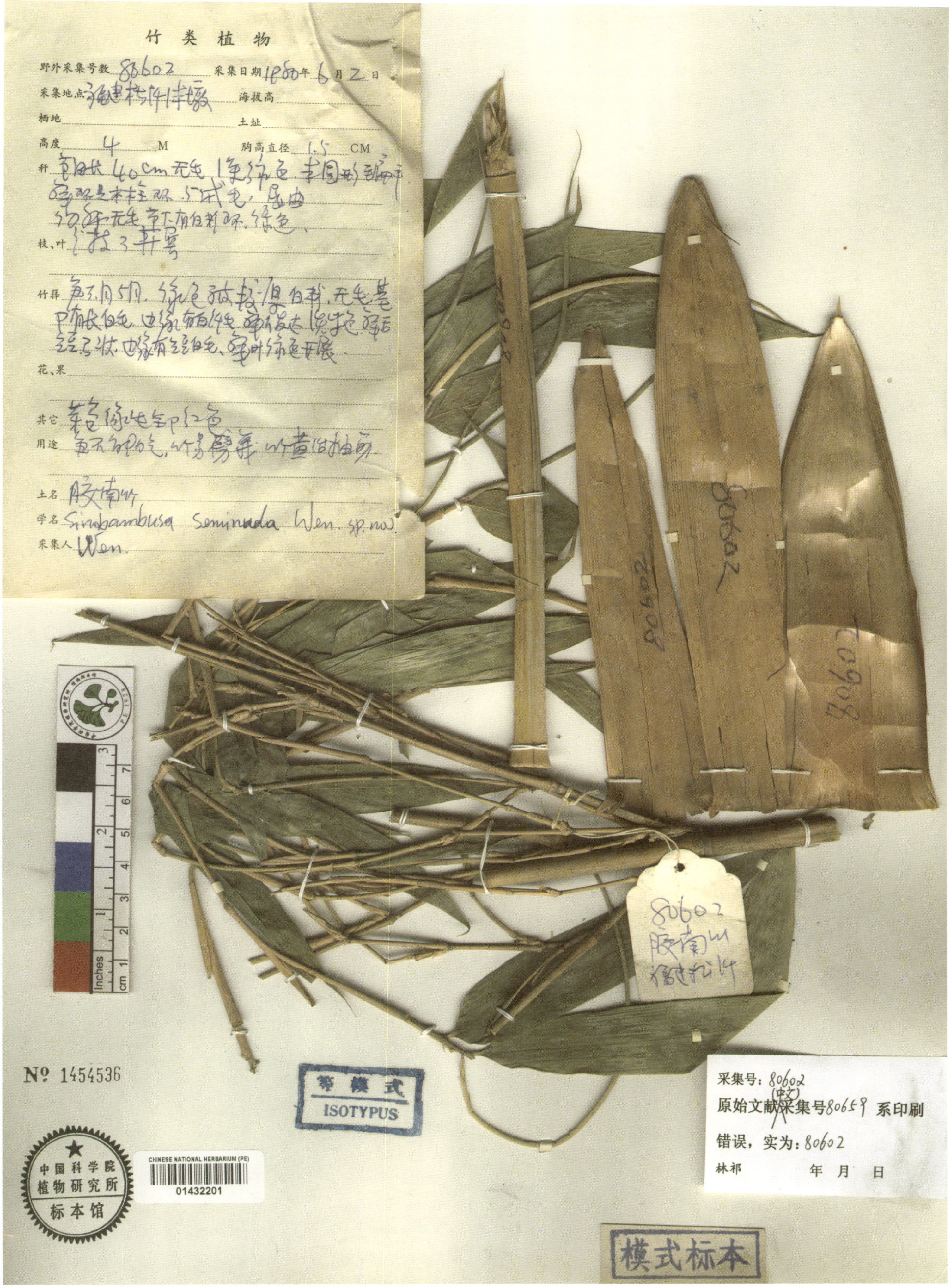

胶南竹 ***Sinobambusa seminuda*** T. F. Wen in Journ. Bamboo Res. 1(2): 18, f. 7. 1982. **Isotype:** China. Fujian: Songxi, 1980-06-02, T. H. Wen 80602.

大花大油芒 ***Spodiopogon sibiricus*** Trin. var. ***grandiflorus*** L. Liu ex S. M. Phillips & S. L. Chen, Fl. China 22: 576. 2006. **Holotype:** China. Sichuan: Barkam, alt. 2600 m, 1957-09-01, X. Li 71733.

越西大油芒 ***Spodiopogon yuexiensis*** S. L. Zhong in Journ. SW Agric. Coll. 1982(4): 77, f. 2. 1982. **Isotype:** China. Sichuan: Yuexi, alt. 1600 m, 1976-09-03, Sichuan Exped. 14256.

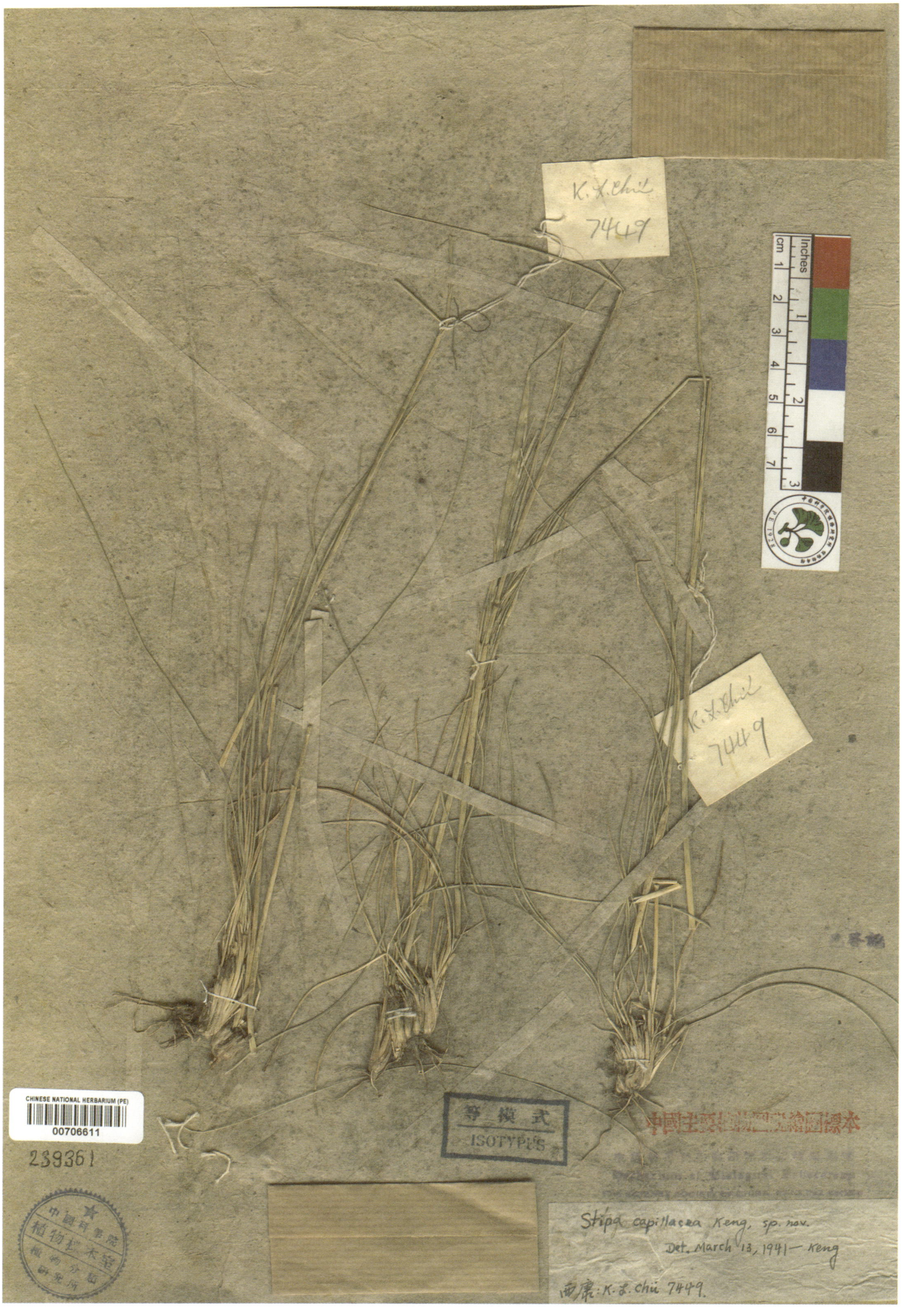

丝颖针茅 ***Stipa capillacea*** Keng in Sunyatsenia 6(2): 100, pl. 15. 1941. **Isotype:** China. Sichuan: Taining, 1940-07-22, K. L. Chu 7449.

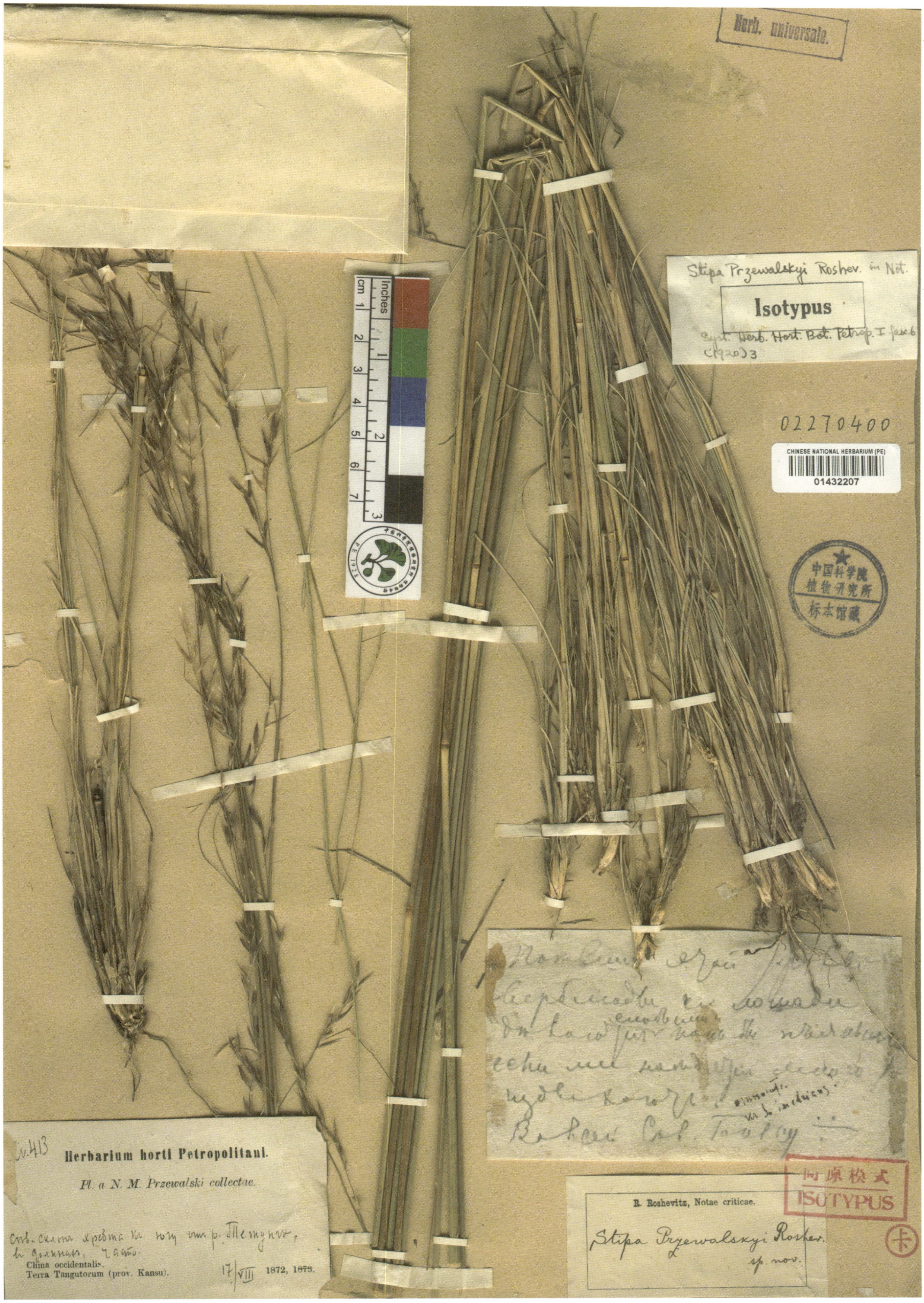

甘青针茅 ***Stipa przewalskyi*** R. J. Roshevitz in Bot. Mat. Gerb. Glavn. Bot. Sasa RSFSR 1(6): 3. 1920. **Isotype:** China. Gansu: Terra Tangutorum(= Qilianshan), Przewalsky 413.

硬毛蕊草 ***Thrixgyne dura*** Keng in Sunyatsenia 6(2): 82, pl. 13. 1941. **Isotype:** China. Sichuan: Taining, 1940-07-07, K. L. Chu 7268.

贫花三毛草 ***Trisetum pauciflorum*** Keng, Fl. Tsinling. 1(1): 441. 1976. **Isotype:** China. Shaanxi: Yangxian, alt. 1600 m, 1952-07-29, K. T. Fu 5281.

高原鼠茅 ***Vulpia alpina*** L. Liu, Fl. Reip. Pop. Sin. 9(2): 87, 405, pl. 26: 5, 5a. 2002. **Holotype:** China. Xizang: Lhasa, alt. 4000 m, 1965-08-16, Y. T. Chang & K. Y. Lang 1597.

Cyperaceae

莎草科

shacaoke

甘西薹草 ***Carex allivescens*** V. Krecz. in Not. Syst. Herb. Inst. Bot. Nom. Kom. Acad. Sci. URSS. 9(1): 190. 1941.
Paratype: China. W Kansu: Lantshzho-lunva, 1885-05-14, G. Potanin s. n.

安宁薹草 ***Carex anningensis*** F. T. Wang & Tang in Acta Phytotax. Sin. 37(2): 171, f. 10. 1999. **Holotype:** China. Yunnan: Anning, 1940-11-01, T. N. Liu 17217.

长芒鳞薹草 ***Carex aristatisquamata*** Tang & F. T. Wang ex L. K. Dai in Acta Phytotax. Sin. 32(2): 179, f. 4: 1-5. 1994. **Holotype:** China. Sichuan: Maowen, alt. 3500~3800 m, 1959-06-10, S. Jiang & C. L. Jin 711.

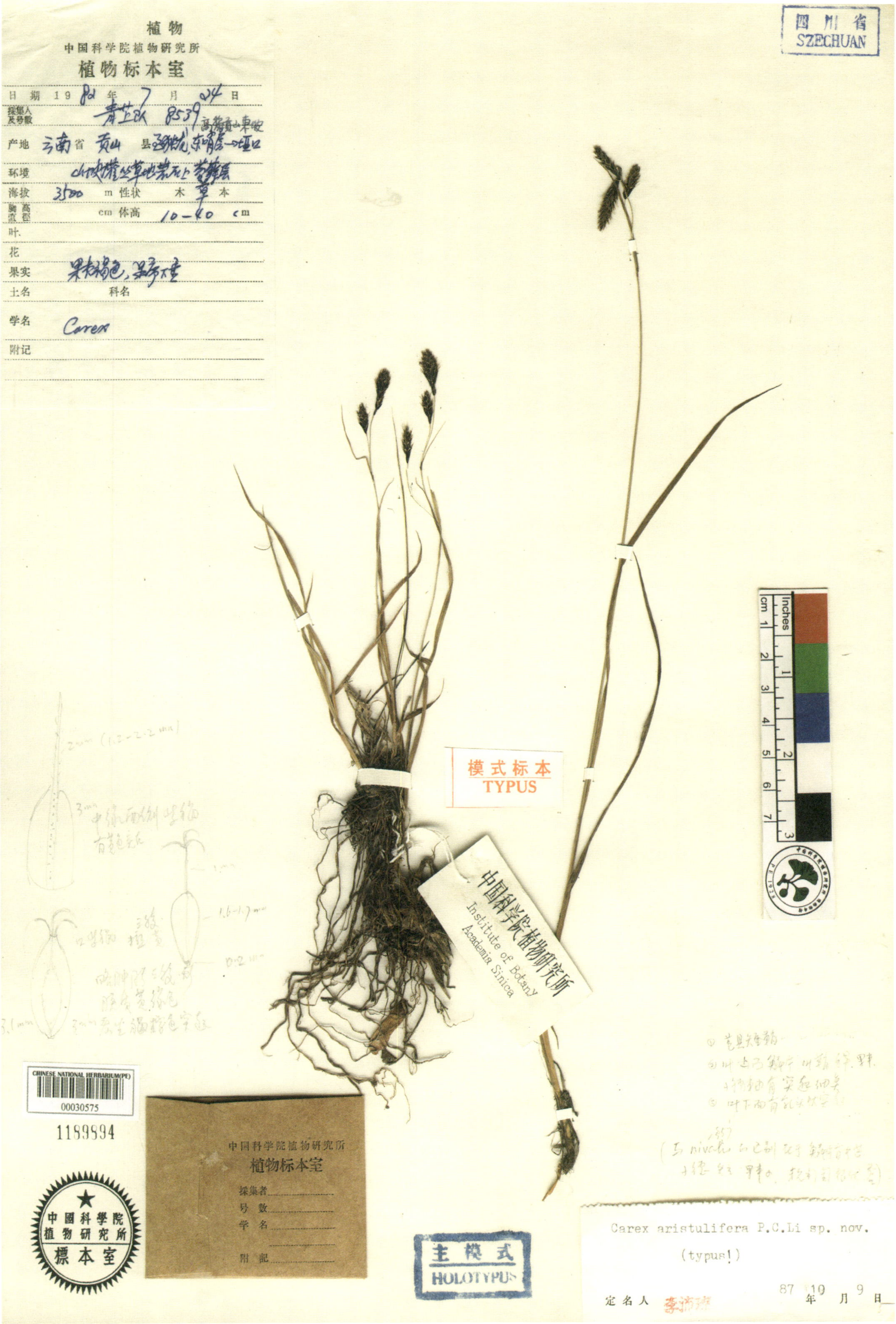

长芒薹草 ***Carex aristulifera*** P. C. Li in Acta Bot. Yunnan. 12(2): 141, f. 6. 1990. **Holotype:** China. Yunnan: Gongshan, alt. 3500 m, 1982-07-24, Qinghai-Xizang. Exped. 8539.

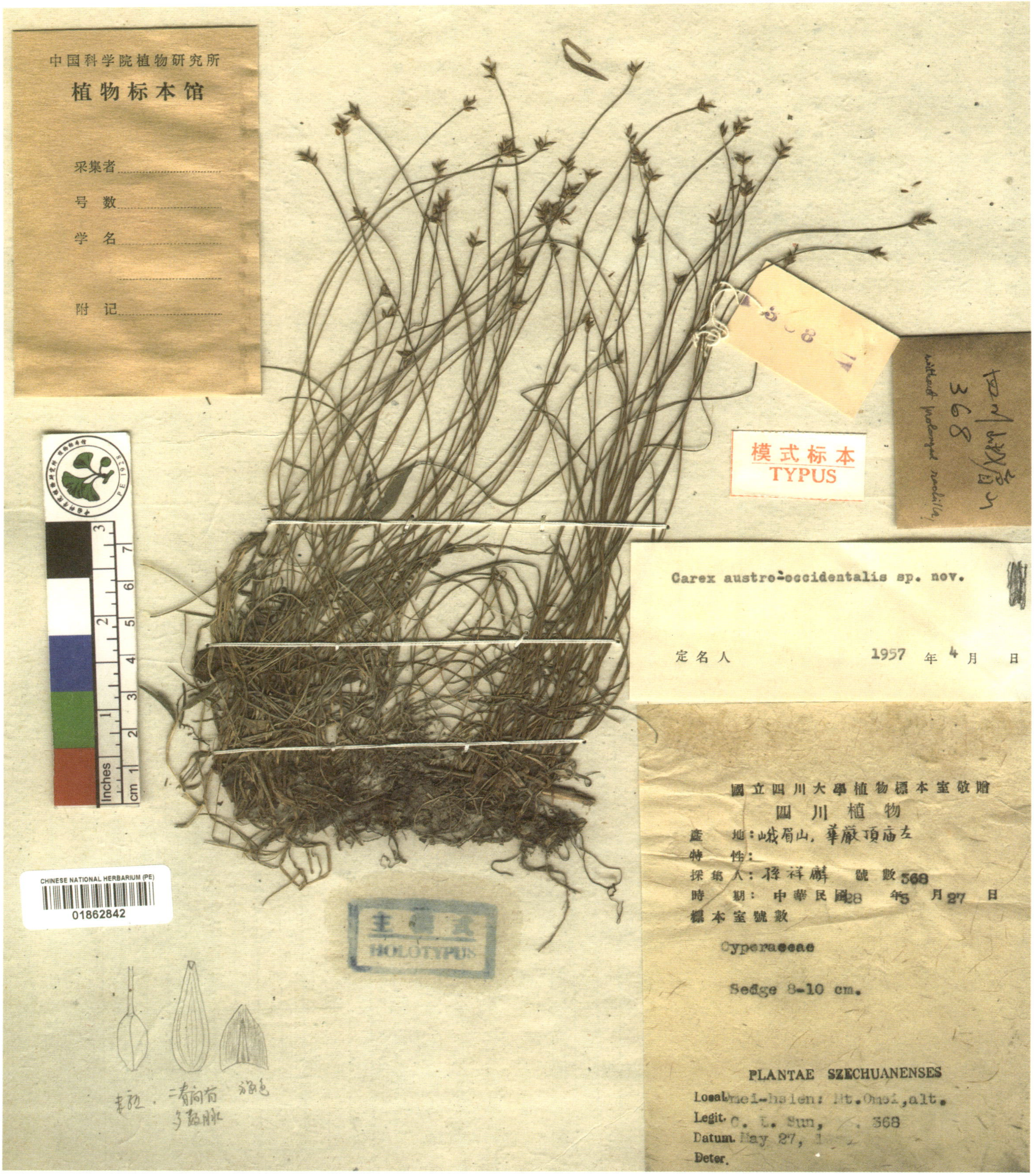

西南薹草 ***Carex austro-occidentalis*** F. T. Wang & Tang, Fl. Reip. Pop. Sin. 12: 527. 2000. **Holotype:** China. Sichuan: Emei, Emeishan, 1939-05-27, C. L. Sun 368.

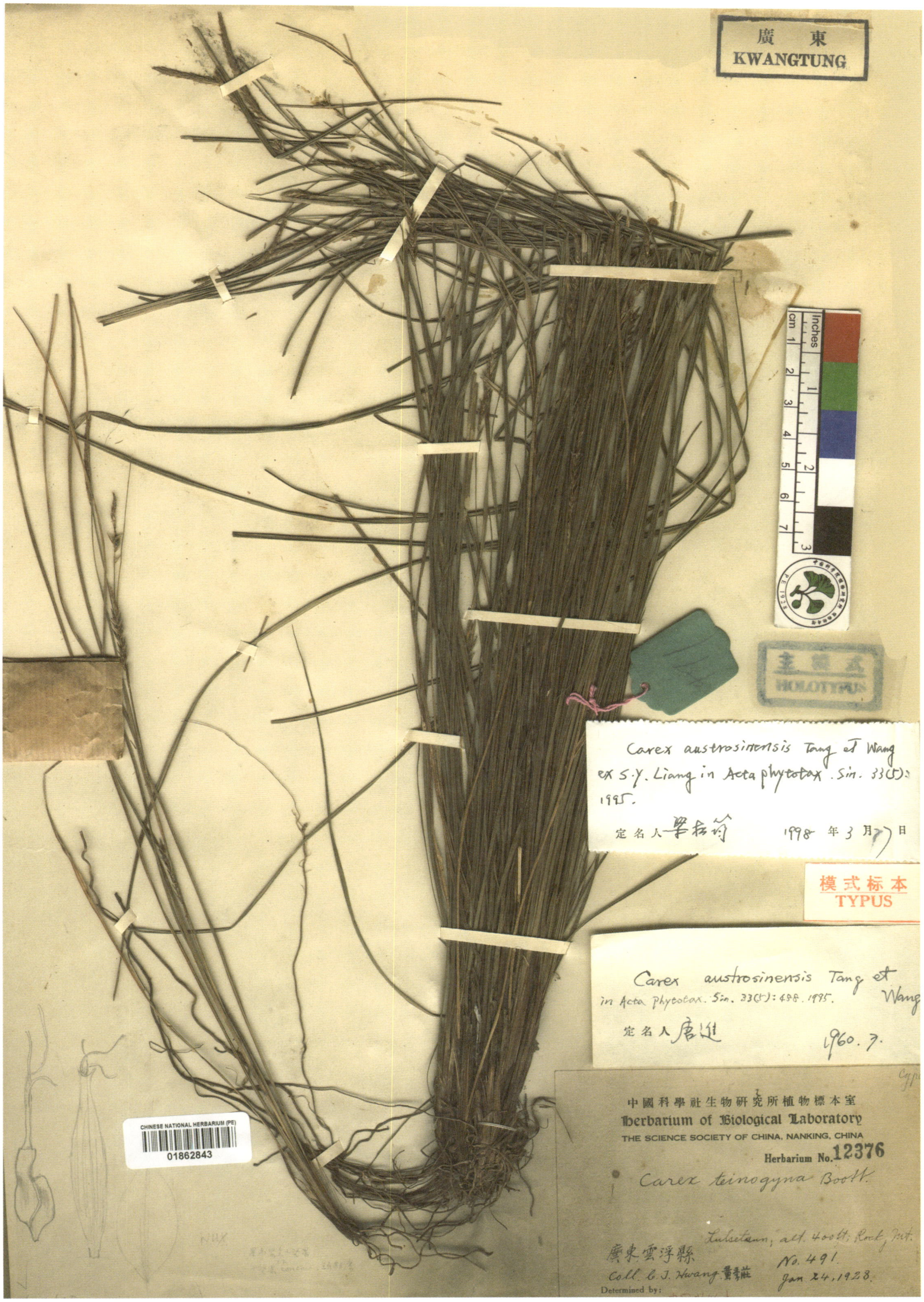

华南薹草 ***Carex austrosinensis*** F. T. Wang & Tang in Acta Phytotax. Sin. 33(5): 488, f. 1: 12-17. 1995. **Holotype:** China. Guangdong: Yunfu, alt. 1030 m, 1928-01-24, C. J. Hwang 491.

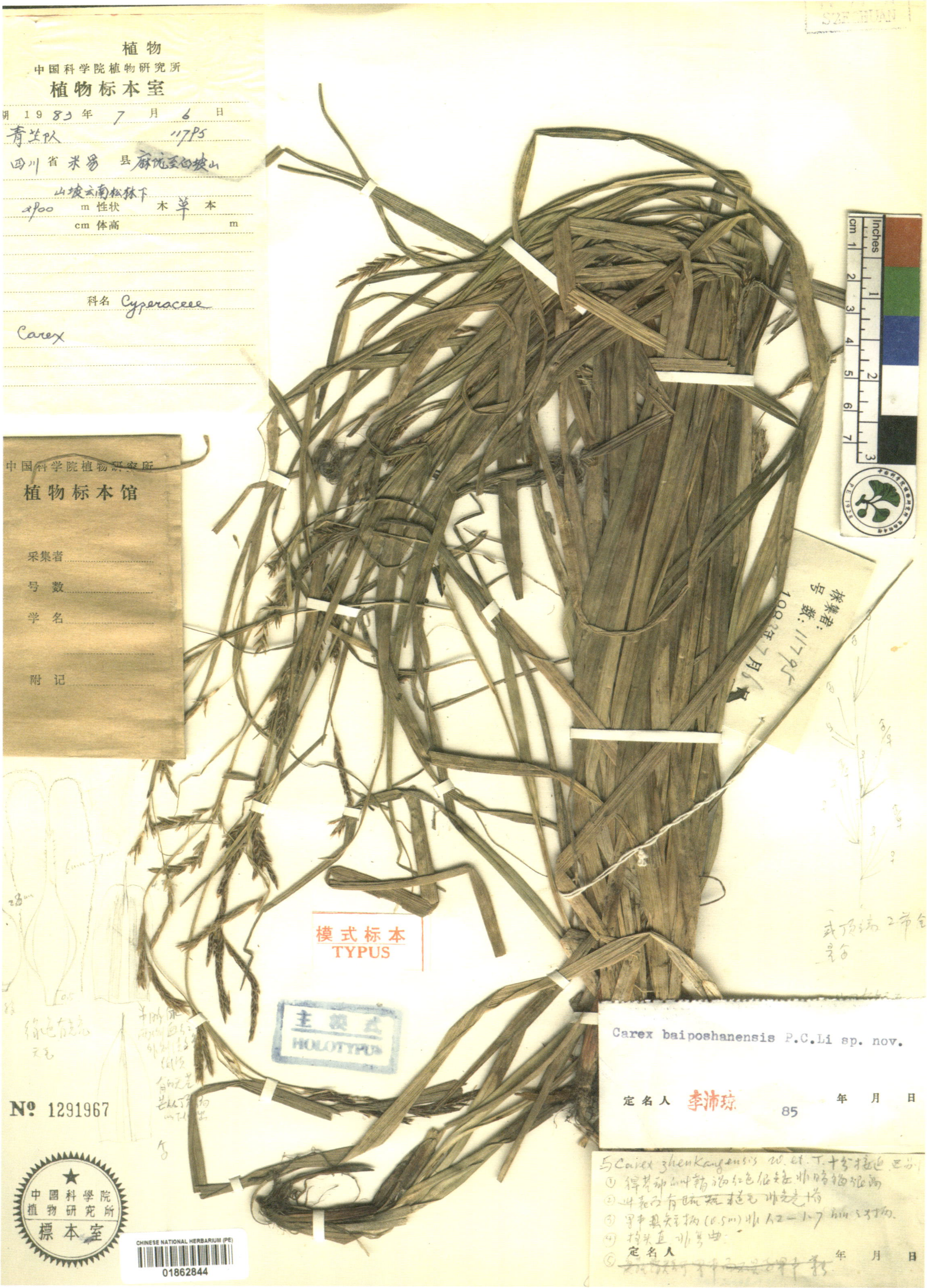

百坡山薹草 ***Carex baiposhanensis*** P. C. Li in Acta Bot. Yunnan. 12(2): 145, f. 12. 1990. **Holotype:** China. Sichuan: Miyi, alt. 2900 m, 1983-07-06, Qinghai-Xizang Exped. 11795.

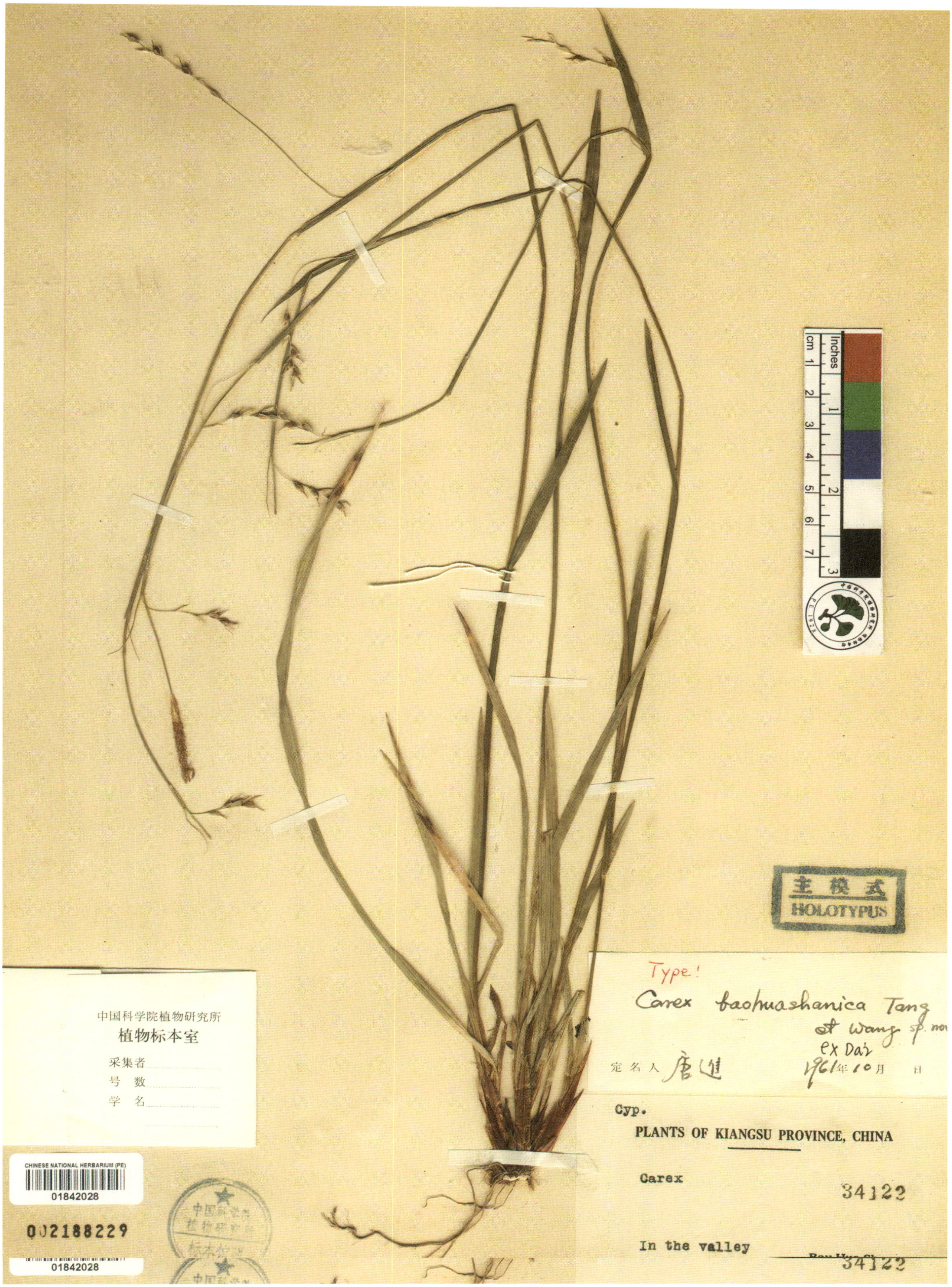

宝华山薹草 ***Carex baohuashanica*** Tang & F. T. Wang ex L. K. Dai in Acta Phytotax. Sin. 32(2): 182, f. 5: 5-8. 1994.
Holotype: China. Jiangsu: Baohuashan, 1932-04-16, H. C. Cheo 3527.

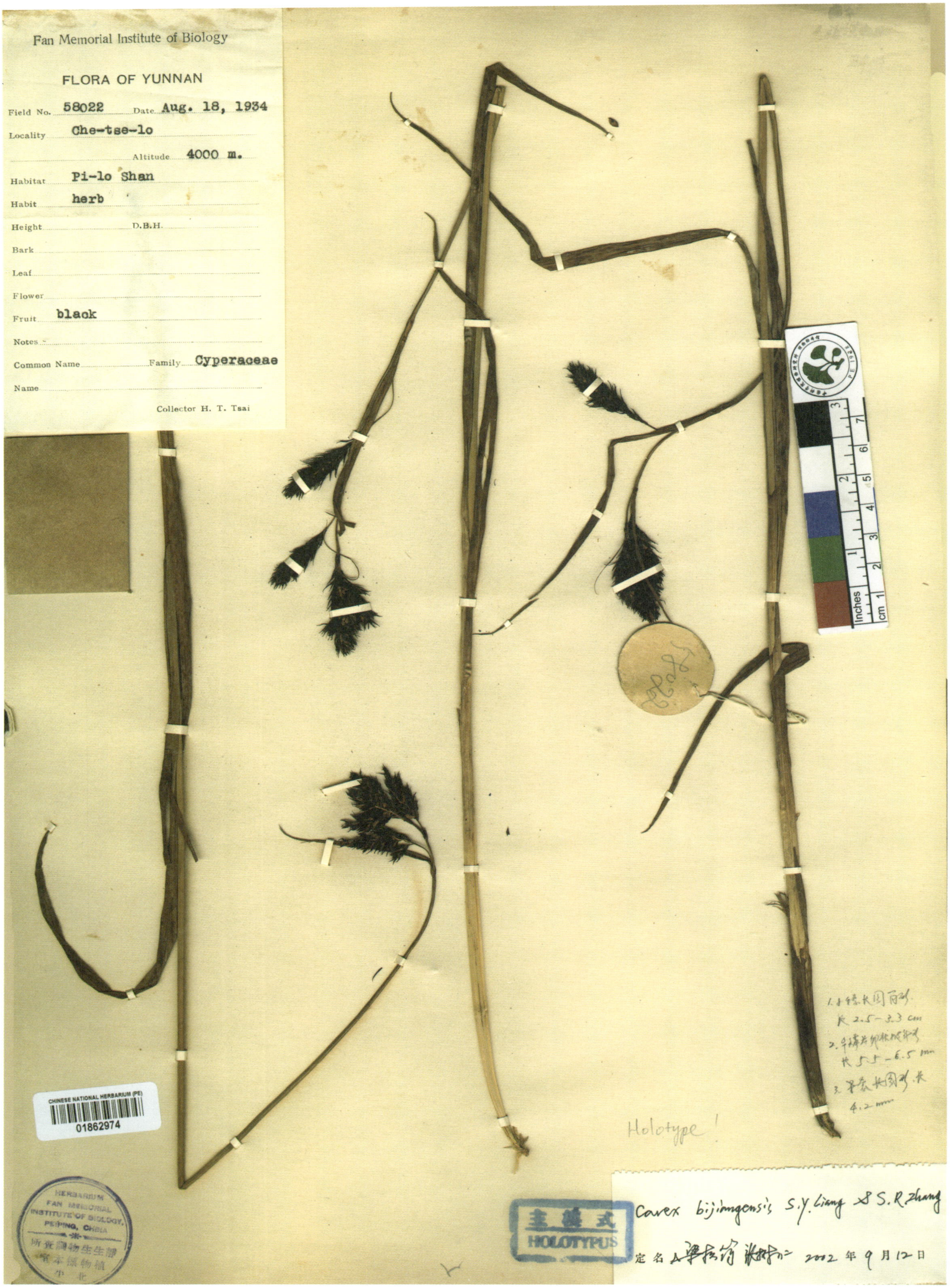

碧江薹草 ***Carex bijiangensis*** S. Y. Liang & S. R. Zhang in Novon 16: 364, f. 1. 2006. **Holotype:** China. Yunnan: Chetselo (= Bijing), alt. 4000 m, 1934-08-18, H. T. Tsai 58022.

垂穗薹草 ***Carex brachyathera*** Ohwi in Japan. Journ. Bot. 7: 190. 1934. **Isotype:** China. Taiwan: Mt. Nanko-taisan, 1933-07-??, J. Ohwi 4067.

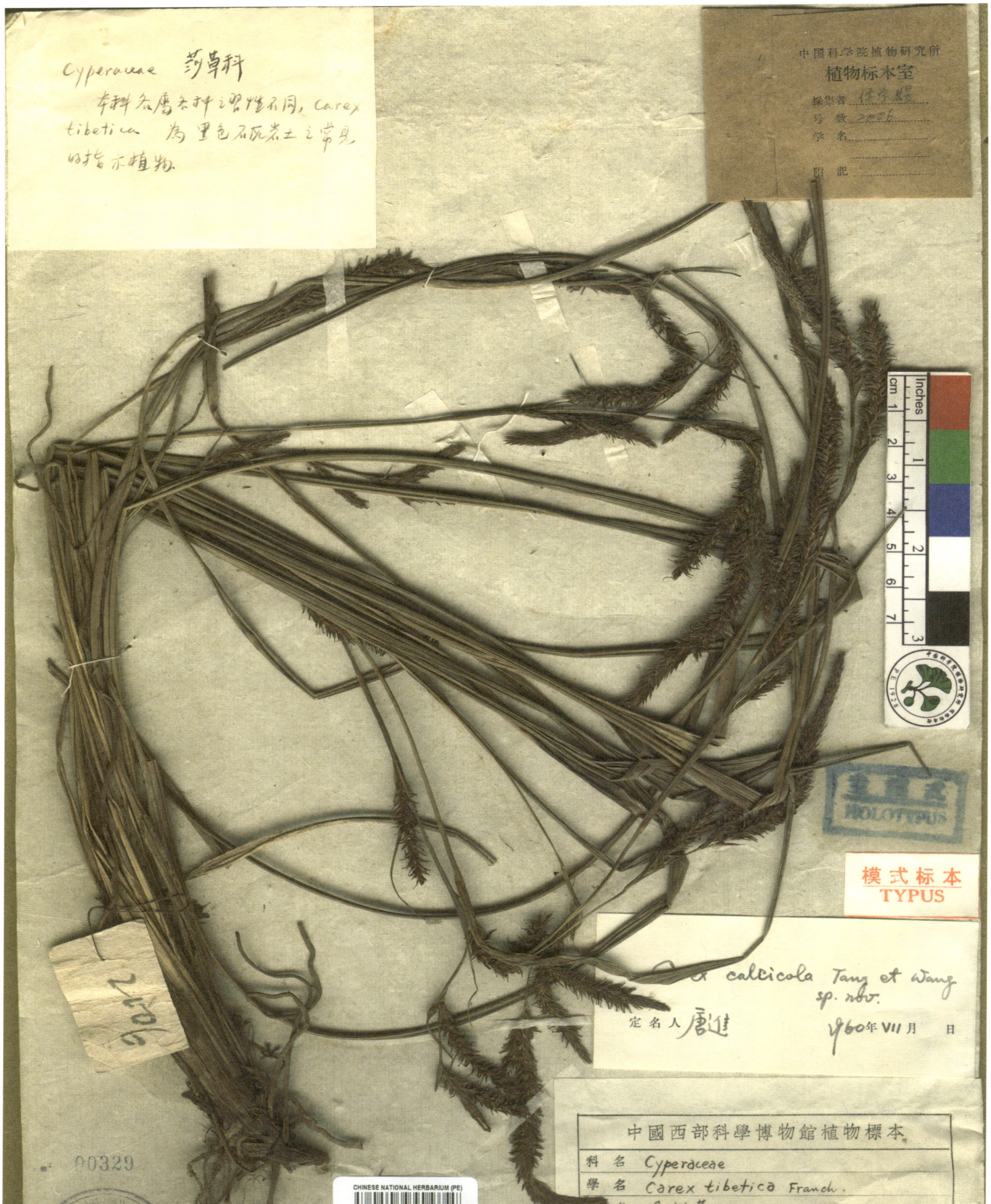

灰岩生薹草 ***Carex calcicola*** Tang & F. T. Wang in Acta Phytotax. Sin. 24(3): 244, f. 4. 1986. **Holotype:** China. Guizhou: Libo, alt. 800 m, 1932-02-24, H. Y. Hou 2006.

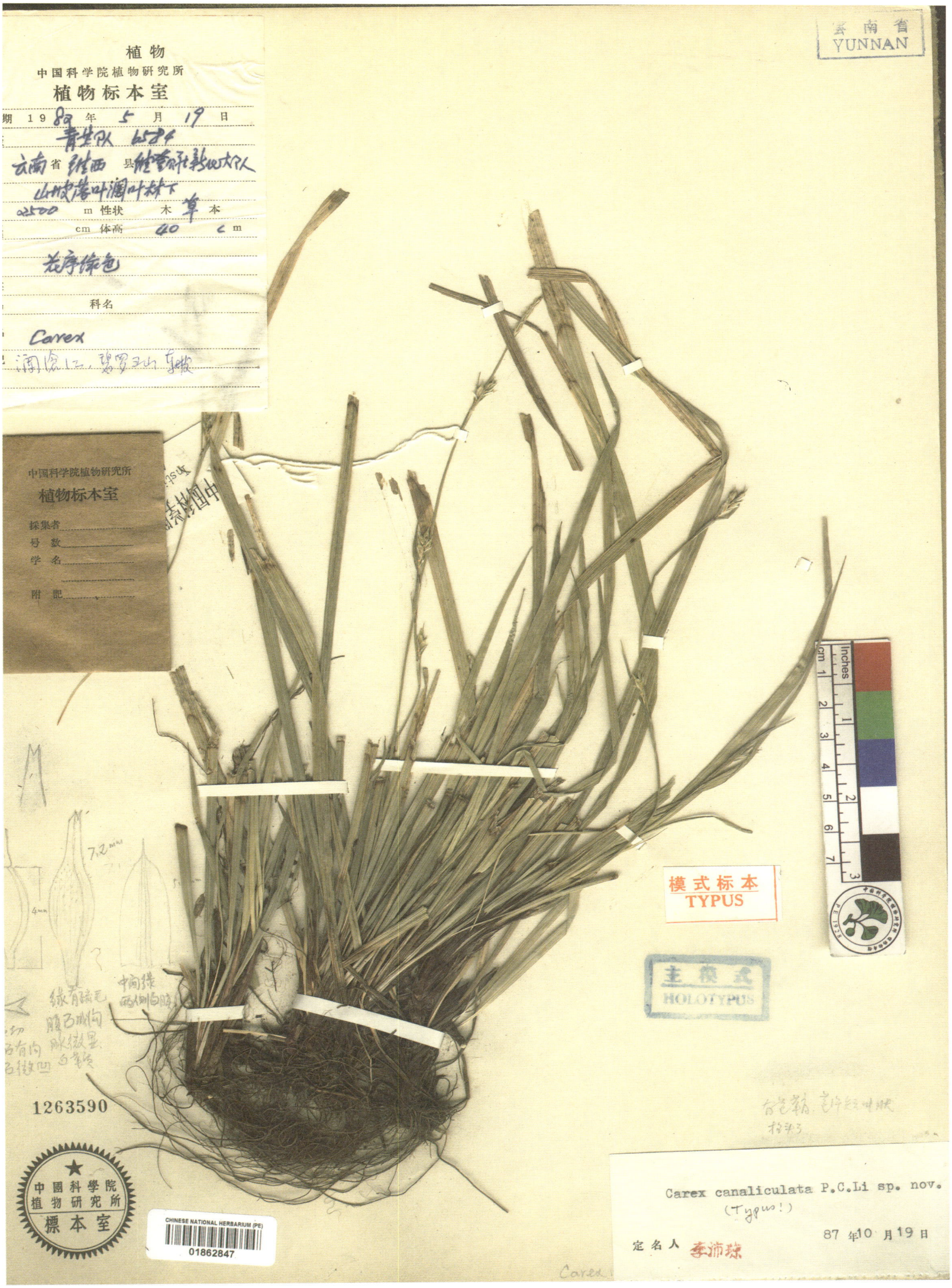

沟囊薹草 ***Carex canaliculata*** P. C. Li in Acta Bot. Yunnan. 12(2): 148, f. 16. 1990; S. Y. Jin, Catal. Type Speeim (Corm.) Herb. China 67. 1999. **Holotype:** China. Yunnan: Weixi, alt. 2500 m, 1982-05-19. Qinghai-Xizang Exped. 6584.

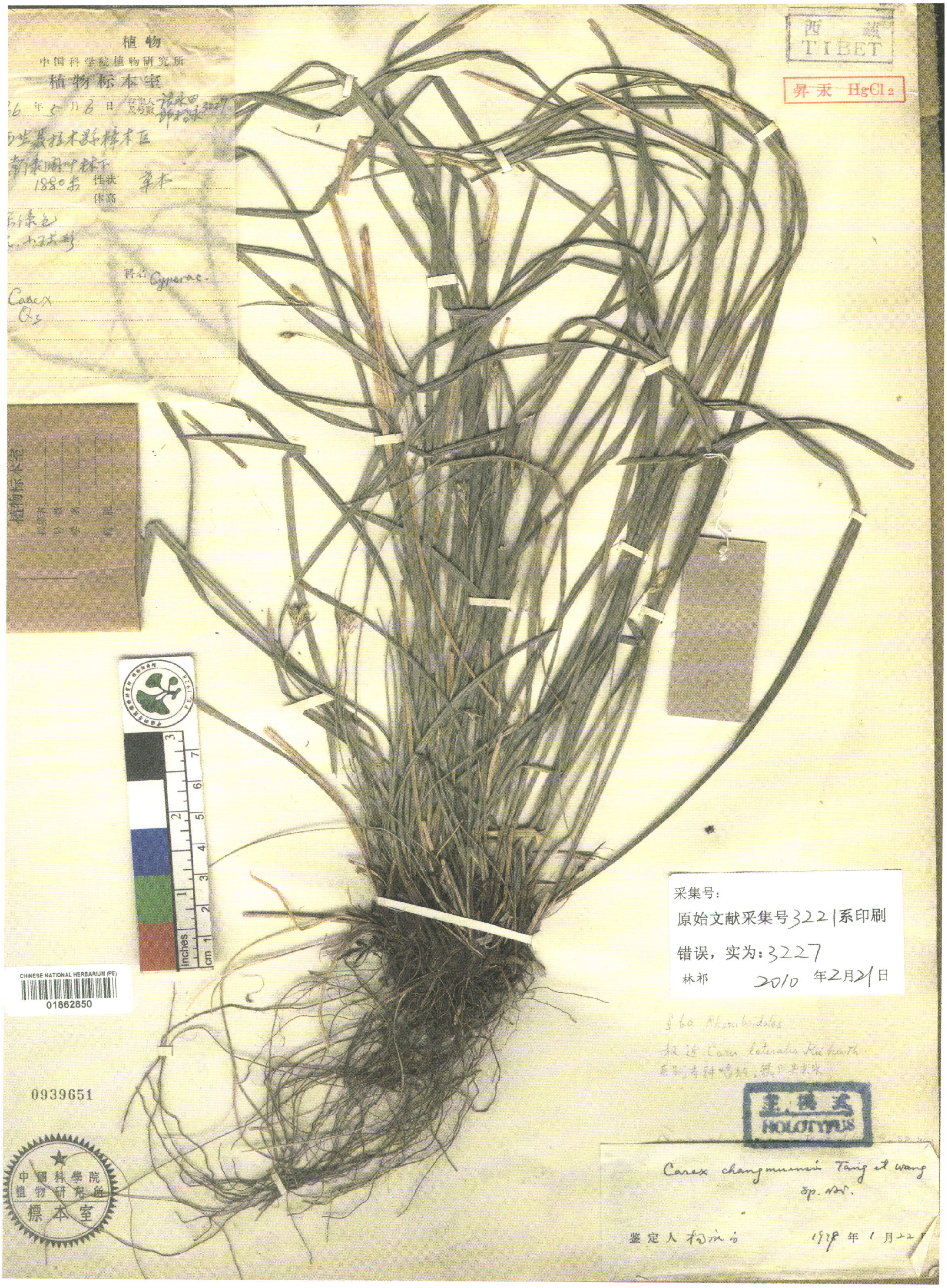

樟木薹草 ***Carex changmuensis*** Tang & F. T. Wang ex Y. C. Yang, Fl. Xizang. 5: 410, f. 234. 1987. **Holotype:** China. Xizang: Nyalam, alt. 1880 m, 1966-05-06, Y. T. Cheng & K. Y. Lang 3227.

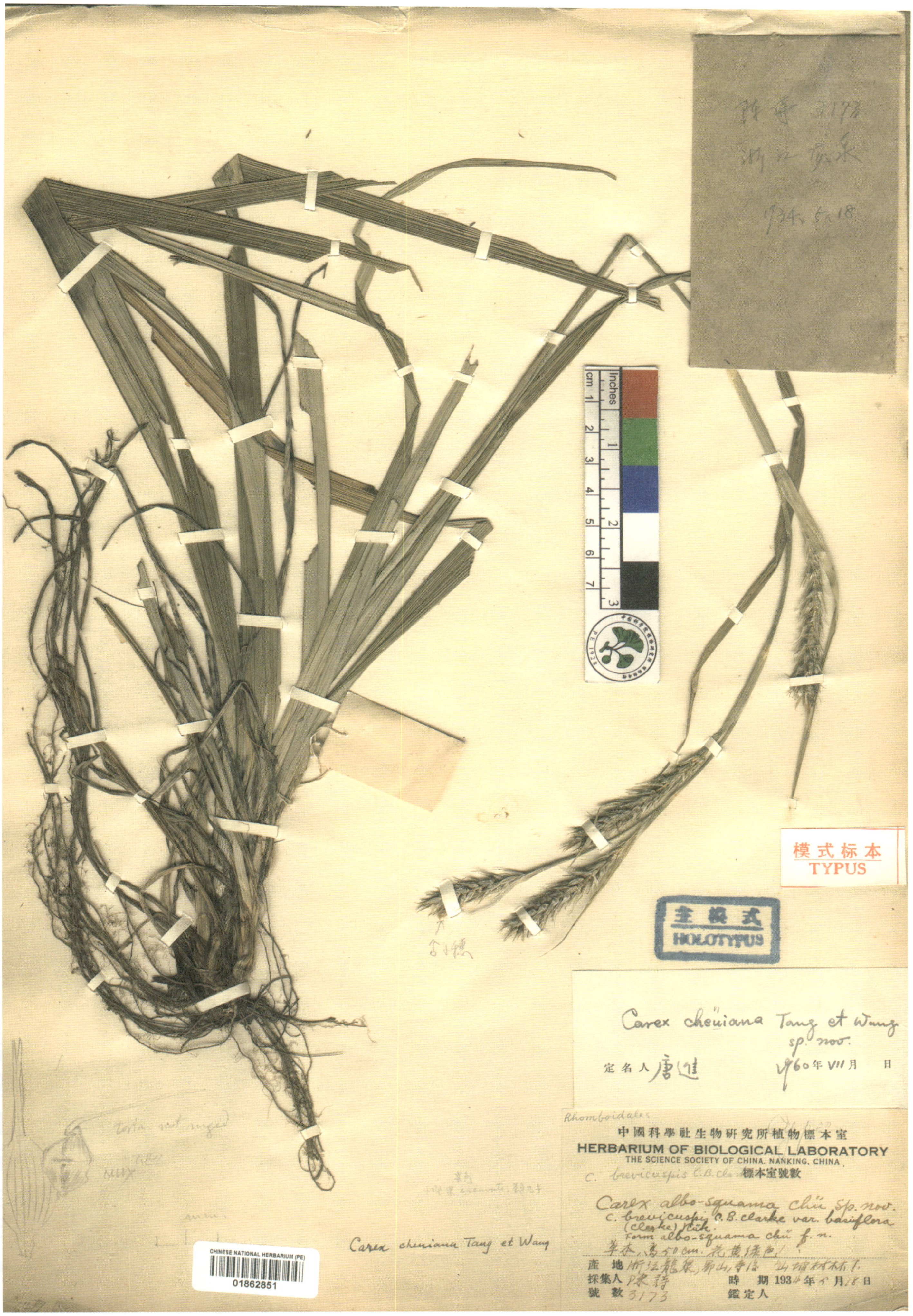

陈氏薹草 ***Carex cheniana*** Tang & F. T. Wang in Acta Phytotax. Sin. 36(6): 532, f. 2: 1-5. 1998. **Holotype:** China. Zhejiang, Longquan, 1934-05-18, S. Chen 3173.

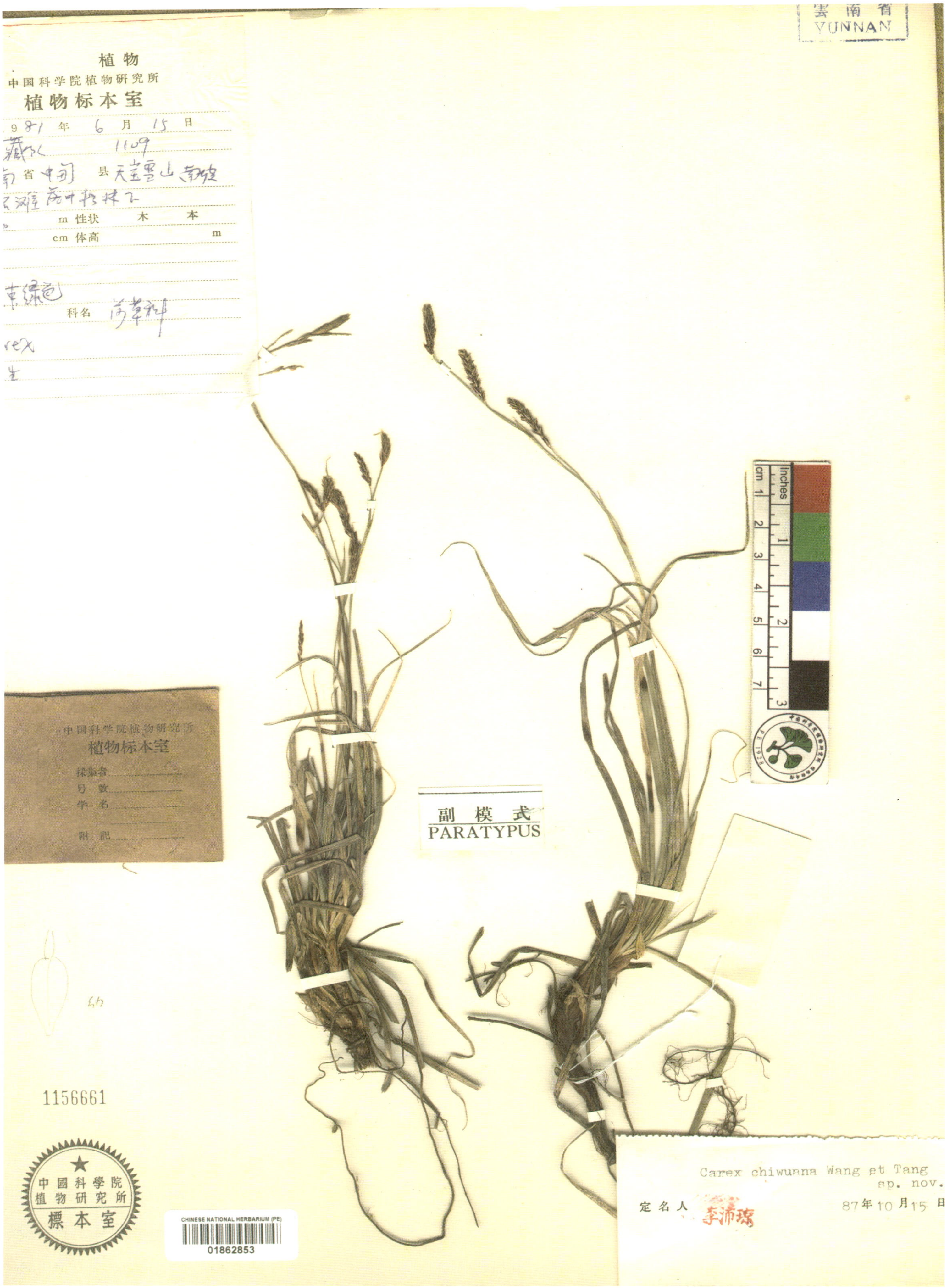

启无薹草 ***Carex chiwuana*** F. T. Wang & Tang in Acta Bot. Yunnan. 12(2): 143, f. 9. 1990. **Paratype:** China. Yunnan: Zhongdian, alt. 3600 m, 1981-06-15, Qinghai-Xizang Exped. 1109.

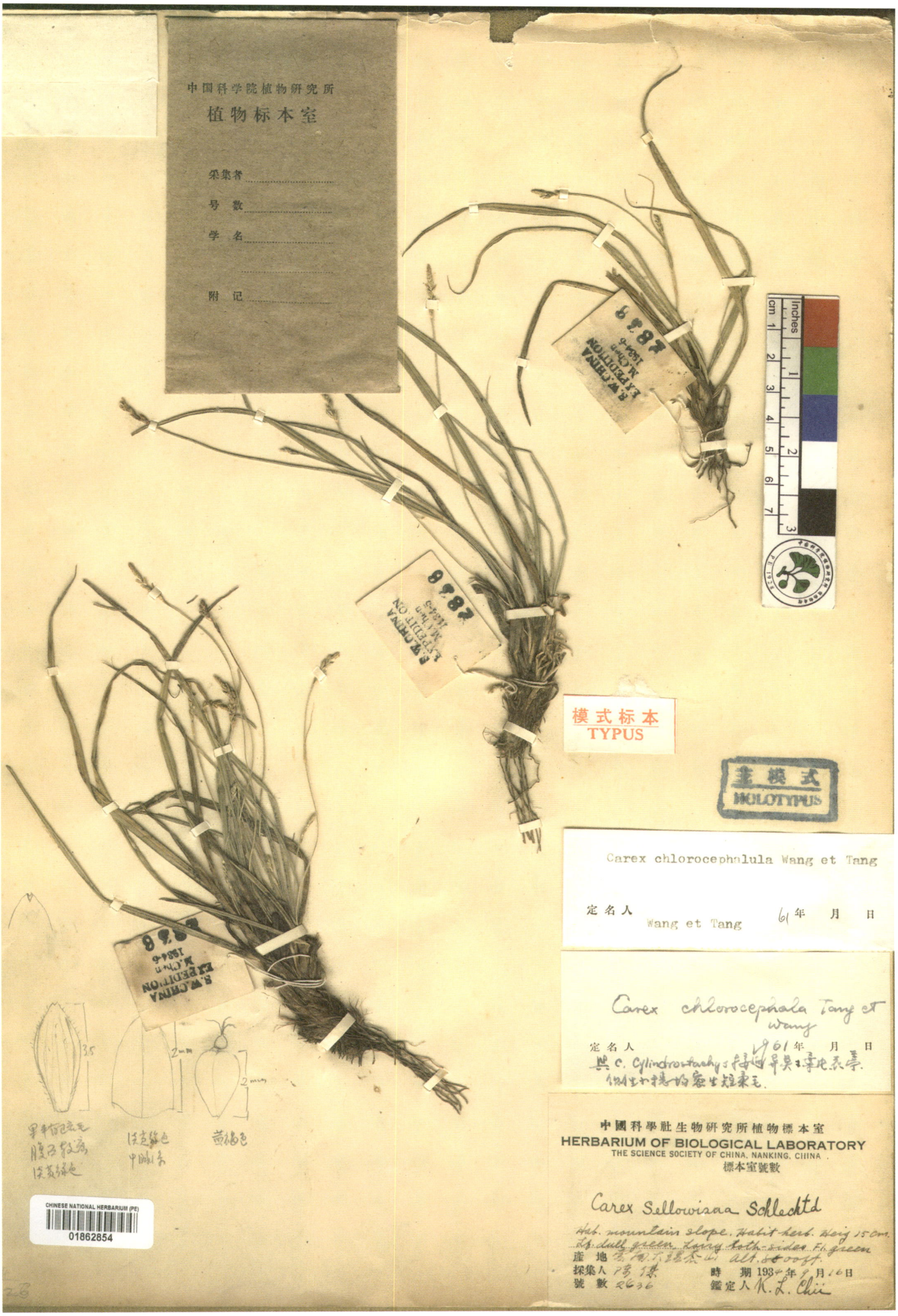

绿头薹草 ***Carex chlorocephalula*** F. T. Wang & Tang in Acta Bot. Yunnan. 12(2): 143, f. 8. 1990. **Holotype:** China. Yunnan: Dali, alt. 2900 m, 1934-09-16, M. Chen 2836.

无喙绿穗薹草 ***Carex chlorostachys*** Steven var. ***conferta*** Tang & F. T. Wang, Fl. Reip. Pop. Sin. 12: 259, 526, pl. 49: 15-16. 1999. **Holotype:** China. Qinghai: Datong, alt. 1960 m, 1936-07-28, K. M Liou 6326.

桂龄薹草 ***Carex chuiana*** F. T. Wang & Tang in Acta Phytotax. Sin. 37(2): 167, f. 6. 1999. **Holotype:** China. Sichuan: Taining, 1940-08-03, K. L. Chu 7490.

曲氏薹草 *Carex chuii* Nelmes in Kew Bull. 1948(1): 108. 1948. **Isotype:** China. Sichuan: Baoxing, alt. 2450 m, 1936-06-30, K. L. Chu 3022.

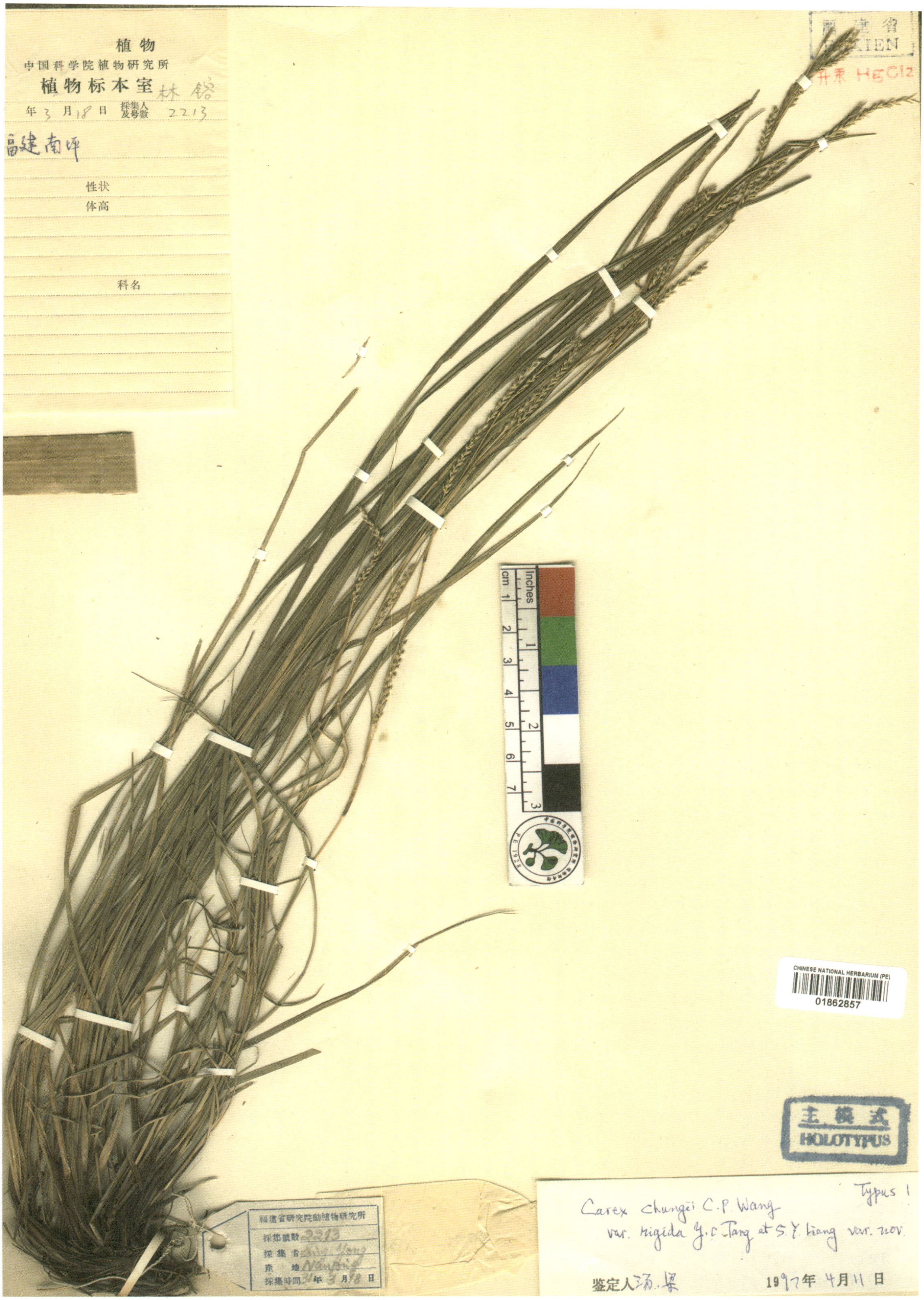

坚硬薹草 ***Carex chungii*** C. P. Wang var. ***rigida*** Y. C. Tang & S. Y. Liang, Fl. Reip. Pop. Sin. 12: 522. 2000. **Holotype:** China. Fujian, Nanping, 1931-03-18, Y. Ling 2213.

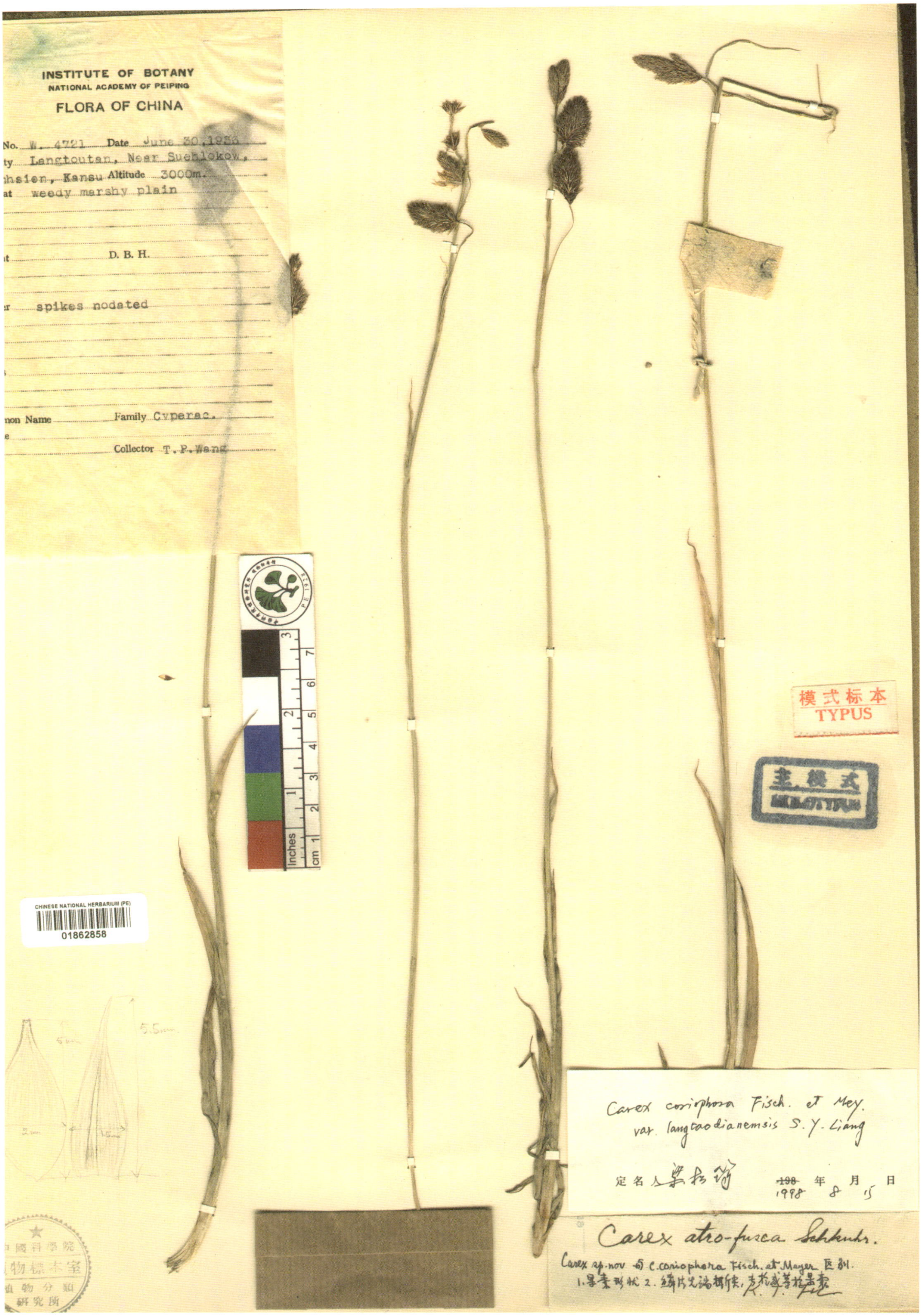

浪淘殿薹草 ***Carex coriophora*** Fisch. & C. A. Mey. subsp. ***langtaodianensis*** S. Y. Liang, Fl. Reip. Pop. Sin. 12: 524. 2000. **Holotype:** China. Gansu: Min Xian, alt. 3000 m, 1936-06-30, T. P. Wang 4721.

鹤果薹草 ***Carex cranaocarpa*** Nelmes in Bull. Misc. Inform. Kew 1939: 184. 1939. **Isotype:** China. Hebei: Tientsin, 1929-08-09, Licent 9002.

密生薹草 ***Carex crebrae*** V. Krecz. in Not. Syst. Herb. Inst. Bot. Nom. Kom. Acad. Sci. URSS 9(1): 190. 1941. **Isotype:** China. W Gansu: Dshoni, 1885-06-06, G. Potanin s. n.

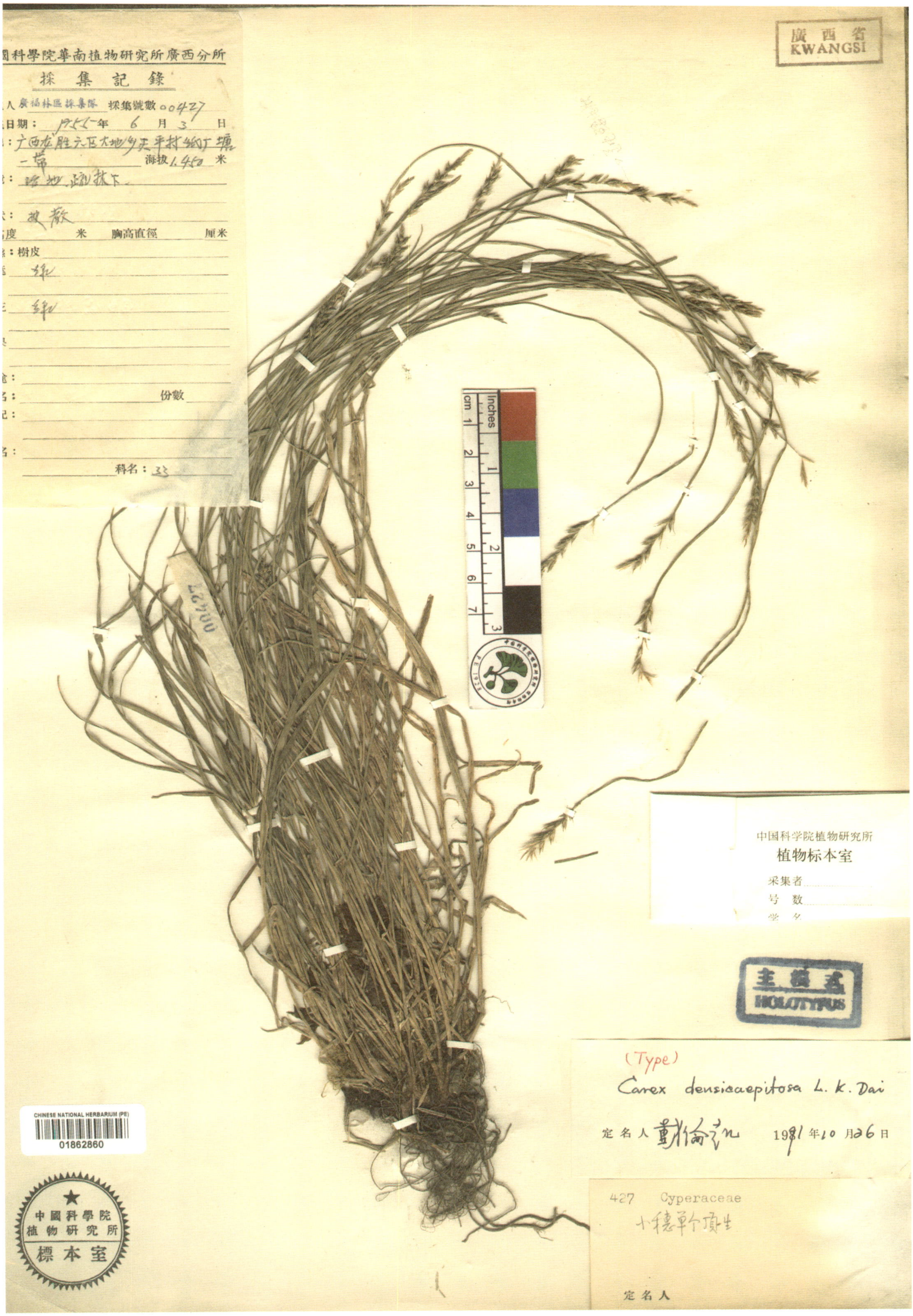

密丛薹草 ***Carex densicaespitosa*** L. K. Dai in Acta Phytotax. Sin. 32(2): 173, f. 1: 1-4. 1994. **Holotype:** China. Guangxi: Longsheng, alt. 1450 m, 1955-06-03, Guangfu Forest Exped. 427.

密缘毛薹草 ***Carex densifimbriata*** Tang & F. T. Wang in Acta Phytotax. Sin. 24(3): 240, f. 1. 1986. **Paratype:** China. Hunan: Qiangyang, 1953-04-18, Anjiang Agric. School 275.

德钦薹草 ***Carex deqinensis*** L. K. Dai in Acta Phytotax. Sin. 32(2): 182, f. 5:1-4. 1994. **Holotype:** China. Yunnan: Dêqên, alt. 2900~3300 m, 1940-07-18, K. M. Feng 5607.

秦岭薹草 ***Carex diplodon*** Nelmes in Bull. Misc. Inform. Kew 1939: 184. 1939. **Isotype:** China. Gansu: Koanhiaho, 1919-04-16, Licent 5035.

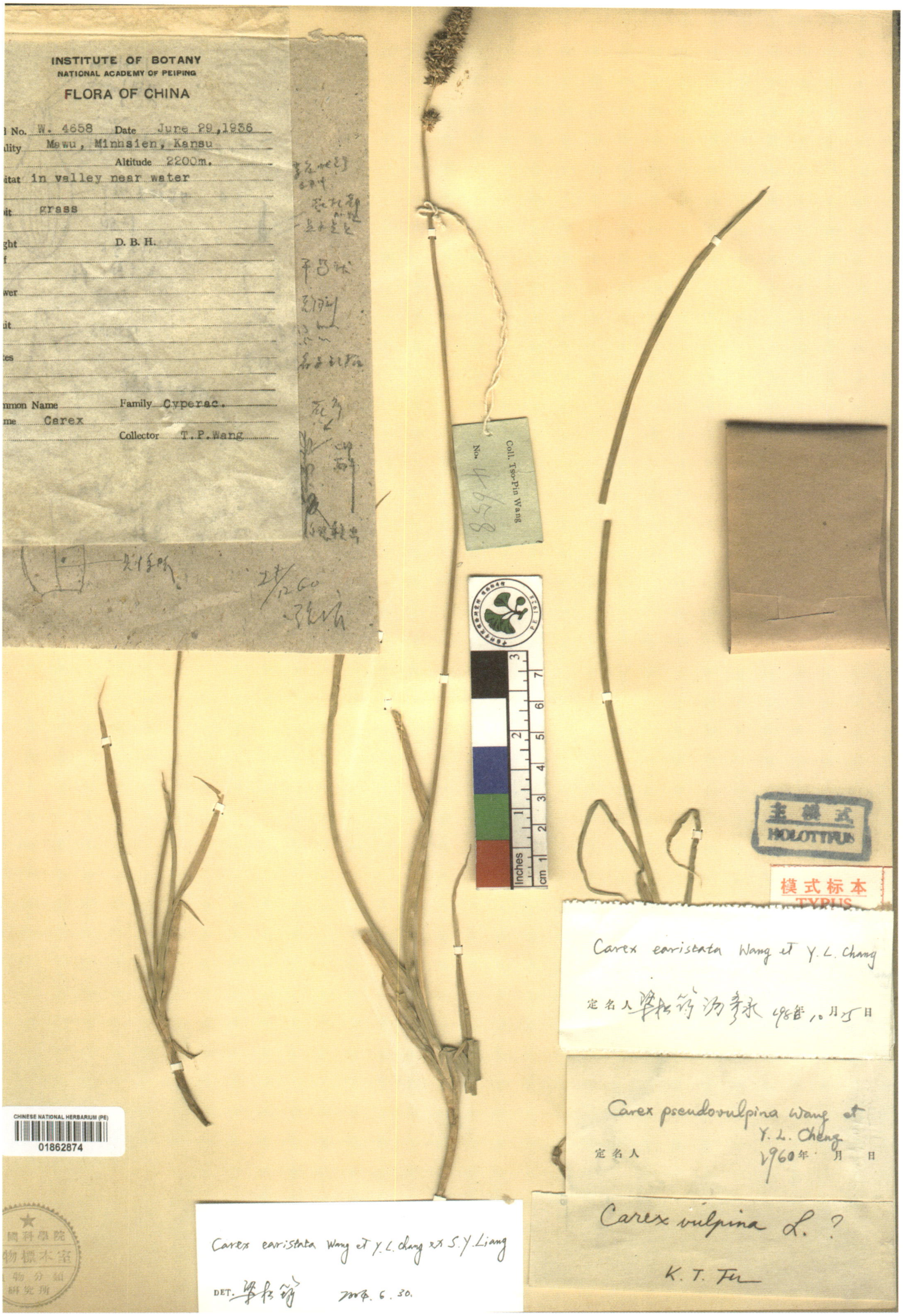

无芒薹草 ***Carex earistata*** F. T. Wang & Y. L. Chang ex S. Y. Liang in Acta Phytotax. Sin. 28(2): 153, f. 1. 1990. **Holotype:** China. Gansu: Min Xian, alt. 2200 m, 1936-06-29, T. P. Wang 4658.

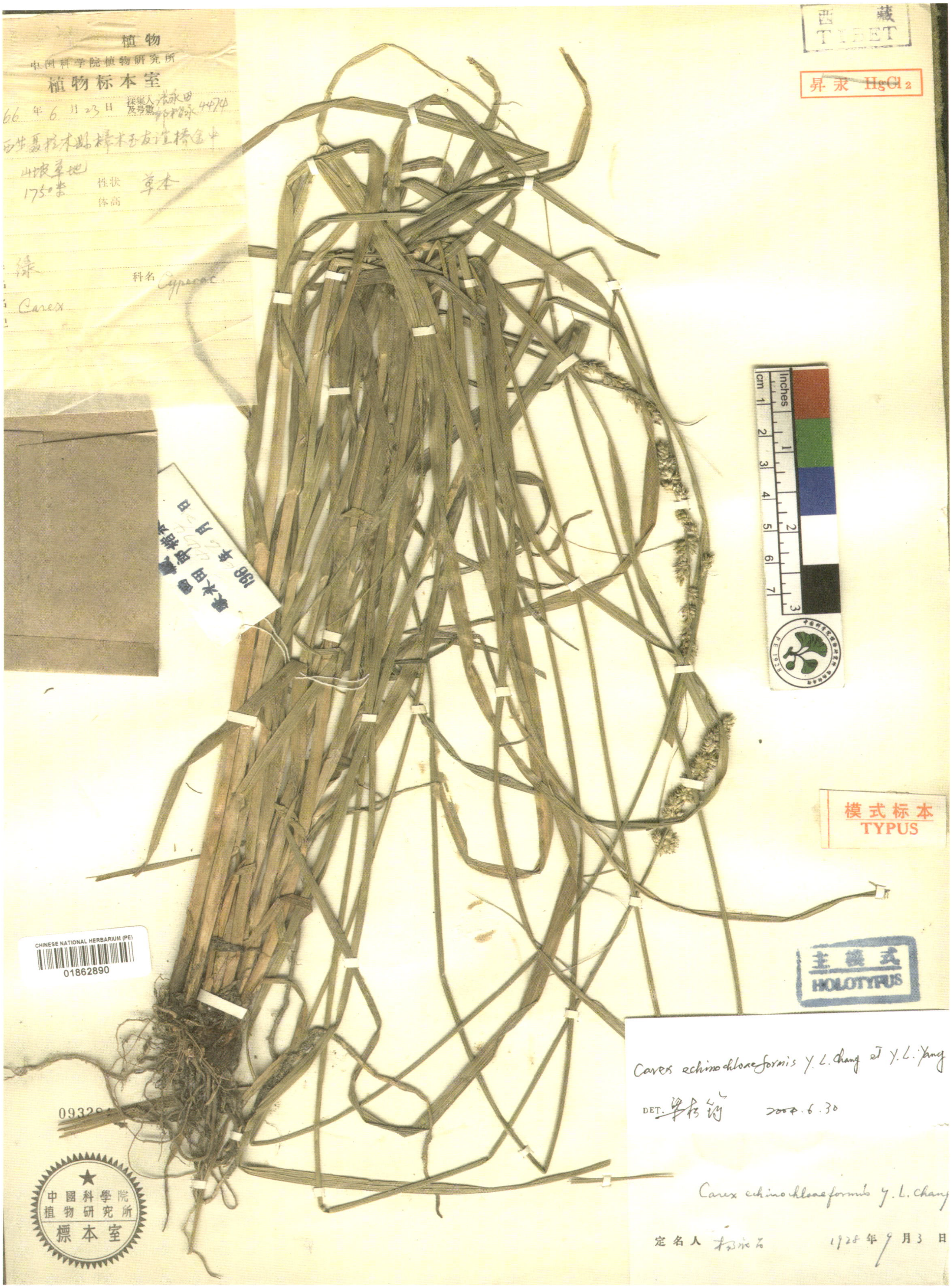

类稗薹草 ***Carex echinochloaeformis*** Y. L. Chang & Y. C. Yang, Fl. Xizang. 5: 436, f. 254. 1987. **Holotype:** China. Xizang: Nyalam, Zham, alt. 1750 m, 1966-06-23, Y. T. Chang & K. Y. Lang 4474.

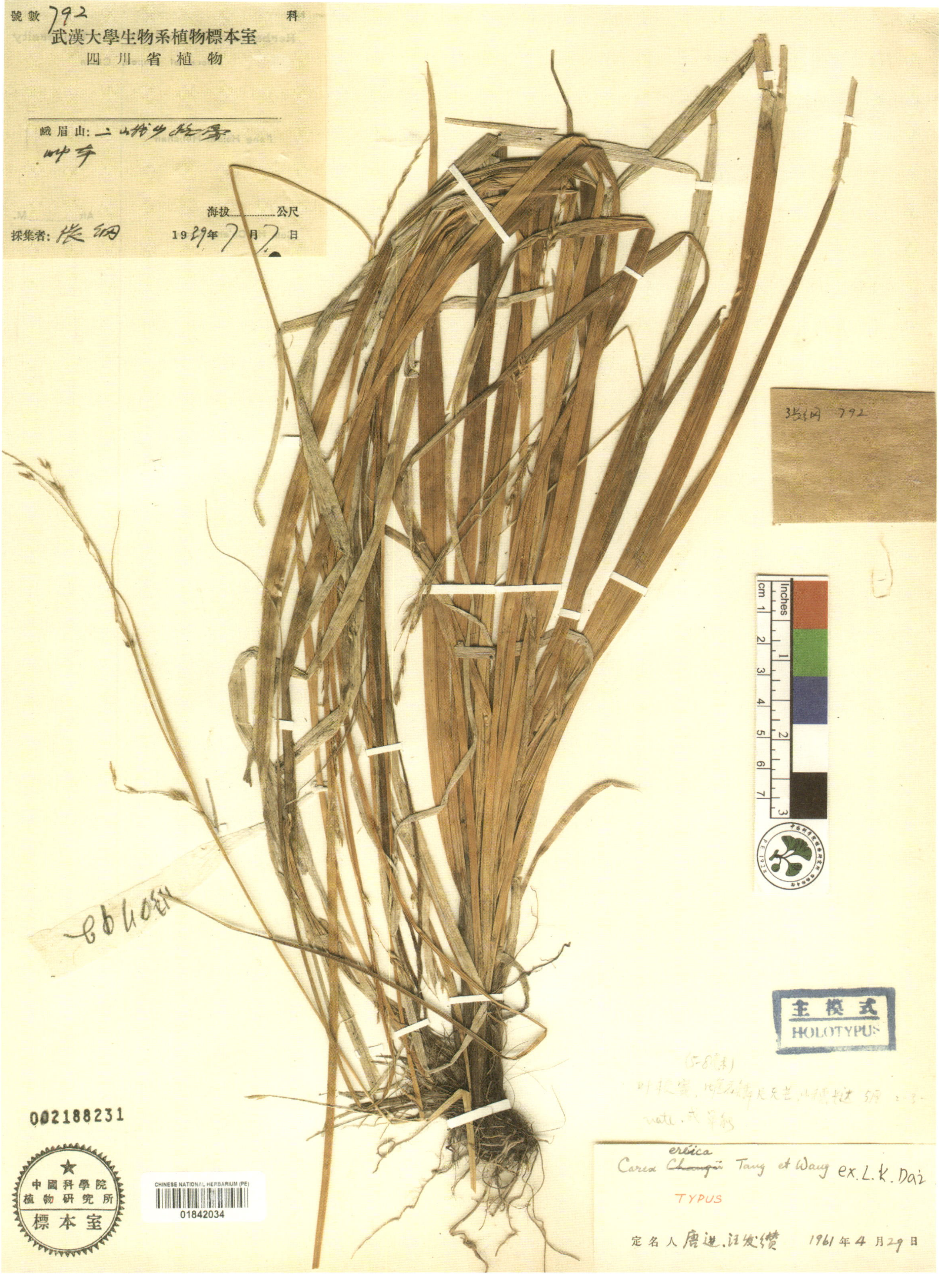

二峨薹草 ***Carex ereica*** Tang & F. T. Wang ex L. K. Dai in Acta Phytotax. Sin. 32(2): 177, f. 3: 9-11. 1994. **Holotype:** China. Sichuan: Emei, Emeishan, 1939-07-07, G. Zhang 792.

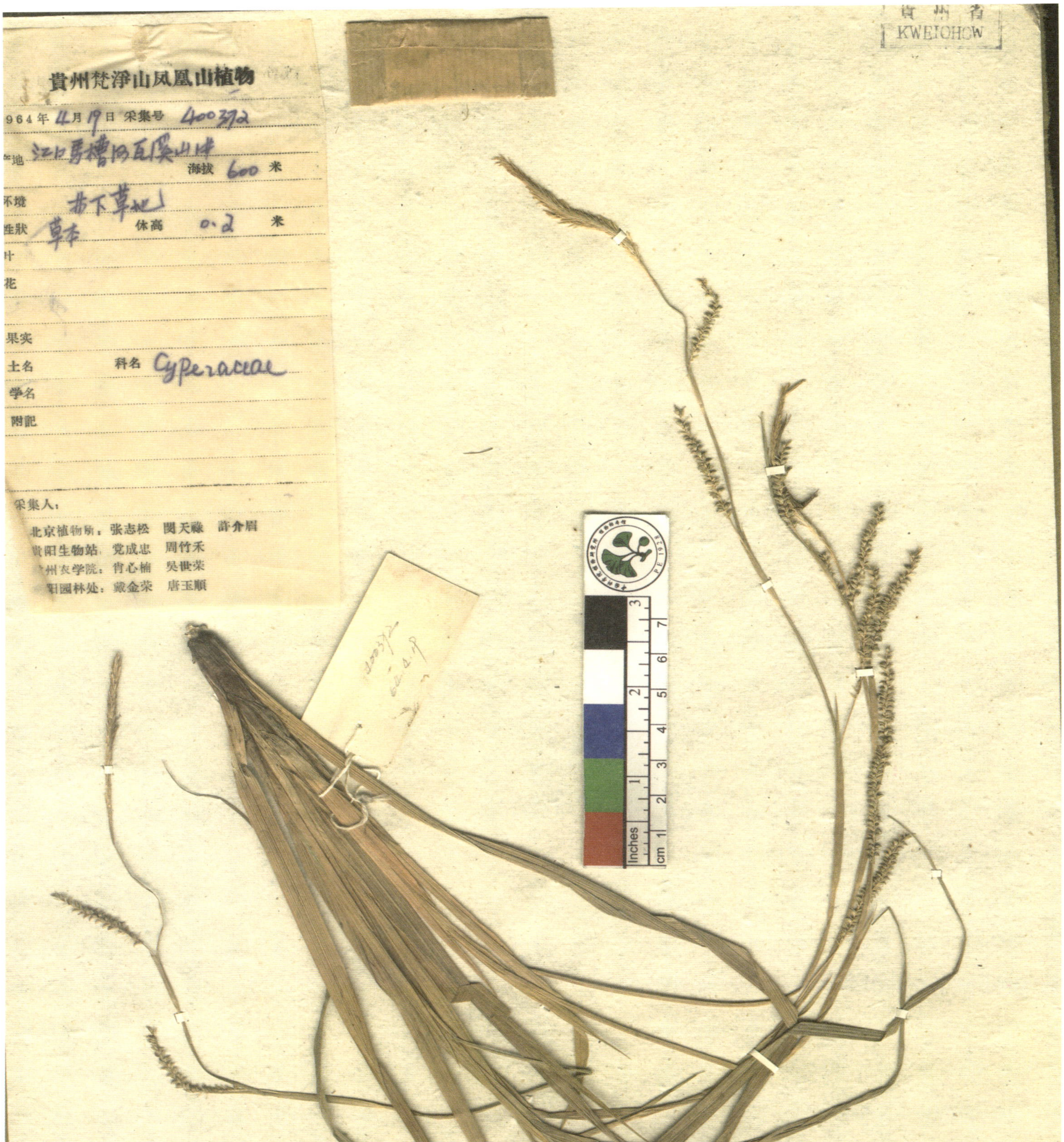

拟穿孔薹草 ***Carex foraminatiformis*** Y. C. Tang & S. Y. Liang, Fl. Reip. Pop. Sin. 12: 521. 2000. **Holotype:** China. Guizhou: Jiangkou, alt. 600 m, 1964-04-19, Z. S. Chang & al. 400372.

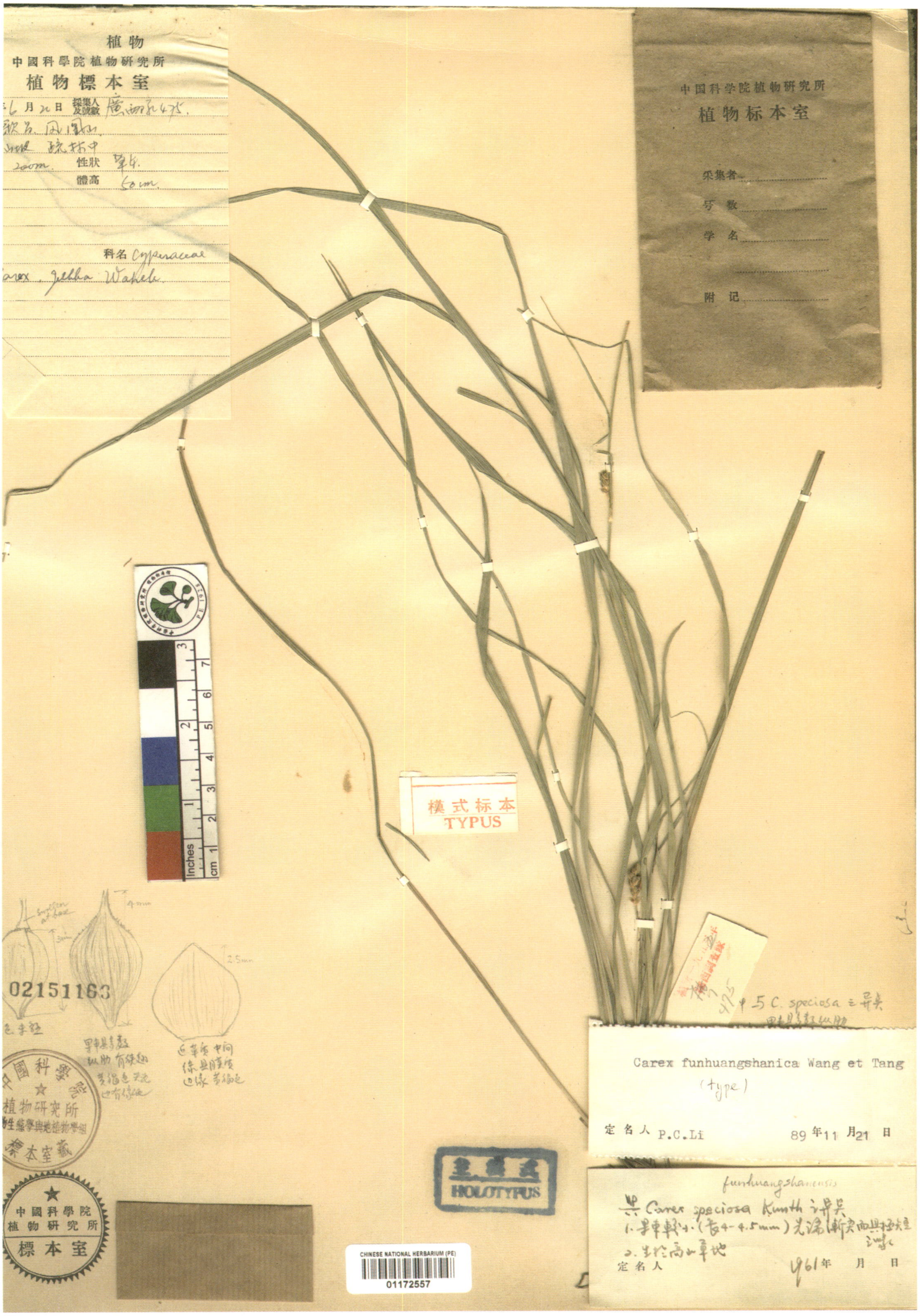

凤凰山薹草 *Carex funhuangshanica* F. T. Wang & Tang ex P. C. Li in Acta Phytotax. Sin. 37(2): 168, f. 7. 1999. **Holotype:** China. Guangxi: Gechang, Fenghuangshan, alt. 2000 m., 1955-06-20, Guangxi Exped. 475.

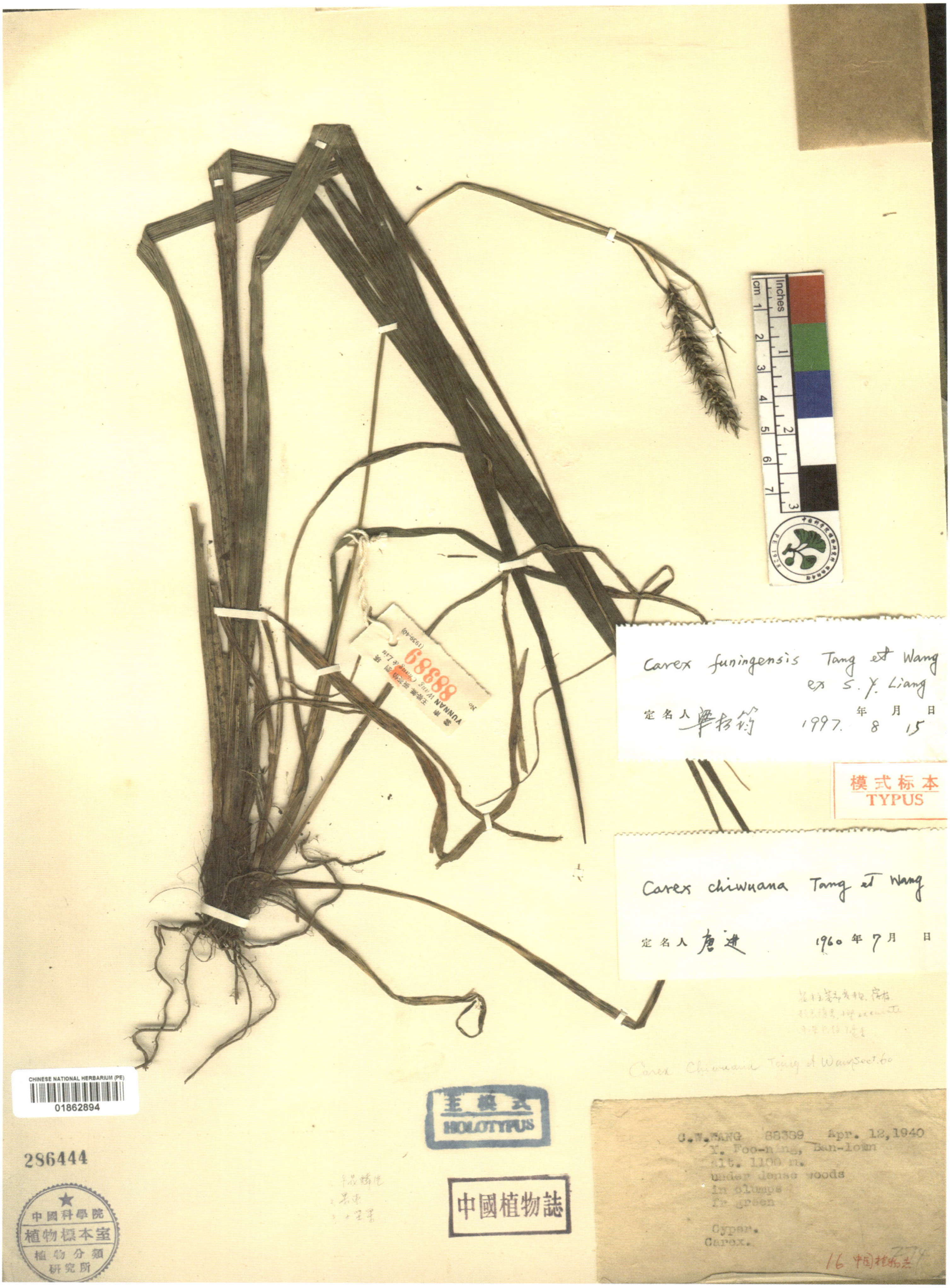

富宁薹草 ***Carex funingensis*** Tang & F. T. Wang ex S. Y. Liang in Acta Phytotax. Sin. 36(6): 534, f. 2: 6-8. 1998. **Holotype:** China. Yunnan: Funing, alt. 1100 m, 1940-04-12, C. W. Wang 88389.

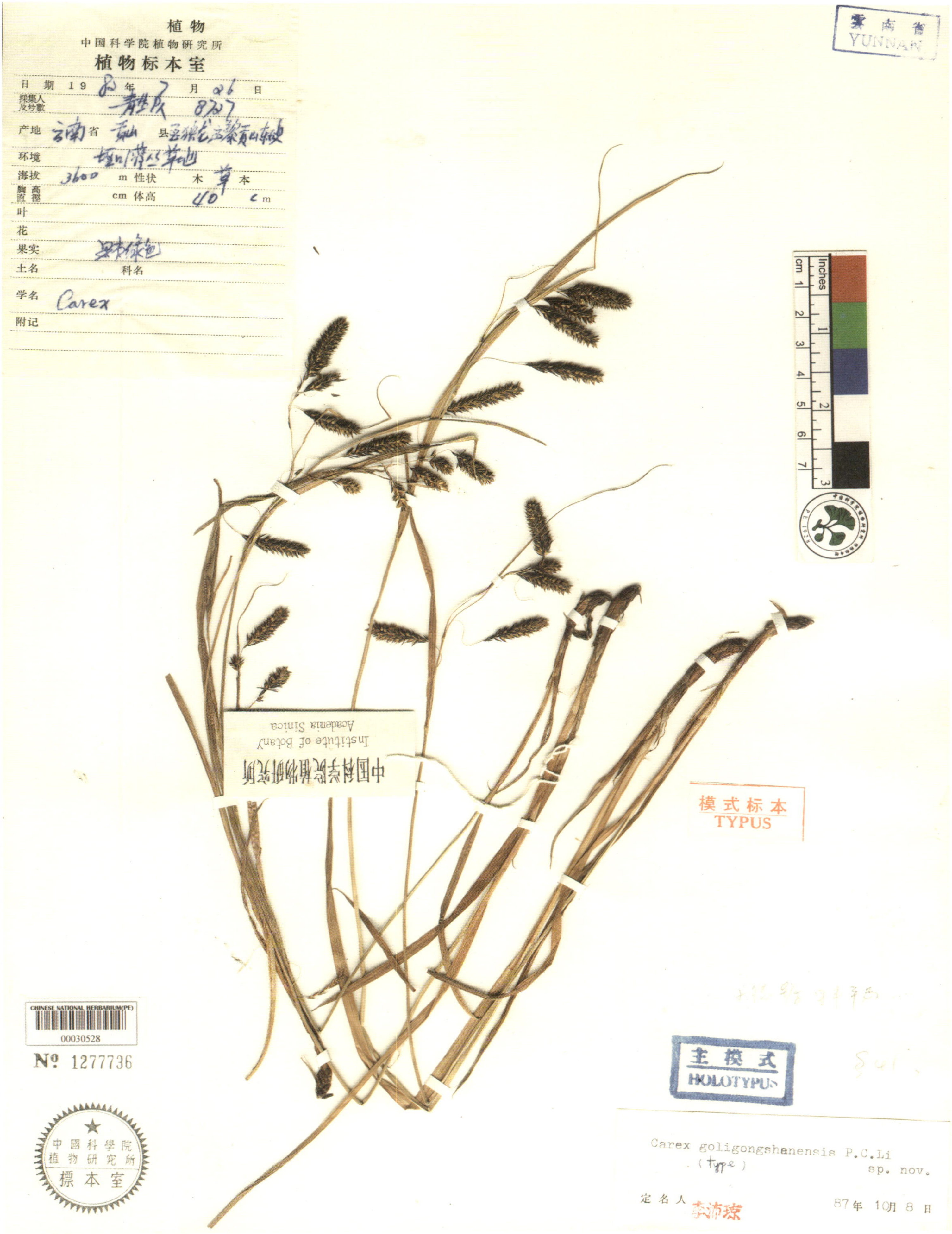

高黎贡山薹草 ***Carex gaoligongshanensis*** P. C. Li. in Acta Bot. Yunnan. 12(2): 140, f. 4. 1990. **Holotype:** China. Yunnan: Gongshan, alt. 3600 m, 1982-07-26, Qinghai-Xizang Exped. 8727.

宽叶亲族薹草 ***Carex gentilis*** Franch. var. ***intermedia*** Tang & F. T. Wang ex L. K. Dai in Acta Phytotax. Sin. 37(2): 187. 1999. **Holotype:** China. Chongqing: Chengkou, alt. 1400 m, 1958-09-02, T. L. Dai 104274.

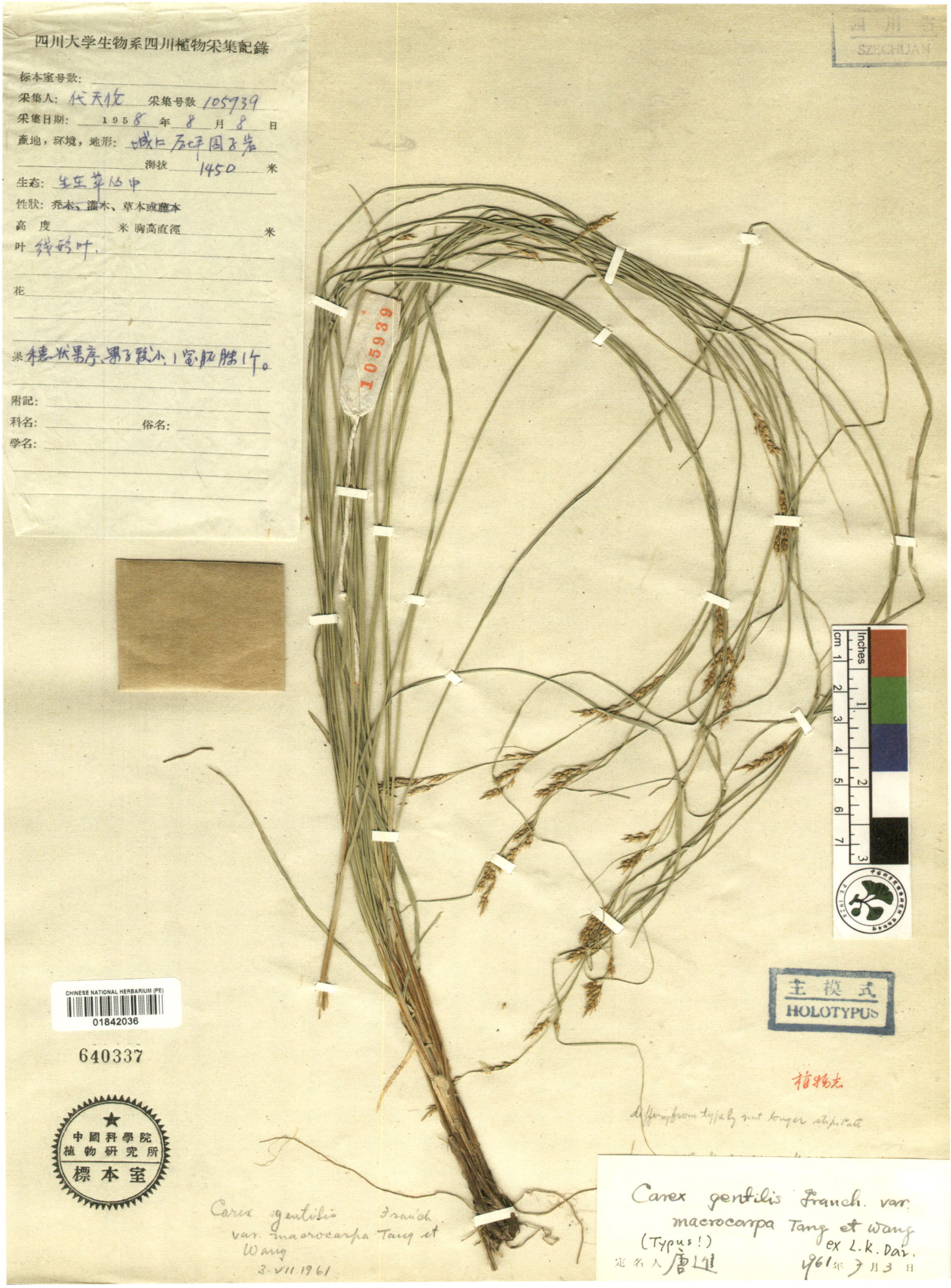

大果亲族薹草 ***Carex gentilis*** Franch. var. ***macrocarpa*** Tang & F. T. Wang ex L. K. Dai in Acta Phytotax. Sin. 37(2): 188. 1999. **Holotype:** China. Chongqing: Chengkou, alt. 1450 m, 1958-08-08, T. L. Dia 105939.

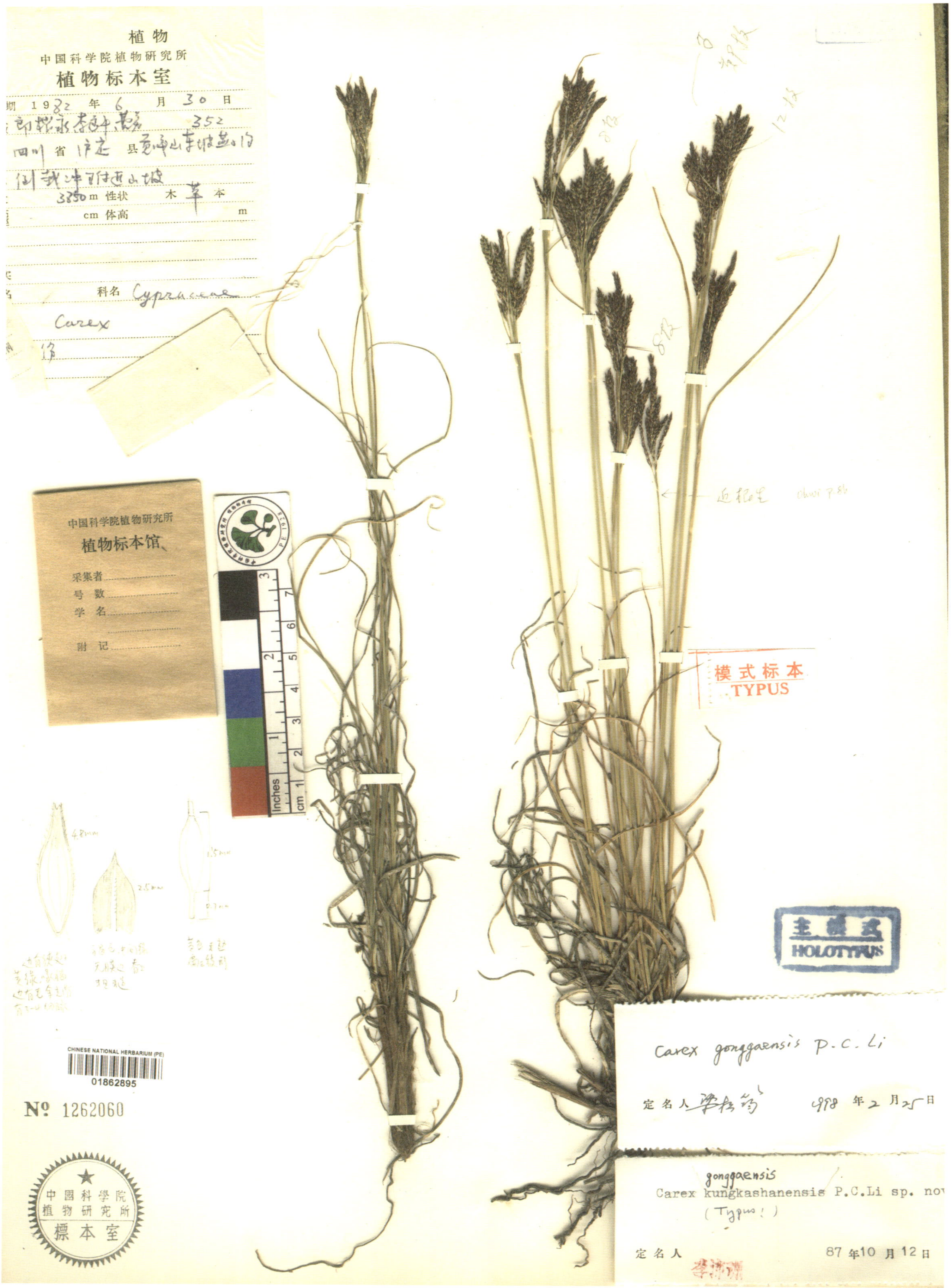

贡嘎薹草 ***Carex gonggaensis*** P. C. Li in Acta Bot. Yunnan 12(2): 147, f. 14. 1990. **Holotype:** China. Sichuan: Luding, alt. 3850 m， 1982-06-30, K. Y. Lang & al. 352.

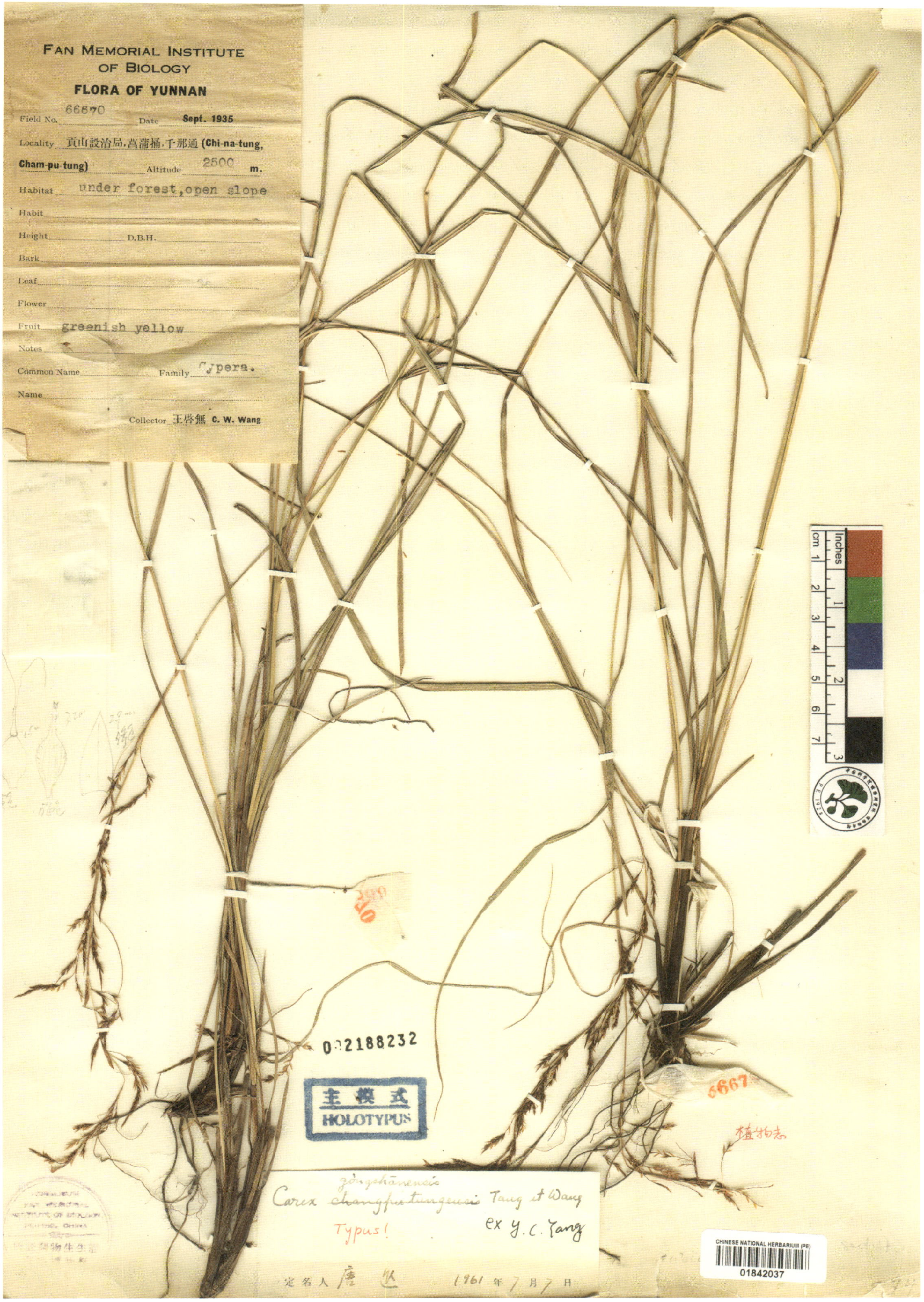

贡山薹草 ***Carex gongshanensis*** Tang & F. T. Wang, Fl. Xizang. 5: 404, f. 229. 1987. **Holotype:** China. Yunnan: Gongshan, alt. 2500 m，1935-09-??, C. W. Wang 66670.

红棕薹草 ***Carex haematostoma*** Nees var. ***digyna*** Kükenth. in Acta Hort. Gothob. 5: 45. 1929. **Isotype:** China. Sichuan: Merge, alt. 3500 m, 1922-09-03, H. Smith 4318.

Carex handelii Kükenth., Symbolae Sinicae 7(5): 1268, pl. 38, f. 4. 1936.

Locality: China: Yunnan: Tannenwaldern, 3700-4100 m, 18 June 1916; Collector and Number: Handel-Mazzetti 8972

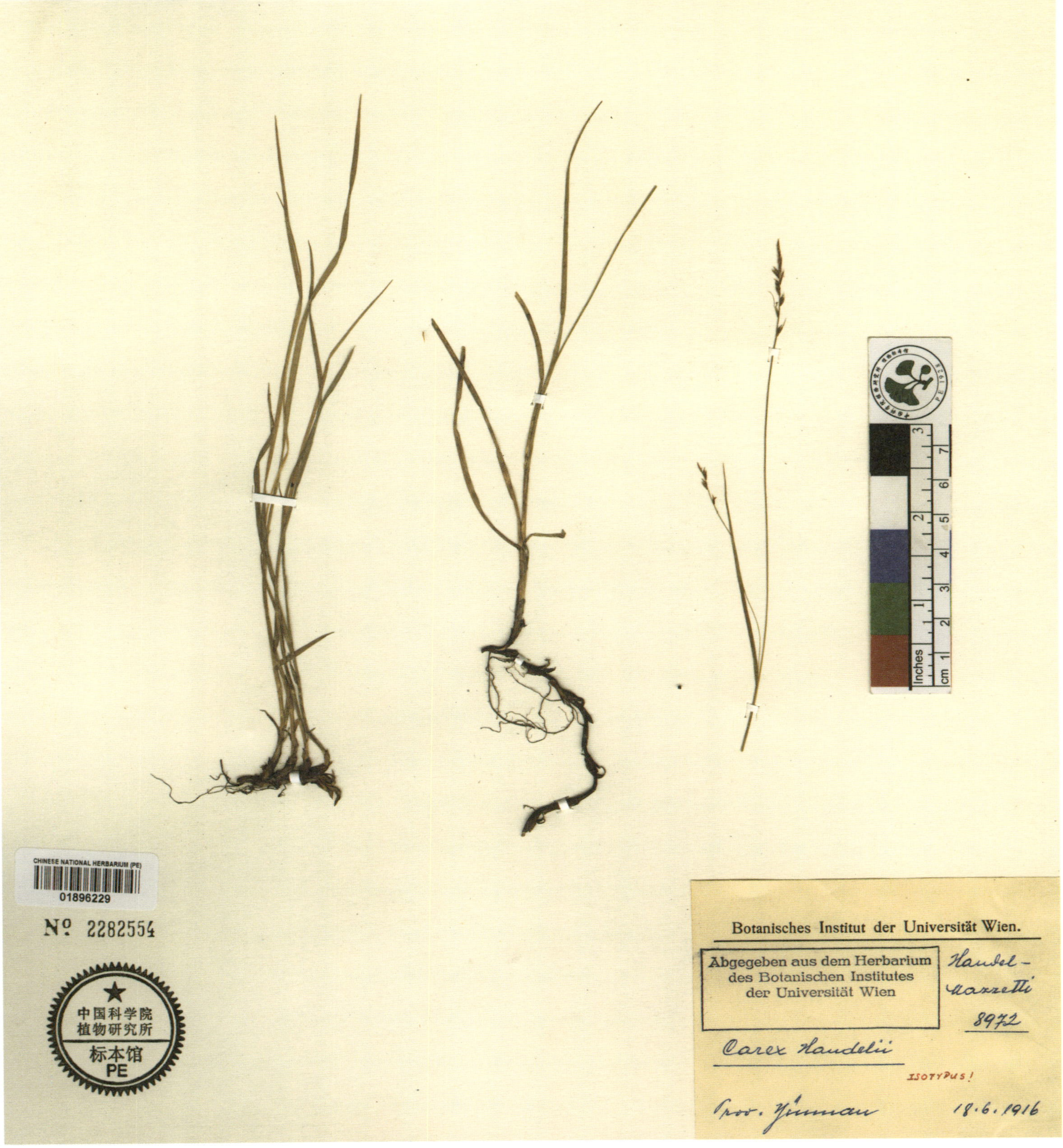

双脉囊薹草 ***Carex handelii*** Kükenth. in Hand.-Mazz., Symb. Sin. 7: 1268. 1936. **Isotype:** China. Yunnan: Tannenwaldern, alt. 3700~4100 m, 1916-06-18, Handel-Mazzetti 8972.

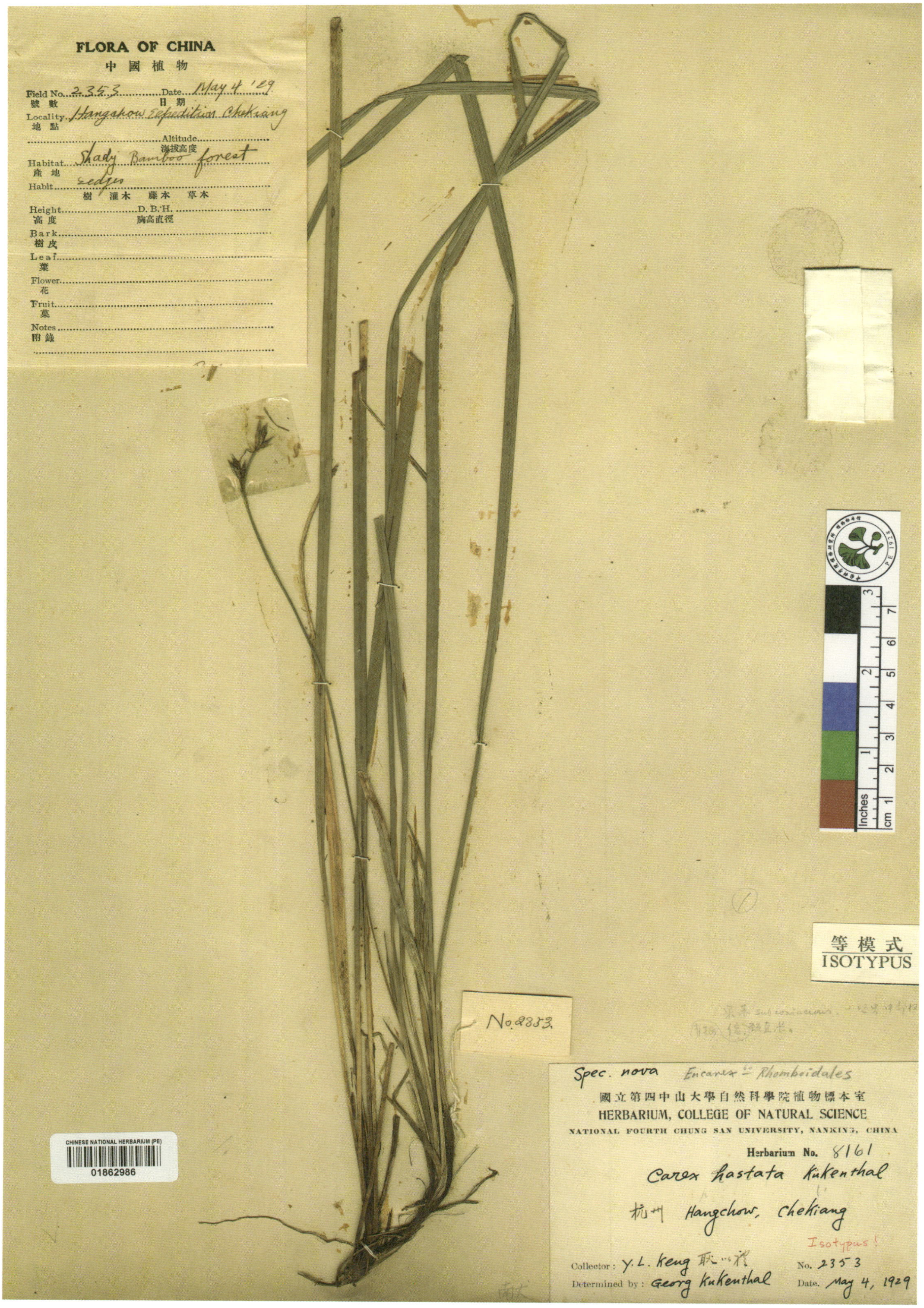

戟叶薹草 ***Carex hastata*** Kükenth. in Fedde, Rep. Spec. Nov. Reg. Veg. 27: 110. 1929. **Isotype:** China. Zhejiang: Hangzhou, 1929-05-04, Y. L. Keng 2353.

和硕薹草 ***Carex heshuoensis*** S. Y. Liang, Fl. Reip. Pop. Sin. 12: 525. 2000. **Holotype:** China. Xinjiang: Heshuo, alt. 2540 m. 1958-08-11, A. R. Li & J. N. Zhu 7680.

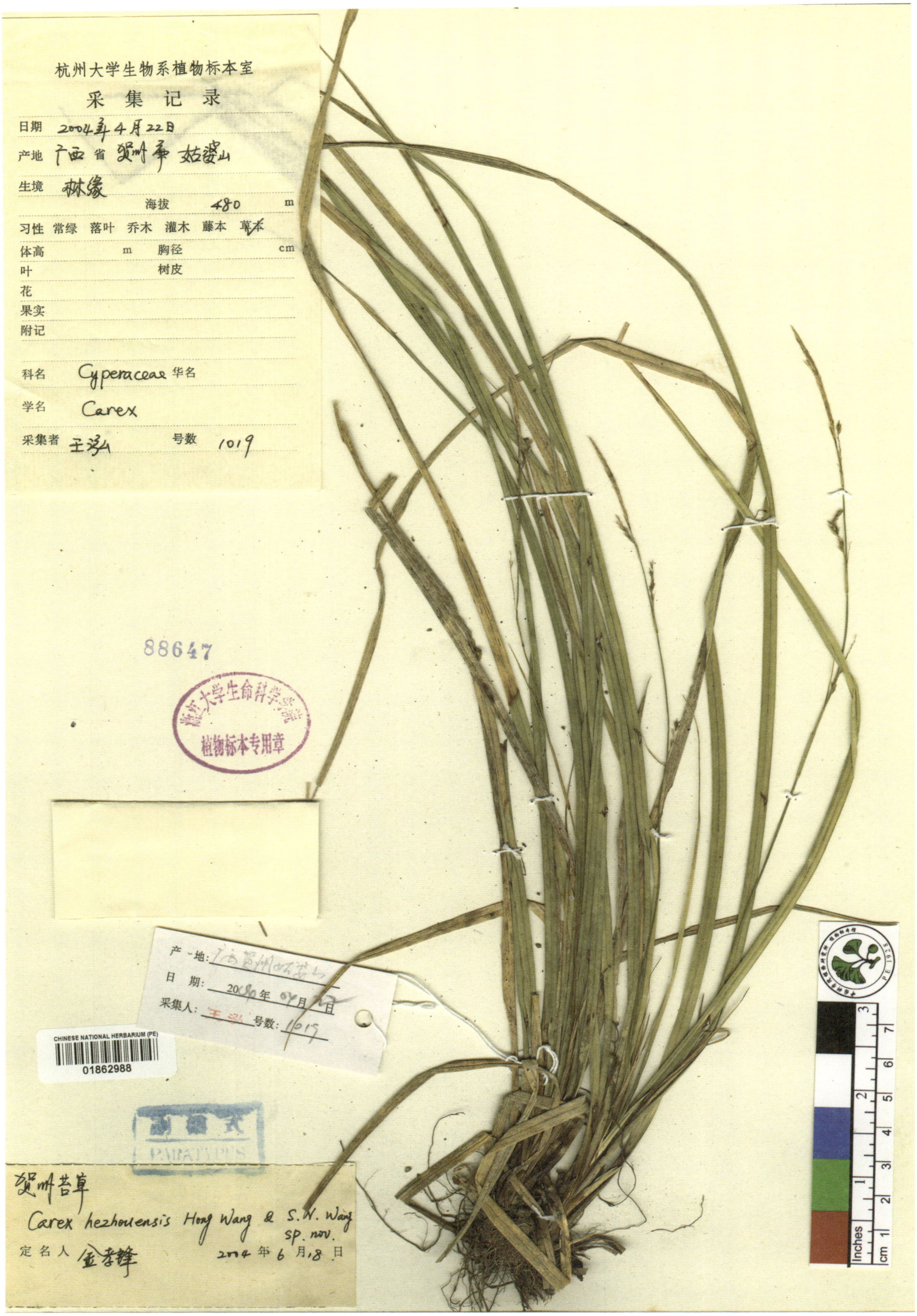

贺州薹草 ***Carex hezhouensis*** H. Wang & S. N. Wang in Guihaia 25(2): 105, pl. 1. 2005.**Paratype:** China. Guangxi: Hezhou, alt. 480 m, 2004-04-22, H. Wang 1019.

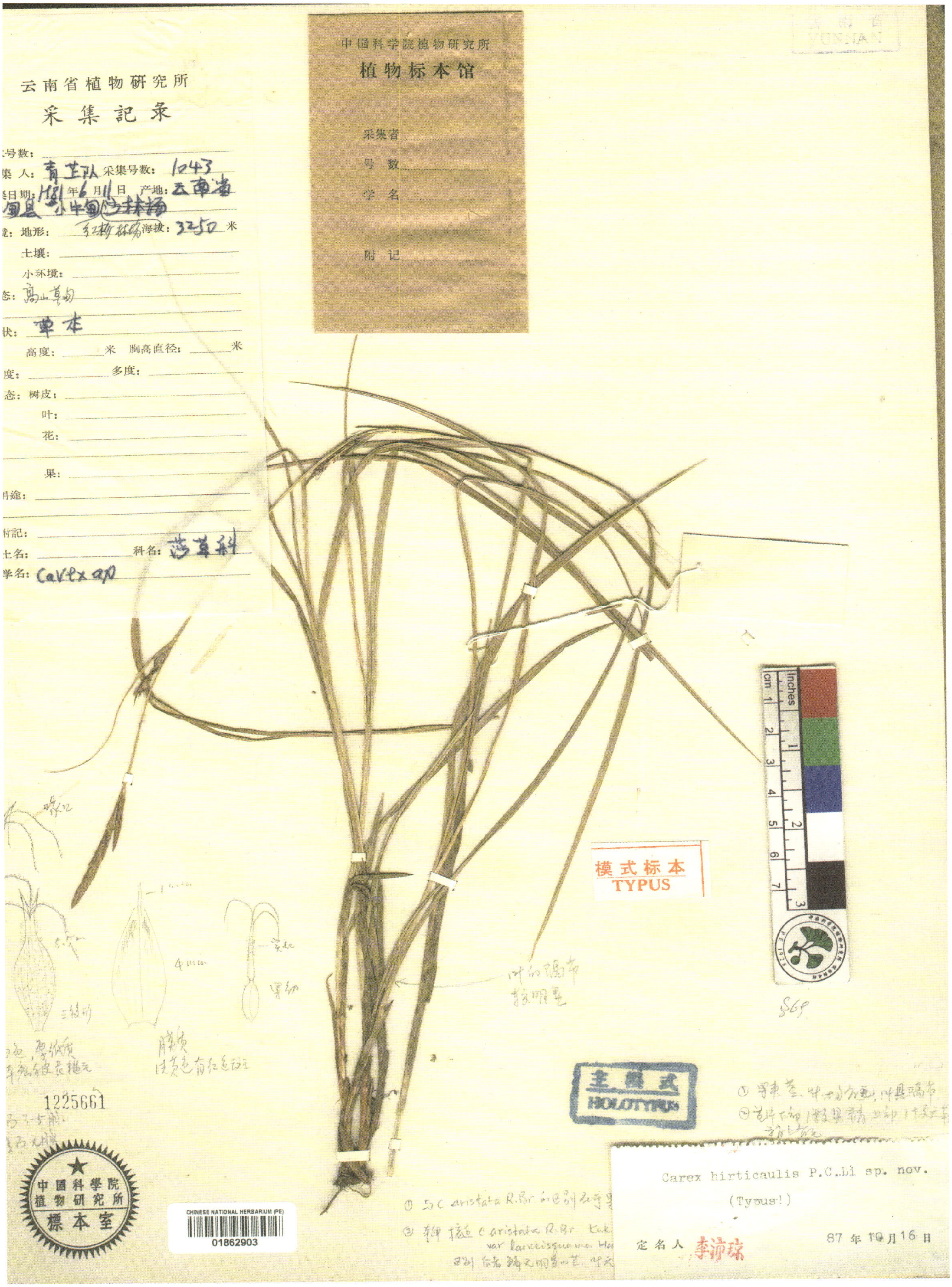

密毛薹草 ***Carex hirticaulis*** P. C. Li in Acta Bot. Yunnan 12(2): 149, f. 18. 1990. **Holotype:** China. Yunnan: Zhongdian, alt. 3250 m, 1981-06-11, Qinghai-Xizang Exped. 1043.

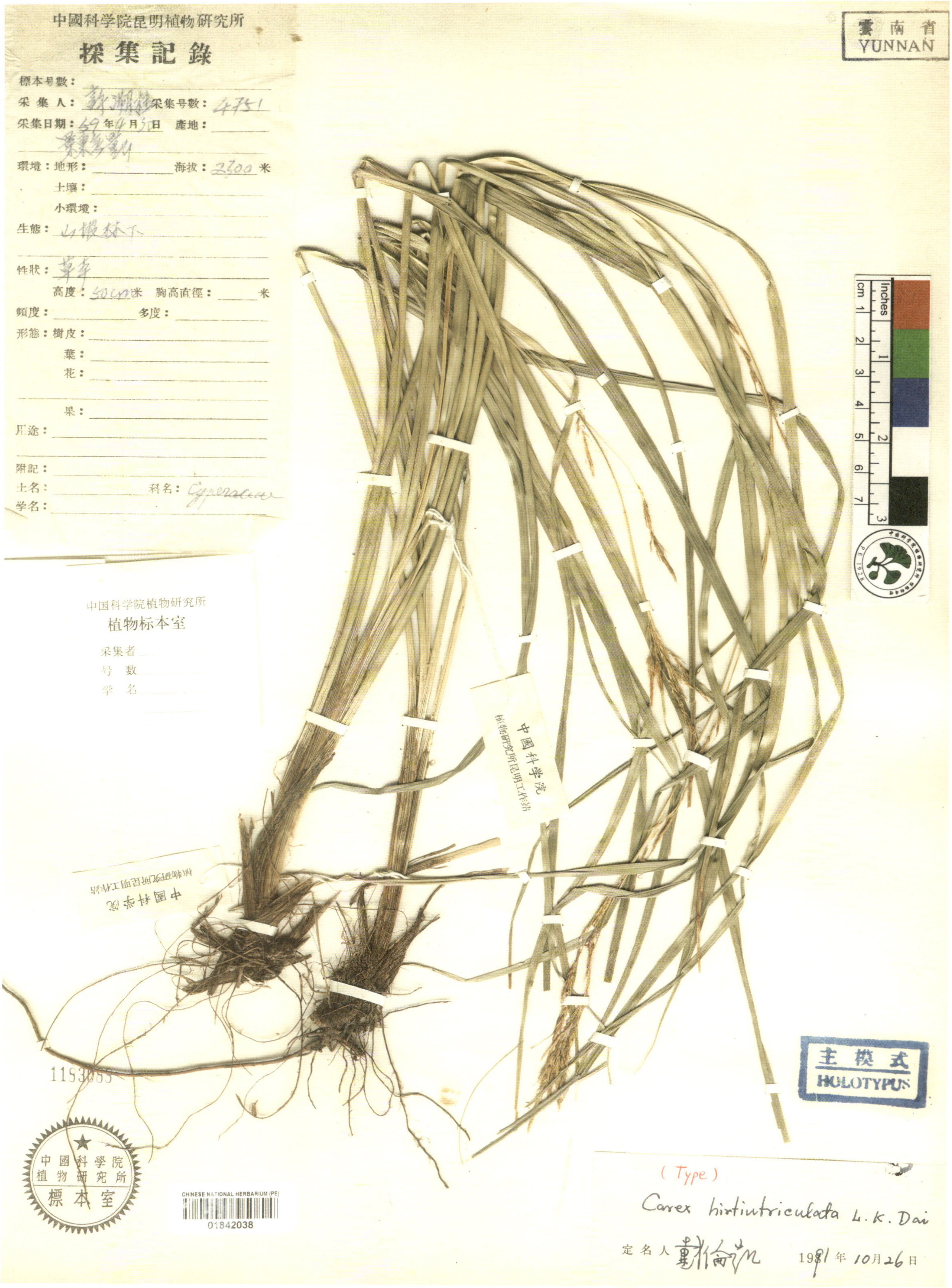

糙毛囊薹草 ***Carex hirtiutriculata*** L. K. Dai in Acta Phytotax. Sin. 32(2): 181, f. 1: 5-8. 1994. **Holotype:** China. Yunnan: Jingdong, alt. 2300 m, 1959-04-30, S. G. Xu 4751.

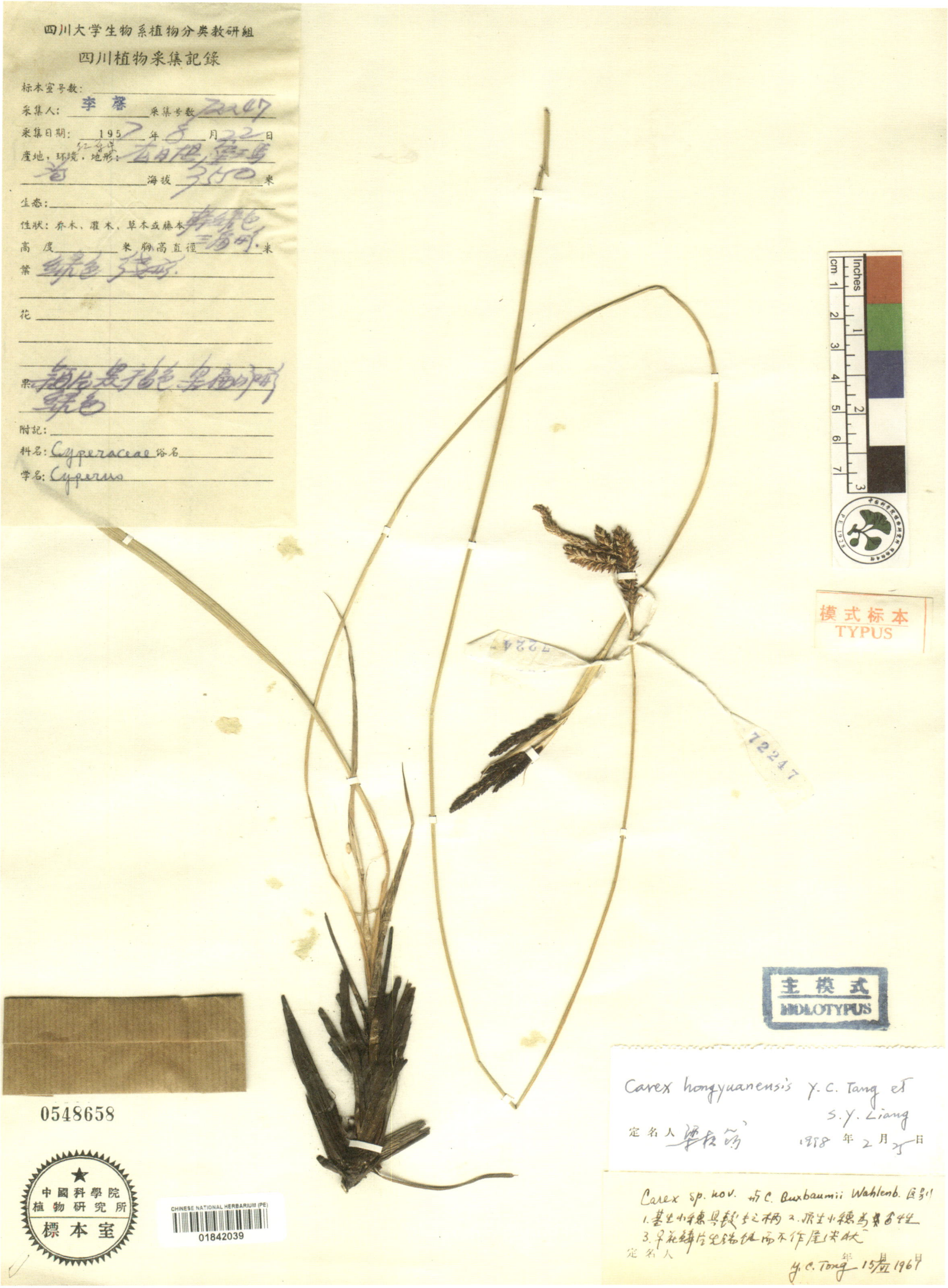

红原薹草 ***Carex hongyuanensis*** Y. C. Tang & S. Y. Liang, Fl. Reip. Pop. Sin. 12:103, 520, pl. 19: 1-4. 2000. **Holotype:** China. Sichuan: Hongyuan, alt. 3550 m, 1957-08-22, X. Li 72247.

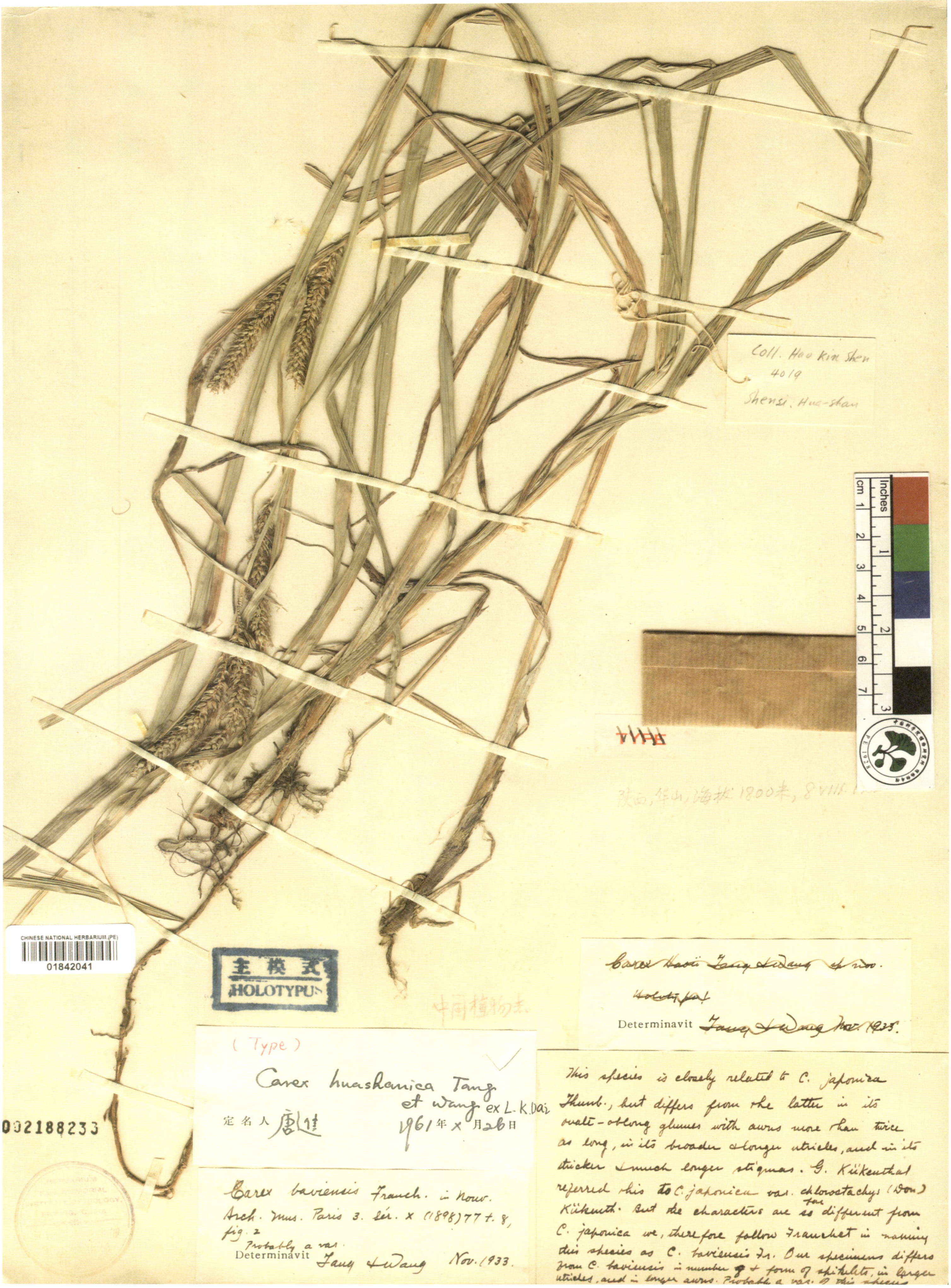

华山薹草 ***Carex huashanica*** Tang & F. T. Wang ex L. K. Dai in Acta Phytotax. Sin. 32(2): 186, f. 6: 1-4. 1994. **Holotype:** China. Shaansi: Huashan, alt. 1800 m. 1932-08-08, K. S. Hao 4019.

印度型薹草 ***Carex indicaeformis*** Tang & F. T. Wang ex P. C. Li in Acta Phytotax. Sin. 37(2): 156, f. 1. 1999. **Holotype:** China. Hainan: Yaihsien (=Sanya), 1935-05-(04-24), S. K. Lau 6219.

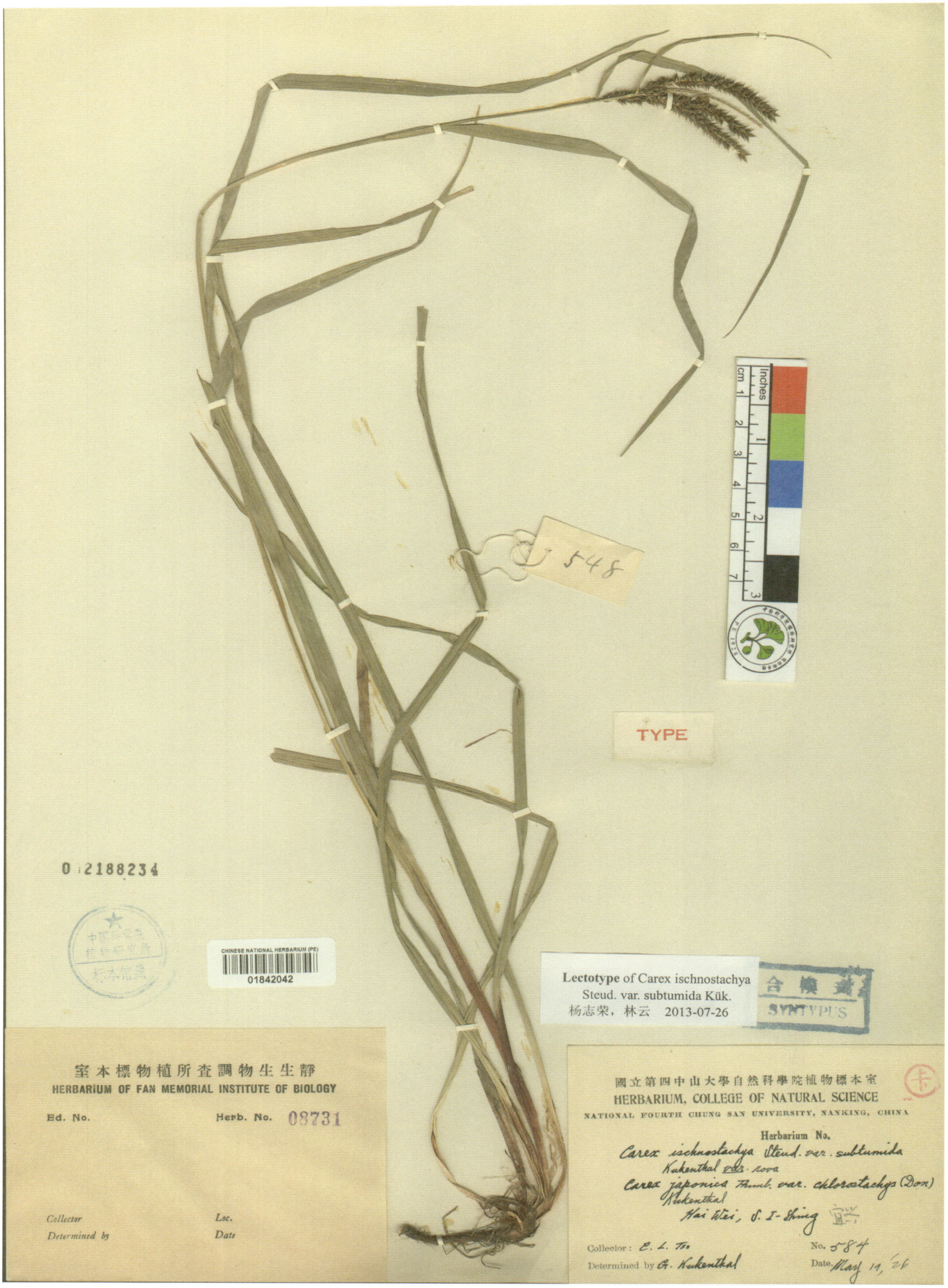

肿胀果薹草 ***Carex ischnostachya*** Steud. var. ***subtumida*** Kükenth. in Repert. Spec. Nov. Regni Veg. 27(726-733): 109. 1929. **Lectotype** (designated by Y. Lin & al. in Acta Bot. Bor.-Occ. Sin. 34: 412. 2014.): China. Jiangsu: Yixing, 1926-05-17, R. C. Ching & C. L. Tso 548.

胶东薹草 ***Carex jiaodongensis*** Y. M. Zhang & X. D. Chen in Acta Phytotax. Sin. 31(4): 381, f. 1. 1993. **Isotype:** China. Shandong: Muping, 1990-05-08, Y. M. Zhang 90015.

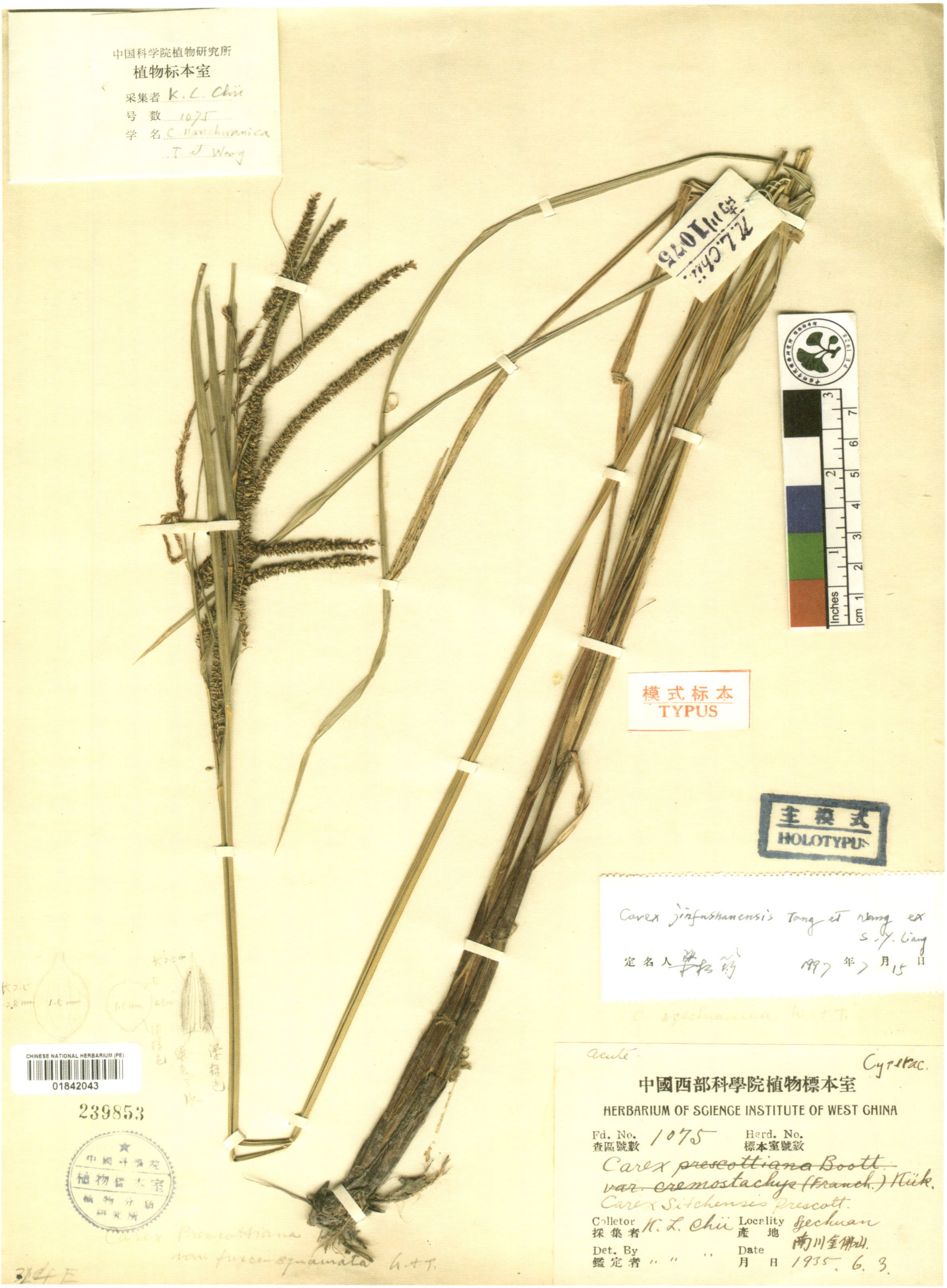

金佛山薹草 ***Carex jinfoshanensis*** Tang & F. T. Wang ex S. Y. Liang in Acta Phytotax. Sin. 37(2): 191, f. 1:5-7. 1999.
Holotype: China. Chongqing: Nanchuan, alt. 1200 m, 1935-06-03, K. L. Chu 1075.

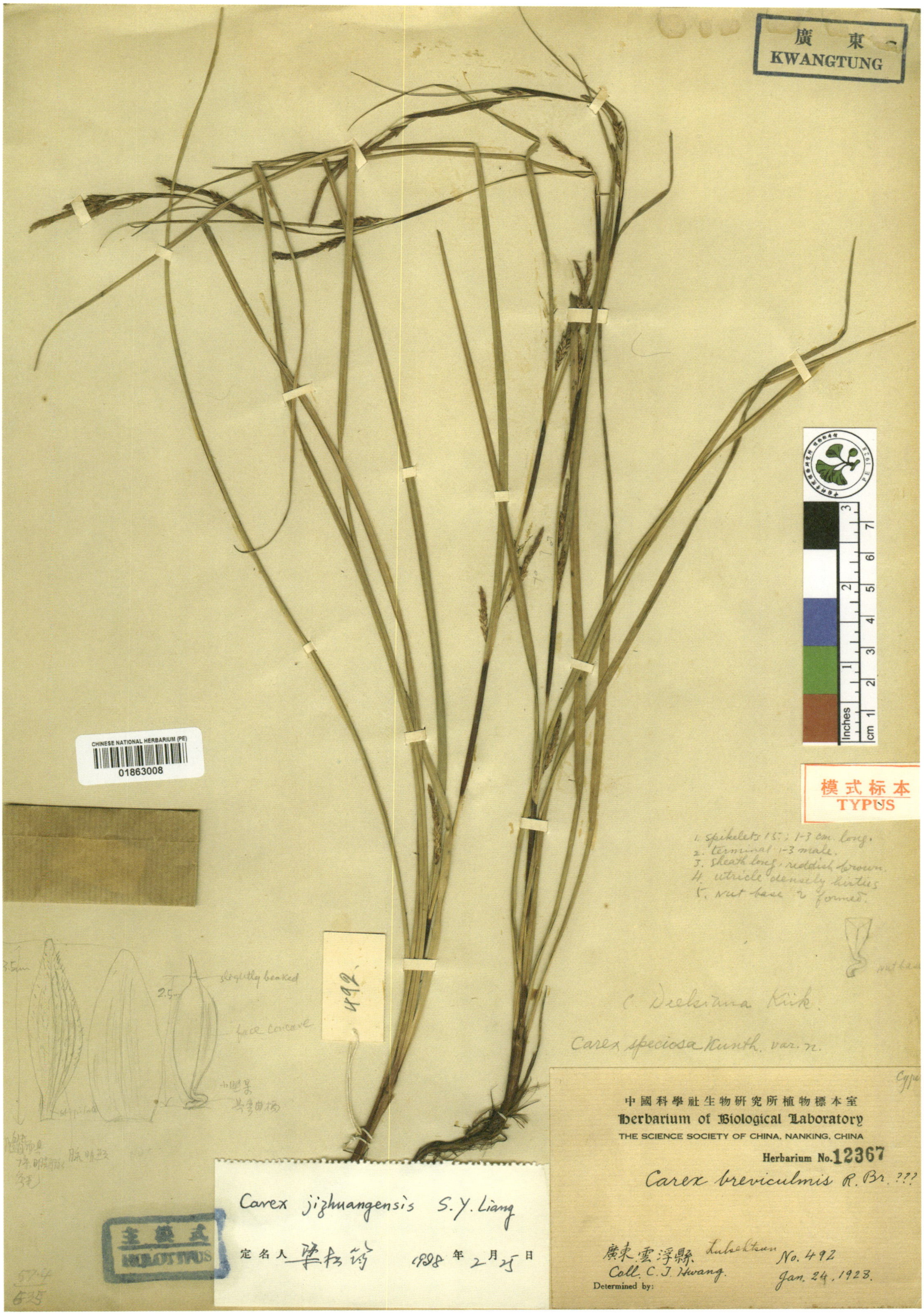

季庄薹草 ***Carex jizhuangensis*** S. Y. Liang, Fl. Reip. Pop. Sin. 12: 222, 522, pl. 44: 1-4. 2000. **Holotype:** China. Guangdong: Yunfu, 1928-01-24, C. J. Hwang 492.

甘肃薹草 ***Carex kansuensis*** Nelmes in Bull. Misc. Inform. Kew 1939: 201. 1939. **Isotype:** China. Gansu: Lanzhou, alt. 3000 m, 1918-07-10, Licent 4232.

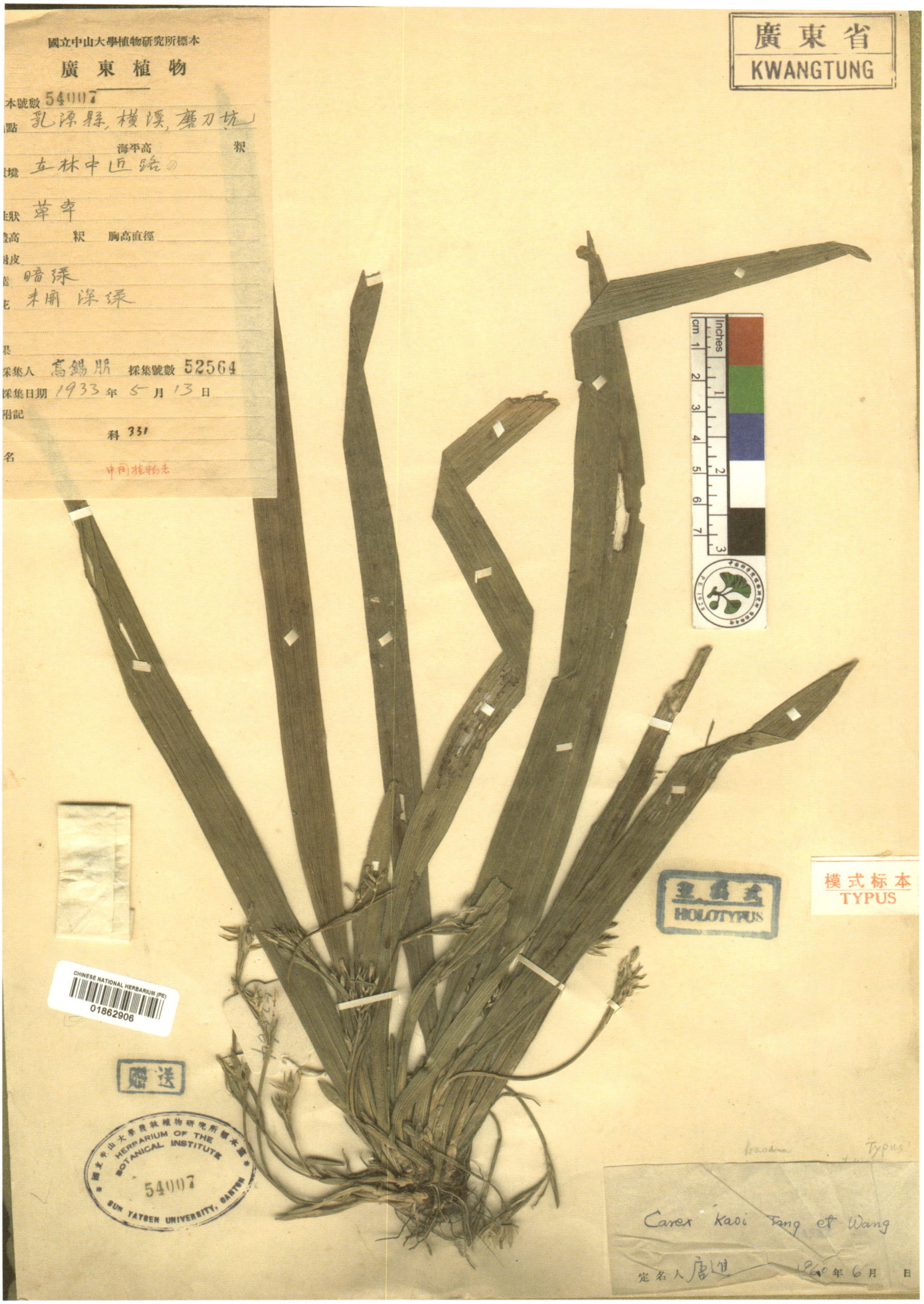

高氏薹草 ***Carex kaoi*** Tang & F. T. Wang ex S. Y. Liang in Acta Phytotax. Sin. 36(6): 530, f. 1: 1-5. 1998. **Holotype:** China. Guangdong: Ruyuan, 1933-05-13, S. P. Ko 52564.

卡郎薹草 ***Carex karlongensis*** Kukenth. in Acta Hort. Gothob. 5: 47. 1929. **Isotype:** China. Sichuan: Karlong, alt. 3300 m, 1922-08-04, H. Smith 3037.

吉林薹草 ***Carex kirinensis*** F. T. Wang & Y. L. Chang, Fl. Pl. Herb. China Bor.-Orient. 11: 174, 207, pl. 80: 6-11. 1976.
Isotype: China. Jilin: Hunjiang, alt. 950 m, 1950-06-15, T. N. Liou 904.

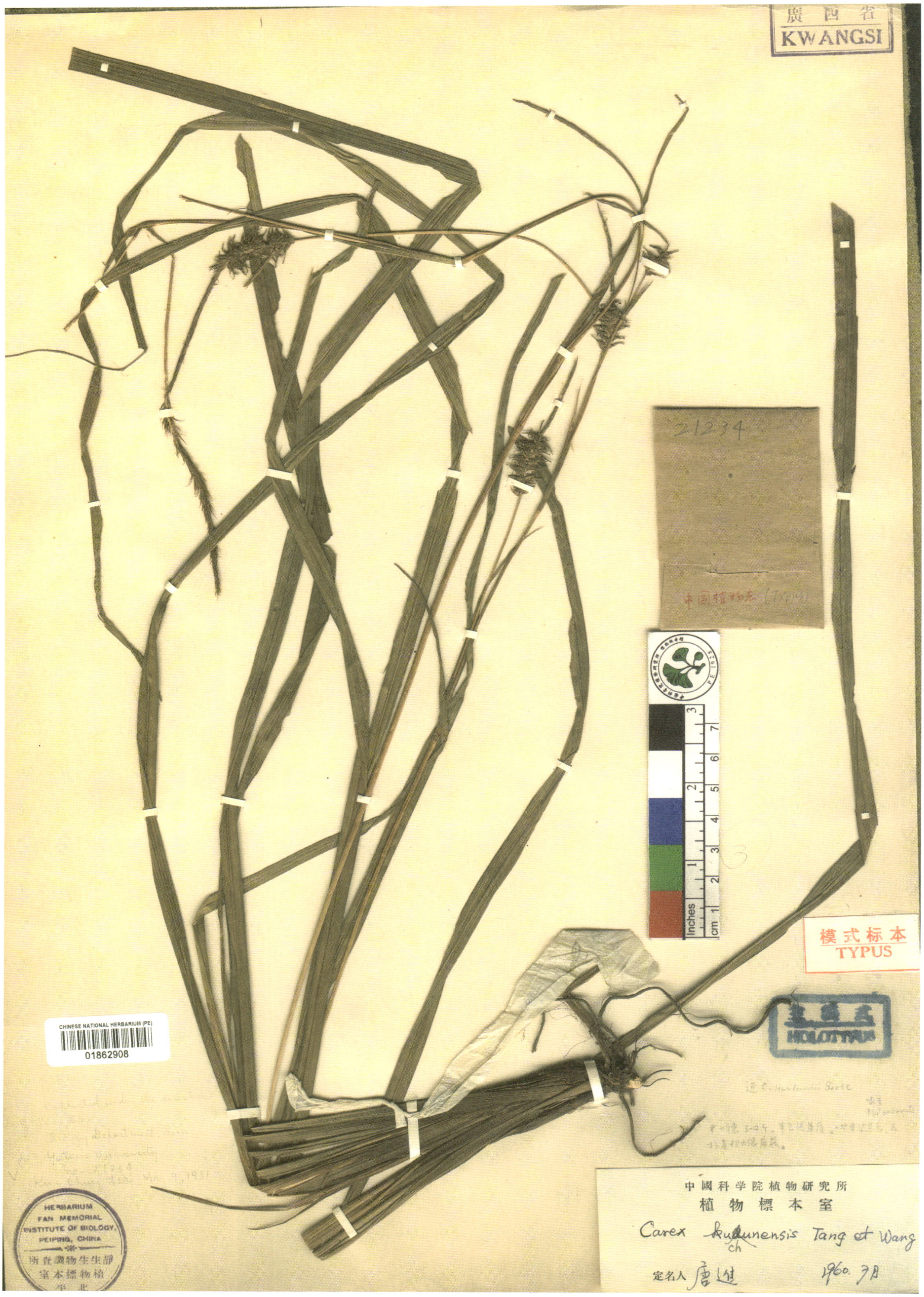

古城薹草 ***Carex kuchunensis*** Tang & F. T. Wang ex S. Y. Liang in Acta Phytotax. Sin. 36(6): 536, f. 3: 6-10. 1998.

Holotype: China. Guangxi: Precise locality not known, alt. 900 m. Anonymous 21234.

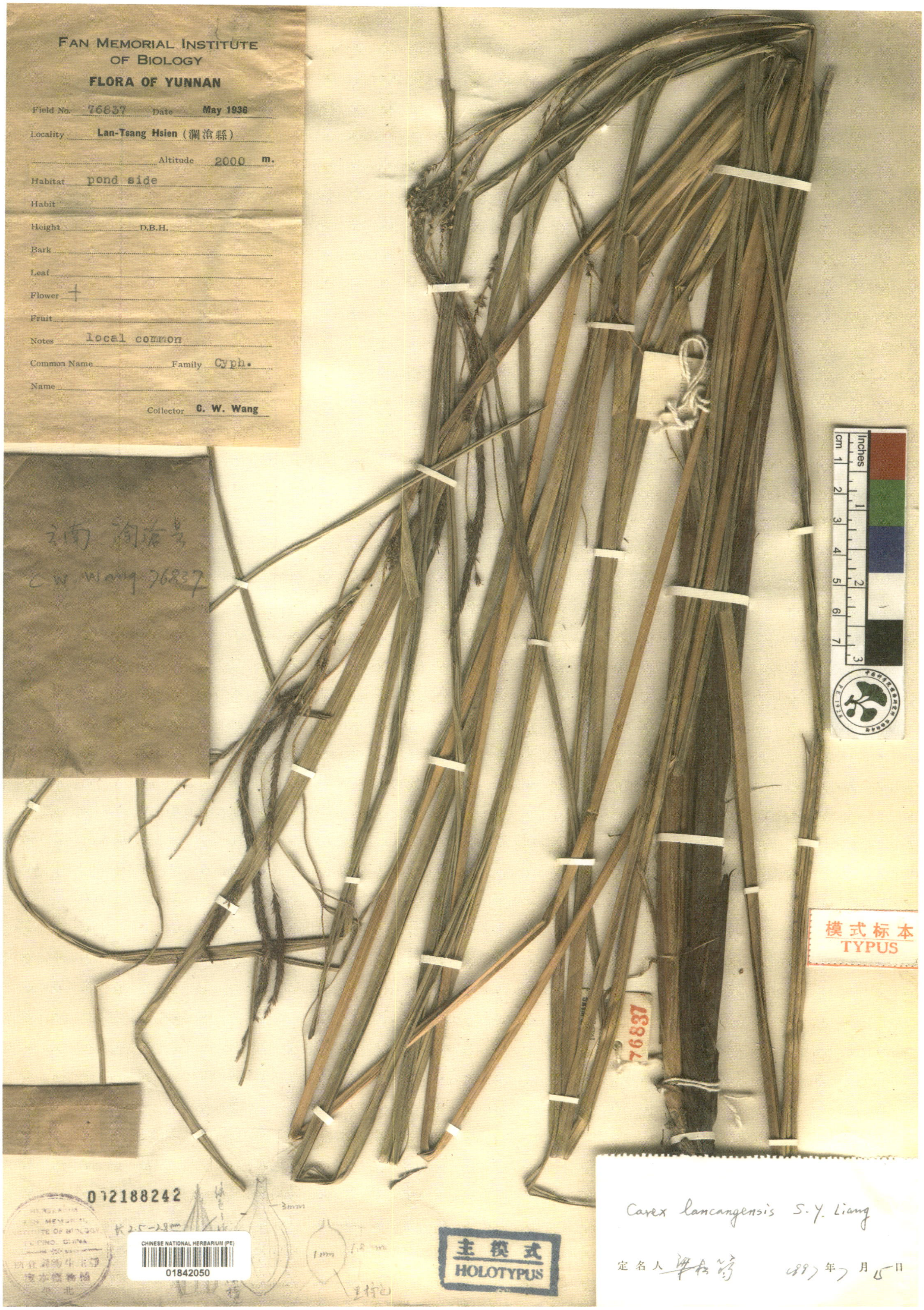

澜沧薹草 ***Carex lancangensis*** S. Y. Liang in Acta Phytotax. Sin. 37(2): 189, f. 1: 1-4. 1999. **Holotype:** China. Yunnan: Lancang, alt. 2000 m, 1936-05-??, C. W. Wang 76837.

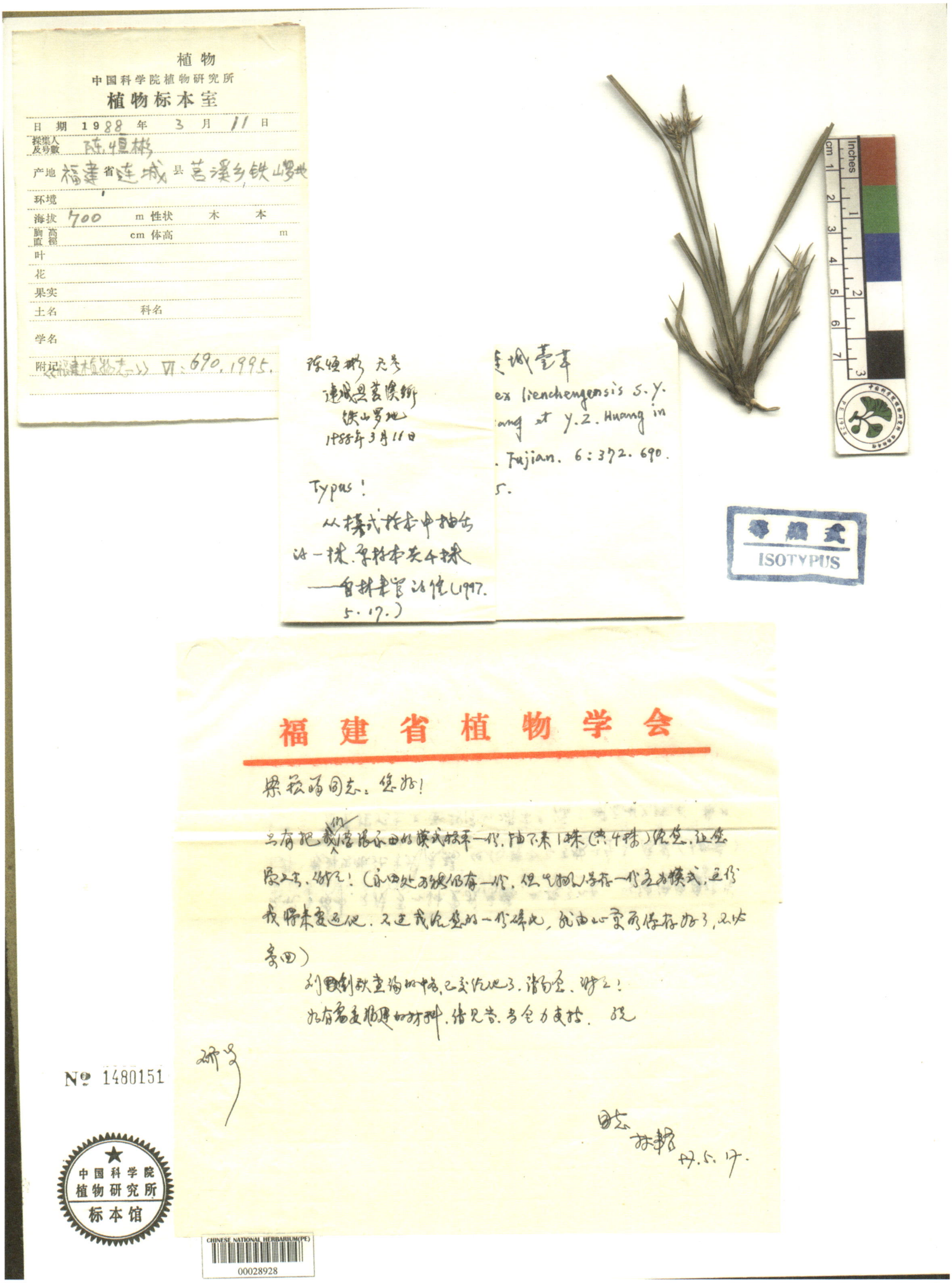

连城薹草 ***Carex lianchengensis*** S. Y. Liang & Y. Z. Huang, Fl. Fujian. 6: 372, 690, f. 340. 1995. **Isotype:** China. Fujian: Liancheng, alt. 700 m, 1988-03-11, H. B. Chen s. n.

小粒薹草 ***Carex licentii*** Nelmes in Bull. Misc. Inform. Kew 1939: 185. 1939. **Isotype:** China. Shanxi: Precise locality not known, 1916-06-06, Licent 2007.

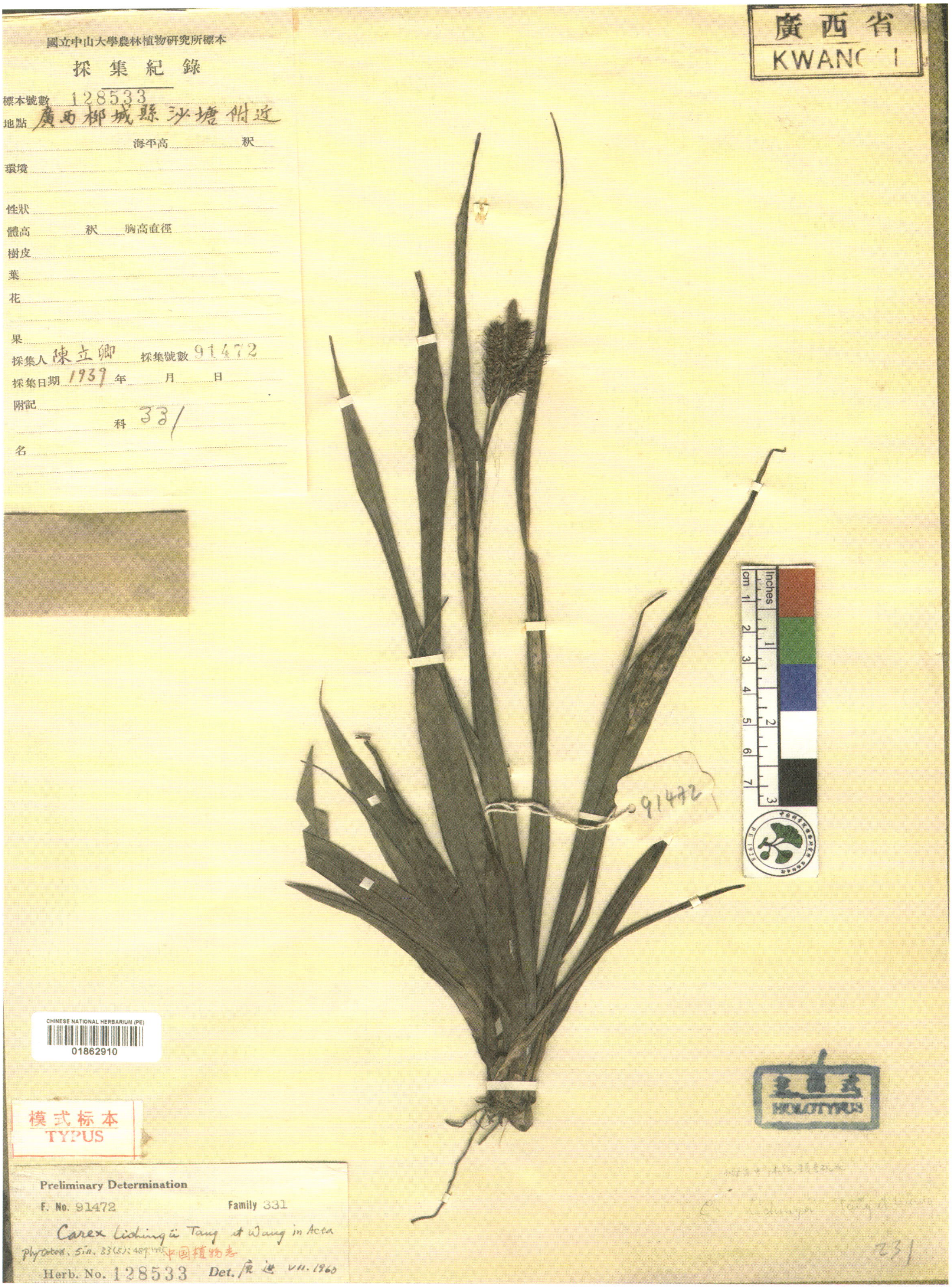

立卿薹草 ***Carex liqingii*** Tang & F. T. Wang ex S. Y. Liang in Acta Phytotax. Sin. 33(5): 489, f. 2: 6-10. 1995. **Holotype:** China. Guangxi: Liucheng, 1939-??-??, L. Q. Chen 91472.

城湾薹草 ***Carex longerostrata*** C. A. Mey. var. ***hoi*** K. L. Chu ex S. Y. Liang in Acta Phytotax. Sin. 36(6): 537. 1998.

Holotype: China. Zhejiang: Precise locality not known, 1932-04-19, Y. Y. Ho 943.

无柄穗薹草 ***Carex longipes*** D. Don var. ***sessilis*** Tang & F. T. Wang ex L. K. Dai in Acta Phytotax. Sin. 32(2): 179, f. 3: 5-8. 1994. **Holotype:** China. Yunnan: Jingdong, alt. 2550 m, 1956-11-16, B. Y. Qiu 53758.

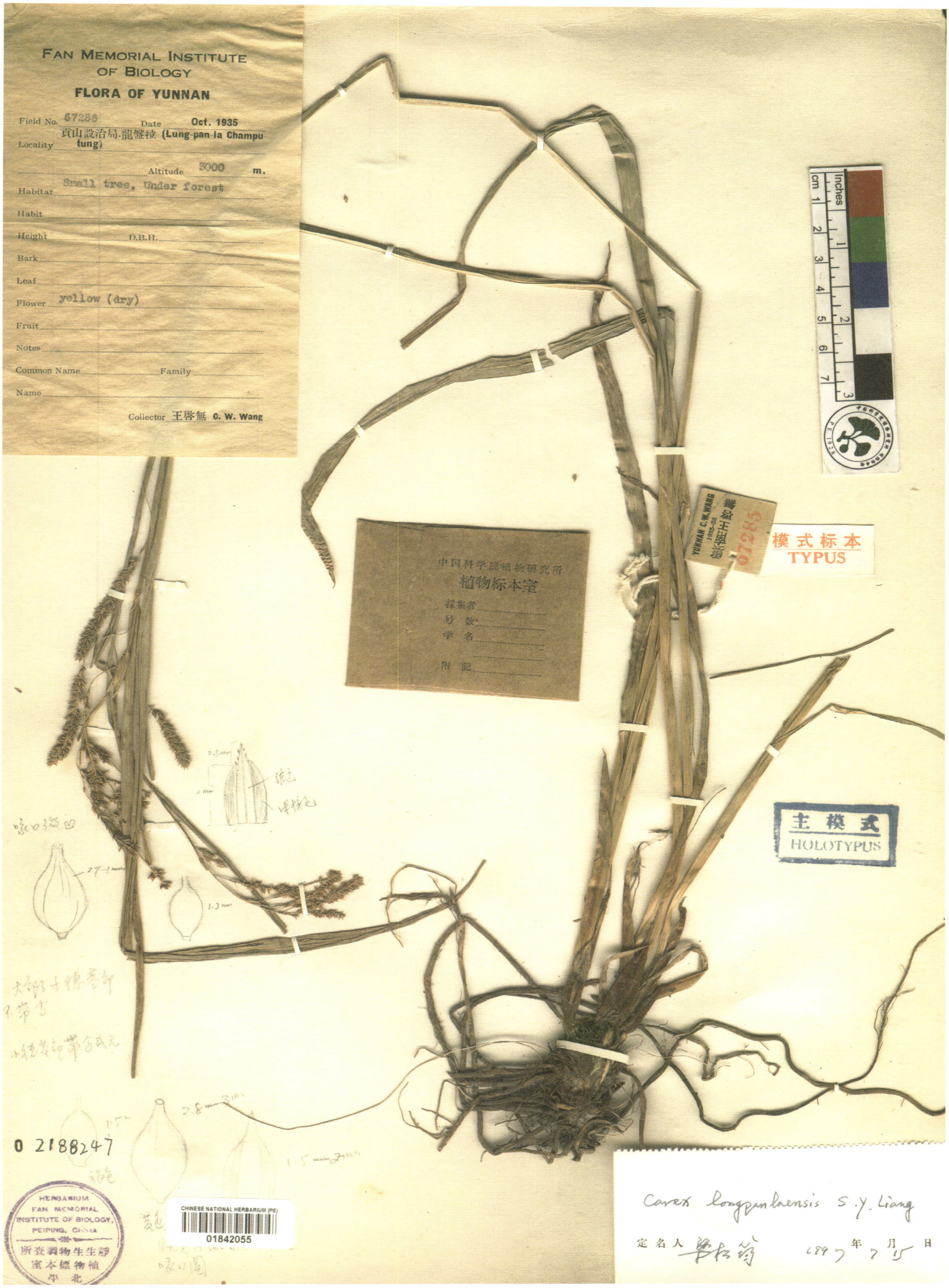

龙盘拉薹草 ***Carex longpanlaensis*** S. Y. Liang in Acta Phytotax. Sin. 37(2): 191, f. 1: 8-10. 1999. **Holotype:** China. Yunnan: Gongshan, alt. 3000 m, 1935-10-??. C. W. Wang 67285.

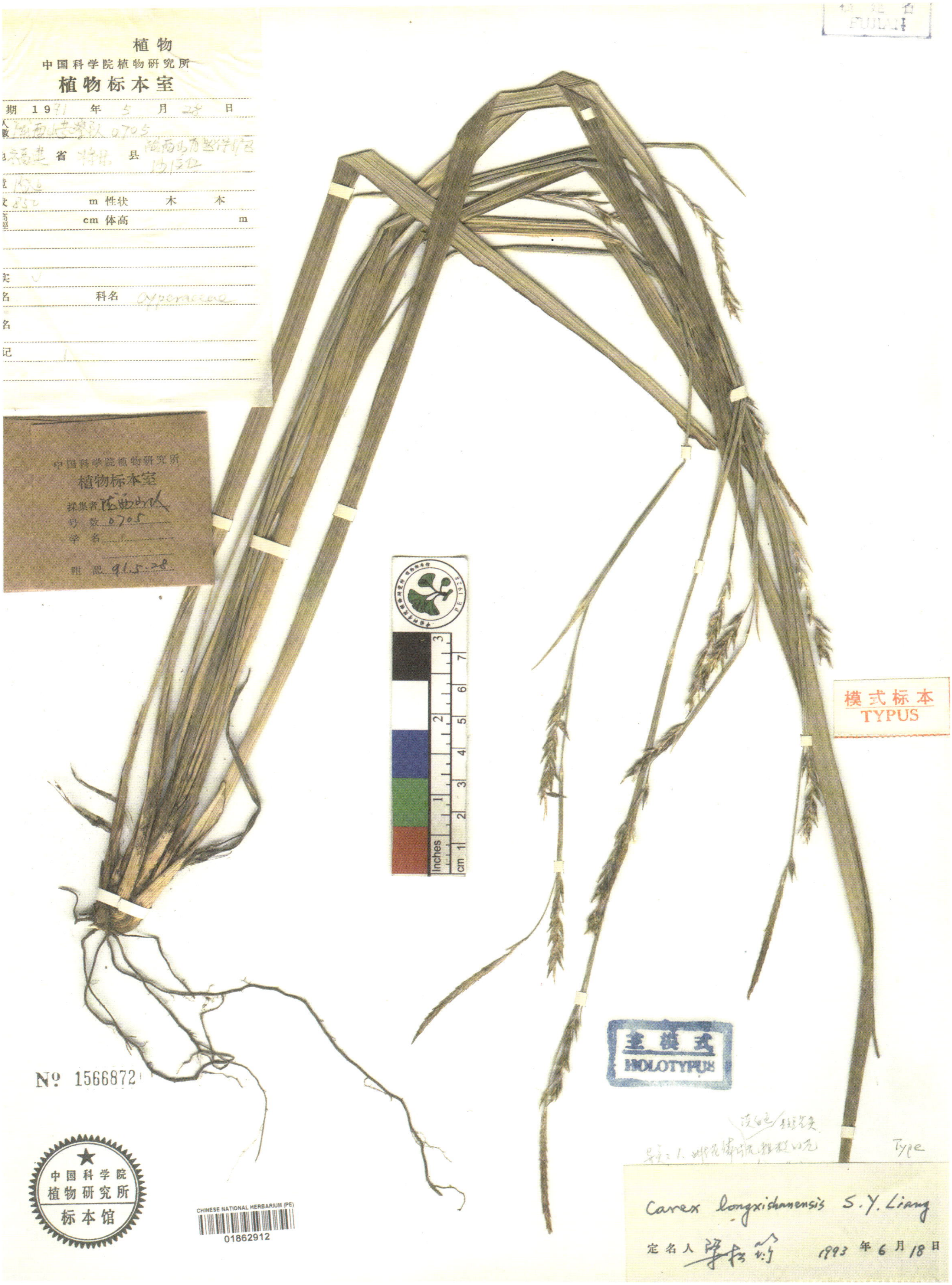

龙栖山薹草 ***Carex longqishanensis*** S. Y. Liang, Pl. Longqi Mount., Fujian, China 603, f. 8-3. 1994. **Holotype:** China. Fujian: Jiangle, alt. 850 m, 1991-05-28, Longqishan Exped. 705.

龙胜薹草 ***Carex longshengensis*** Y. C. Tang & S. Y. Liang, Fl. Reip. Pop. Sin. 12: 219, 522, pl. 43: 1-4. 2000. **Holotype:** China. Guangxi: Longsheng, 1955-05-31, Guangfu Forest Exped. 363.

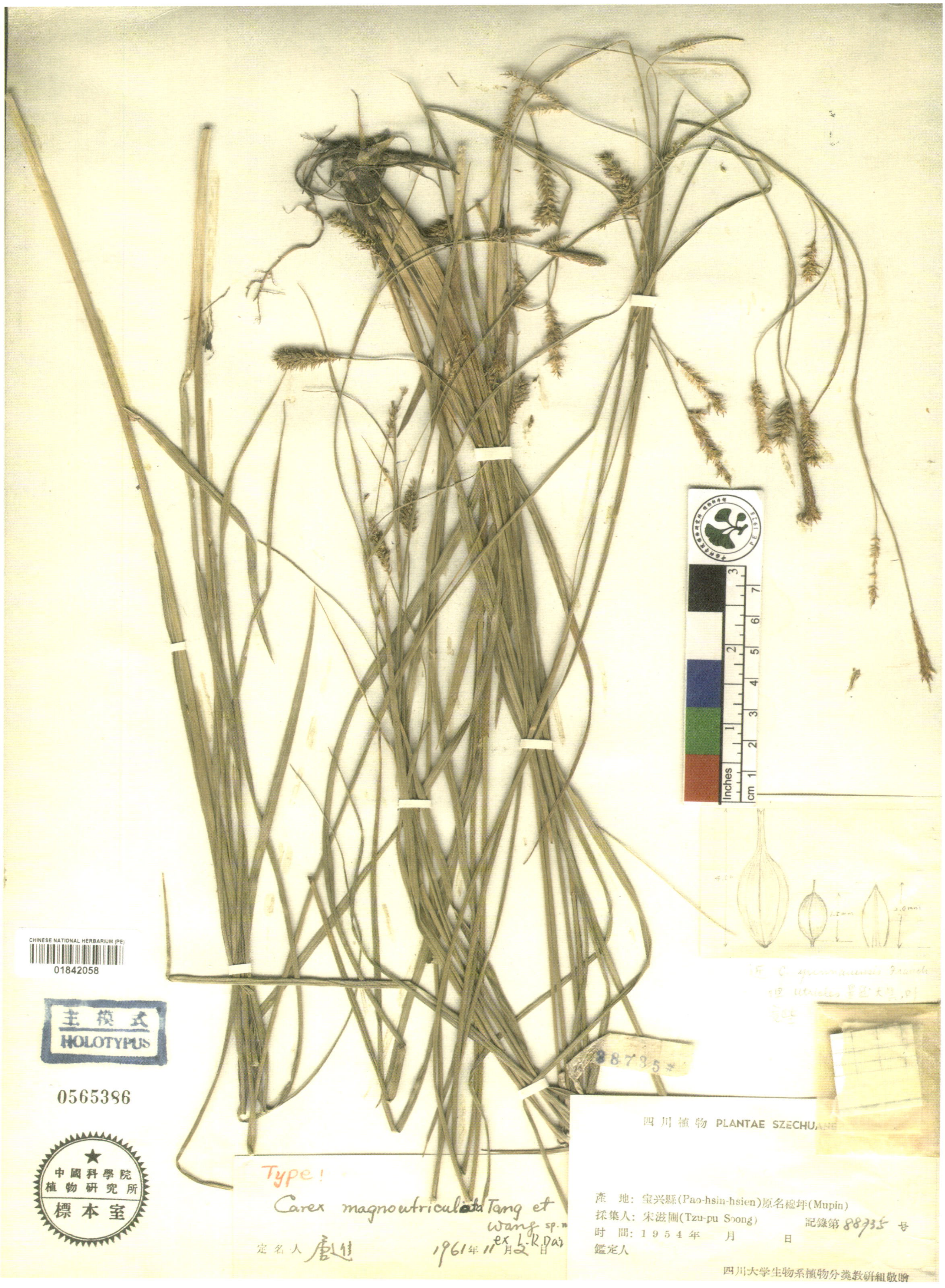

大果囊薹草 ***Carex magnoutriculata*** Tang & F. T. Wang ex L. K. Dai in Acta Phytotax. Sin. 32(2): 179, f. 4: 6-9. 1994.
Holotype: China. Sichuan: Baoxing, 1954-??-??, T. P. Soong 88735.

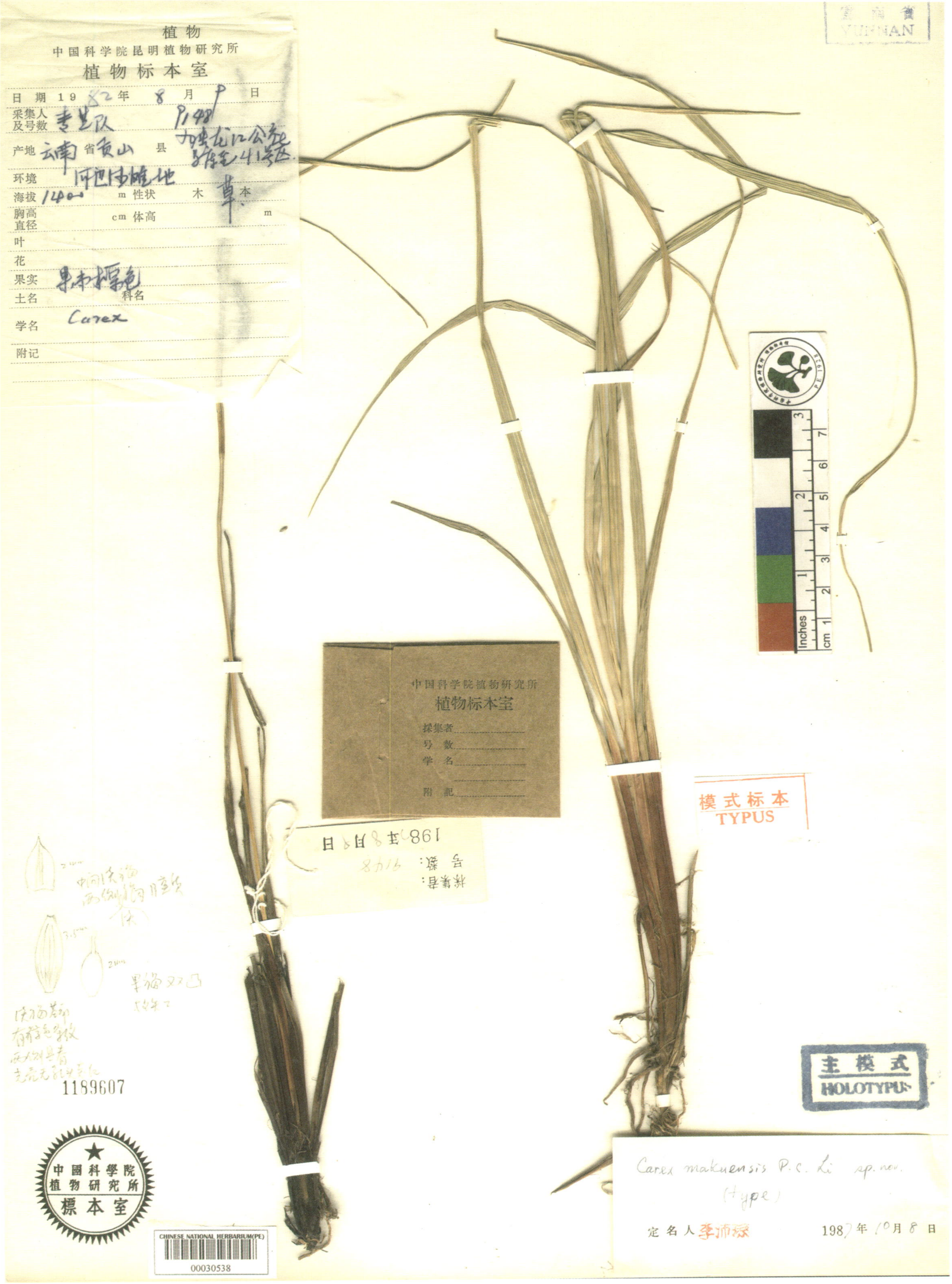

马库薹草 ***Carex makuensis*** P. C. Li in Acta Bot. Yunnan. 12(2): 139, f. 3. 1990. **Holotype:** China. Yunnan: Gongshan, alt. 1400 m, 1982-08-09, Qinghai-Xizang Exped. 9148.

昌宁薹草 ***Carex melinacra*** Franch. var. ***changningensis*** S. Y. Liang in Acta Phytotax. Sin. 37(2): 193. 1999. **Holotype:** China. Yunnan: Shunning (=Fengqing), alt. 3000 m, 1938-06-10, T. T. Yu 16205.

岷县薹草 ***Carex minxianensis*** S. Y. Liang, Fl. Reip. Pop. Sin. 12: 107, 520, pl. 19: 5-8. 2000. **Holotype:** China. Gansu: Min Xian, alt. 3000 m, 1936-07-01, T. P. Wang 4805.

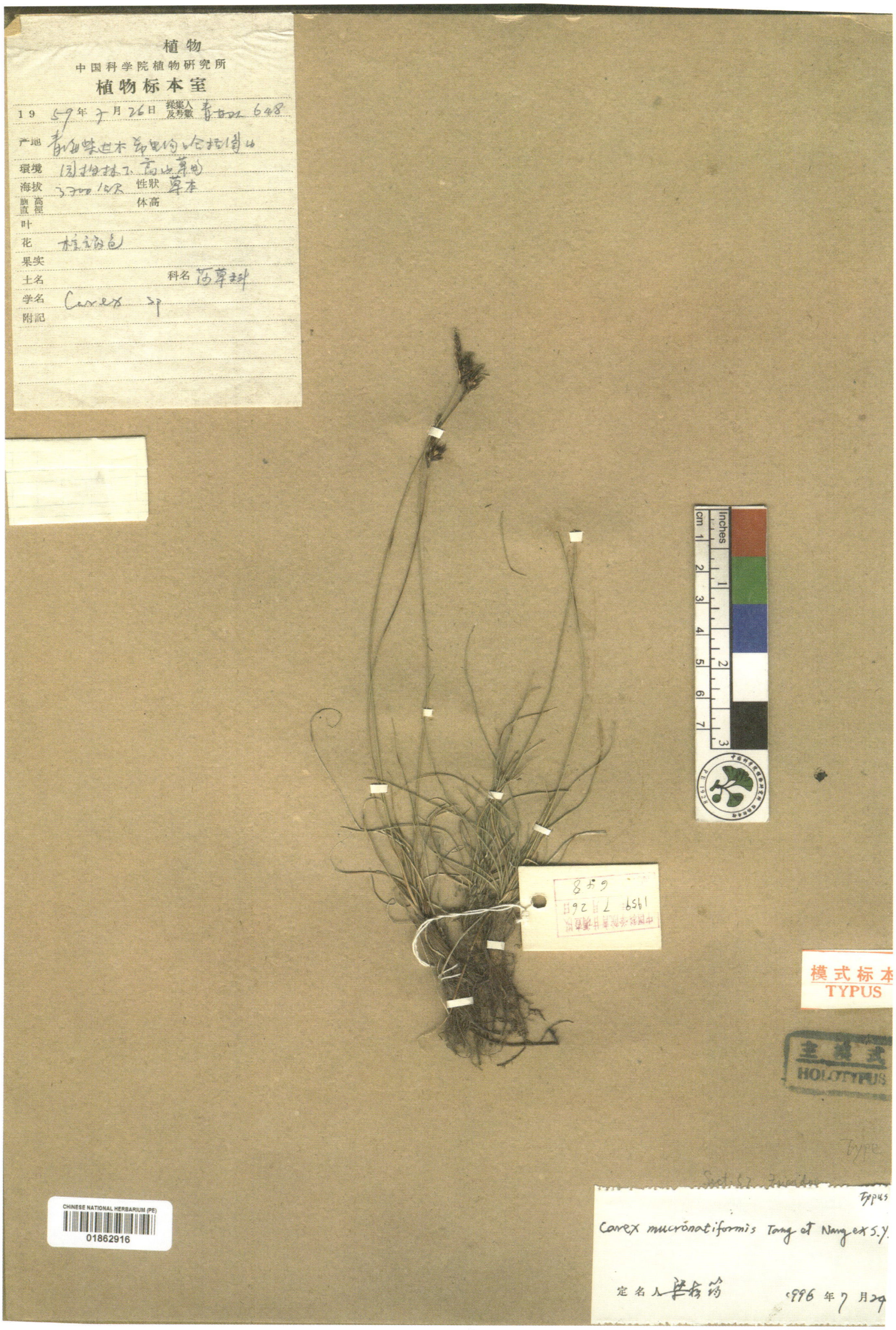

类短尖薹草 ***Carex mucronatiformis*** Tang & F. T. Wang ex S. Y. Liang, Fl. Reip. Pop. Sin. 12: 523. 2000. **Holotype:** China. Qinghai: Chaidamu, alt. 3700 m, 1959-07-26, Qinghai-Gansu Exped. 648.

南川薹草 ***Carex nanchuanensis*** K. L. Chu ex S. Y. Liang in Acta Phytotax. Sin. 36(6): 532, f. 2: 6-10. 1998. **Holotype:** China. Chongqing: Nanchuan, alt. 2050 m, 1935-06-12, K. L. Chu 1298.

双柱薹草 ***Carex neodigyna*** P. C. Li in Acta Bot. Yunnan. 12(2): 146, f. 13. 1990. **Holotype:** China. Sichuan: Sertar, alt. 3900 m, 1983-09-07, Z. X. Tang 1899.

新多穗薹草 ***Carex neopolycephala*** Tang & F. T. Wang ex L. K. Dai in Acta Phytotax. Sin. 32(2): 175, f. 2: 1-4. 1994.
Holotype: China. Yunnan: Kunming, Kunming Station 146715.

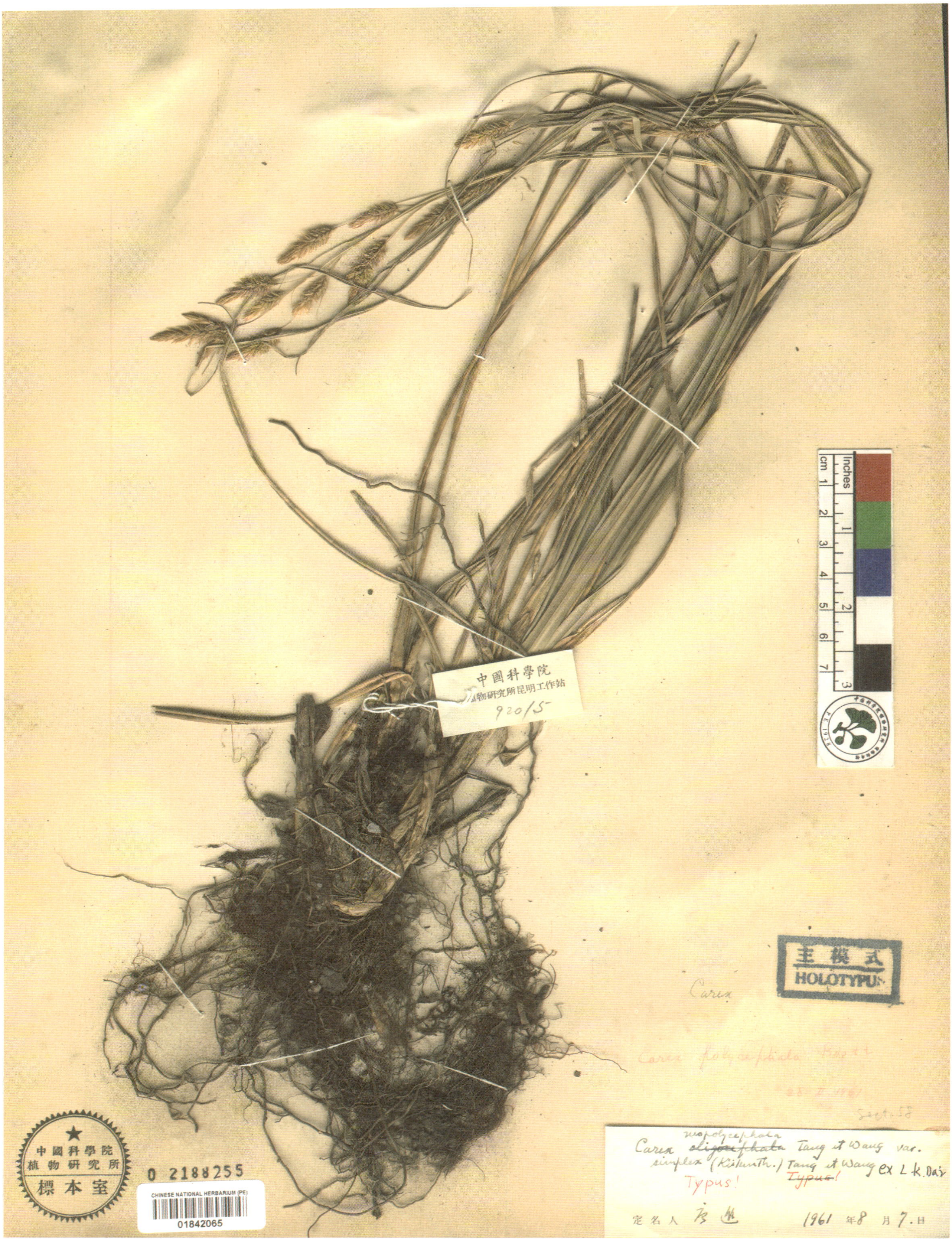

单穗薹草 ***Carex neopolycephala*** Tang & F. T. Wang ex L. K. Dai var. ***simplex*** Tang & F. T. Wang ex L. K. Dai in Acta Phytotax. Sin. 32(2): 175, f. 2: 5-8. 1994. **Holotype:** China. Yunnan: Kunming, Kunming Station 92015.

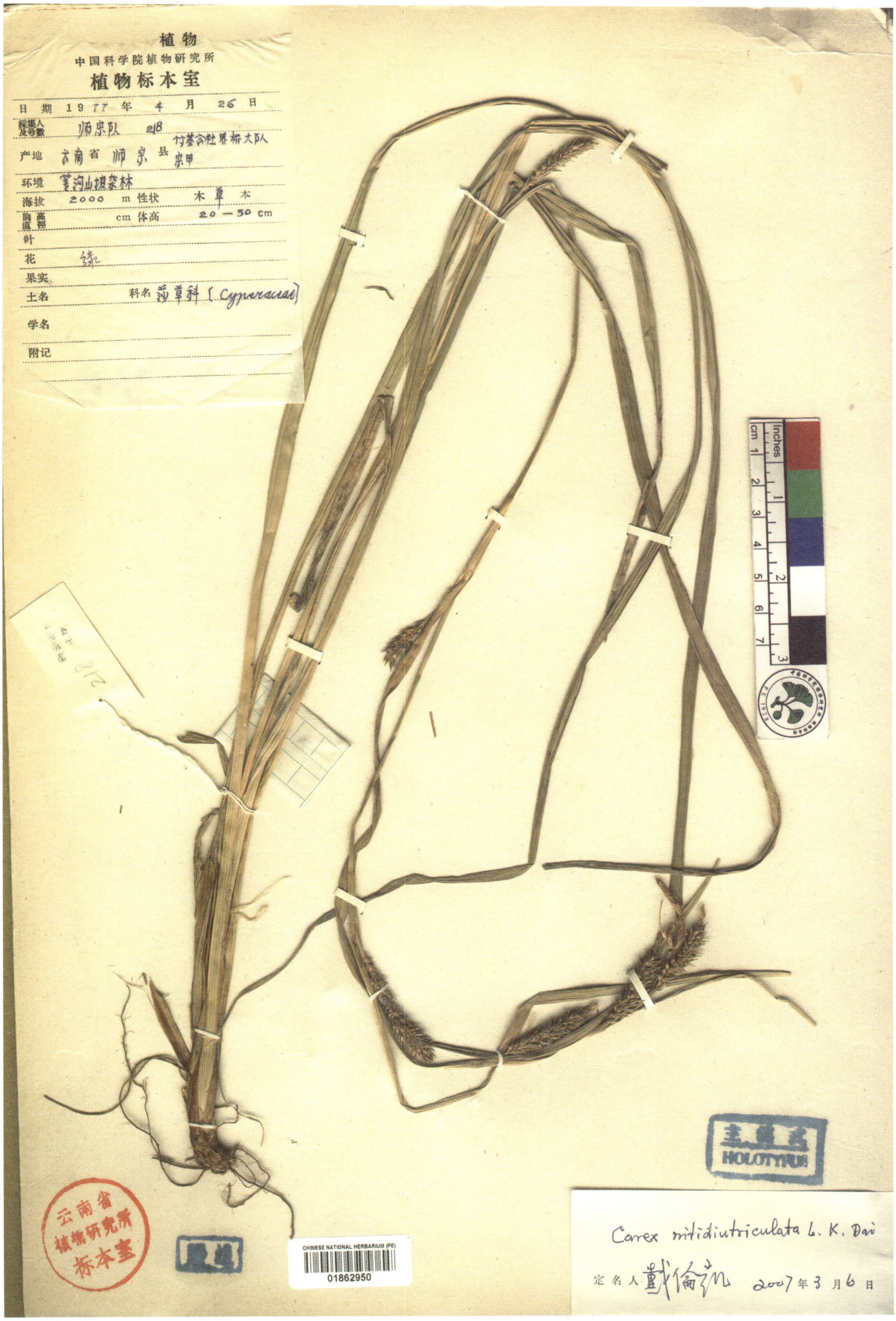

亮果薹草 ***Carex nitidiutriculata*** L. K. Dai in Acta Phytotax. Sin. 37(2): 186, f. 4: 1-4. 1999. **Holotype:** China. Yunnan: Shizong, alt. 2000 m, 1977-04-26, Shizong Exped. 218.

斜口薹草 ***Carex obliquitruncala*** Y. C. Tang & S. Y. Liang in Novon 16(3): 364, f. 2. 2006. **Holotype:** China. Yunnan: Gongshan，alt. 4000 m, 1938-09-30, T. T. Yu 22740.

德浚薹草 ***Carex obovatosquamata*** T. Wang & Y. L. Chang ex. P. C. Li in Acta Phytotax. Sin. 37(2): 175, f. 14. 1999.
Holotype: China.Yunnan, Dêqên, alt. 4100 m, 1937-07-25, T. T. Yu 9162.

峨眉薹草 ***Carex omeiensis*** Tang in Bull. Fan Mem. Inst. Biol. 3: 361. 1932. **Holotype:** China. Sichuan: Emei, Emeishan, alt. 2400 m, 1931-07-19, F. T. Wang 23391.

柔弱峨眉薹草 ***Caerx omeiensis*** Tang f. ***debilis*** Tang in Bull. Fan Mem. Inst. Biol. 3: 362. 1932. **Holotype:** China. Sichuan: Ebian, 1929-09-06, T. Y. Lu & al. 31.

针叶薹草 ***Carex onoei*** Franch. & Savat. var. ***macrogyna*** Tang in Bull. Fan Mem. Inst. Biol. 3: 137. 1932. **Holotype:** China. Hebei: Liulipingtzeshan, alt. 400~500 m, 1931-06-09, T. Tang 1950.

圆坚果薹草 ***Carex orbicularinucis*** L. K. Dai in Acta Phytotax. Sin. 37(2): 183, f. 3: 1-4. 1999. **Isotype:** China. Yunnan: Kunming, 1955-07-25, Kunming Station 50894.

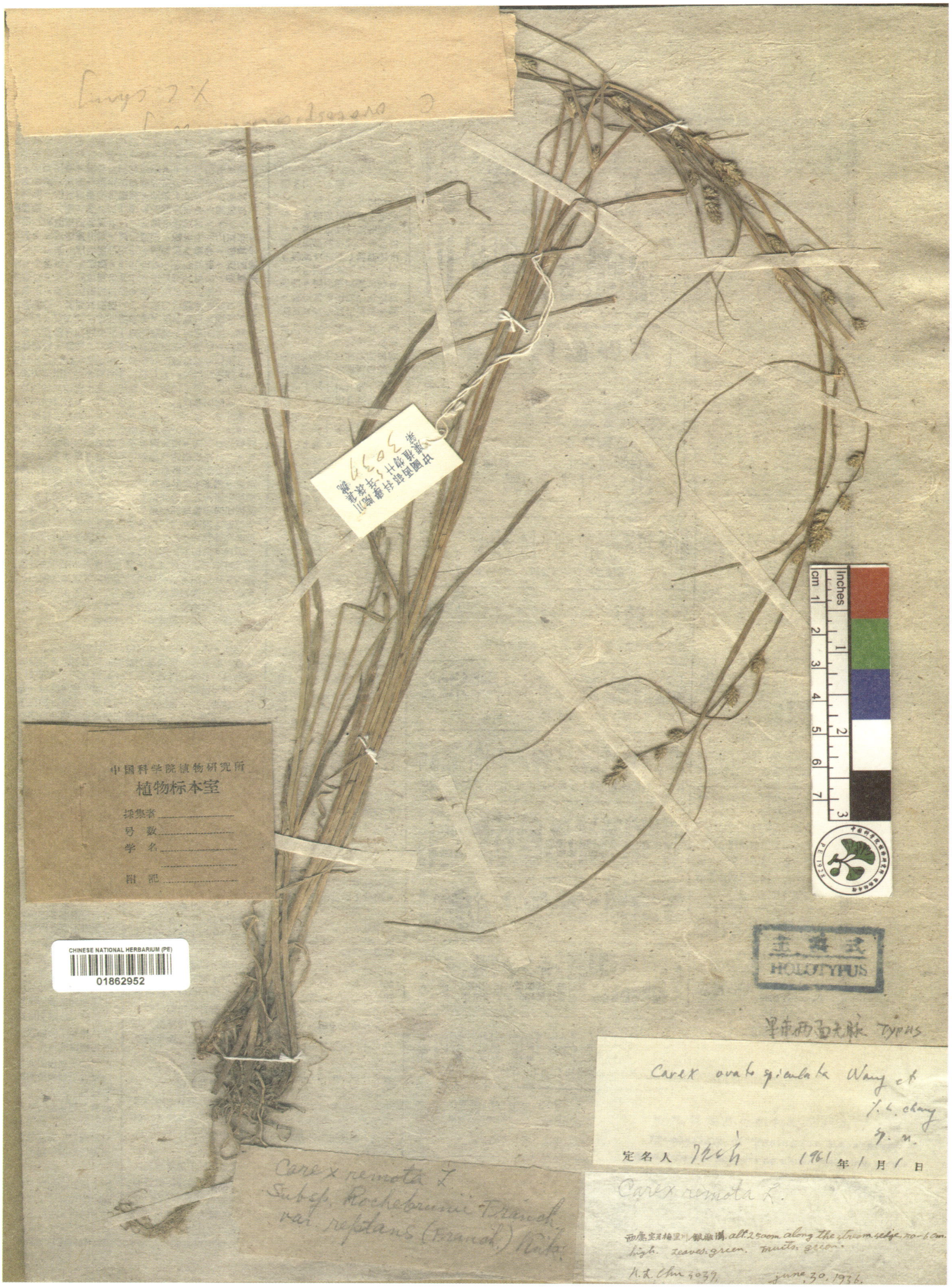

卵穗薹草 ***Carex ovatispiculata*** T. Wang & Y. L. Chang ex S. Y. Liang in Acta Phytotax. Sin. 28(2): 155, f. 2. 1990. **Holotype:** China. Sichuan: Baoxing, alt. 2500 m, 1936-06-30, K. L. Chu 3037.

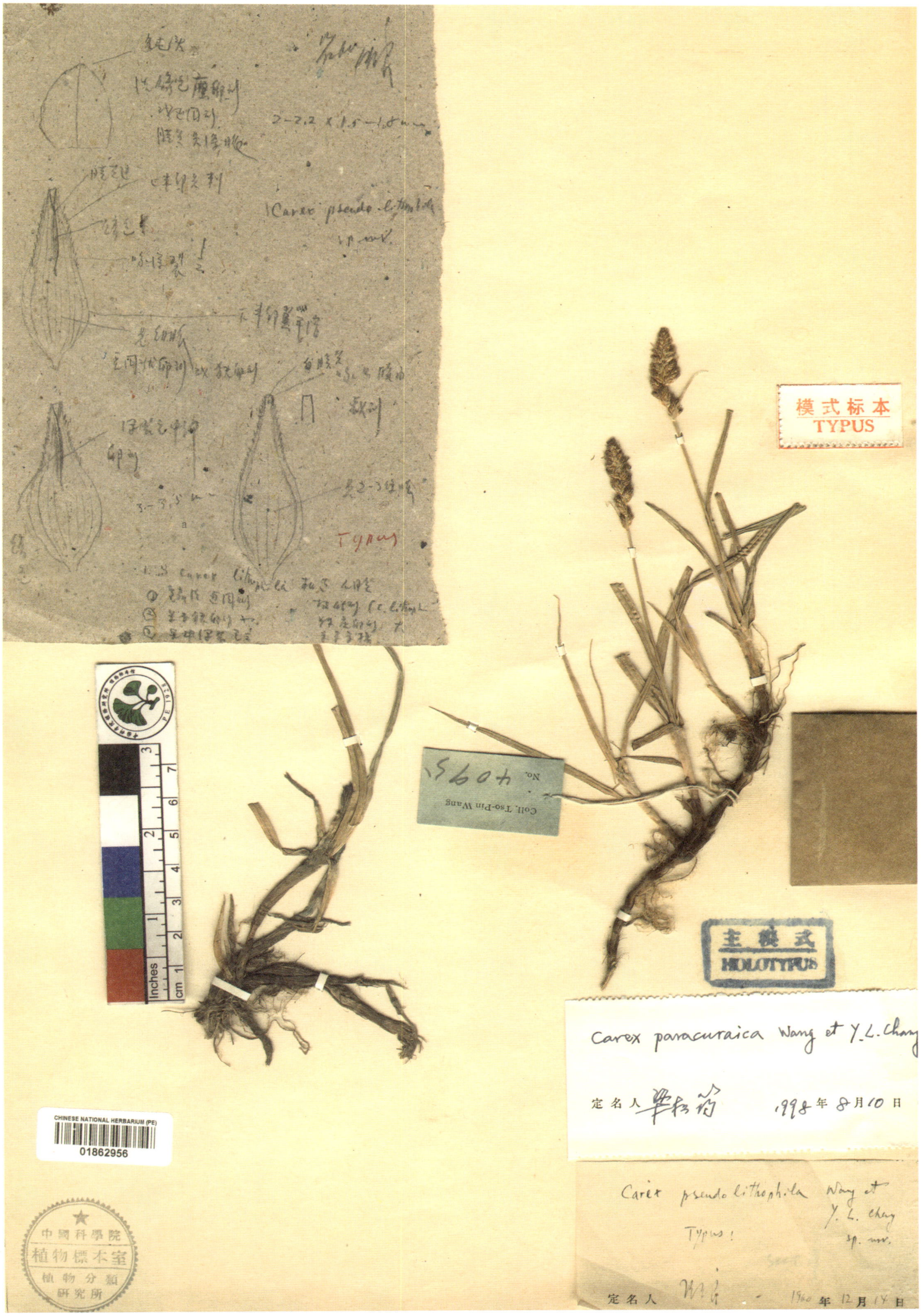

陇县薹草 ***Carex paracuraica*** F. T. Wang & Y. L. Chang ex S. Y. Liang, Fl. Reip. Pop. Sin. 12: 494, 527. 2000. **Holotype:** China. Shaanxi: Long Xian, alt. 1900 m, 1936-06-02, T. P. Wang 4095.

裴氏薹草 ***Carex peiana*** F. T. Wang & Tang in Contr. Inst. Bot. Nat. Acad. Peiping 6(2): 55. 1949. **Holotype:** China. Sichuan: Tianquan, alt. 1200 m, 1938-03-18, C. Pein 8256.

纤细薹草 ***Carex pergracilis*** Nelmes in Kew Bull. 1948(1): 107. 1948. **Isotype:** China. Sichuan: Tianquan, alt. 2450 m, 1936-04-27, K. L. Chu 3435.

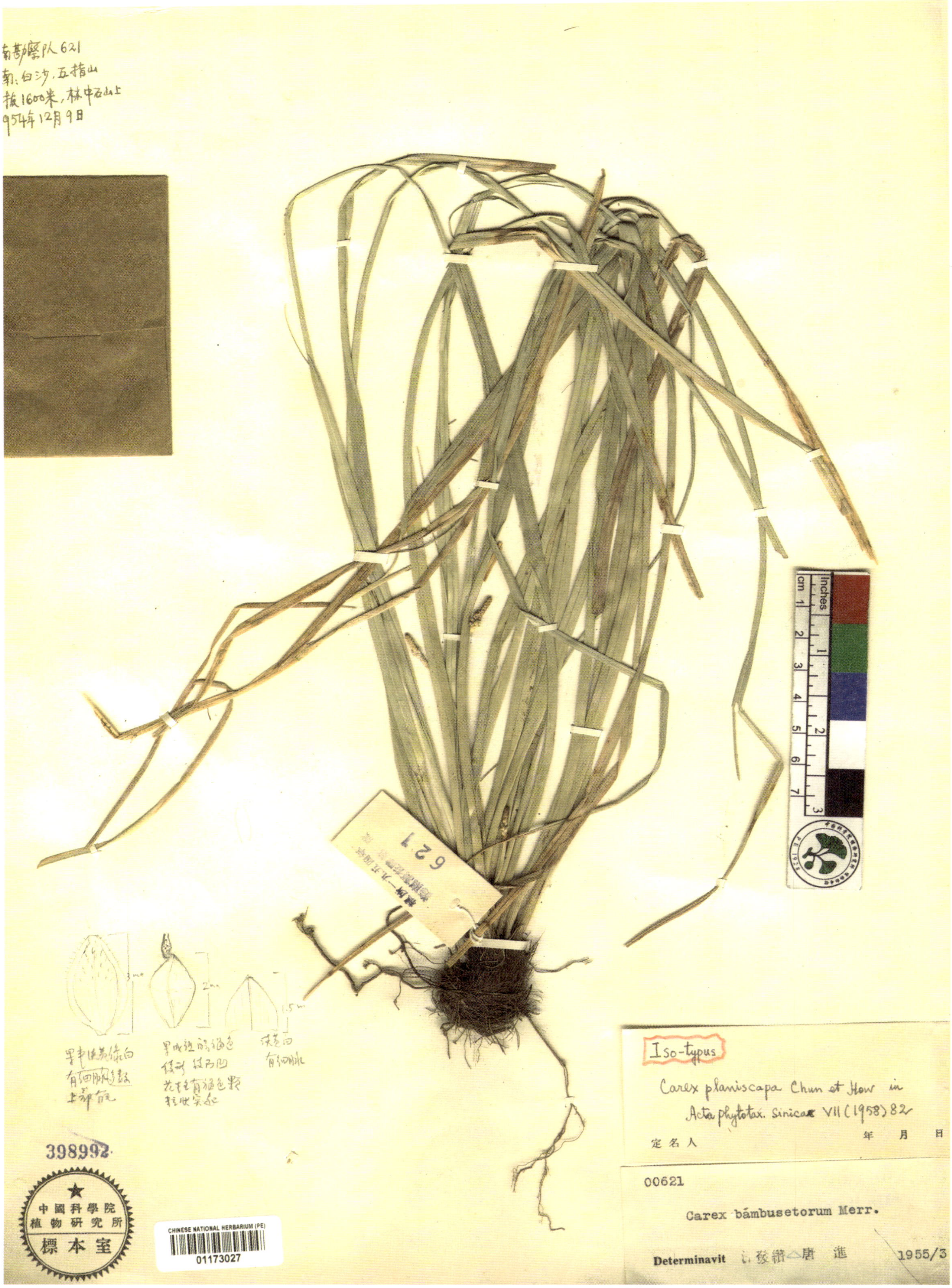

扁茎薹草 ***Carex planiscapa*** Chun & How in Acta Phytotax. Sin. 7(1): 82, pl. 25: 2. 1958. **Isotype:** China. Hainan: Baisha, Wuzhishan, alt. 1600 m, 1954-12-09, Hainan Exped. 621.

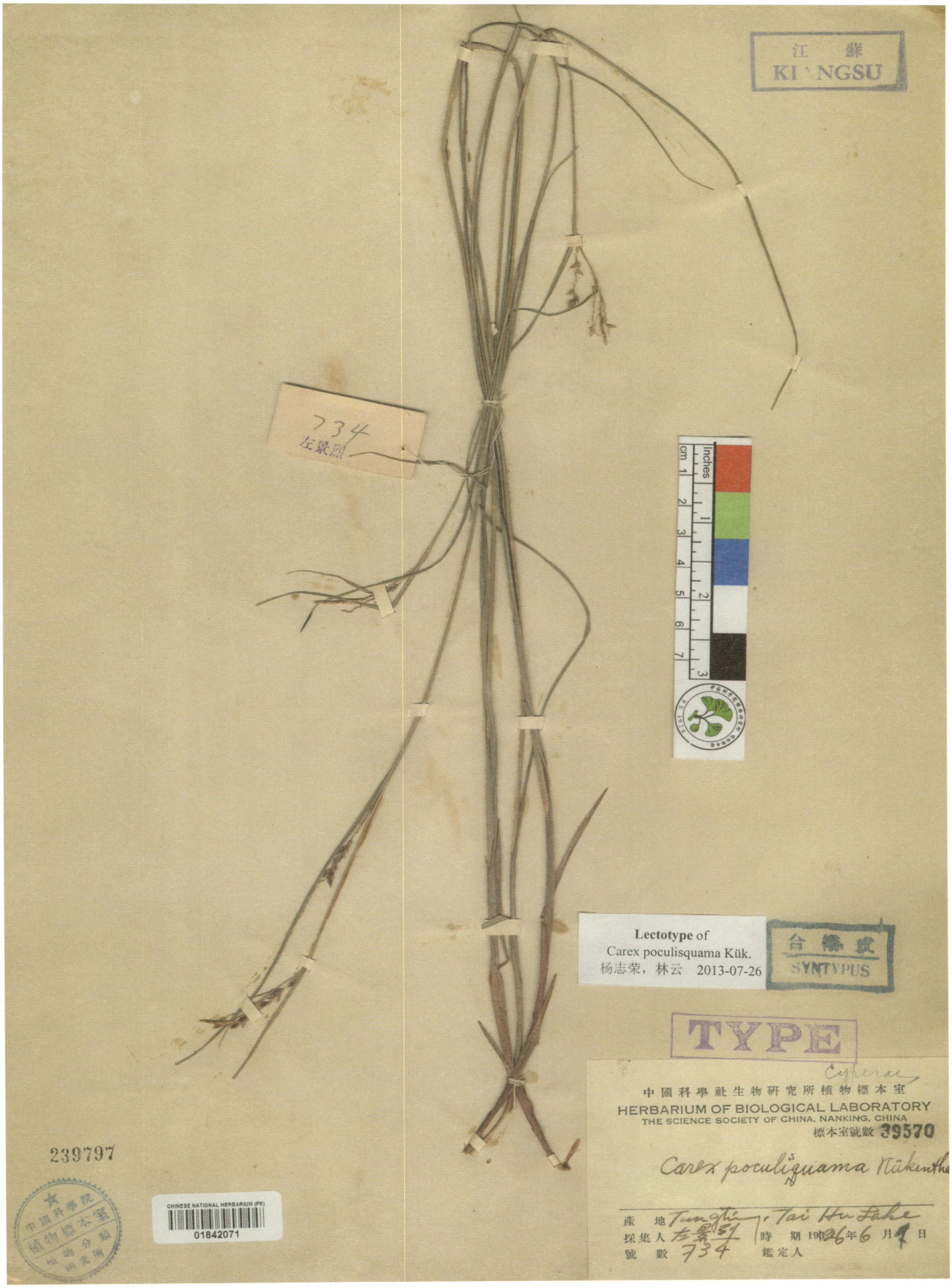

杯鳞薹草 ***Carex poculisquama*** Kükenth. in Repert. Sepc. Nov. Regni Veg. 27(726-733): 111. 1929. **Lectotype** (designated by Y. Lin & al. in Acta Bot. Bor.-Occ. Sin. 34: 412. 2014.): China. Jiangsu: Suzhou, 1926-06-07, R. C. Ching & C. L. Tso 734.

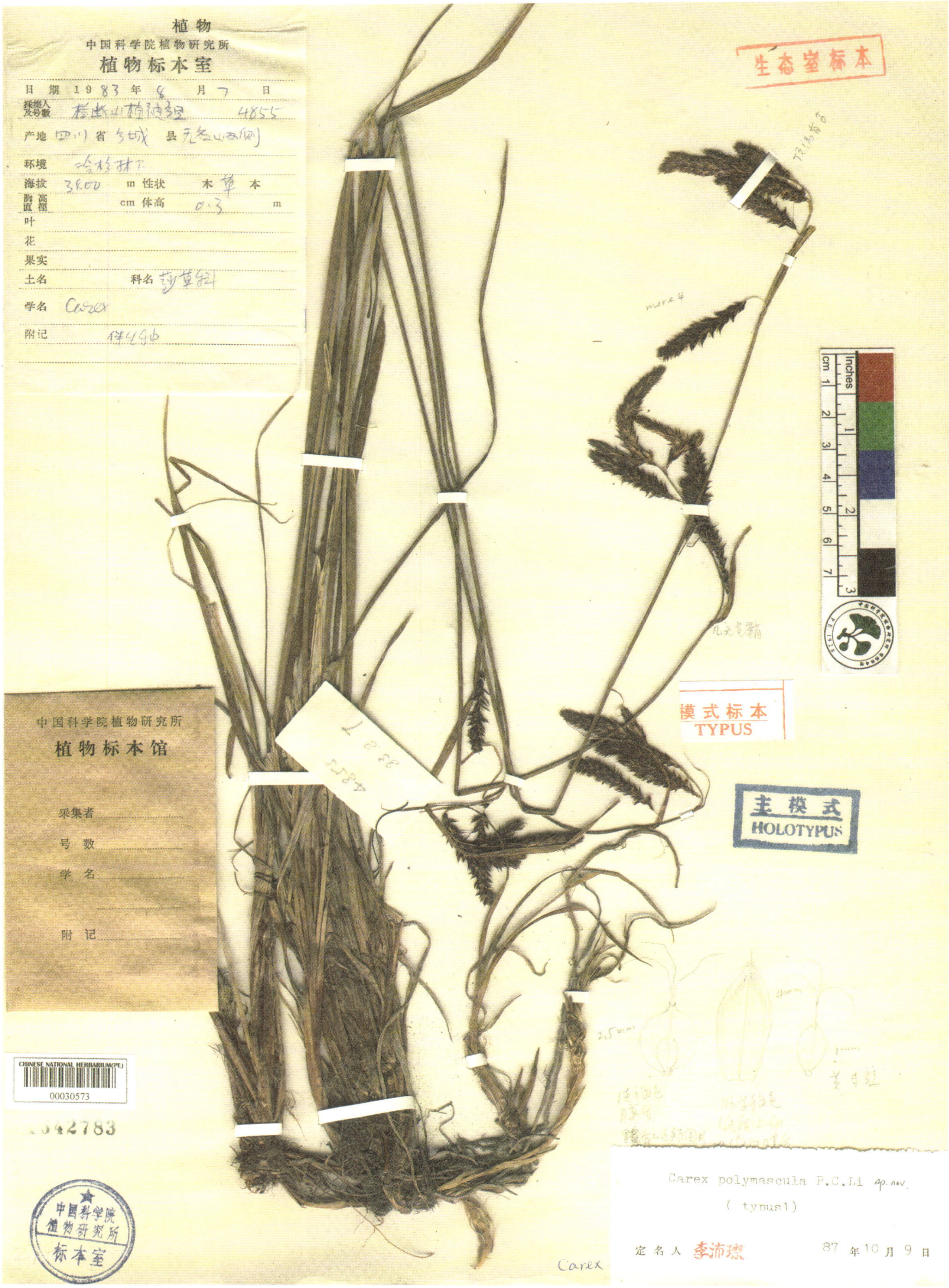

多雄薹草 ***Carex polymascula*** P. C. Li in Acta Bot. Yunnan. 12(2): 140, f. 5. 1990. **Holotype:** China. Sichuan: Xiangcheng, alt. 3800 m, 1983-08-07, Hengduanshan Exped. 4855.

弥勒山薹草 ***Carex pseudo-laticeps*** Tang & F. T. Wang ex S. Y. Liang in Acta Phytotax. Sin. 36(6): 536, f. 3: 1-5. 1998.
Holotype: China. Hongkong: Mileshan, 1929-04-20, C. T. Tso 20216.

锈点薹草 ***Carex pseudoligulata*** L. K. Dai in Acta Phytotax. Sin. 32(2): 187, f. 7: 5-8. 1994. **Holotype:** China. Hunan: Shimen, alt. 700~800 m, 1987-06-21, Hupingshan Exped. 474.

假长嘴薹草 ***Carex pseudo-longerostrata*** Y. L. Chang & Y. L. Yang, Fl. Pl. Herb. China Bor.-Orient. 11: 131, 206, pl. 59: 1-5. 1976. **Isotype:** China. Jilin: Antu, Changbaishan, alt. 2600 m, 1963-07-28, W. Wang 2287.

假头序薹草 ***Carex pseudo-phyllocephala*** L. K. Dai in Acta Phytotax. Sin. 32(2): 187, f. 7: 1-4. 1994. **Holotype:** China. Hunan: Yongshun, 1988-06-20, Beijing Exped. 1767.

高山薹草 ***Carex pseudosupina*** Y. C. Tang ex L. K. Dai in Acta Phytotax. Sin. 37(2): 177, f. 1: 5-8. 1999. **Holotype:** China. Sichuan: Kangding, Haizishan, alt. 3700 m, 1934-08-28, C. S. Liu 1389.

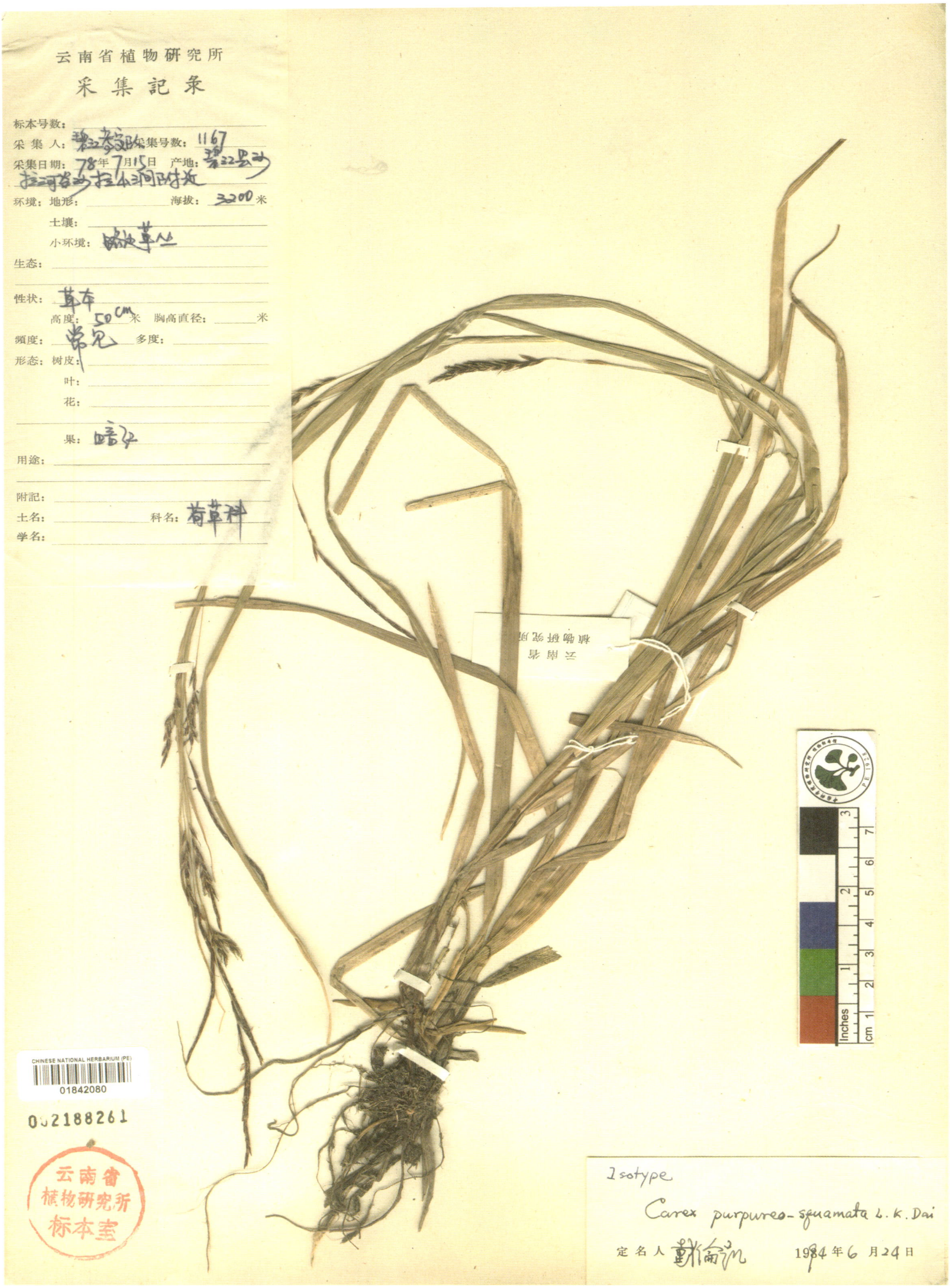

紫鳞薹草 ***Carex purpureosquamata*** L. K. Dai in Acta Phytotax. Sin. 37(2): 183, f. 3: 5-9. 1999. **Isotype:** China. Yunnan: Bijiang, alt. 3200 m, 1978-07-15, Bijiang Exped. 1167.

紫红鞘薹草 ***Carex purpureovagina*** F. T. Wang & Y. L. Chang ex S. Y. Liang, Fl. Reip. Pop. Sin. 12: 525. 2000. **Holotype:** China. Guangxi: Longsheng, alt. 1600~1700 m, 1955-05-05, Guangfu Forest Exped. 131.

普陀薹草 ***Carex putuoensis*** S. Y. Liang in Acta Phytotax. Sin. 33(5): 486, f. 2: 1-5. 1995. **Holotype:** China. Zhejiang: Putuoshan, 1936-06-05, H. Migo s. n.

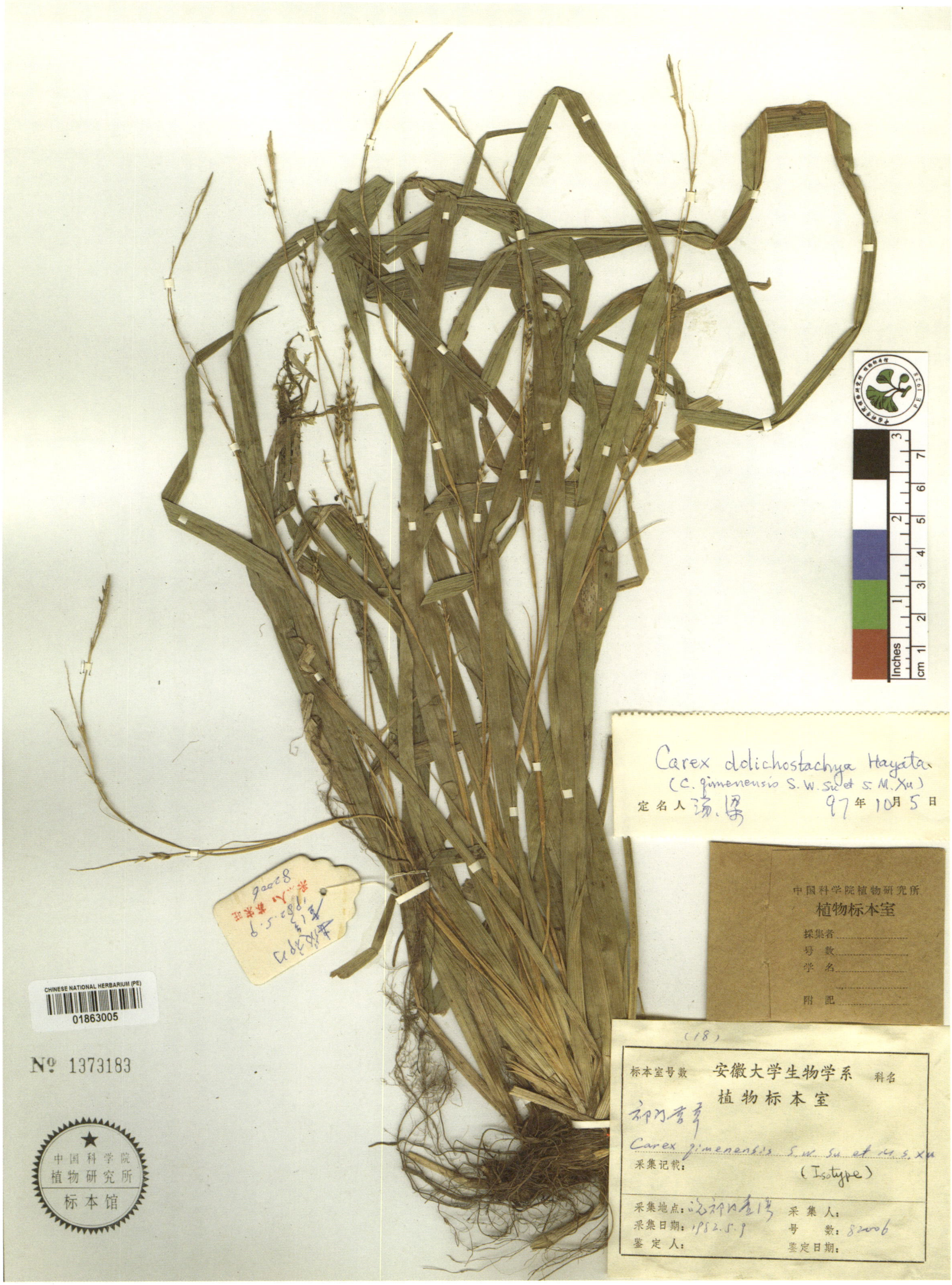

祁门薹草 ***Carex qimenensis*** S. W. Su & S. M. Xu in Bull. Bot. Res., Harbin 9(1): 49, f. 4. 1989. **Isotype:** China. Anhui: Qimen, 1982-05-09, S. W. Su 82006.

青岛薹草 ***Carex qingdaoensis*** F. Z. Li & S. J. Fan in Bull. Bot. Res., Harbin 13(1): 71, f. 1993. **Isotype:** China. Shandong: Qingdao, Changmenyan, 1988-06-29, S. J. Fan 88008.

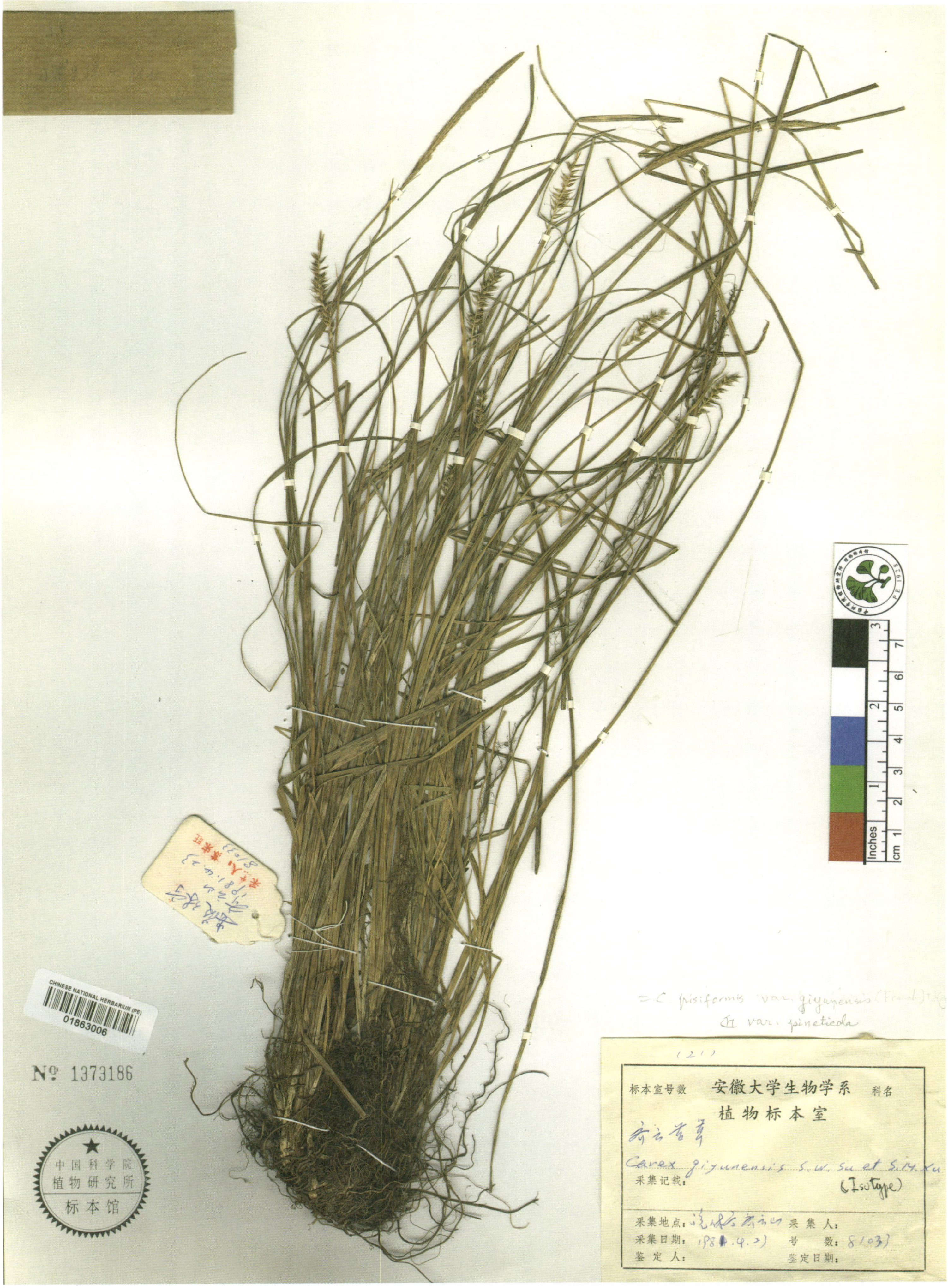

祁云薹草 ***Carex qiyunensis*** S. W. Su & S. M. Xu in Bull. Bot. Res., Harbin 9(1): 50, f. 5. 1989. **Isotype:** China. Anhui: Xiuning, 1981-04-23, S. W. Su 81033.

根花薹草 ***Carex radiciflora*** Dunn in Journ. Linn. Soc. Bot. 38: 371. 1908. **Isotype:** China. Fujian: Fuzhou, 1905-(04-06)-??, Hongkong Herb. 3653.

垂果薹草 ***Carex recurvisaccus*** T. Koyama in Jap. Journ. Bot. 15(2): 166, f. 2. 1956. **Isotype:** China. Guangdong: Zengcheng, Nankunshan, 1932-04-03, W. T. Tsang 20068.

反折果薹草 ***Carex retrofracta*** Kükenth. in Fedde, Rep. Spec. Nov. Reg. Veg. 27: 110. 1930. **Isotype:** China. Zhejiang: Hangzhou, 1929-05-04, Y. L. Keng 2352.

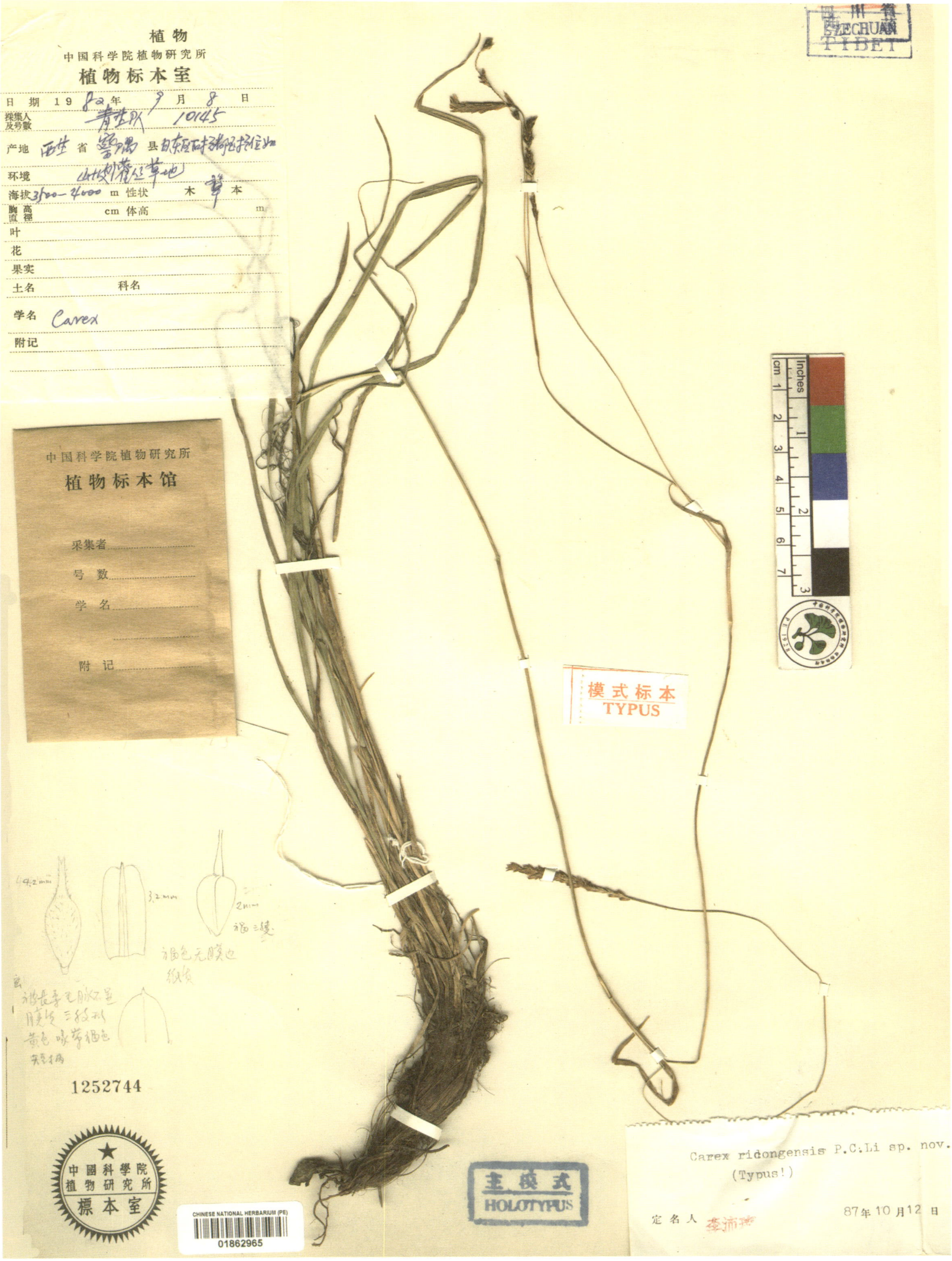

日东薹草 ***Carex ridongensis*** P. C. Li in Acta Bot. Yunnan 12(2): 148, f. 15. 1990. **Holotype:** China. Xizang: Zayü, alt. 3500~4000 m, 1982-09-08, Qinghai-Xizang Exped. 10145.

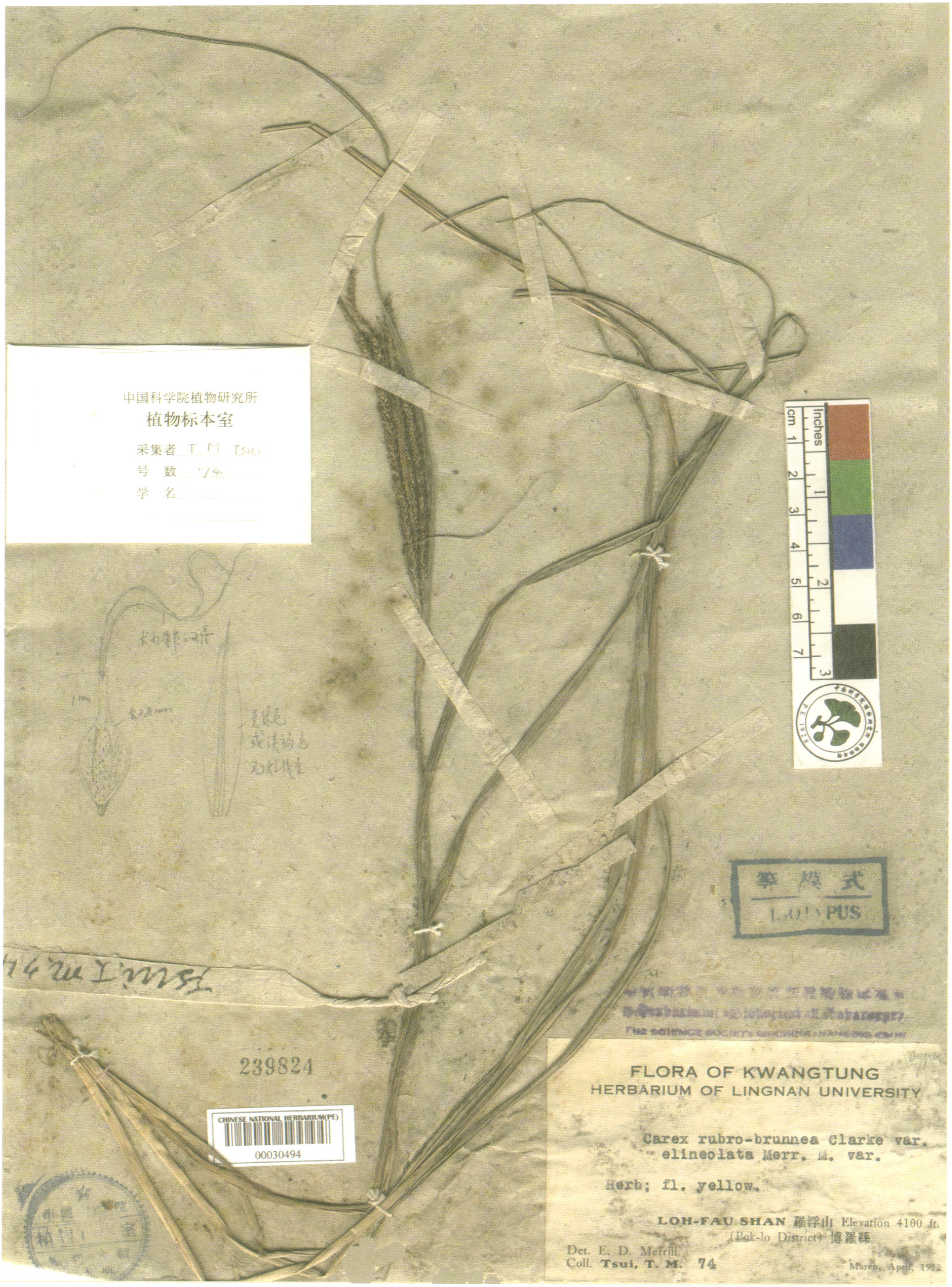

罗浮山薹草 ***Carex rubrobrunnea*** C. B. Clarke var. ***elineolata*** Merr. in Lingnan Sci. Journ. 13(1): 18. 1934. **Isotype:** China. Guangdong: Boluo, Luofoshan, alt. 1700 m, 1932-(03-04)-??, T. M. Tsui 74.

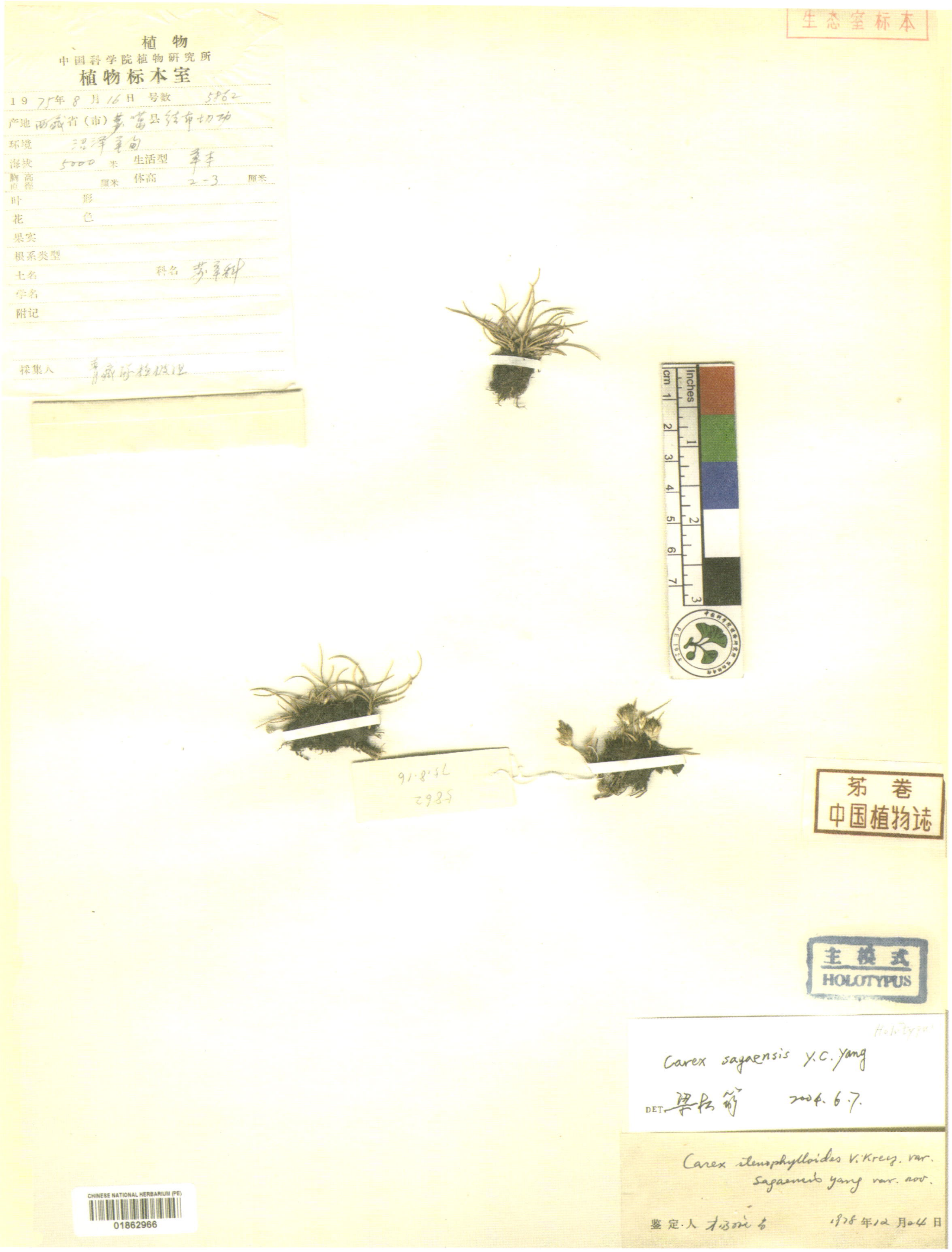

萨噶薹草 ***Carex sagaensis*** Y. C. Yang ex C. Y. Wu, Fl. Xizang. 5: 439, f. 256. 1987. **Holotype:** China. Xizang: Saga, alt. 5000 m, 1975-08-16, Qinghai-Xizang Exped. 5862.

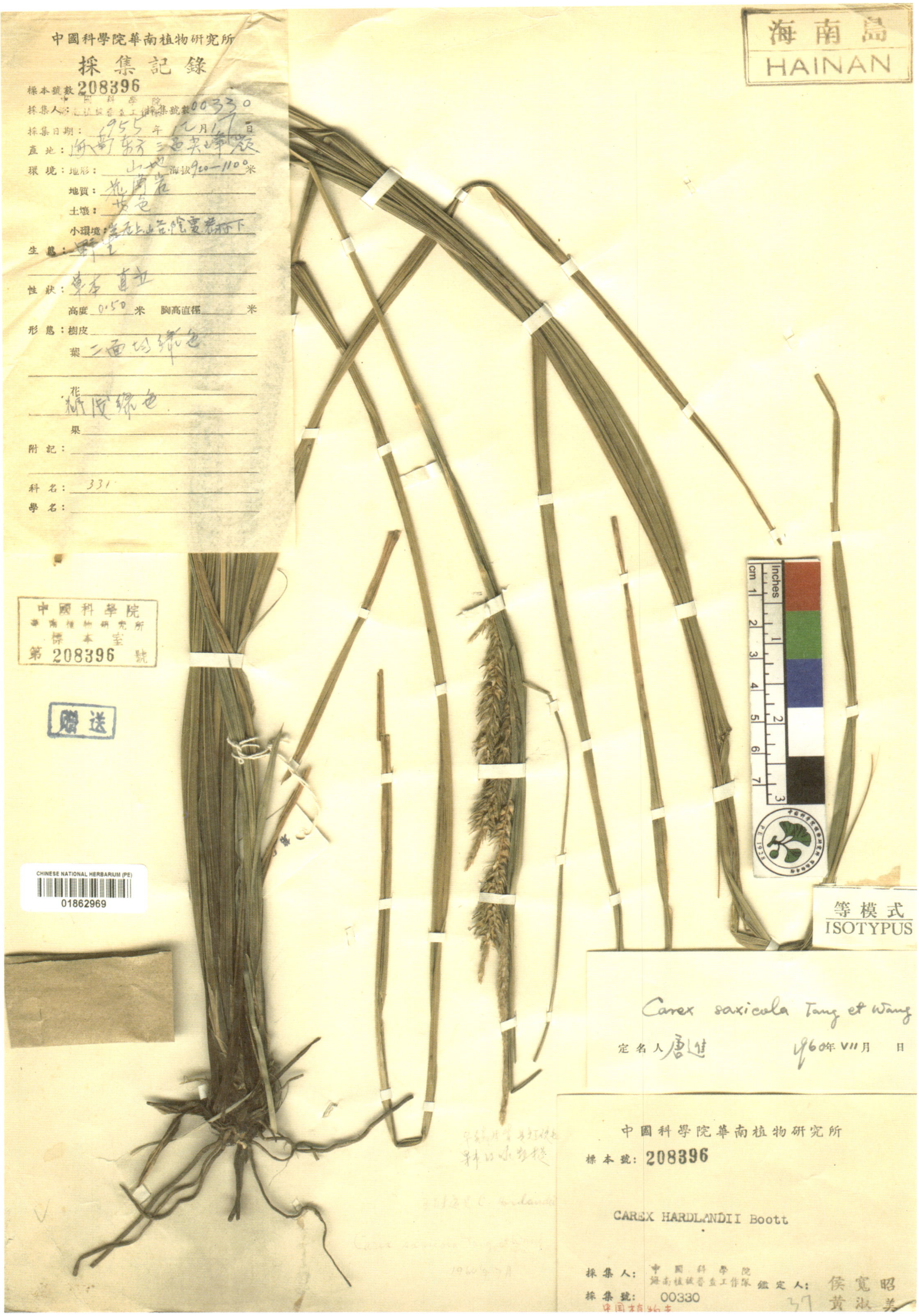

岩生薹草 ***Carex saxicola*** Tang & F. T. Wang, Fl. Hainan. 4: 535. 1977. **Isotype:** China. Hainan: Dongfang, alt. 920~1100 m, 1955-12-17, Hainan Exped. 330.

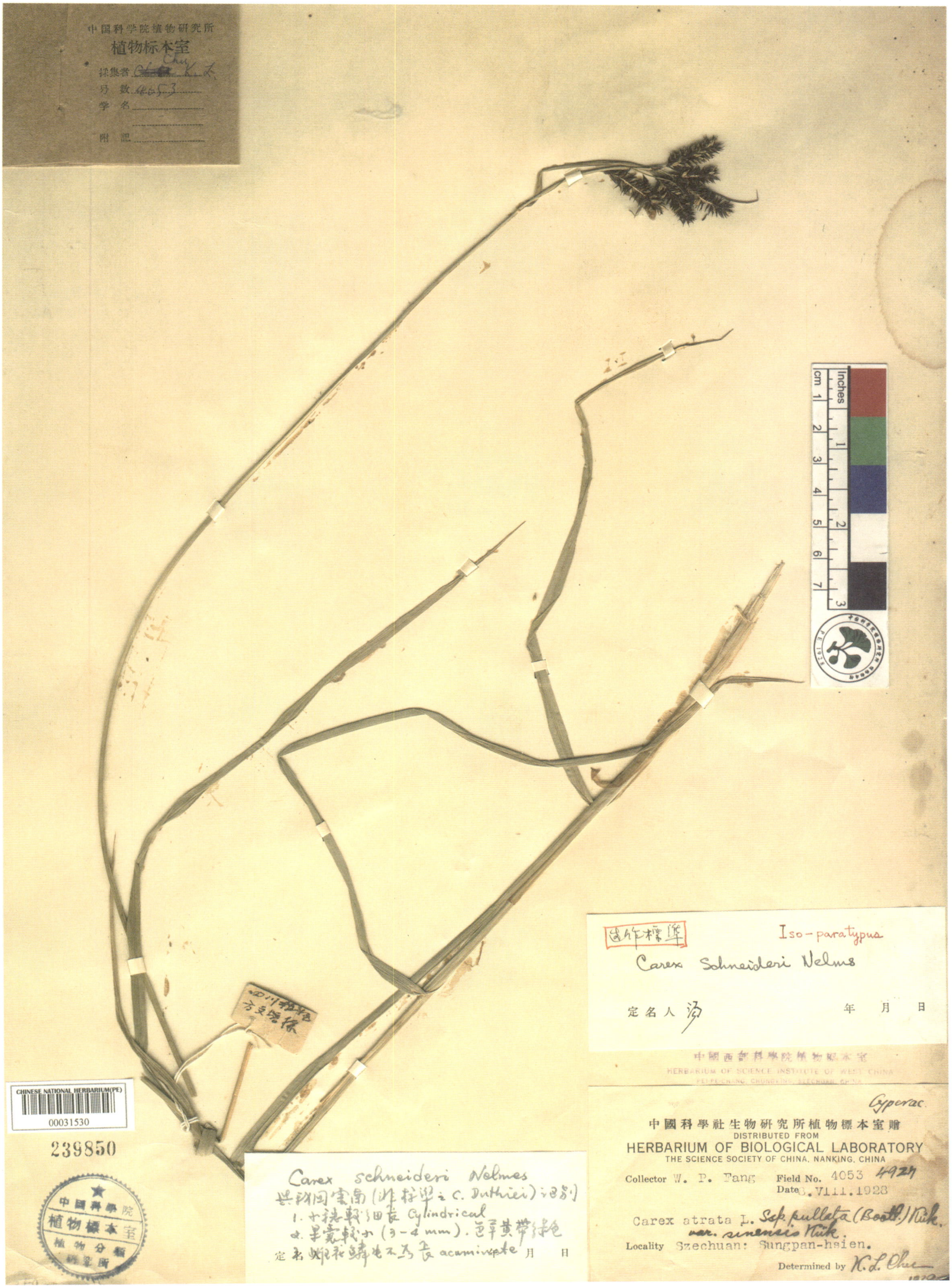

川滇薹草 ***Carex schneideri*** Nelmes in Bull. Misc. Inform. Kew 1939: 201. 1939. **Isoparatype:** China. Sichuan: Songpan, 1928-08-08, W. P. Fang 4053.

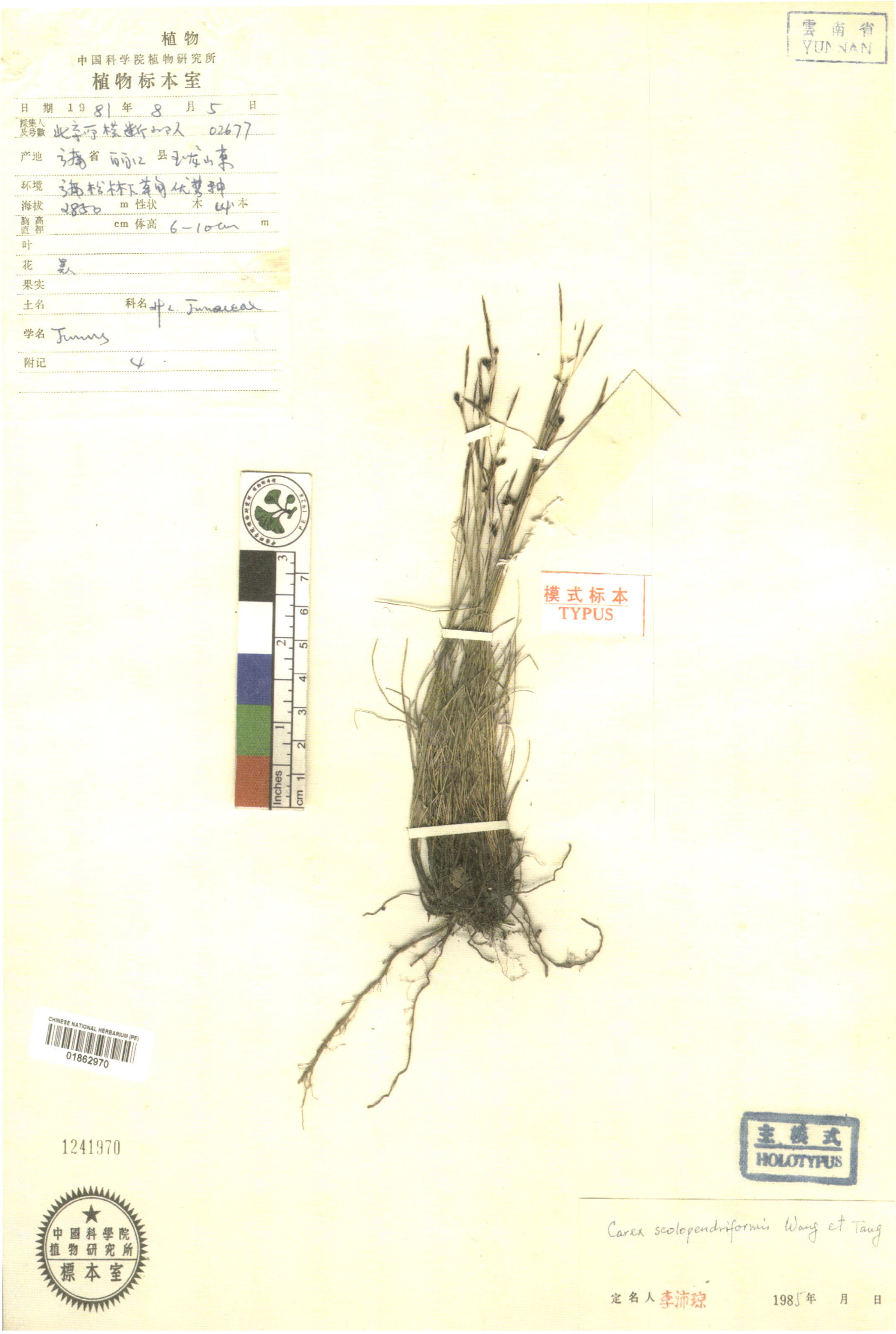

蜈蚣薹草 ***Carex scolopendriformis*** F. T. Wang & Tang ex P. C. Li in Acta Bot. Yunnan. 12(2): 142, f. 7. 1990. **Holotype:** China. Yunnan: Lijiang, alt. 2850 m, 1981-08-05, Hengduanshan Exped. 2677.

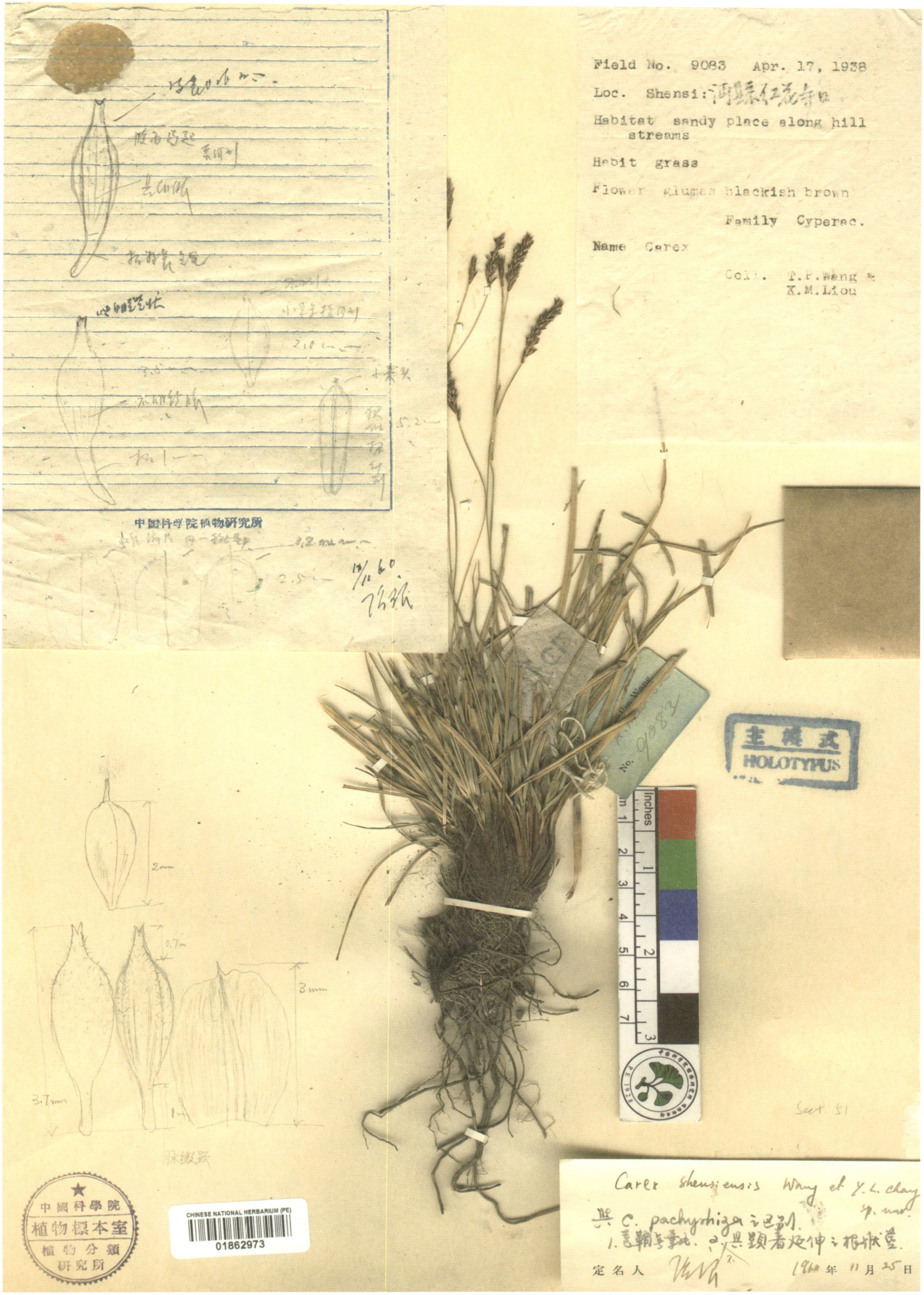

沔县薹草 ***Carex setosa*** Boott var. ***mianxianica*** S. Y. Liang, Fl. Reip. Pop. Sin. 12: 525. 2000. **Holotype:** China. Shaanxi: Mian Xian, 1938-04-17, T. P. Wang & K. M. Liou 9083.

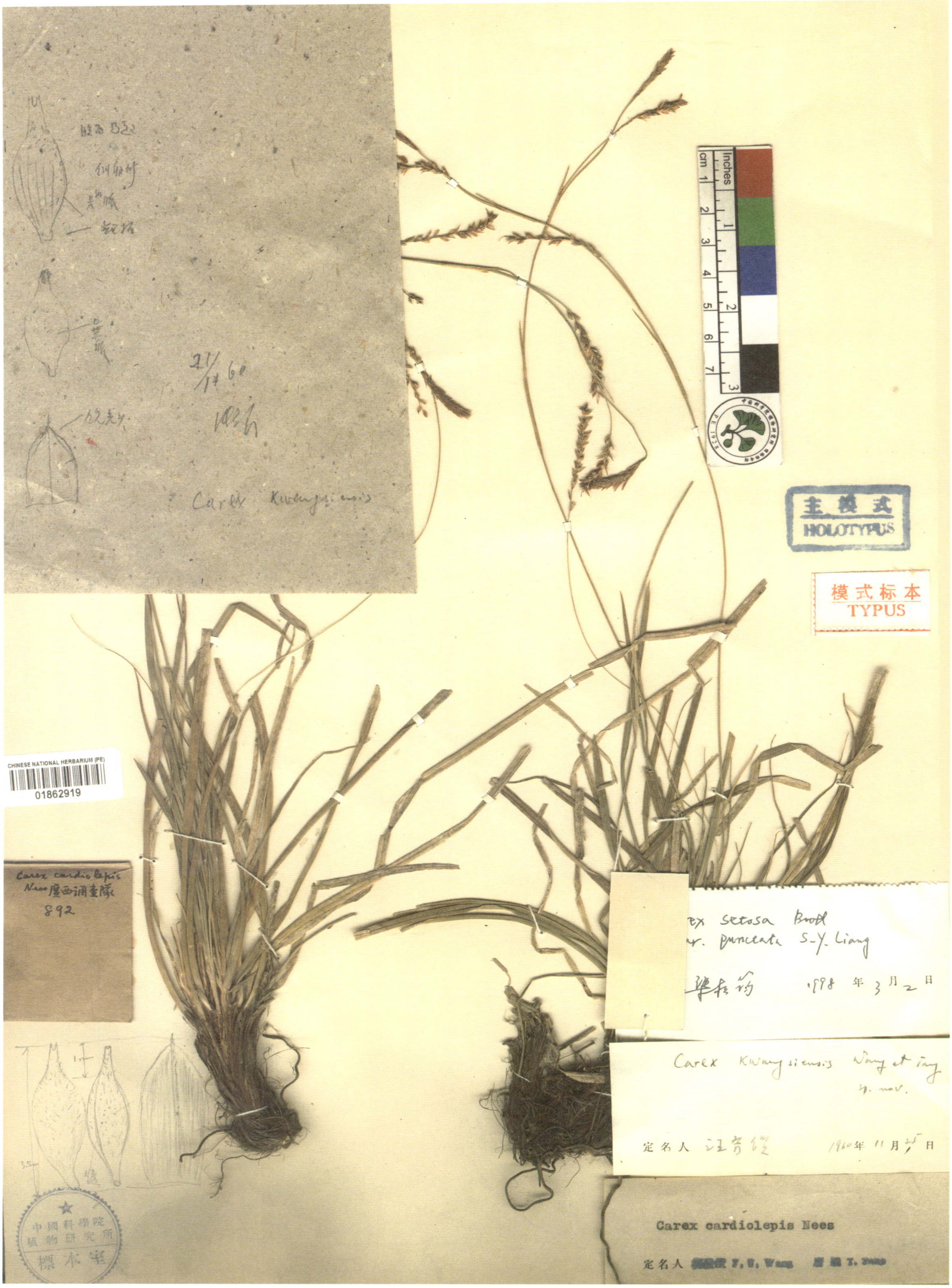

锈点薹草 ***Carex setosa*** Boott var. ***punctata*** S. Y. Liang, Fl. Reip. Pop. Sin. 12: 525. 2000. **Holotype:** China. Guangxi: Xingan, 1953-06-??, Guangxi Exped. 892.

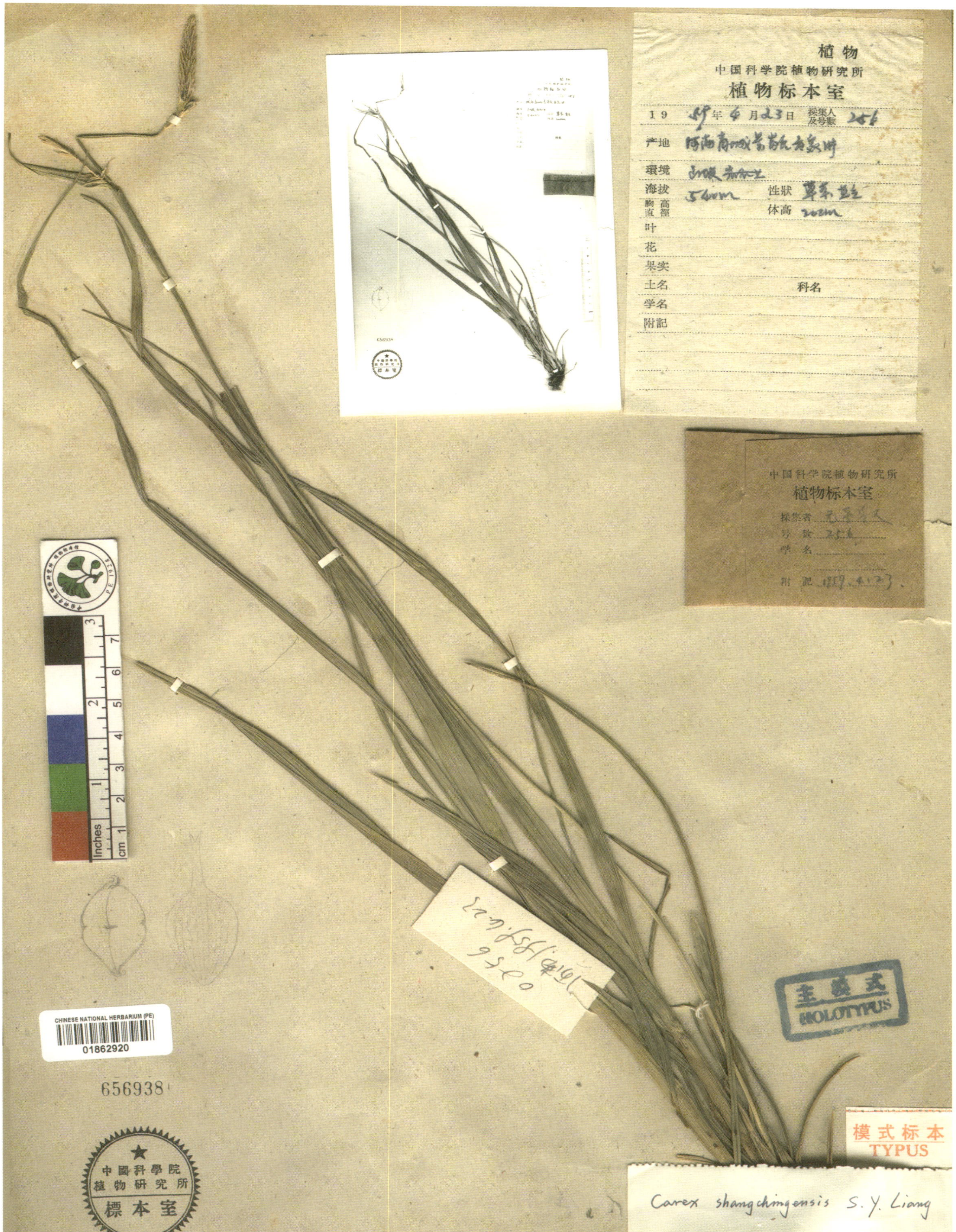

商城薹草 ***Carex shangchengensis*** S. Y. Liang in Acta Phytotax. Sin. 24(3): 242, f. 3. 1986. **Holotype:** China. Henan: Shangcheng, alt. 540 m, 1959-04-23, Anonymous 256.

上海薹草 ***Carex shanghaiensis*** S. X. Qian & Y. Q. Liu in Acta Phytotax. Sin. 37(1): 103, f. 1. 1999. **Holotype:** China. Shanghai: Precise locality not known, 1995-04-27, Y. Q. Liu & S. X. Qian 63.

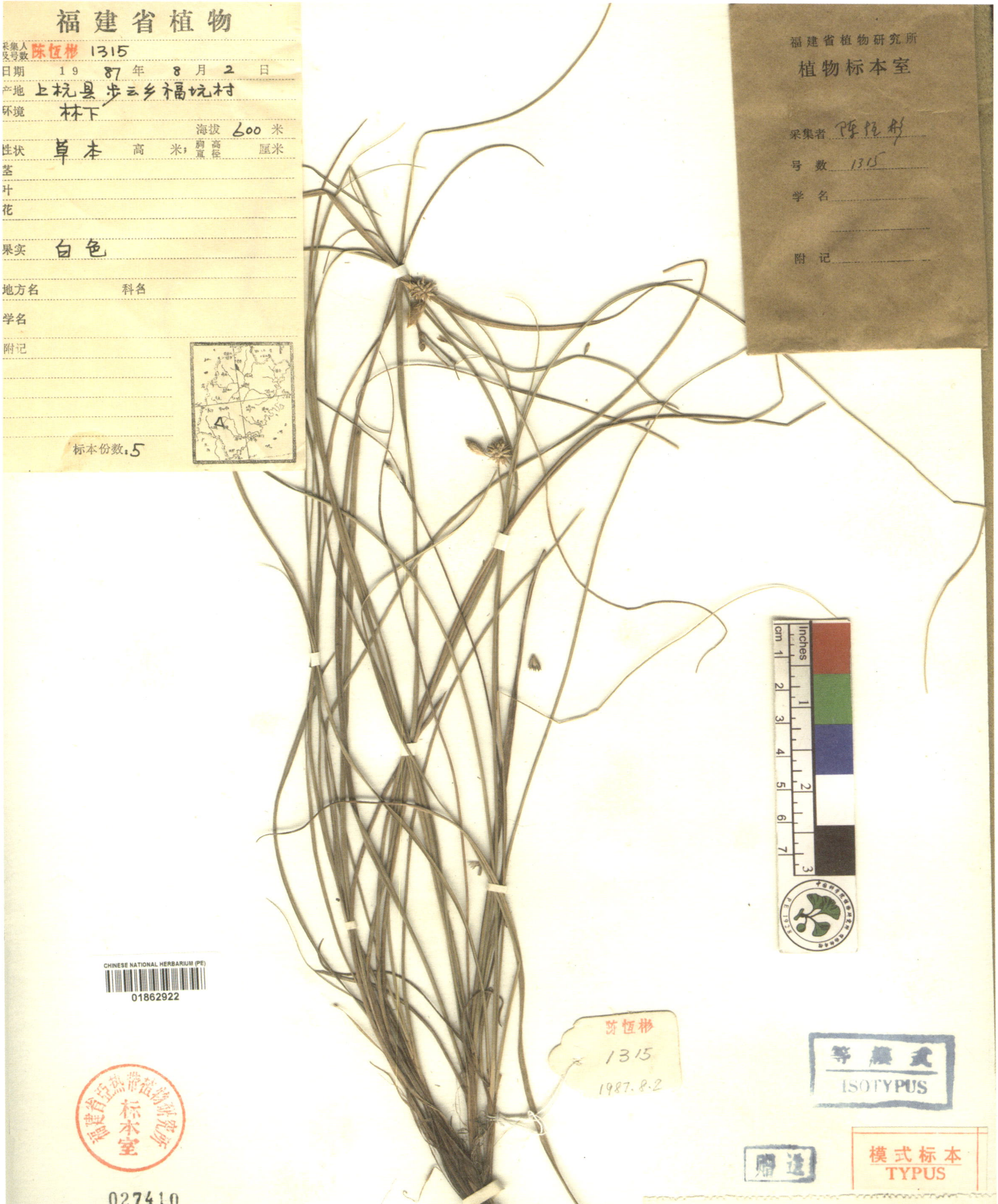

上杭薹草 ***Carex shanghangensis*** S. Y. Liang in Acta Phytotax. Sin. 34(1): 94, f. 2. 1996. **Isotype:** China. Fujian: Shanghang, alt. 600 m, 1987-08-02, H. B. Chen 1315.

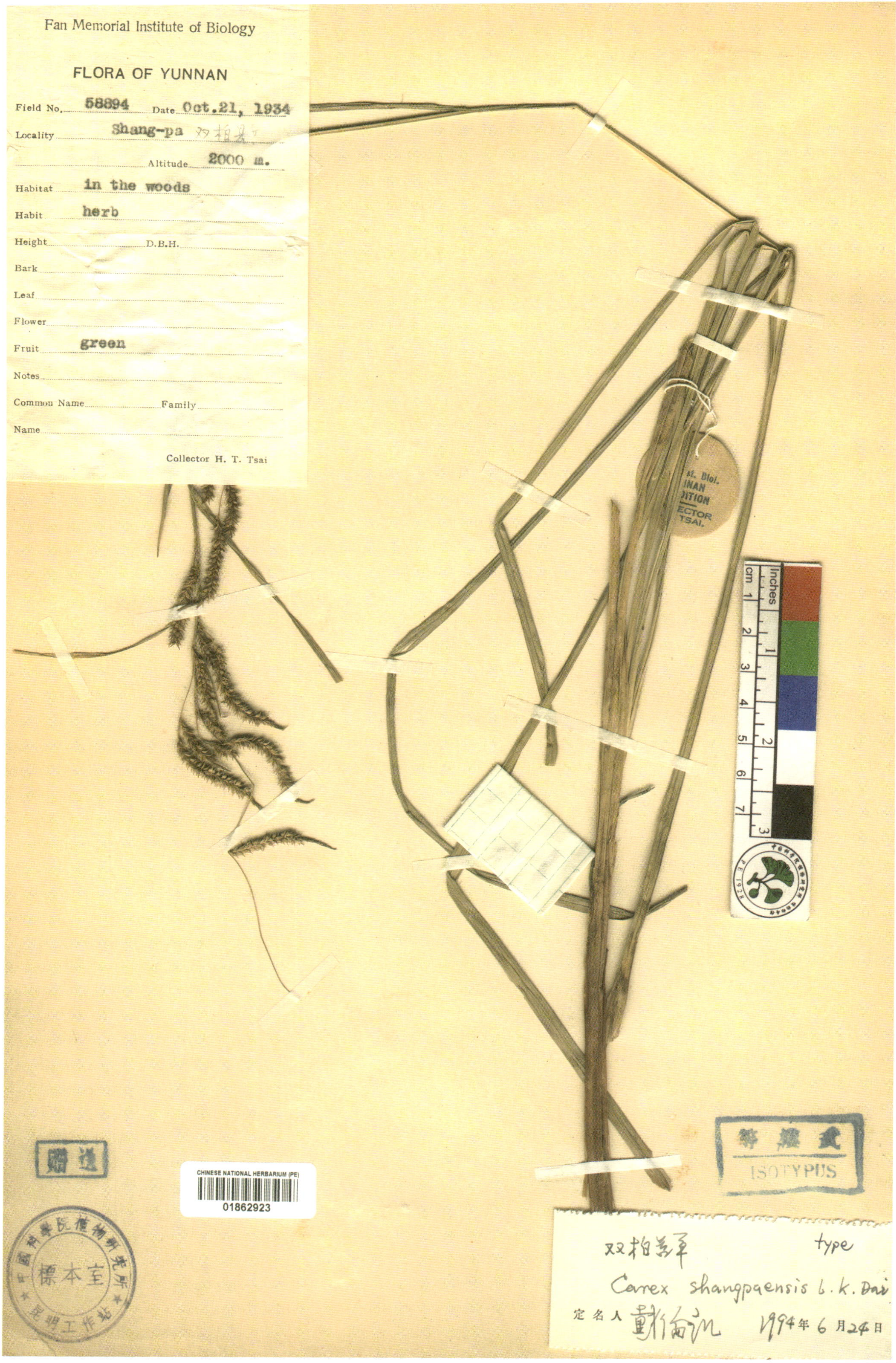

双柏薹草 ***Carex shuangbaiensis*** L. K. Dai in Acta Phytotax. Sin. 37(2): 187, f. 4: 5-8. 1999. **Isotype:** China. Yunnan: Shuangbai, alt. 2000 m, 1934-10-21, H. T. Tsai 58894.

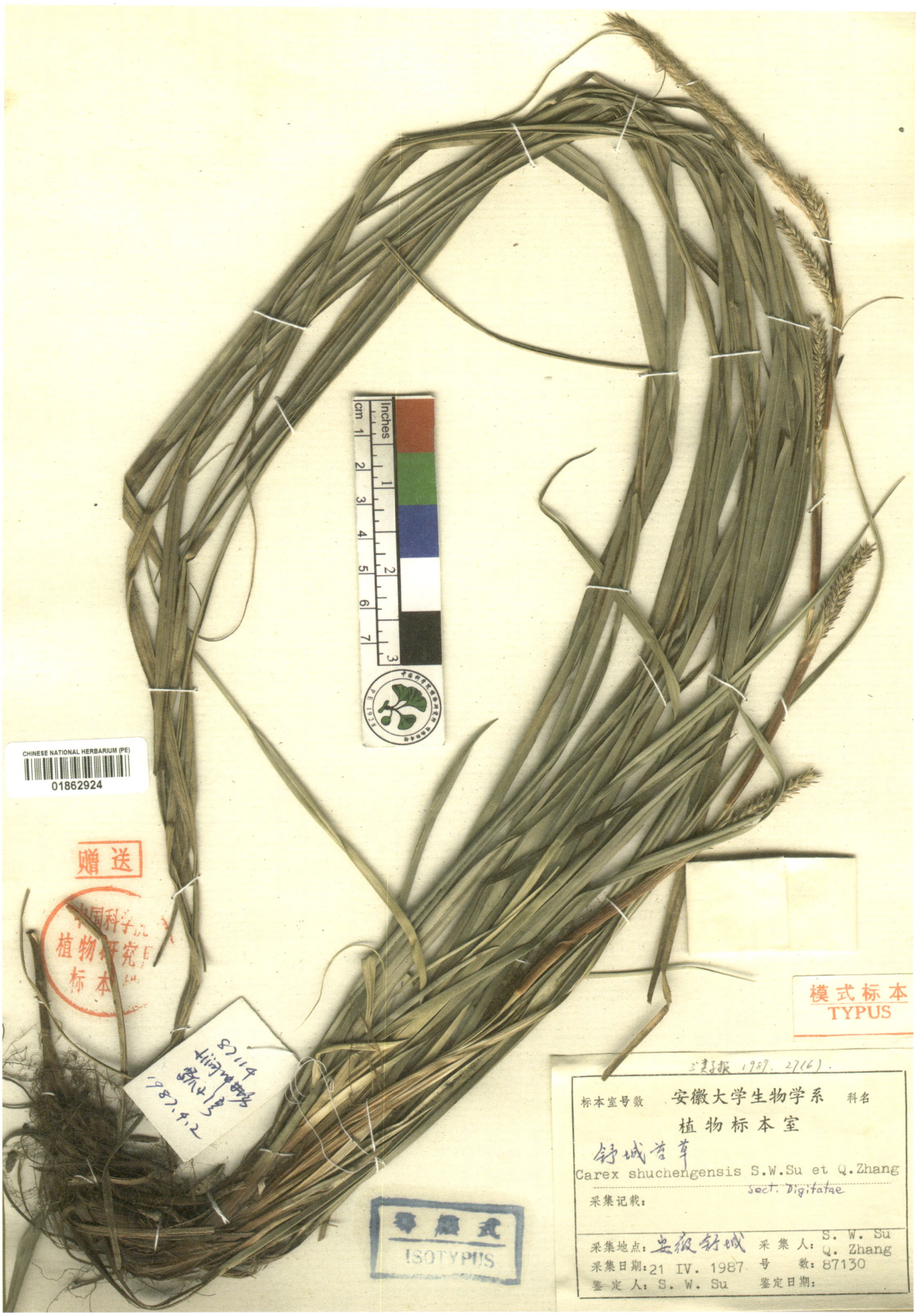

舒城薹草 ***Carex shuchengensis*** S. W. Su & Q. Zhang in Acta Phytotax. Sin. 27(6): 465, f. 1. 1989. **Isotype:** China. Anhui: Shucheng, 1987-04-21, S. W. Su & Q. Zhang 87130.

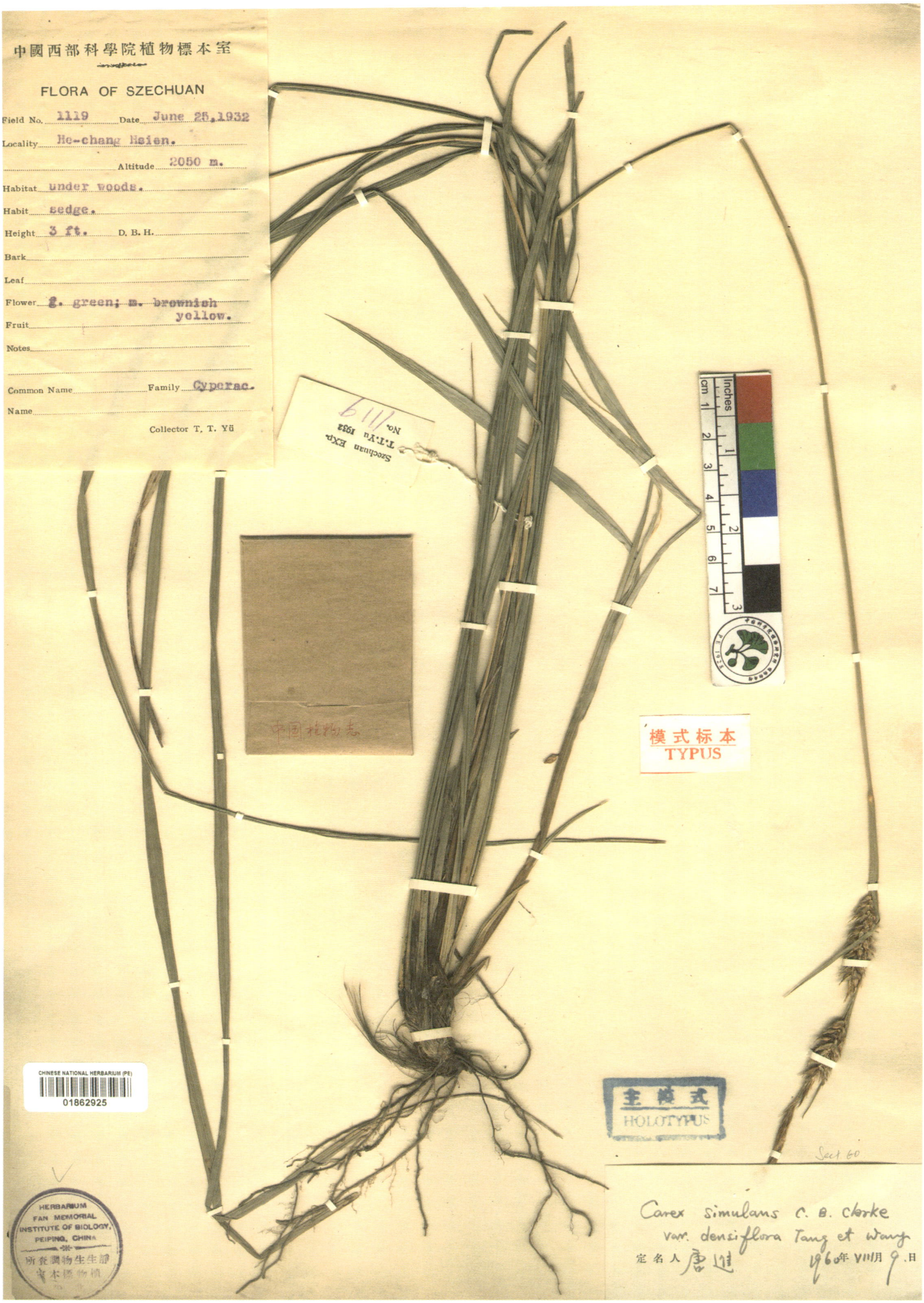

密花相仿薹草 ***Carex simulans*** C. B. Clarke var. ***densiflora*** Tang & F. T. Wang ex S. Y. Liang in Acta Phytotax. Sin. 36(6): 537. 1998. **Holotype:** China. Sichuan: Xichang, alt. 2050 m, 1932-06-25, T. T. Yu 1119.

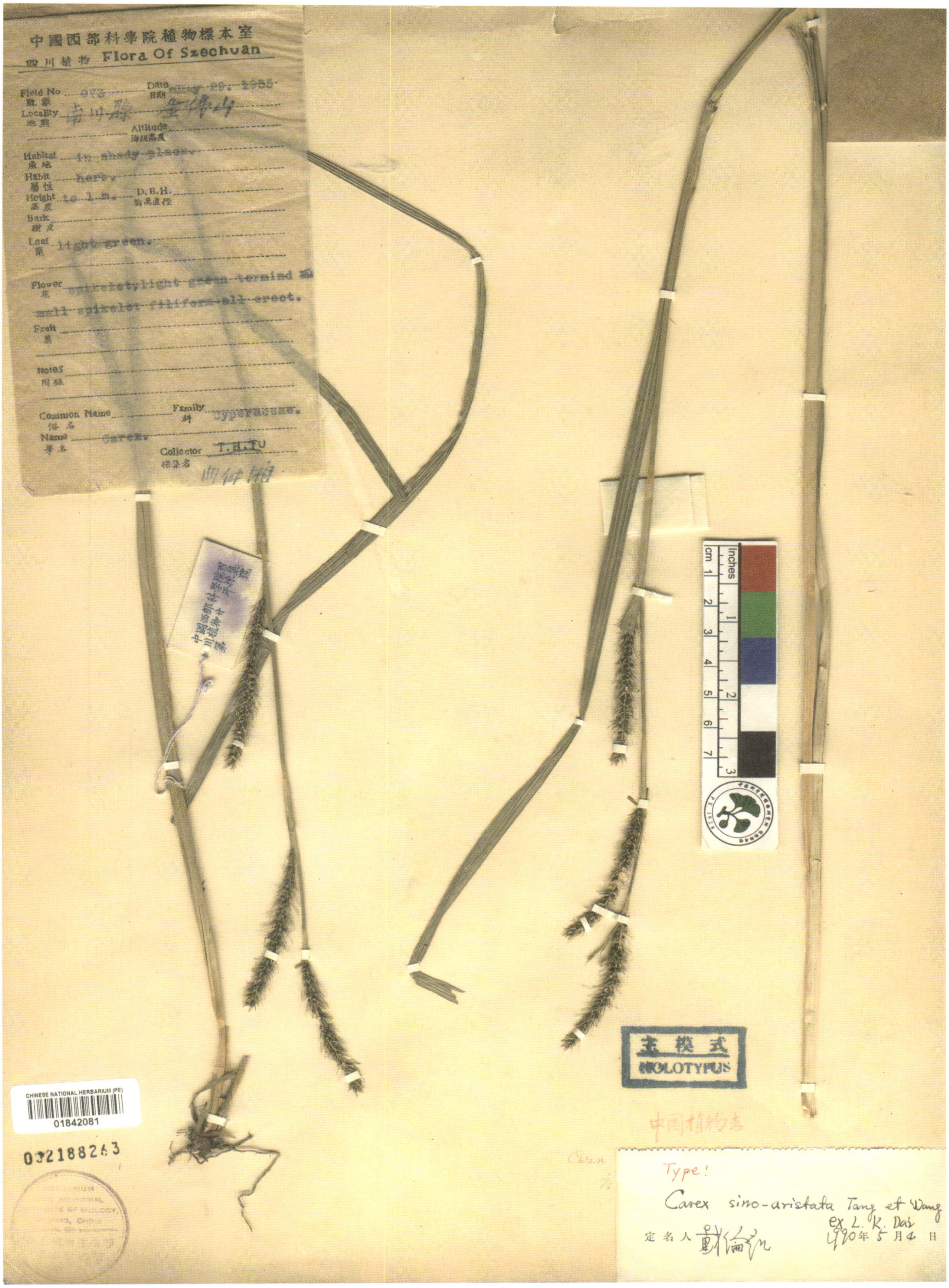

芒鳞苔草 ***Carex sino-aristata*** Tang & F. T. Wang ex L. K. Dai in Acta Phytotax. Sin. 32(2): 184, f. 6: 9-13. 1994.

Holotype: China. Chongqing: Nanchuan, 1935-05-29, C. H. Chu (=K. L. Chu) 973.

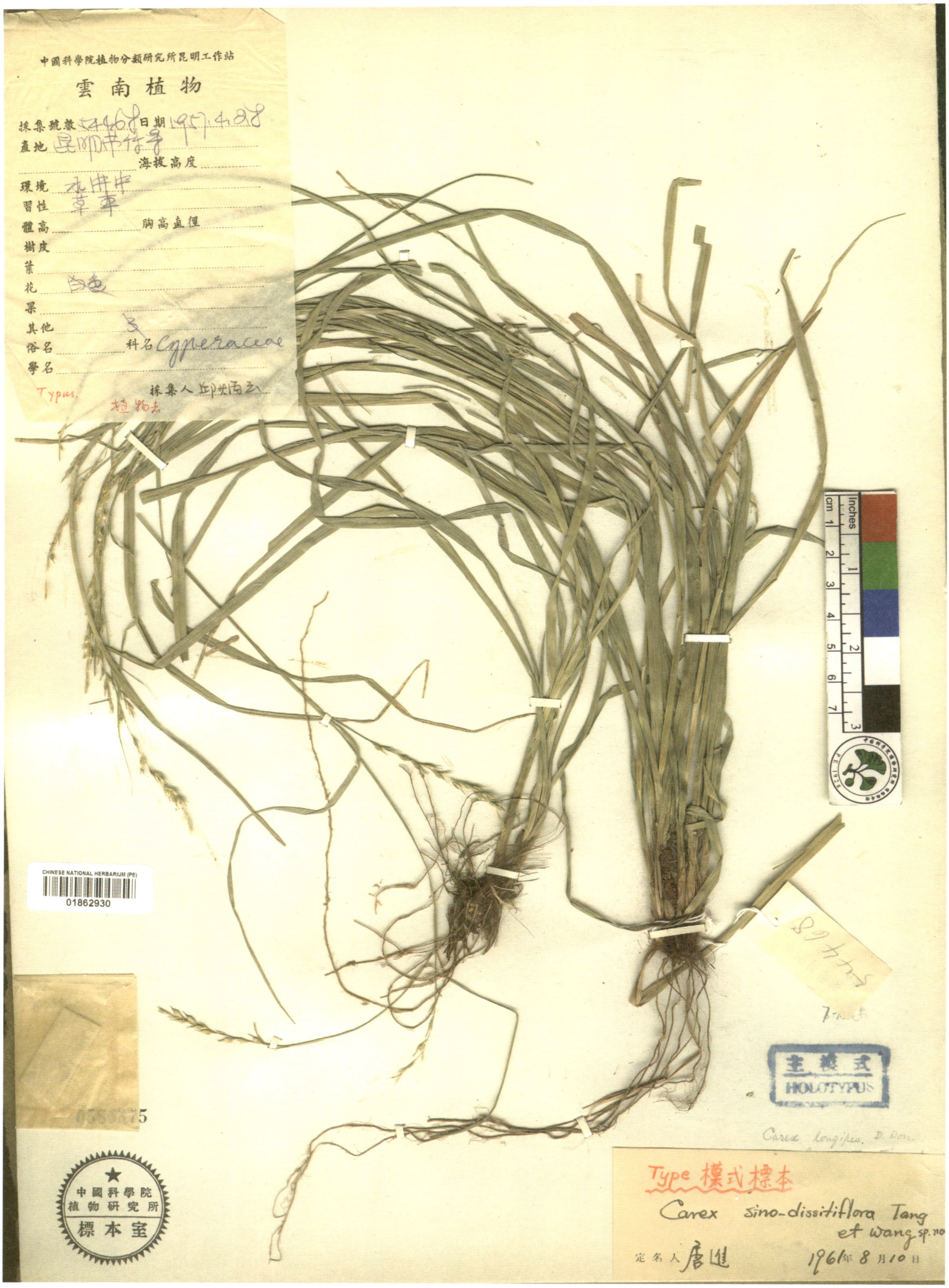

远穗薹草 ***Carex sinodissitiflora*** Tang & F. T. Wang ex L. K. Dai in Acta Phytotax. Sin. 32(2): 177, f. 3: 1-4. 1994.
Holotype: China. Yunnan: Kunming, 1957-04-28, B. Y. Qiu 54468.

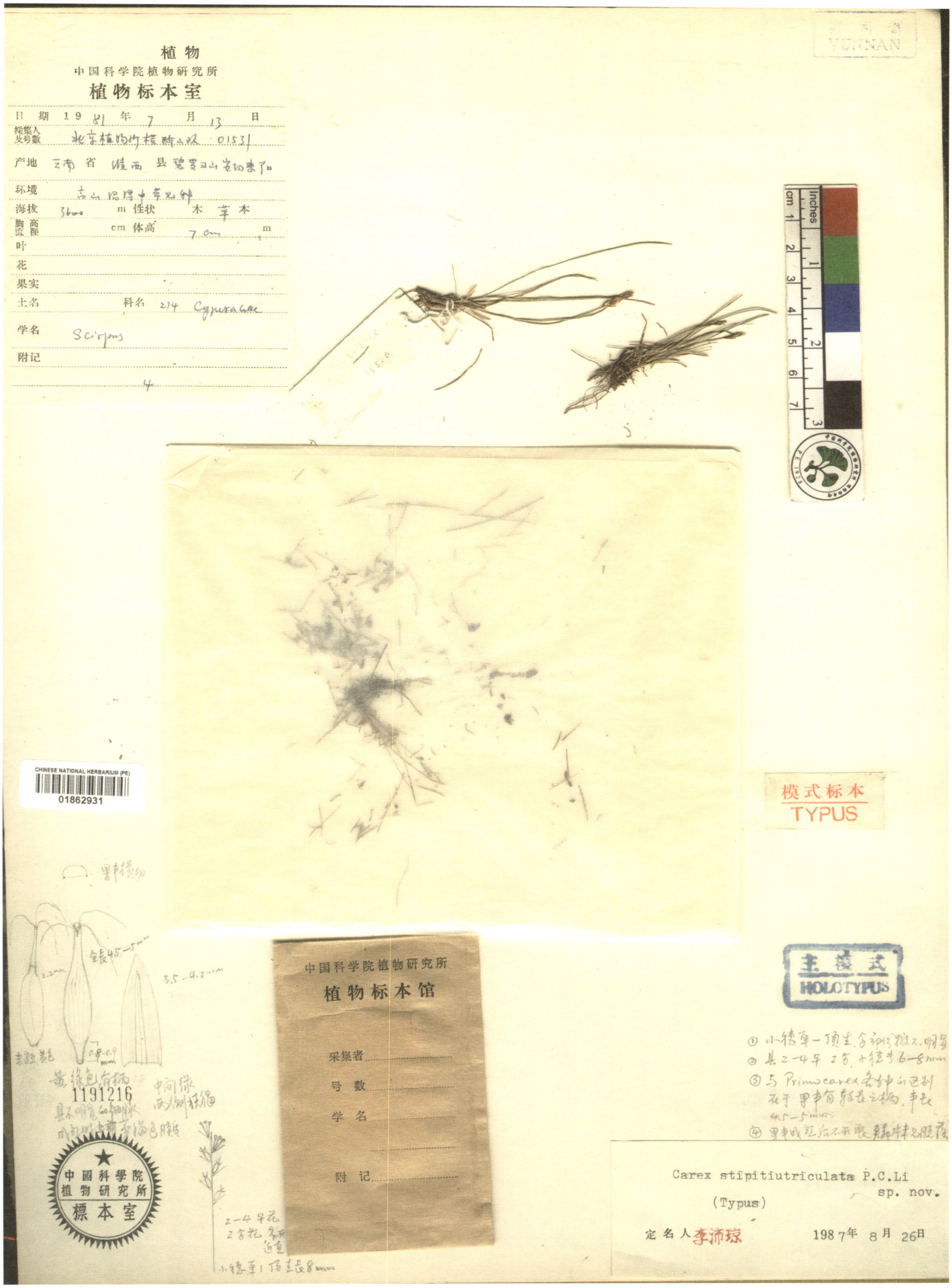

柄囊薹草 ***Carex stipitiutriculata*** P. C. Li in Acta Bot. Yunnan. 12(2): 137, f. 1. 1990. **Holotype:** China. Yunnan: Weixi, alt. 3600 m, 1981-07-13, Hengduanshan Exped. 1531.

拟柔果薹草 ***Carex submollicula*** Tang & F. T. Wang ex L. K. Dai in Acta Phytotax. Sin. 32(2): 186, f. 6: 5-8. 1994.
Holotype: China. Guangdong: Boluo, Luofushan, 1936-05-(24-26), Metcalf & Lau Fun 18218.

似矮生薹草 ***Carex subpumila*** Tang & F. T. Wang ex L. K. Dai in Acta Phytotax. Sin. 37(2): 182, f. 2: 14-17. 1999.

Holotype: China. Hebei: Beidaihe, 1931-06-30, K. S. Ho 1619.

太湖薹草 ***Carex taihuensis*** S. W. Su & S. M. Xu in Bull. Bot. Res., Harbin 9(1): 51, f. 6. 1989. **Isotype:** China. Anhui: Taihu, 1983-05-02, S. W. Su 83030.

太白山薹草 ***Carex taipaishanica*** K. T. Fu, Fl. Tsinling. 1(1): 271, 449. 1976. **Isotype:** China. Shaanxi: Taibai, Taibaishan, alt. 1200 m, 1937-04-16, T. P. Wang 6474.

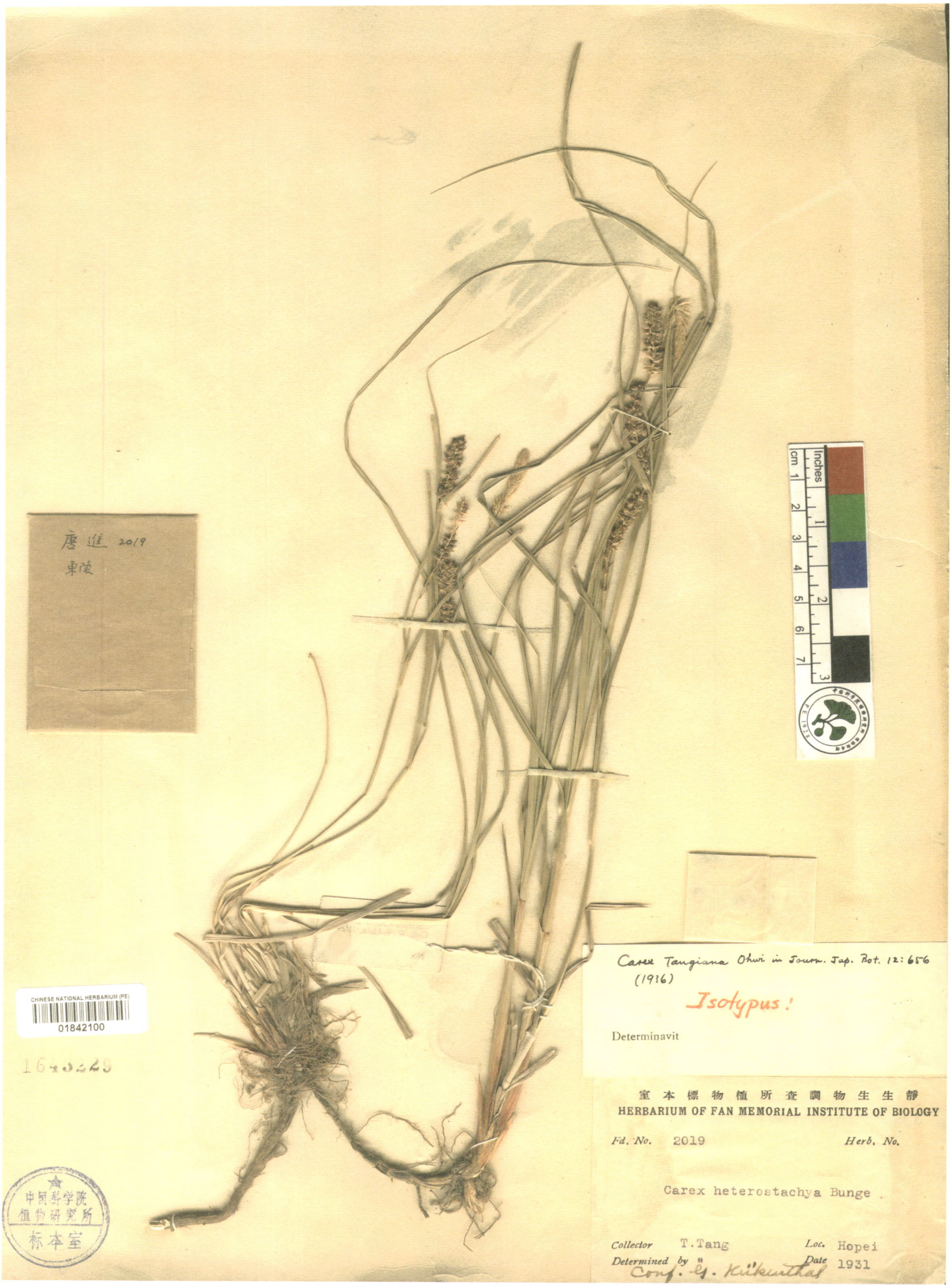

唐进薹草 ***Carex tangiana*** Ohwi in Journ. Jap. Bot. 12(1-12): 656. 1936. **Isotype:** China. Hebei: Precise locality not known, T. Tang 2019.

河北薹草 ***Carex tangii*** Kükenth. in Bull. Fan Mem. Inst. Biol. 3: 144. 1932. **Holotype:** China. Hebei: Dongling, alt. 1600 m, 1931-06-23, T. Tang 2068.

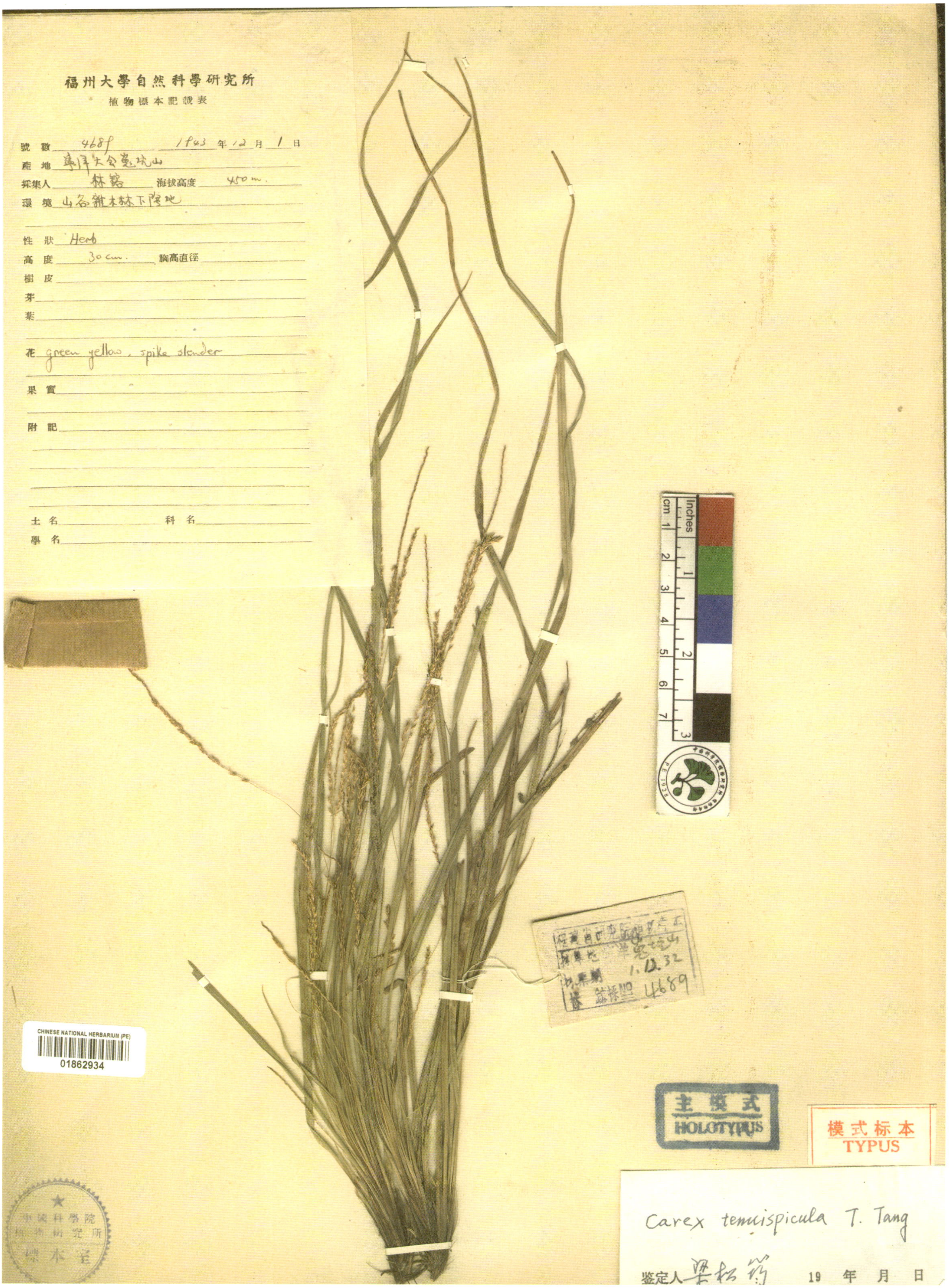

细穗薹草 ***Carex tenuispicula*** Tang ex S. Y. Liang in Acta Phytotax. Sin. 34(1): 92, f. 1. 1996. **Holotype:** China. Fujian: Ningyang, alt. 450 m, 1943-12-01, Y. Ling 4689.

少花藏薹草 ***Carex thibetica*** Franch. var. ***pauciflora*** Tang & F. T. Wang ex S. Y. Liang , Fl. Reip. Pop. Sin. 12: 323, 526. 2000. **Holotype:** China. Yunnan: Yongshan, alt. 2200 m, 1931-06-??, H. T. Tsai 50969.

绿边薹草 ***Carex viridimarginata*** Kükenth. in Acta Hort. Gothob. 5: 113. 1930. **Isotype:** China. Sichuan: Yuanchu, Shuiwangping, alt. 2500 m, 1924-07-21, H. Smith 6703.

瓦屋薹草 ***Carex wawuensis*** Chü ex S. Y. Liang in Acta Phytotax. Sin. 37(2): 192, f. 1: 11. 1999. **Holotype:** China. Sichuan: Hongya, 1939-07-01, C. W. Yao 3995.

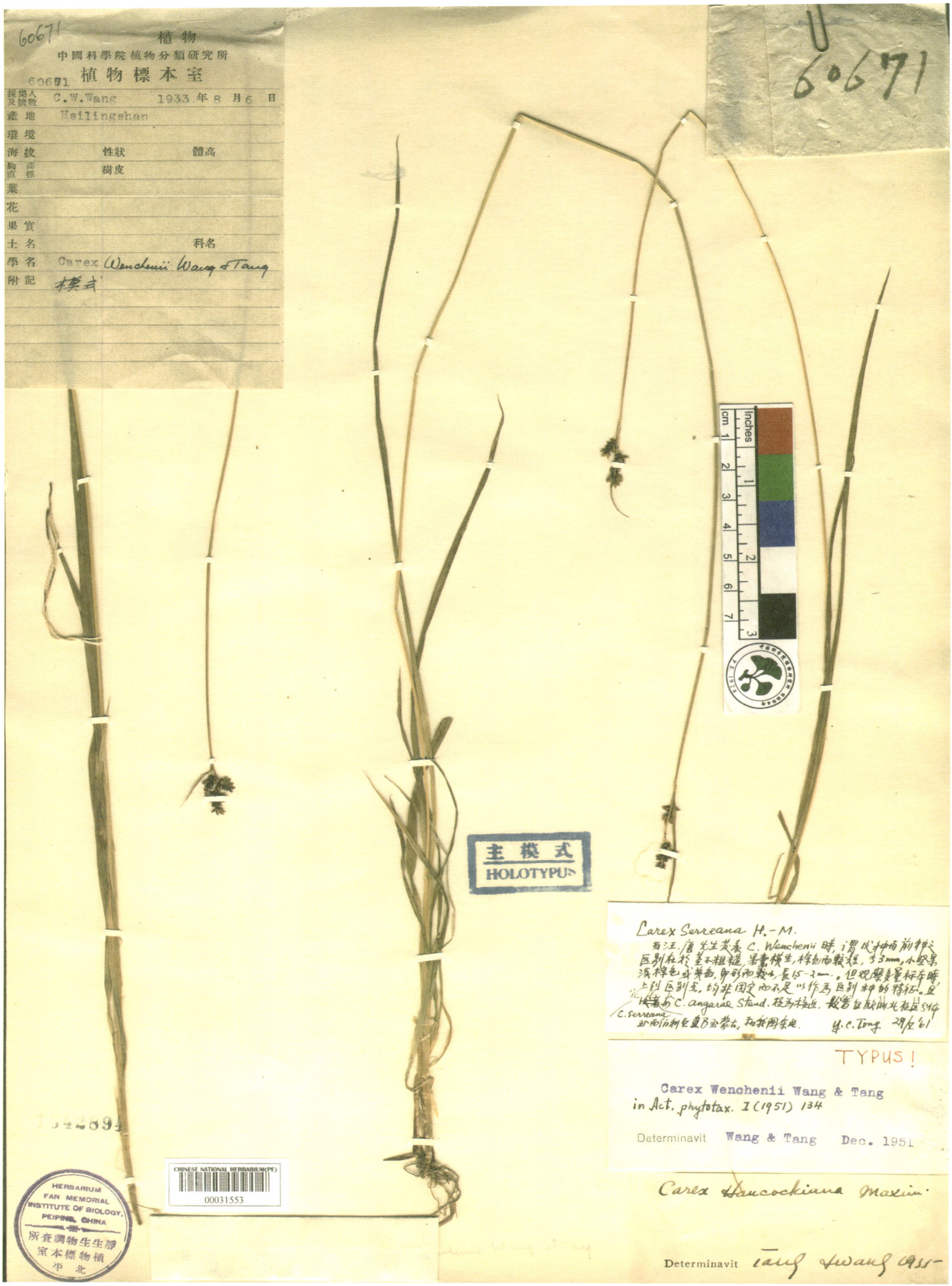

百花山薹草 ***Carex wenchenii*** F. T. Wang & Tang in Acta Phytotax. Sin. 1(1): 134. 1951. **Holotype:** China. Beijing: Baihuashan, 1933-08-06, C. W. Wang 60671.

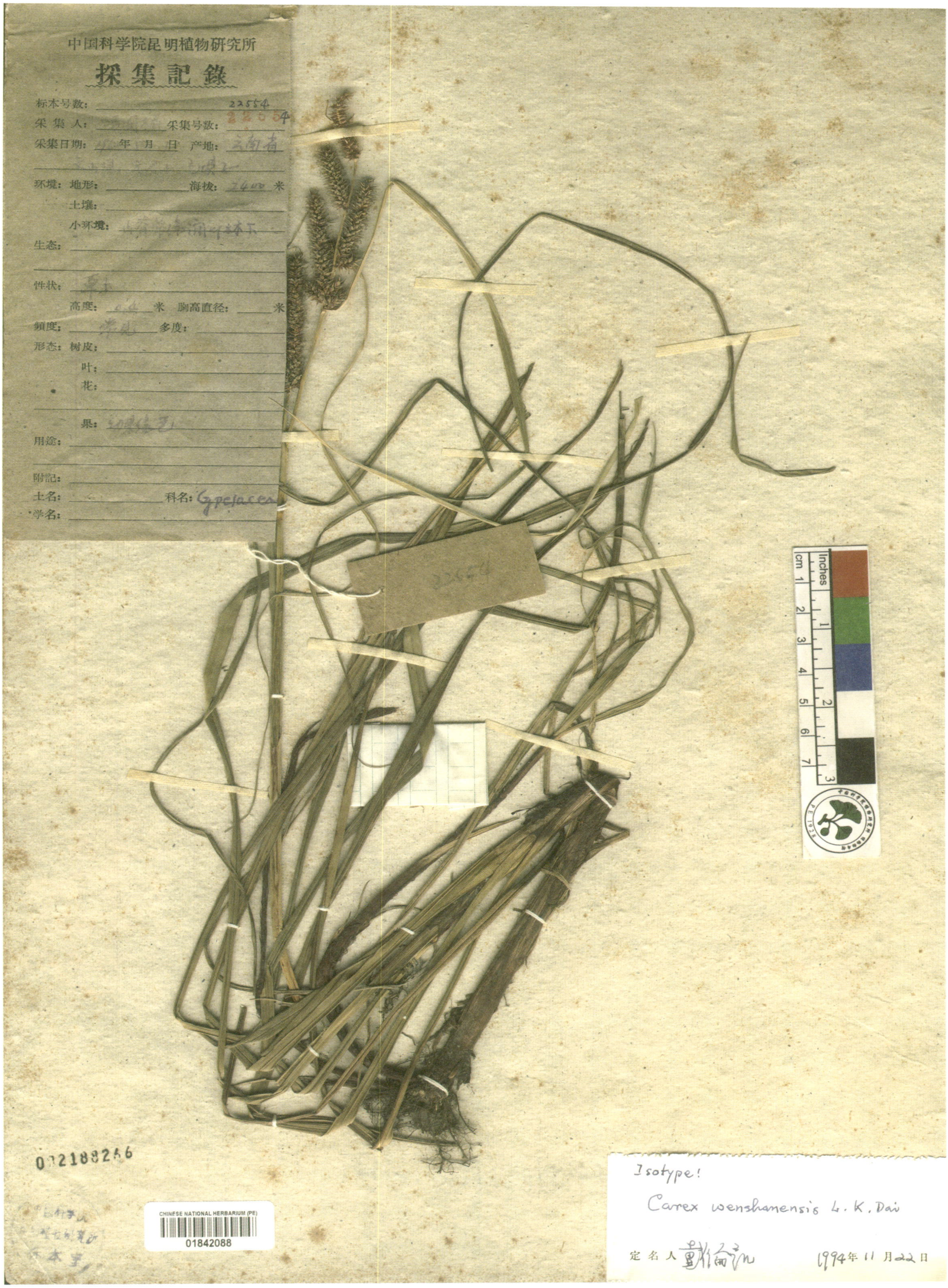

文山薹草 ***Carex wenshanensis*** L. K. Dai in Acta Phytotax. Sin. 37(2): 179, f. 1: 1-4. 1999. **Isotype:** China. Yunnan: Wenshan, alt. 2400 m, 1962-05-17, K. M. Feng 22554.

沙坪薹草 ***Carex wui*** K. L. Chu ex L. K. Dai in Acta Phytotax. Sin. 37(2): 179, f. 2: 1-5. 1999. **Holotype:** China. Sichuan: Ebian, 1938-07-22, C. L. Wu 10443.

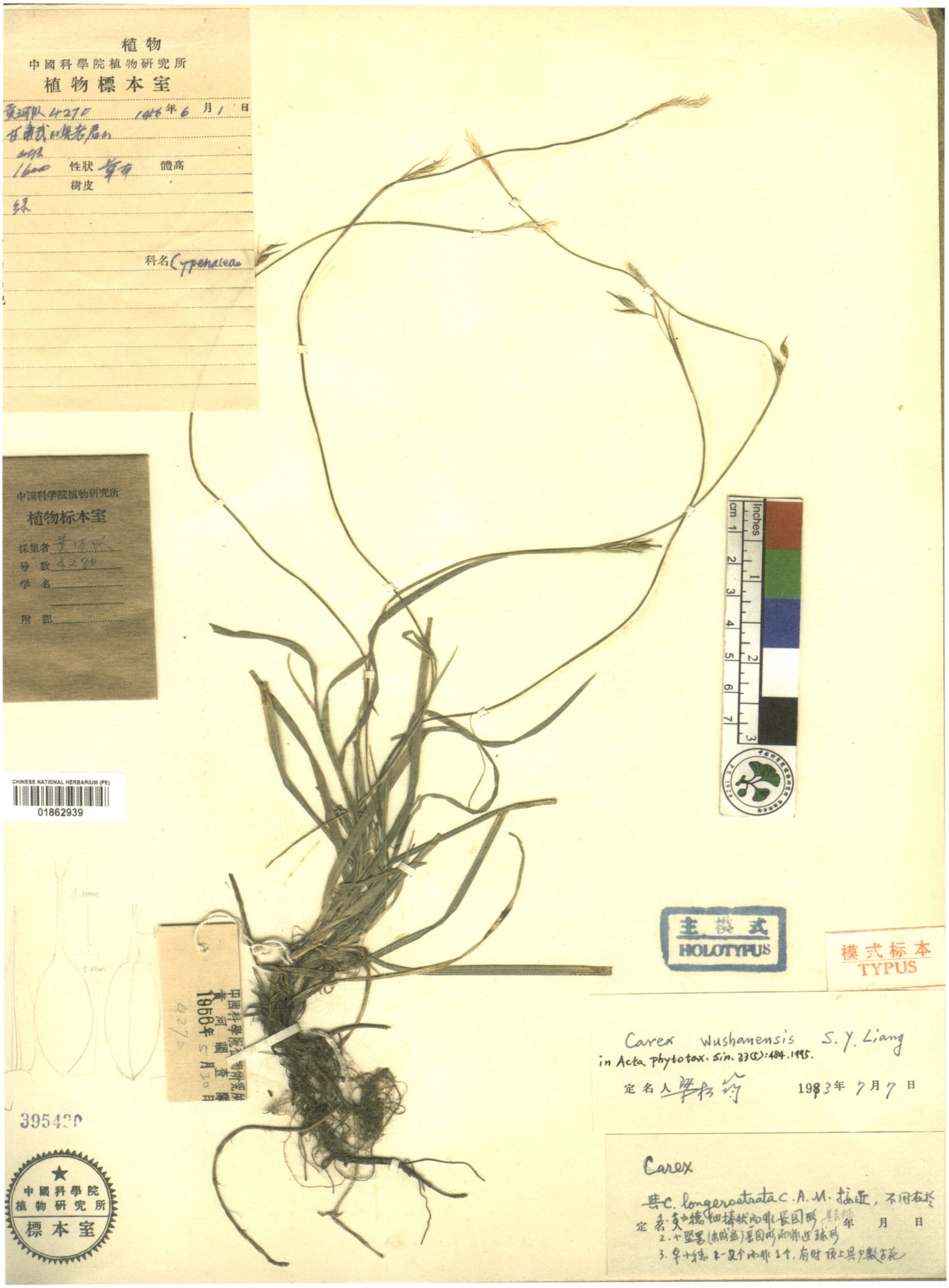

武山薹草 ***Carex wushanensis*** S. Y. Liang in Acta Phytotax. Sin. 33(5): 484, f. 1: 1-5. 1995. **Holotype:** China. Gansu: Wushan, alt. 1600 m, 1956-06-01, Yellow River Exped. 4270.

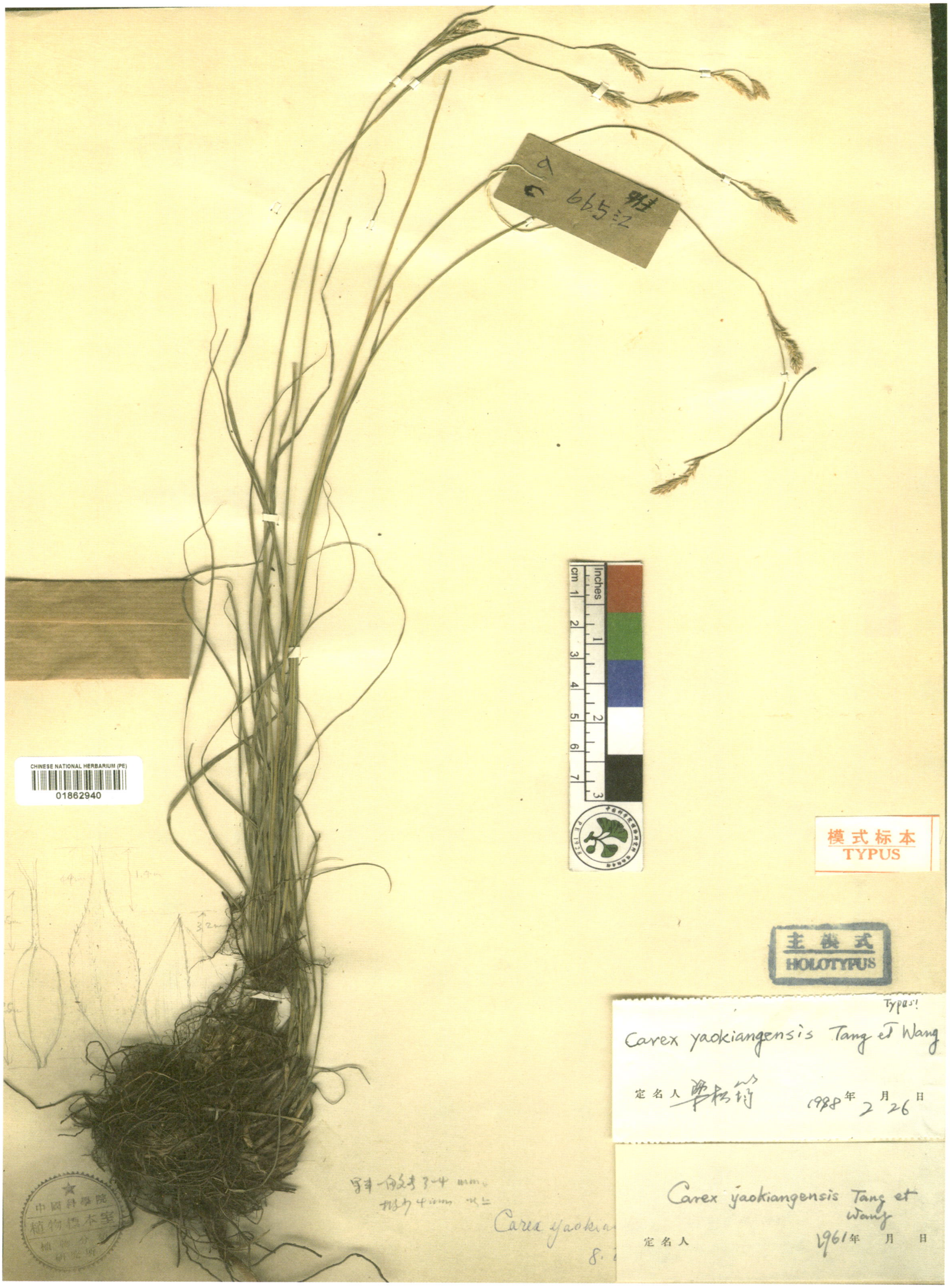

雅江薹草 ***Carex yajiangensis*** Tang & F. T. Wang ex S. Y. Liang, Fl. Reip. Pop. Sin. 12: 240, 524, pl. 46: 1-4. 2000.
Holotype: China. Sichuan: Yajiang, Anonymous 665.

阳朔薹草 ***Carex yangshuoensis*** Tang & F. T. Wang ex S. Y. Liang in Acta Phytotax. Sin. 33(5): 488, f. 1: 6-11. 1995.

Holotype: China. Guangxi: Yangshuo, Shan 750.

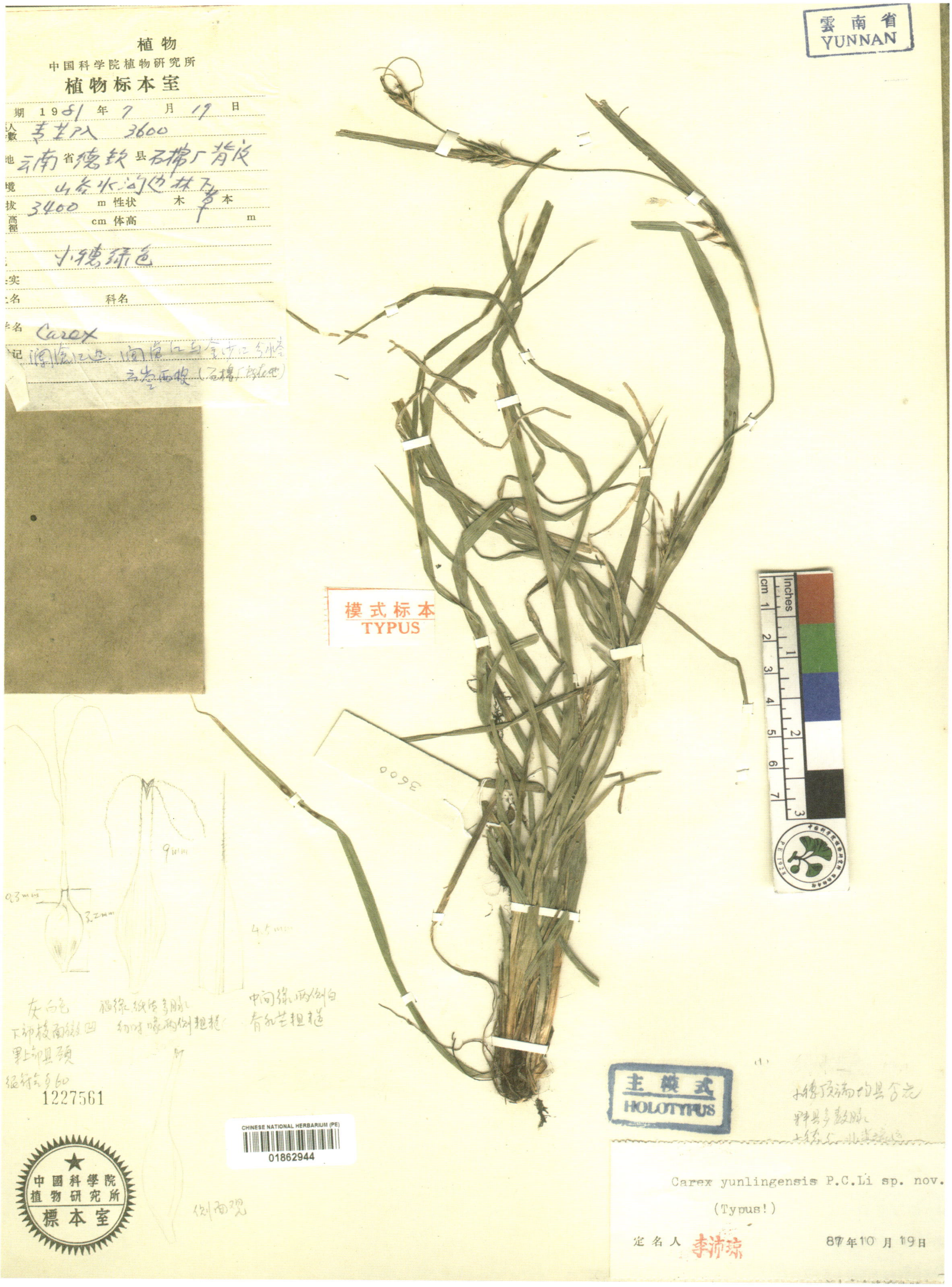

云岭薹草 ***Carex yunlingensis*** P. C. Li in Acta Bot. Yunnan. 12(2): 149, f. 17. 1990. **Holotype:** China. Yunnan: Dêqên, alt. 3400 m, 1981-07-19, Qinghai-Xizang Exped. 3600.

镇康薹草 ***Carex zhenkangensis*** F. T. Wang & Tang ex S. Y. Liang, Fl. Reip. Pop. Sin. 12: 224, 523, pl. 44: 9-12. 2000. **Holotype:** China. Yunnan: Zhenkang, alt. 3500 m, 1938-08-04, T. T. Yu 17191.

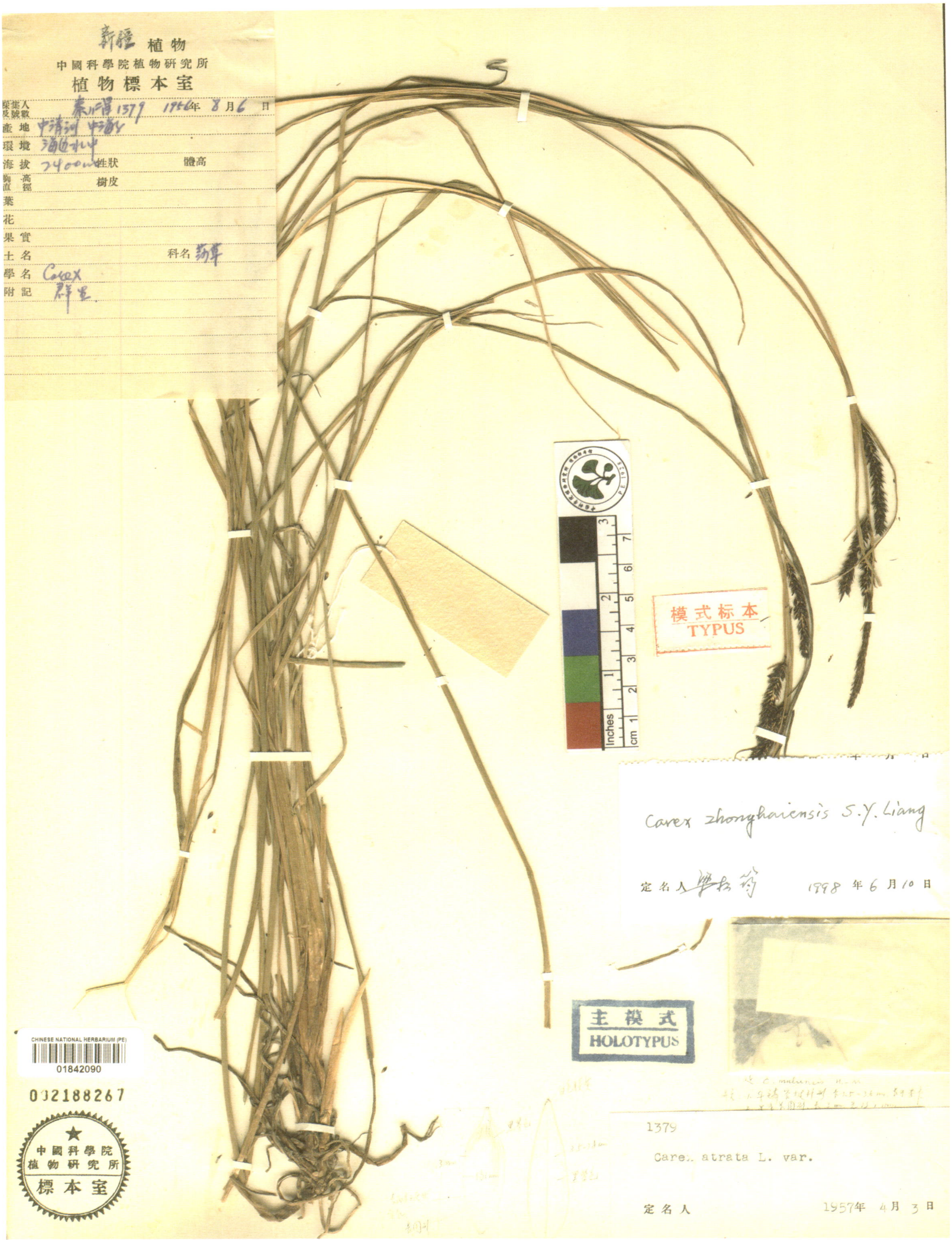

中海薹草 ***Carex zhonghaiensis*** S. Y. Liang, Fl. Reip. Pop. Sin. 12: 403, 526, pl. 84: 1-4. 2000. **Holotype:** China. Xinjiang: Zhongqinghe, alt. 2400 m, 1956-08-06, R. C. Ching 1379.

遵义薹草 ***Carex zunyiensis*** FTang & F. T. Wang ex S. Y. Liang in Acta Phytotax. Sin. 24(3): 241, f. 2. 1986. **Holotype:** China. Guizhou: Zunyi, alt. 1240 m, 1959-04-13, N Guizhou Exped. 263.

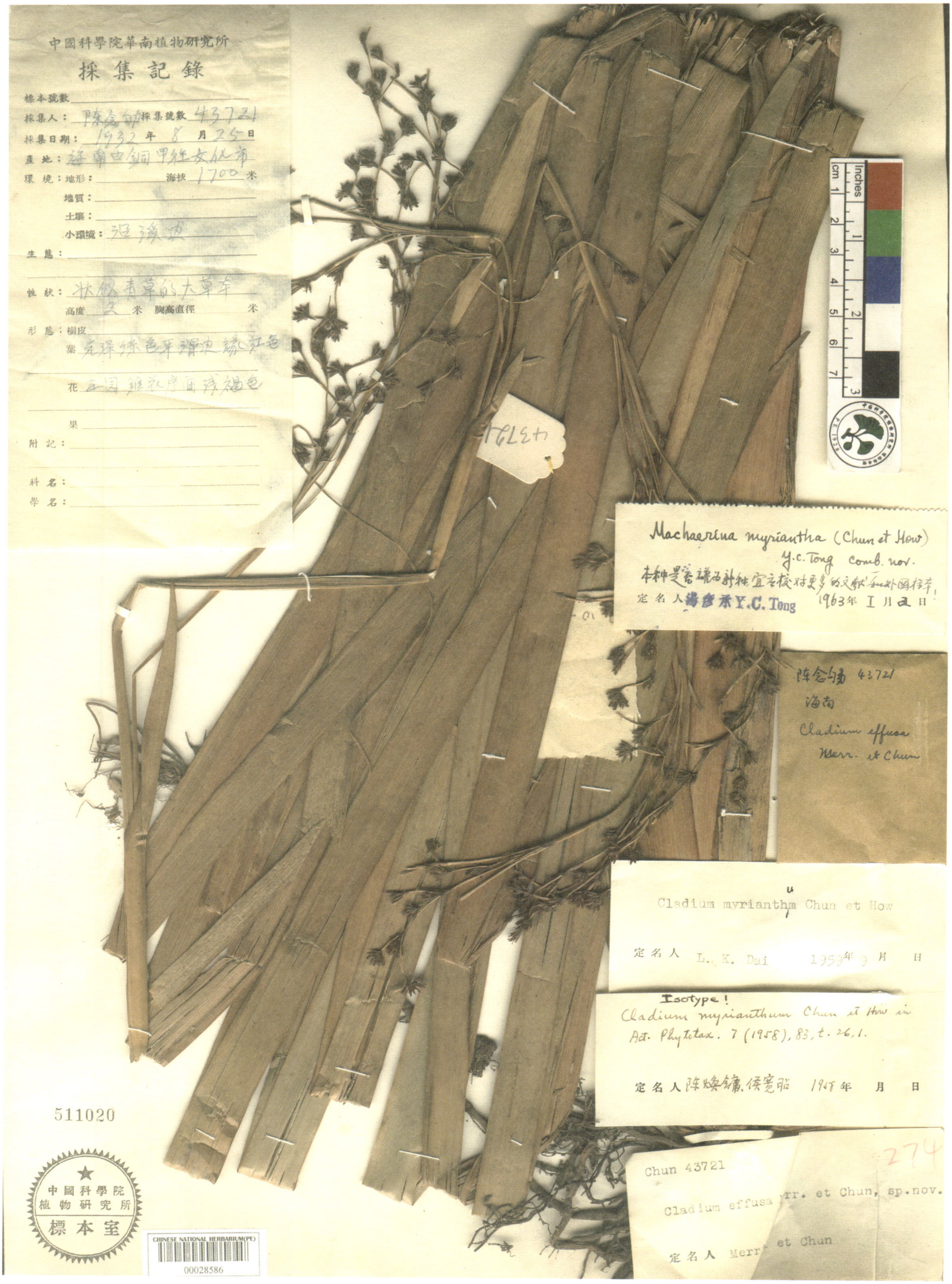

多花一本芒 ***Cladium myrianthum*** Chun & How in Acta Phytotax. Sin. 7(1): 83, pl. 26: 1. 1958. **Isotype:** China. Hainan: Lingshui, alt. 1700 m, 1932-08-25, L. C. Tso & N. K. Chun 43721.

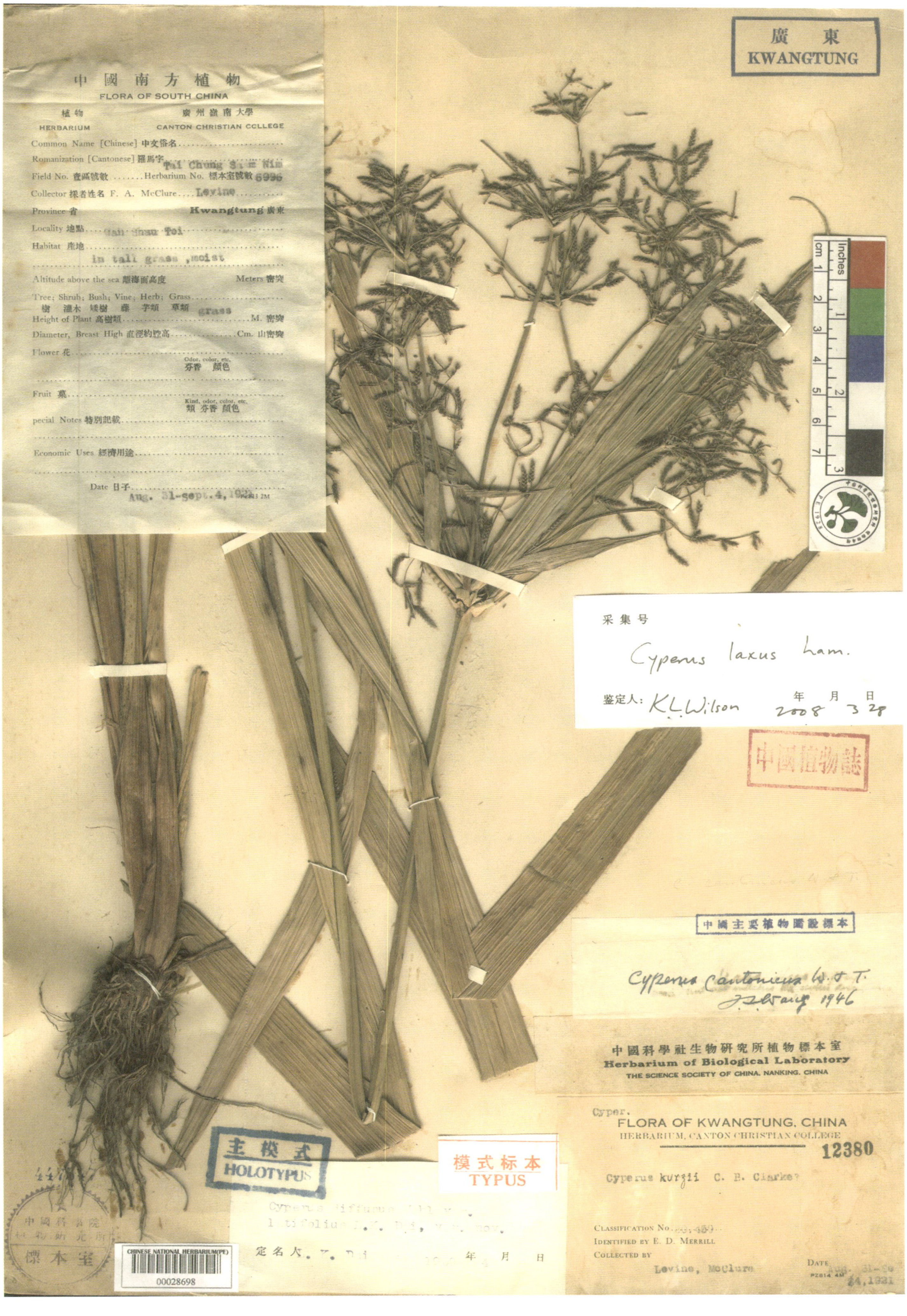

宽叶多脉莎草 ***Cyperus diffusus*** Vahl var. ***latifolius*** L. K. Dai, Fl. Reip. Pop. Sin. 11: 232, pl. 50: 1-7. 1961. **Holotype:** China. Guangdong: Precise locality not known, 1921-08-04, Levine & MaClure 6996.

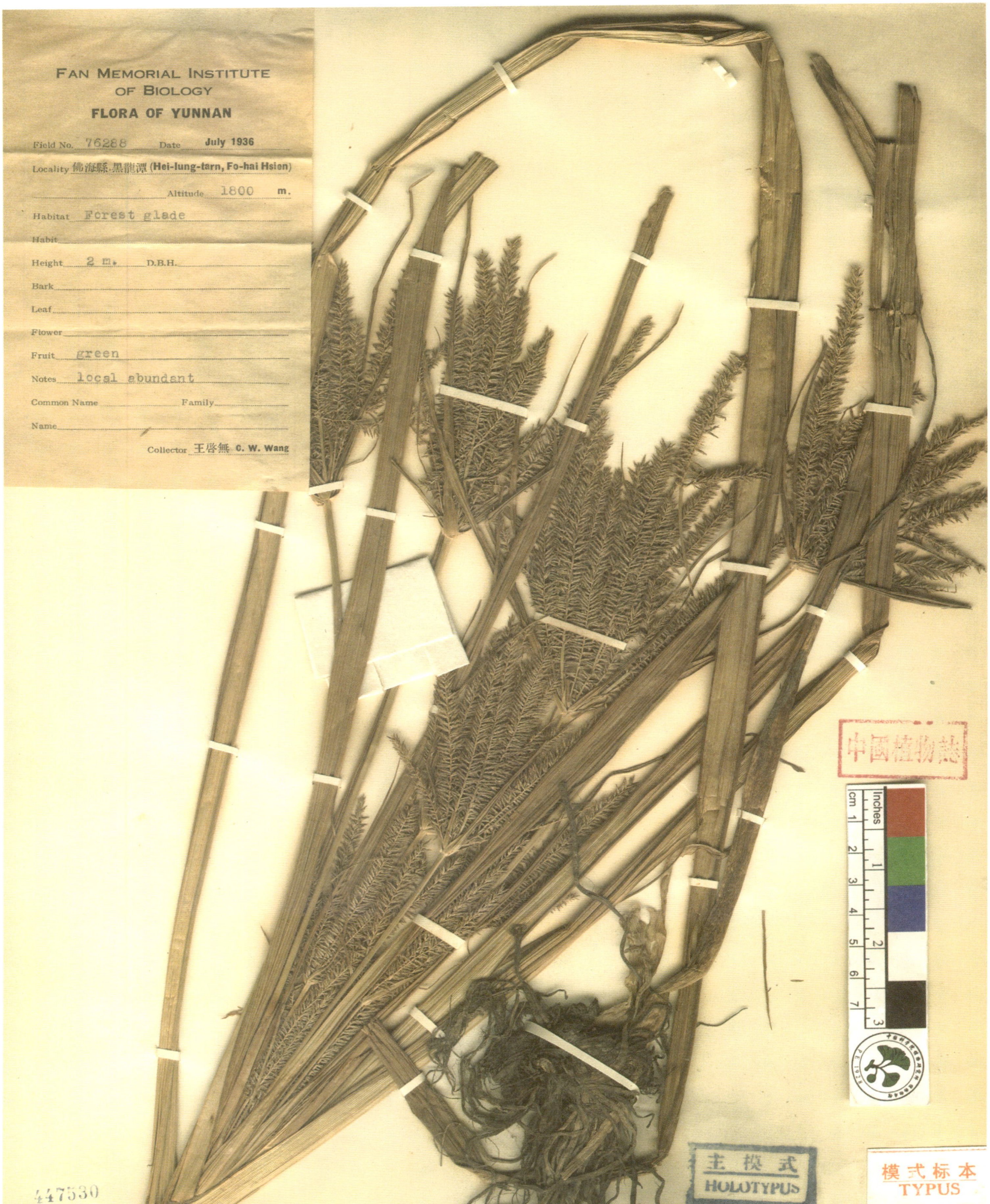

少花穗莎草 ***Cyperus digitatus*** Roxb. var. ***laxiflorus*** L. K. Dai, Fl. Reip. Pop. Sin. 11: 129, 230. 1961. **Holotype:** China. Yunnan: Fohai (=Menghai), alt. 1800 m, 1936-07-??, C. W. Wang 76288.

屏边莎草 ***Cyperus digitatus*** Roxb. var. ***pinbienensis*** L. K. Dai, Fl. Reip. Pop. Sin. 11: 129, 230. 1961. **Holotype:** China. Yunnan: Pingbian, alt. 1500 m, 1939-09-29, C. W. Wang 82189.

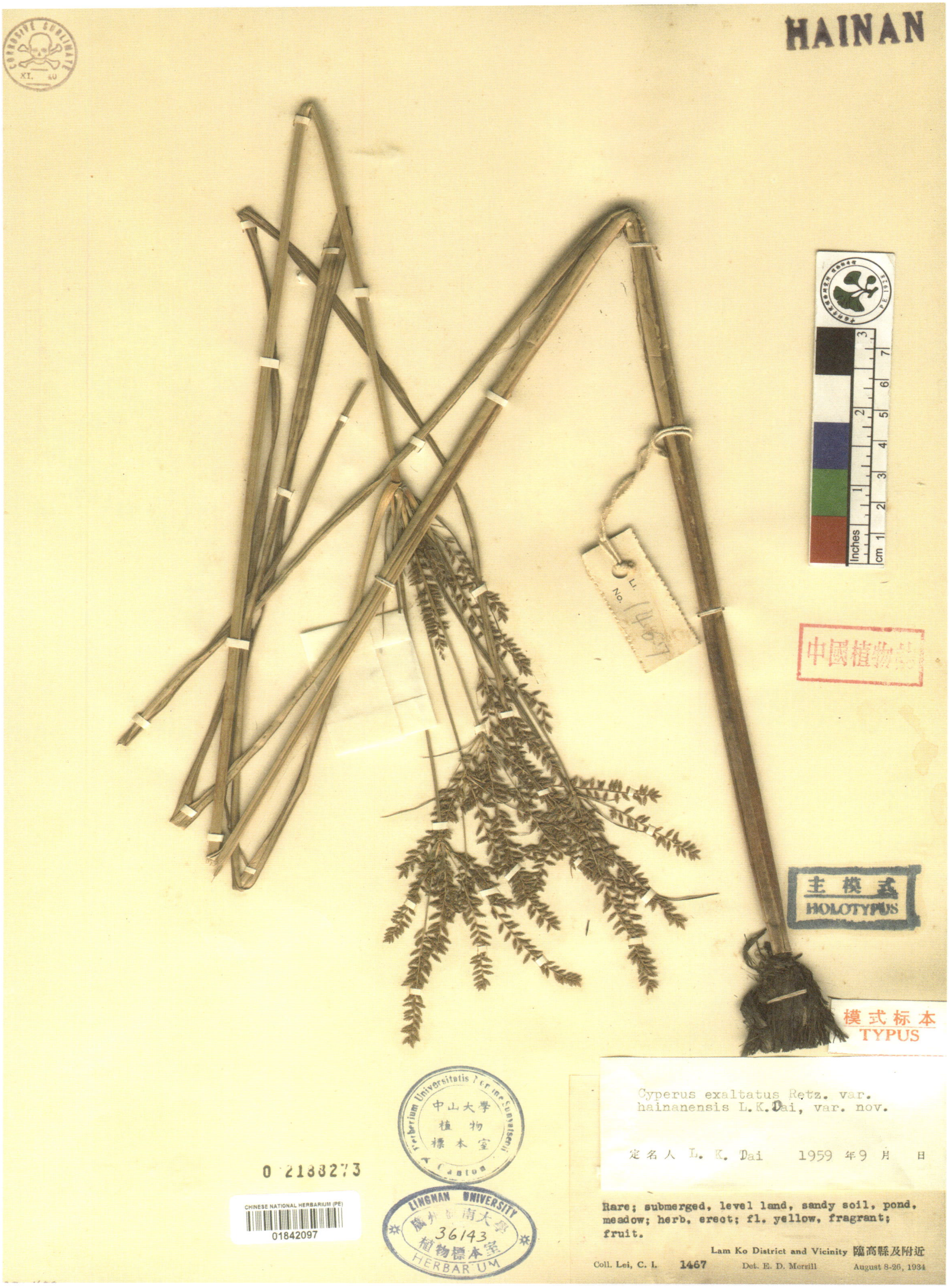

海南高秆莎草 ***Cyperus exaltatus*** Retz. var. ***hainanensis*** L. K. Dai, Fl. Reip. Pop. Sin. 11: 131, 230. 1961. **Holotype:** China. Hainan: Lingao, 1934-08-(08-26), C. L. Lei 1467.

广东高秆莎草 ***Cyperus exaltatus*** Retz. var. ***tenuispicatus*** L. K. Dai, Fl. Reip. Pop. Sin. 11: 131, 230. 1961. **Holotype:** China. Guangdong: Precise locality not known, W. C. Ko 145.

线状穗莎草 ***Cyperus linearispiculatus*** L. K. Dai, Fl. Reip. Pop. Sin. 11: 133, 230, pl. 42: 12-15. 1961. **Holotype:** China. Yunnan: Jinghong, alt. 540 m, 1955-05-07, K. M. Feng 20757.

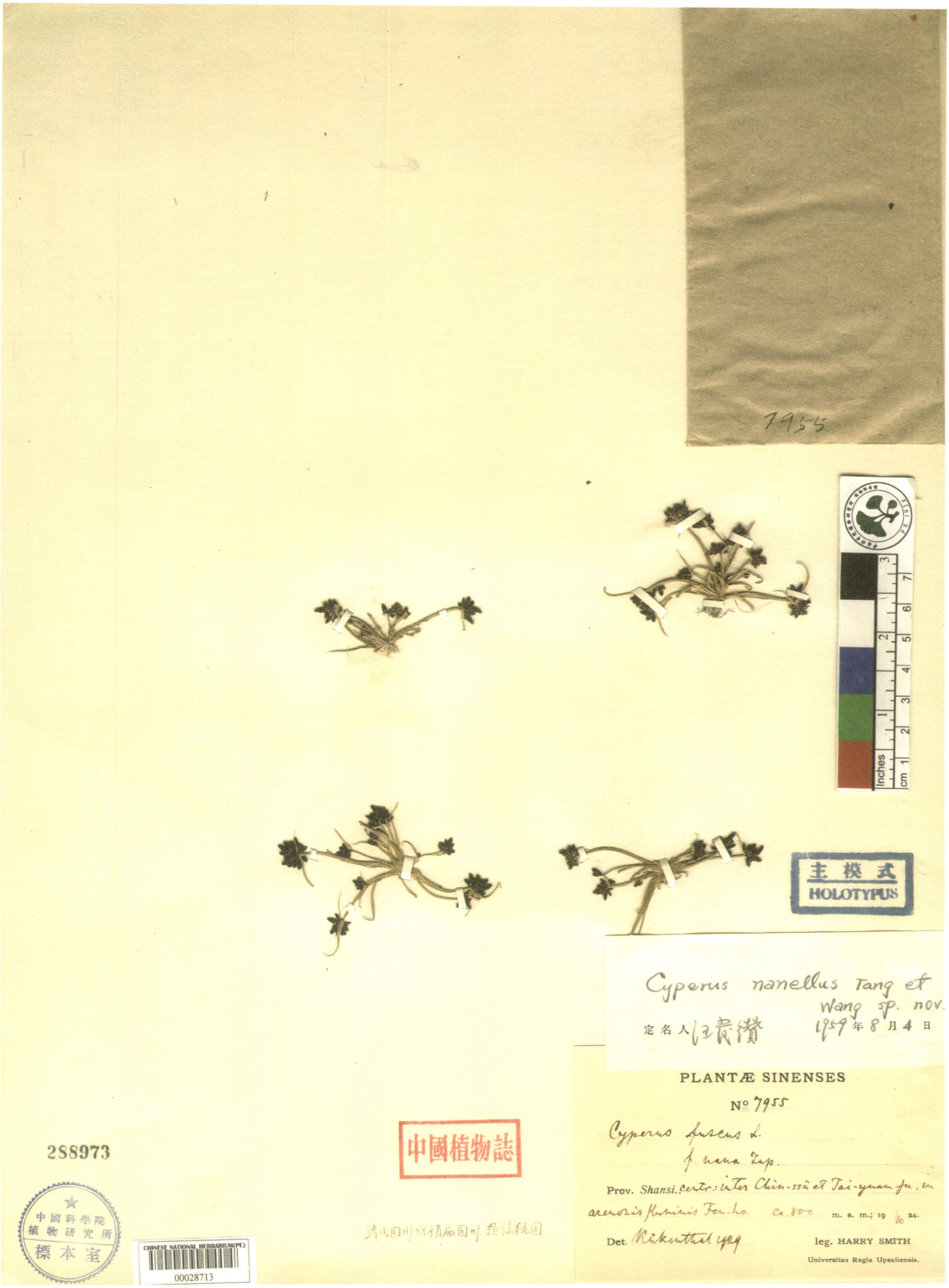

汾河莎草 ***Cyperus nanellus*** Tang & F. T. Wang, Fl. Reip. Pop. Sin. 11: 232, pl. 51: 20-23. 1961. **Holotype:** China. Shanxi: Taiyuan, alt. 800 m, H. Smith 7955.

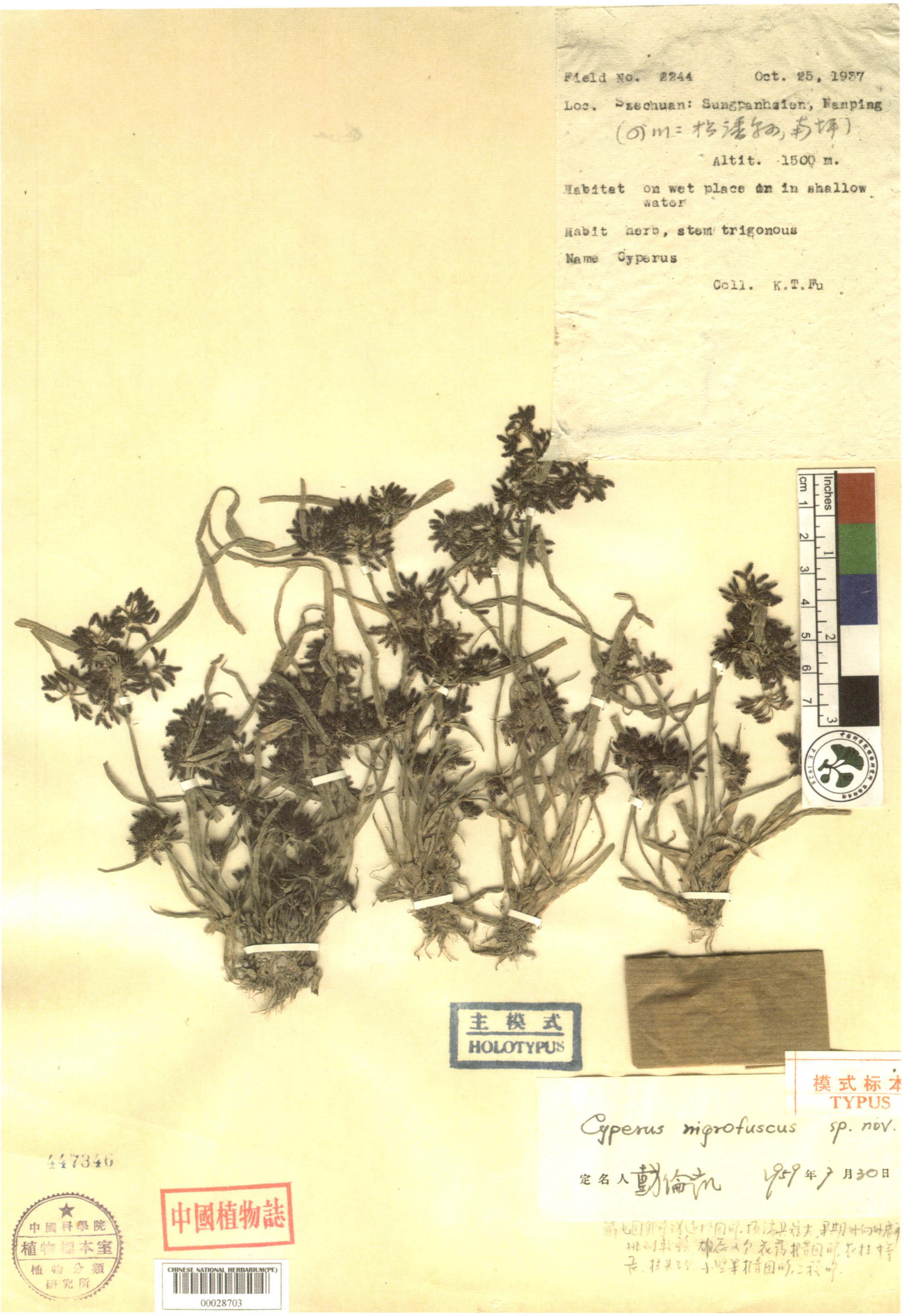

黑穗莎草 ***Cyperus nigrofuscus*** L. K. Dai, Fl. Reip. Pop. Sin. 11: 150, 232, pl. 51: 17-19. 1961. **Holotype:** China. Sichuan: Songpan, alt. 1500 m, 1937-10-25, K. T. Fu 2244.

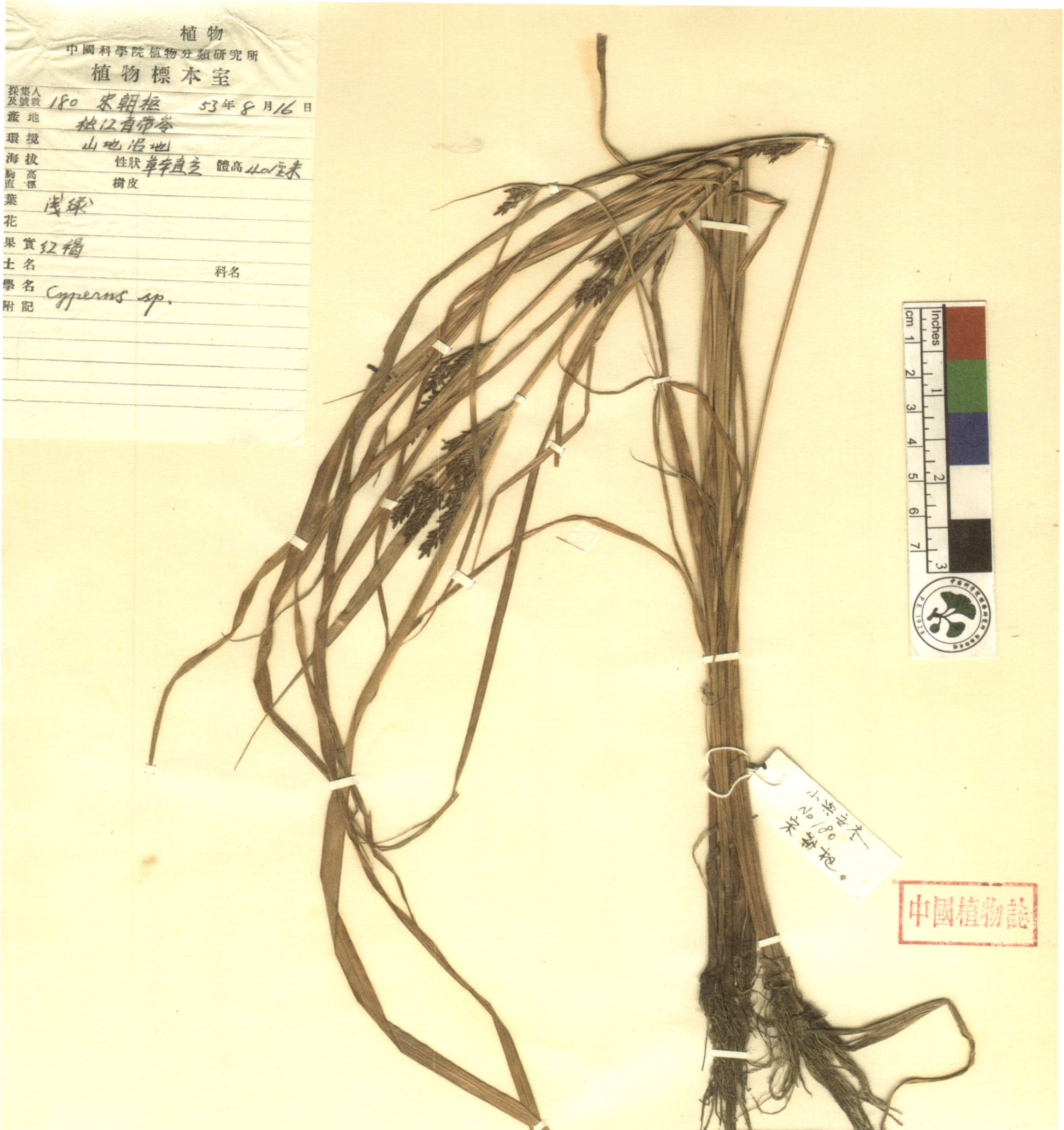

长苞三轮草 ***Cyperus orthostachyus*** Franch. & Savat. var. ***longibracteatus*** L. K. Dai, Fl. Reip. Pop. Sin. 11: 146, 231. 1961. **Holotype:** China. Heilongjiang: Dailing, 1953-08-16, C. S. Song 180.

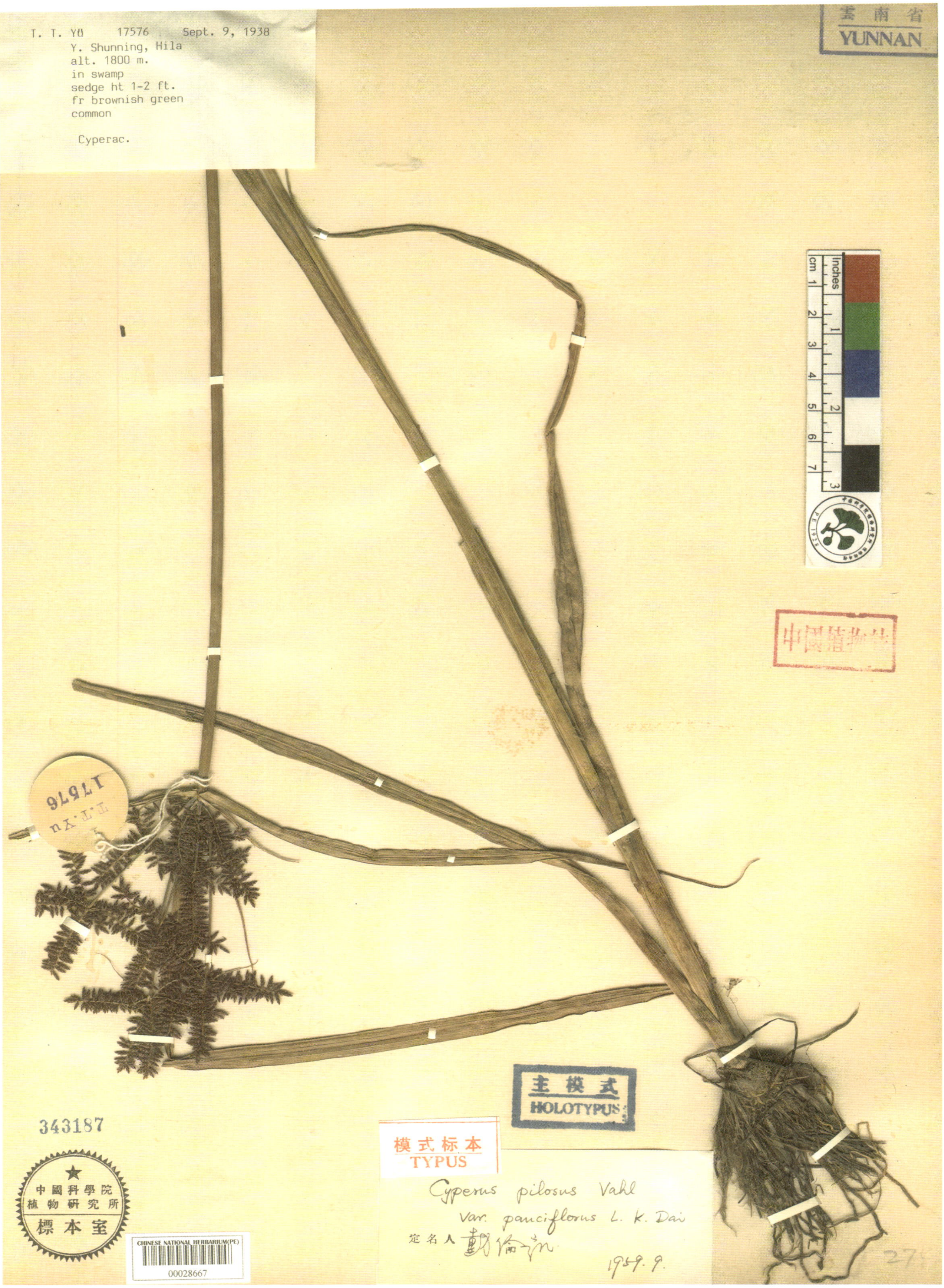

少花毛轴莎草 ***Cyperus pilosus*** Vahl var. ***pauciflorus*** L. K. Dai, Fl. Reip. Pop. Sin. 11: 143, 231. 1961. **Holotype:** China. Yunnan: Shunning (=Fengqing), alt. 1800 m, 1938-09-09, T. T. Yu 17576.

紫穗毛轴莎草 ***Cyperus pilosus*** Vahl var. ***purpurascens*** L. K. Dai, Fl. Reip. Pop. Sin. 11: 143, 231. 1961. **Holotype:** China. Yunnan: Wanyao, alt. 2300 m, 1934-10-14, M. Chen 2937.

金门莎草 ***Cyperus rotundus*** Linn. var. ***quimoyensis*** L. K. Dai, Fl. Reip. Pop. Sin. 11: 135, 231. 1961. **Holotype:** China. Fujian: Xiamen, 1933-08-07, A. N. Steward 3155.

川莎草 ***Cyperus szechuanensis*** T. Koyama in Jap. Journ. Bot. 15(2): 182, f. 10. 1956. **Isotype:** China. Sichuan: Precise locality not known, F. T. Wang 20000D.

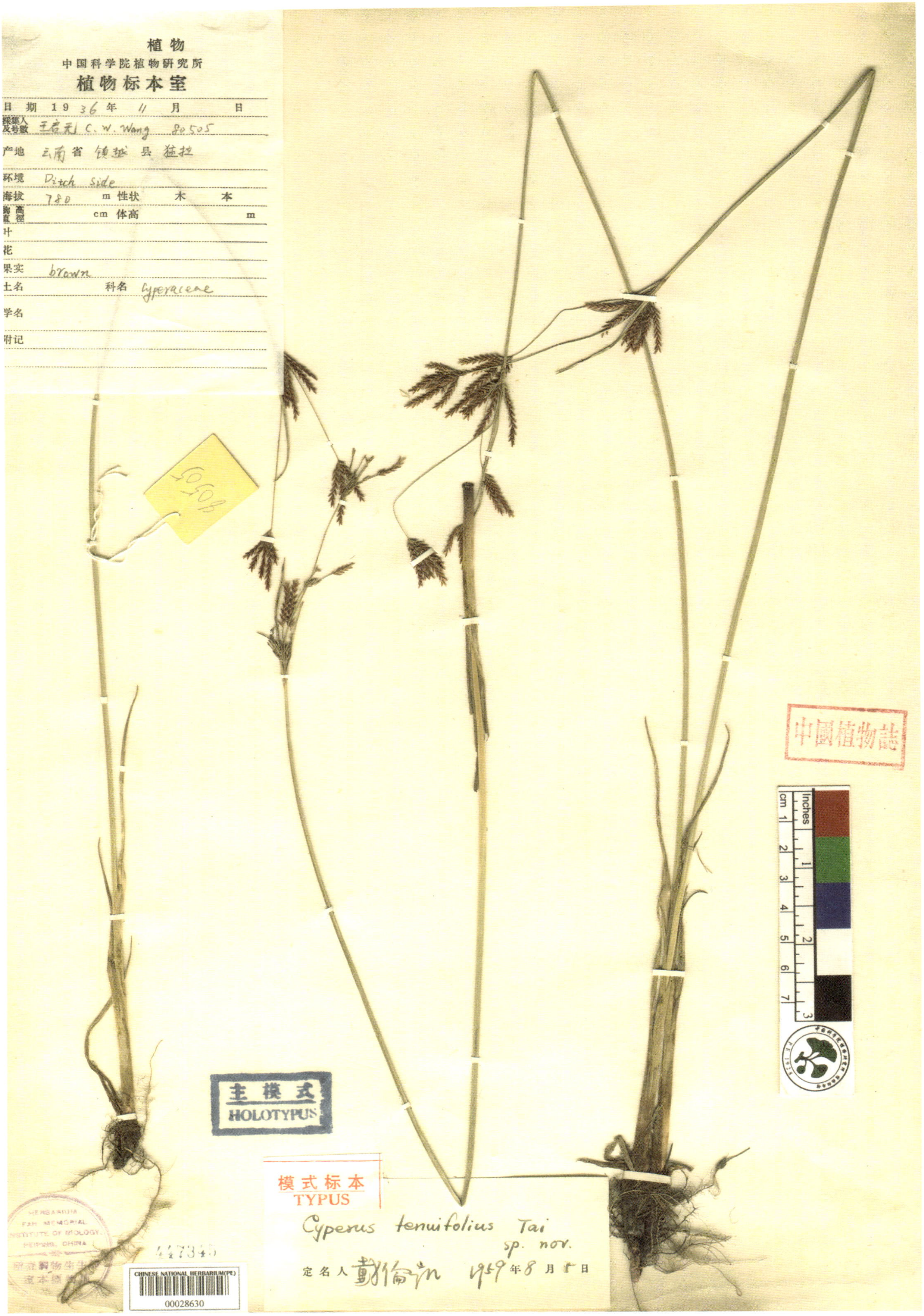

疏鳞莎草 ***Cyperus tenuifolius*** L. K. Dai, Fl. Reip. Pop. Sin. 11: 136, 231. 1961. **Holotype:** China. Yunnan: Zhenyue (=Mengla), alt. 780 m, 1936-11-??, C. W. Wang 80505.

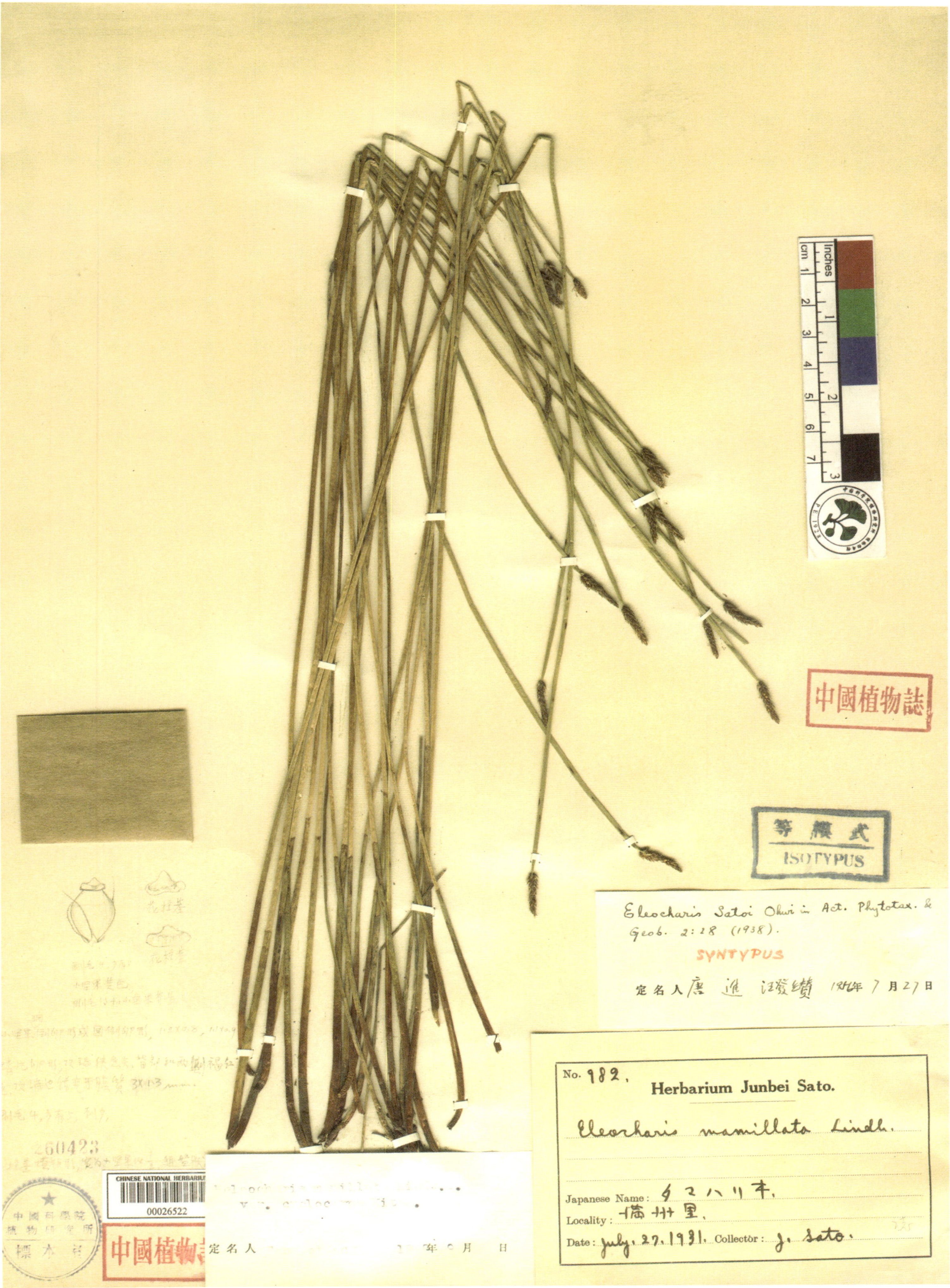

黑龙江荸荠 ***Eleocharis satoi*** Ohwi in Acta Phytotax. Geob. 2: 28. 1933. **Isotype:** China. Heilongjiang, Precise locality not known, 1931-07-27, J. Sato 982.

黔芙兰草 ***Fuirena rhizomatifer*** Tang & F. T. Wang, Fl. Reip. Pop. Sin. 11: 42, 225, pl. 17: 8. 1961. **Holotype:** China. Guizhou: Maotsaoping, alt. 800 m, 1917-07-13, Haudel-Mazzetti 10754.

刘氏荸荠 ***Heleocharis liouana*** Tang & F. T. Wang, Fl. Reip. Pop. Sin. 11: 63, 223. 1961. **Holotype:** China. Yunnan: Kunming, 1941-??-??, T. N. Liou 20112.

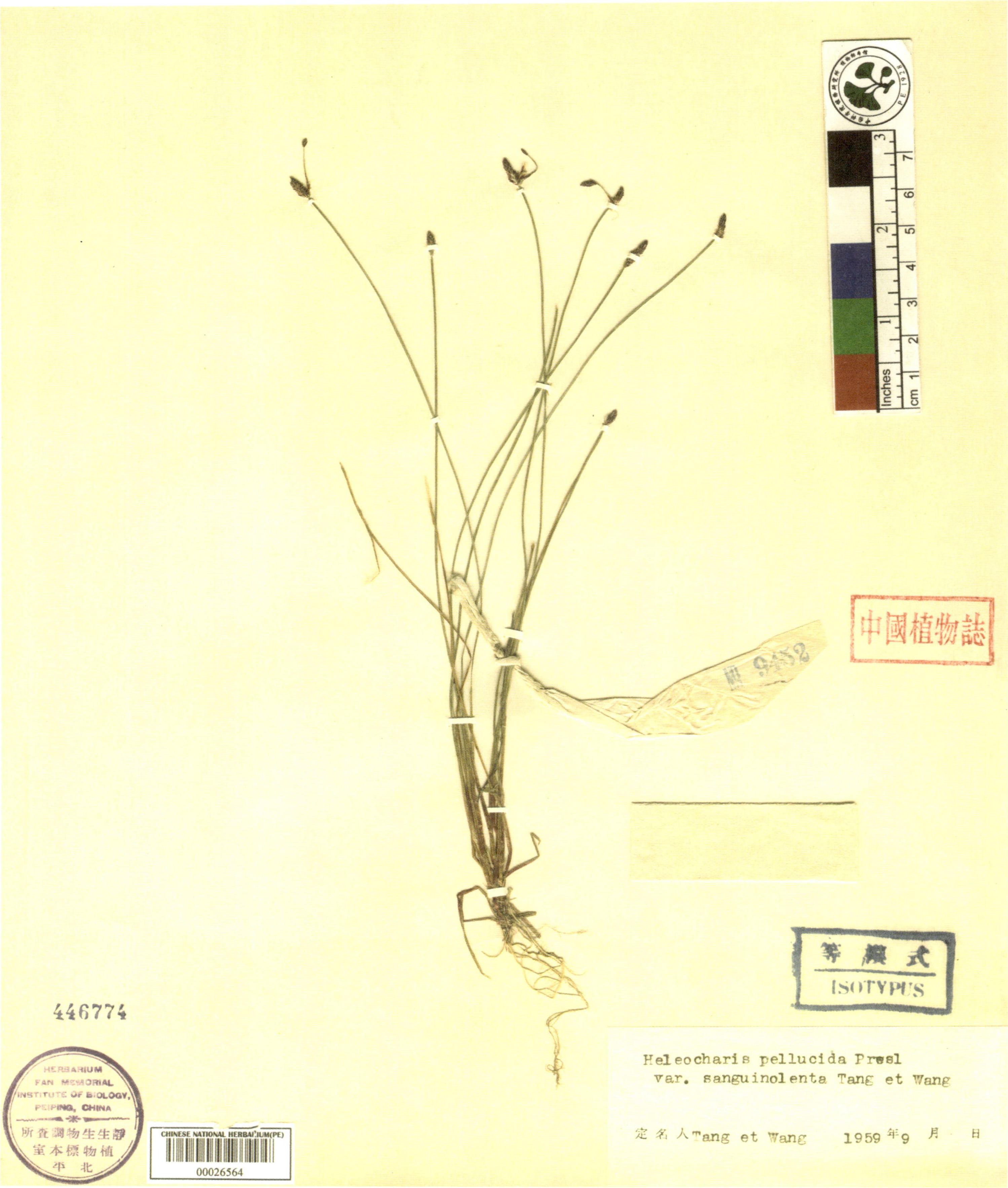

血红穗荸荠 ***Heleocharis pellucida*** Presl var. ***sanguinolenta*** Tang & F. T. Wang, Fl. Reip. Pop. Sin. 11: 58, 226. 1961. **Isotype:** China. Guizhou: Precise locality not known, Anonymous 9452.

海绵基荸荠 ***Heleocharis pellucida*** Presl var. ***spongiosa*** Tang & F. T. Wang, Fl. Reip. Pop. Sin. 11: 58, 226. 1961.
Holotype: China. Jiangxi: Pingxiang, alt. 260 m, 1954-11-13, Jiangxi Exped. 2982.

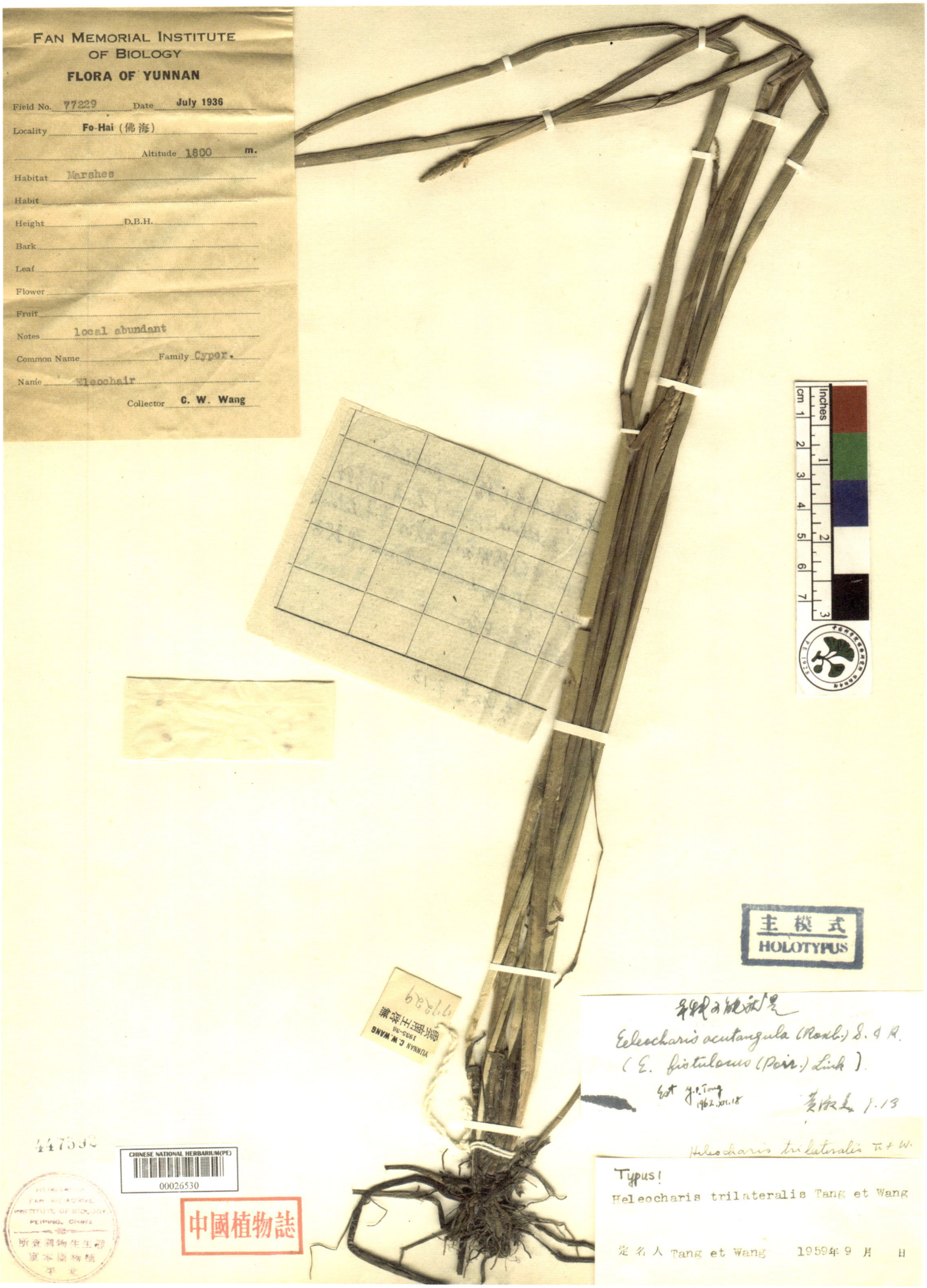

三面秆荸荠 ***Heleocharis trilateralis*** Tang & F. T. Wang, Fl. Reip. Pop. Sin. 11: 50, 225, pl. 19: 4-6. 1961. **Holotype:** China. Yunnan: Fohai (=Menghai), alt. 1800 m, 1936-07-??, C. W. Wang 77229.

三亚割鸡芒 ***Hypolytrum ohwianum*** T. Koyama in Japan. Journ. Bot. 15(2): 183, f. 11. 1956. **Isoparatype:** China. Hainan: Sanya, 1932-07-02, S. K. Lau 197.

少穗割鸡芒 ***Hypolytrum paucistrobiliferum*** Tang & F. T. Wang, Fl. Reip. Pop. Sin. 11: 200, 234. 1961. **Holotype:** China. Hainan: Lingshui, 1932-08-24, N. K. Chun & C. L. Tso 43629.

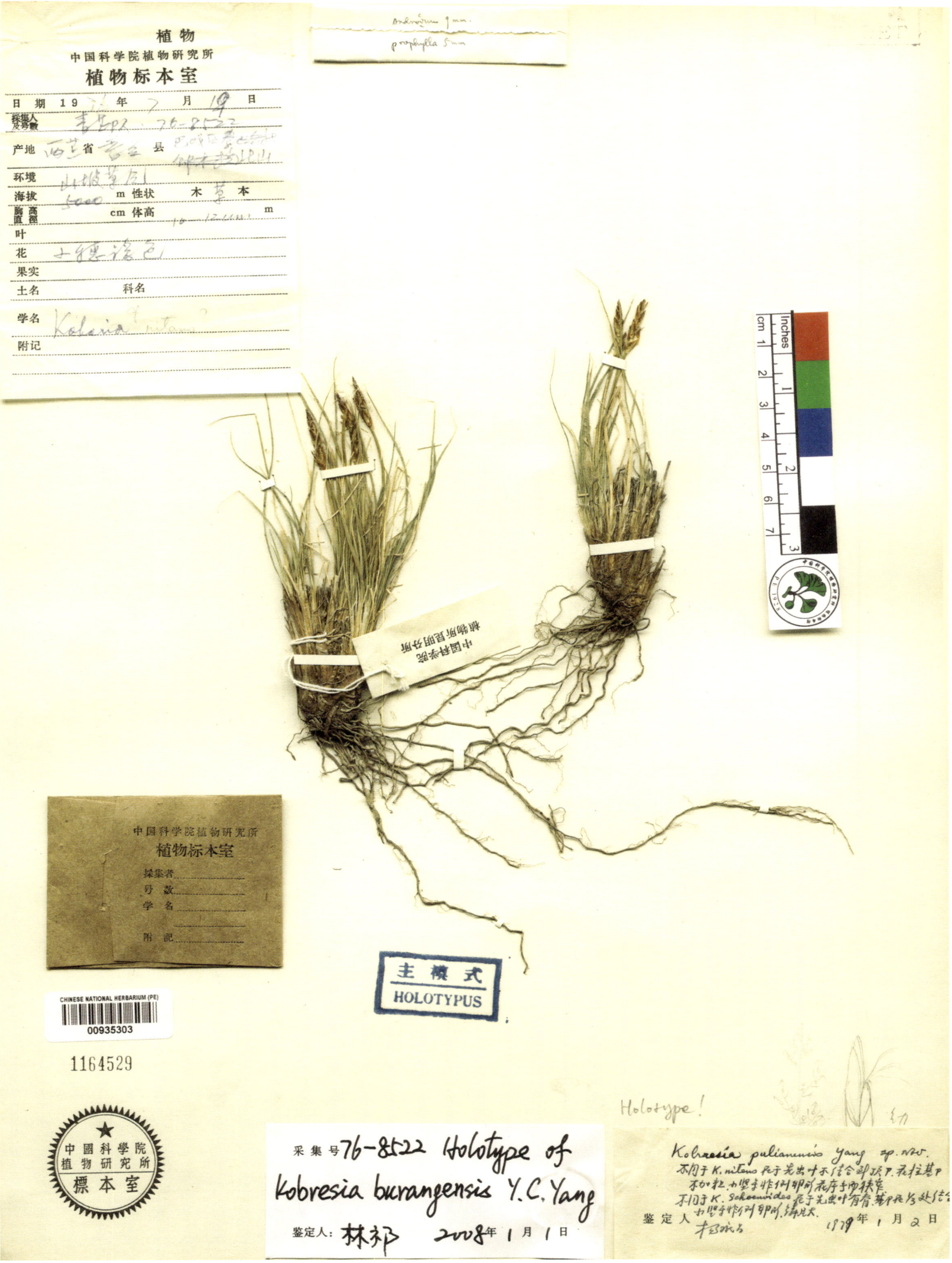

普兰嵩草 ***Kobresia burangensis*** Y. C. Yang, Fl. Xizang 5: 374, f. 208. 1987. **Holotype:** China. Xizang: Burang, alt. 5000 m, 1976-07-18, Qinghai-Xizang Exped. 76-8522.

线穗嵩草 ***Kobresia cercostachya*** (Franch.) C. B. Clarke var. ***capillacea*** P. C. Li in Acta Bot. Yunnan. 12(1): 17. 1990. **Holotype:** China. Xizang: Zayü, alt. 4200 m, 1982-09-09, Qinghai-Xizang Exped. 10239.

镰叶嵩草 ***Kobresia falcata*** F. T. Wang & T. Tang ex P. C. Li in Acta Bot. Yunnan. 12(1): 18, f. 9. 1990. **Holotype:** China. Sichuan: Kangding, 1934-??-??, Z. S. Liou 1336.

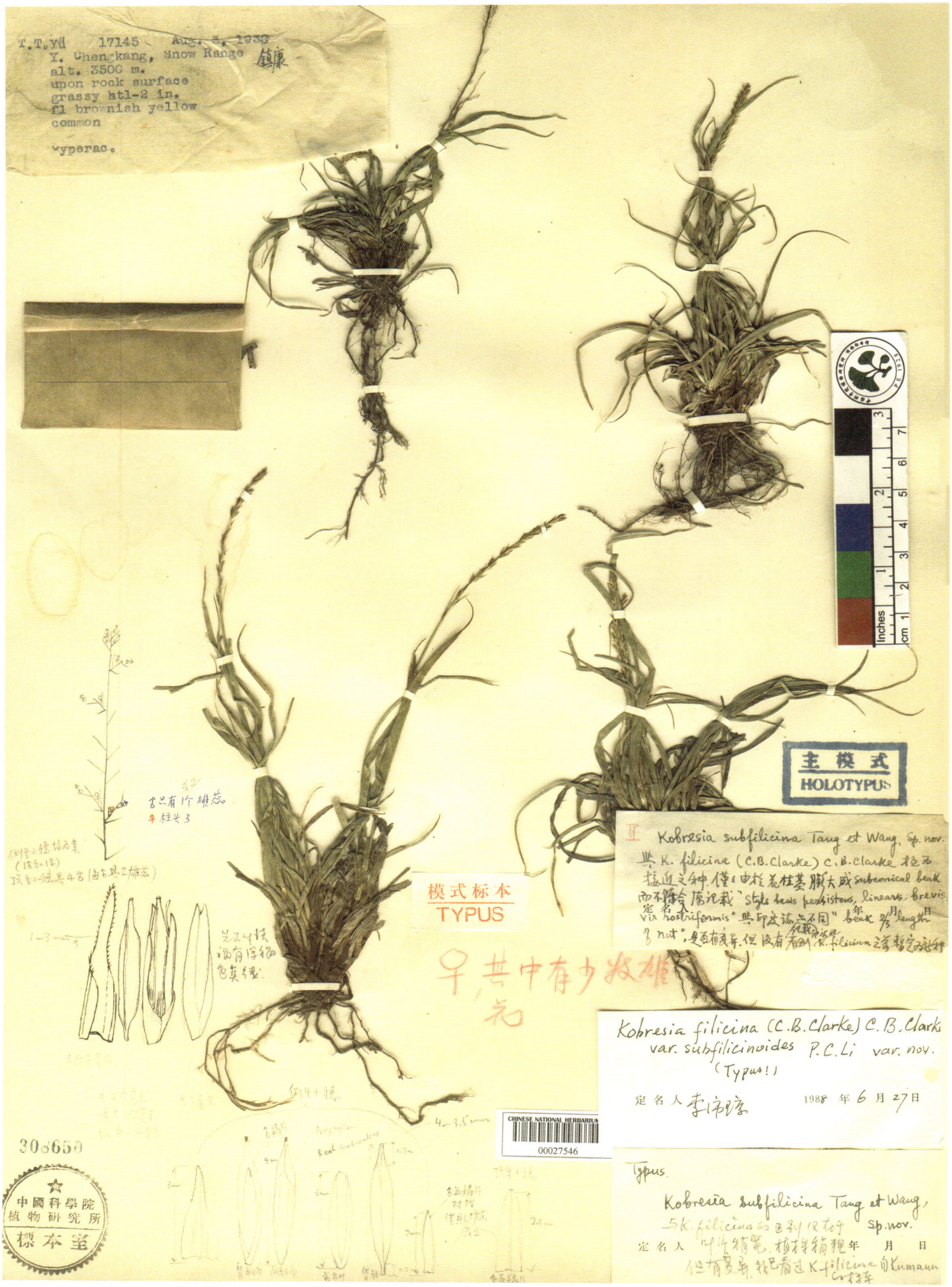

近蕨嵩草 ***Kobresia filicina*** (C. B. Clarke) C. B. Clarke var. ***subfilicinoides*** P. C. Li in Acta Phytotax. Sin. 37(2): 155. 1999. **Holotype:** China. Yunnan: Zhenkang, alt. 3500 m, 1939-08-03, T. T. Yu 17145.

粉绿嵩草 ***Kobresia glaucifolia*** F. T. Wang & Tang ex P. C. Li in Acta Phytotax. Sin. 37(2): 153, f. 1. 1999. **Holotype:** China. Xizang: Nyalam, alt. 4100 m, 1975-06-20, Qinghai-Xizang Exped. 4229.

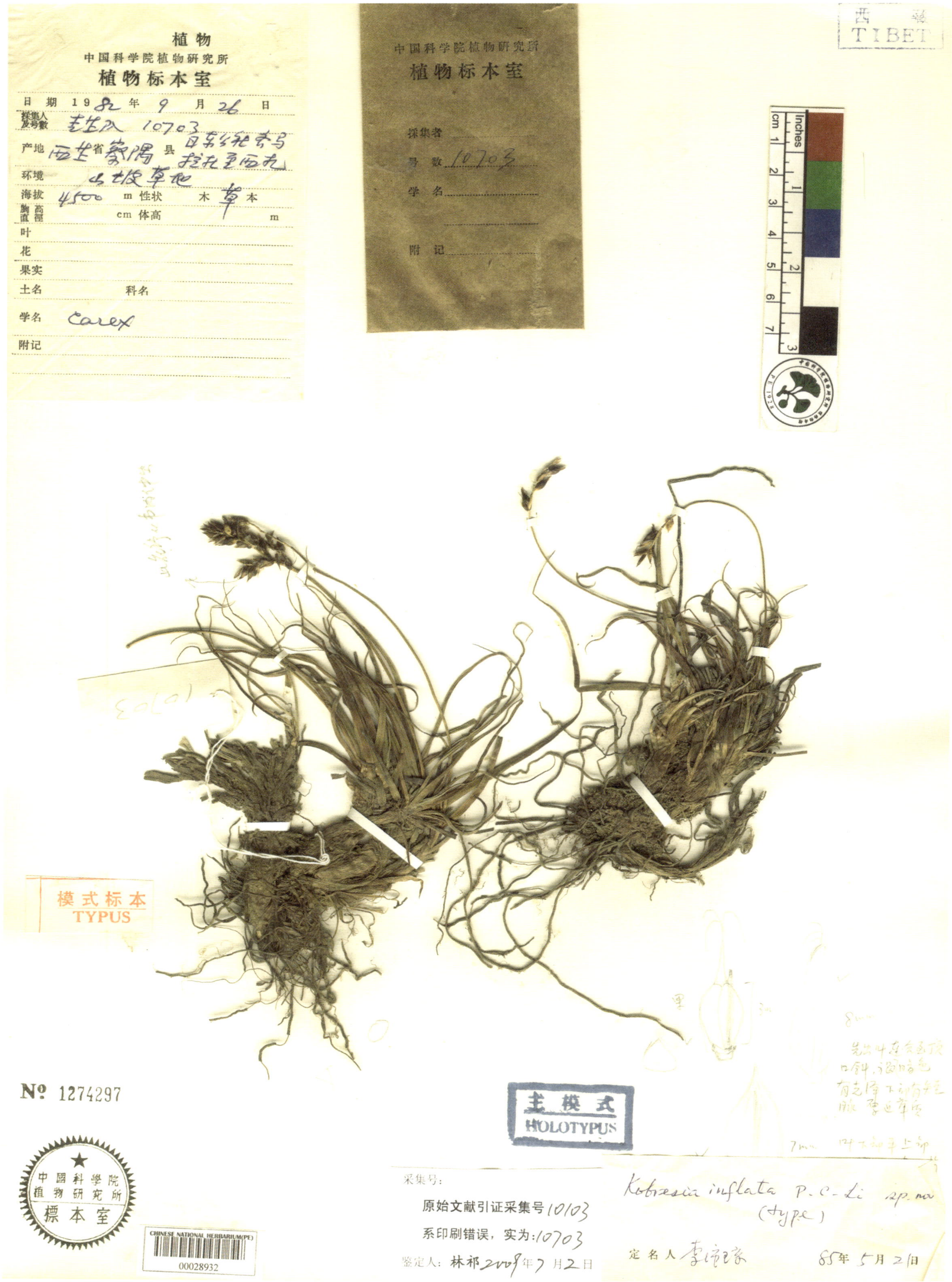

膨囊嵩草 ***Kobresia inflata*** P. C. Li in Acta Bot. Yunnan 12(1): 16, f. 7. 1990. **Holotype:** China. Xizang: Zayü, alt. 4500 m, 1982-09-26, Qinghai-Xizang Exped. 10703.

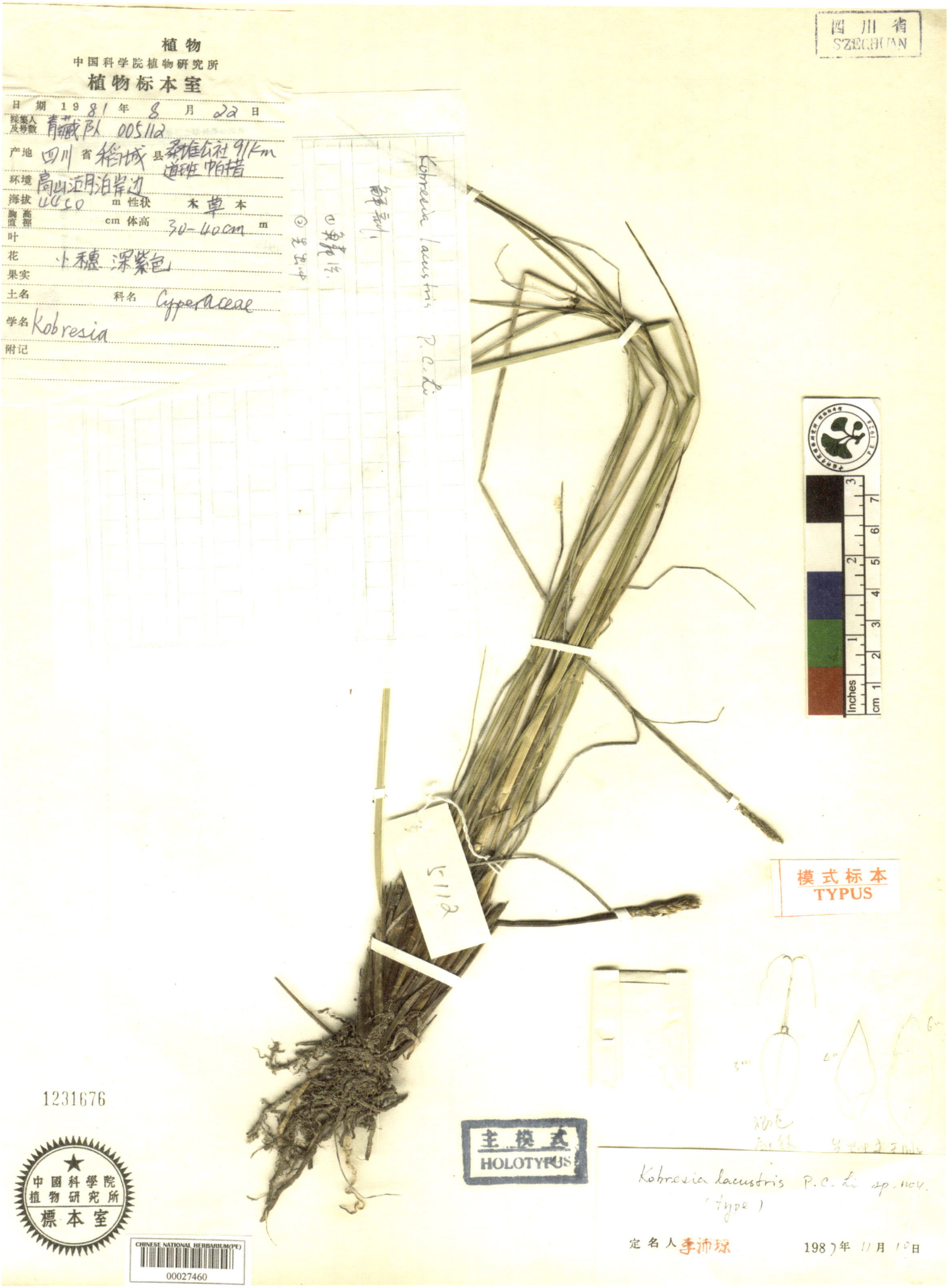

湖滨嵩草 ***Kobresia lacustris*** P. C. Li in Acta Bot. Yunnan 12(1): 14, f. 4. 1990. **Holotype:** China. Sichuan: Daocheng, alt. 4450 m, 1981-08-22, Qinghai-Xizang Exped. 5112.

鳞被嵩草 ***Kobresia lepidochlamys*** F. T. Wang & Tang ex P. C. Li in Acta Bot. Yunnan 12(1): 15, f. 6. 1990. **Holotype:** China. Yunnan: Lijiang, alt. 4300 m, (1923-1924)-??-??, J. F. Rock 11693.

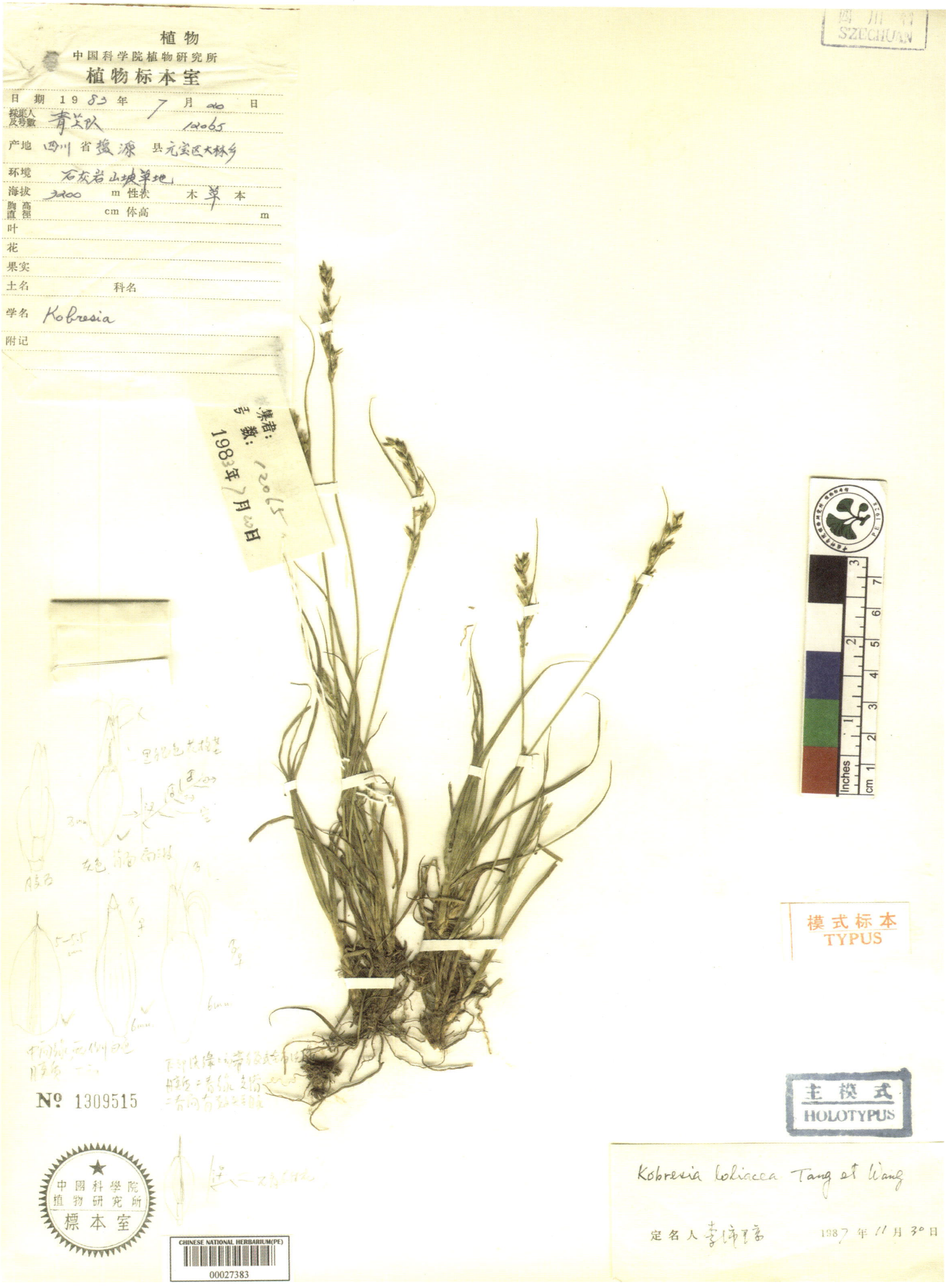

黑麦嵩草 ***Kobresia loliaceae*** F. T. Wang & Tang ex P. C. Li in Acta Bot. Yunnan 12(1): 13, f. 3. 1990. **Holotype:** China. Sichuan: Yanyuan, alt. 3200 m, 1983-07-20, Qinghai-Xizang Exped. 12065.

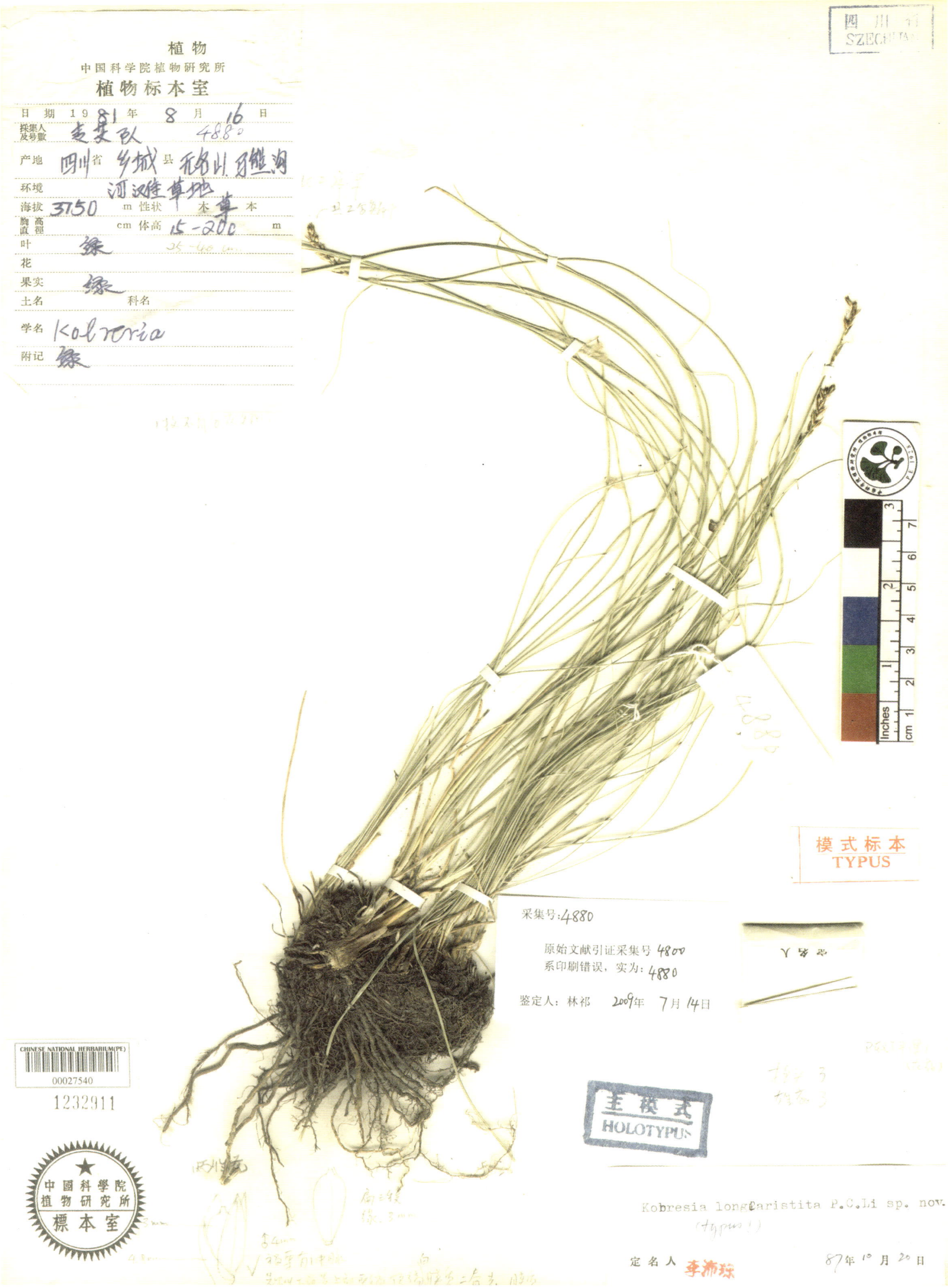

长芒嵩草 ***Kobresia longearistita*** P. C. Li in Acta Bot. Yunnan 12(1): 16, f. 8. 1990. **Holotype:** China. Sichuan: Xiangcheng, alt. 3750 m, 1981-08-16, Qinghai-Xizang Exped. 4880.

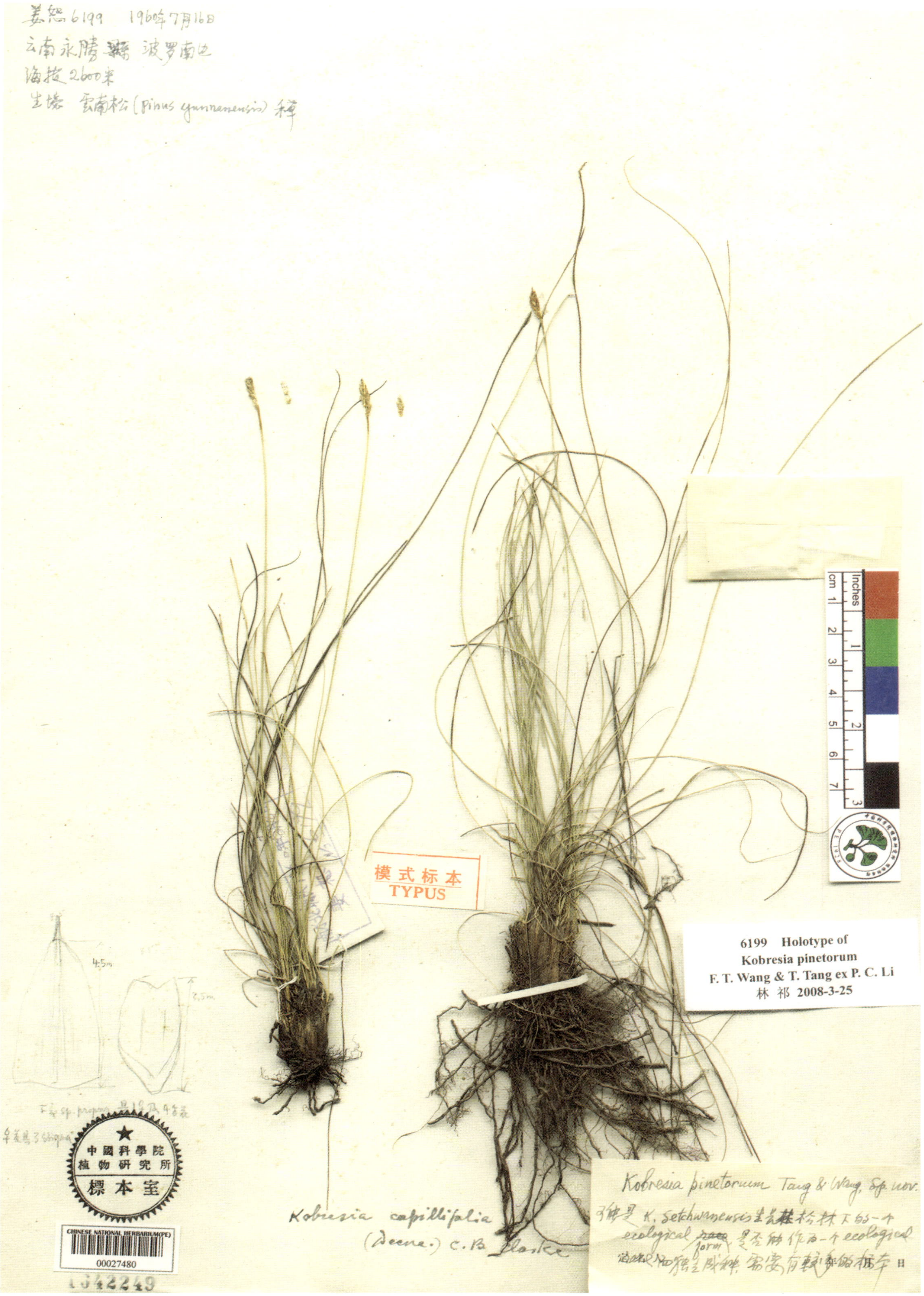

松林嵩草 ***Kobresia pinetorum*** F. T. Wang & T. Tang ex P. C. Li in Acta Bot. Yunnan 12(1): 14, f. 5. 1990. **Holotype:** China. Yunnan: Yongsheng, alt. 2600 m, 1960-07-16, S. Jiang 6199.

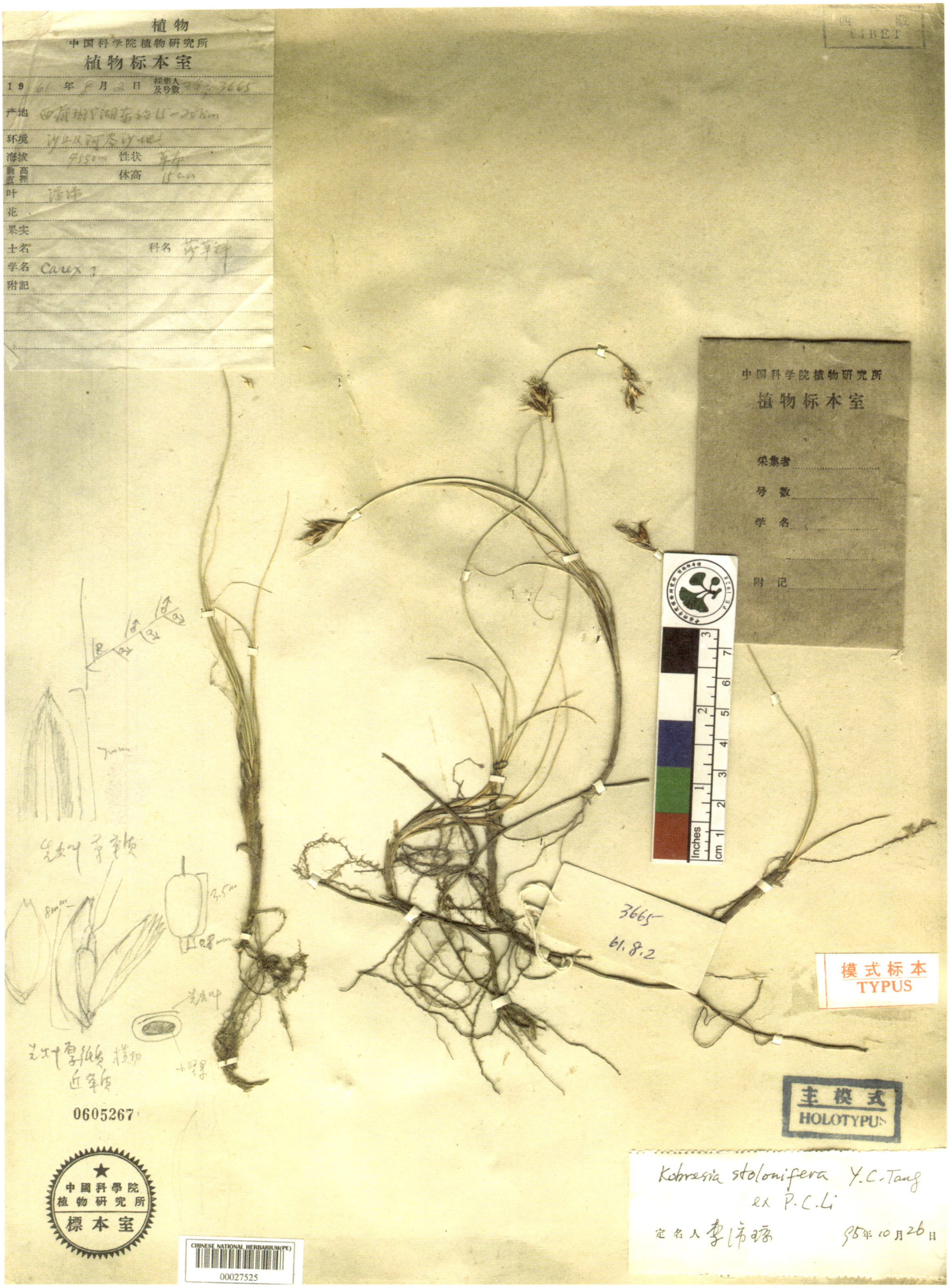

匍茎嵩草 ***Kobresia stolonifera*** Y. C. Tang ex P. C. Li in Acta Phytotax. Sin. 37(2): 154, f. 2. 1999. **Holotype:** China. Xizang: Baingoin, alt. 4550 m, 1961-08-2, J. T. Wang 3665.

长秆擂鼓艻 ***Mapania dolichopoda*** Tang & F. T. Wang, Fl. Reip. Pop. Sin. 11: 197, 233, pl. 63: 5-9. 1961. **Holotype:** China. Guangdong: Yangchun, 1935-11-17, C. Wang 38790.

海南割鸡芒 ***Mapania hainanensis*** Merr. in Lingnan Sci. Journ. 9(1-2): 35. 1930. **Isotype:** China. Hainan: Taamchau (=Dan Xian), 1928-05-25, W. T. Tsang 17244.

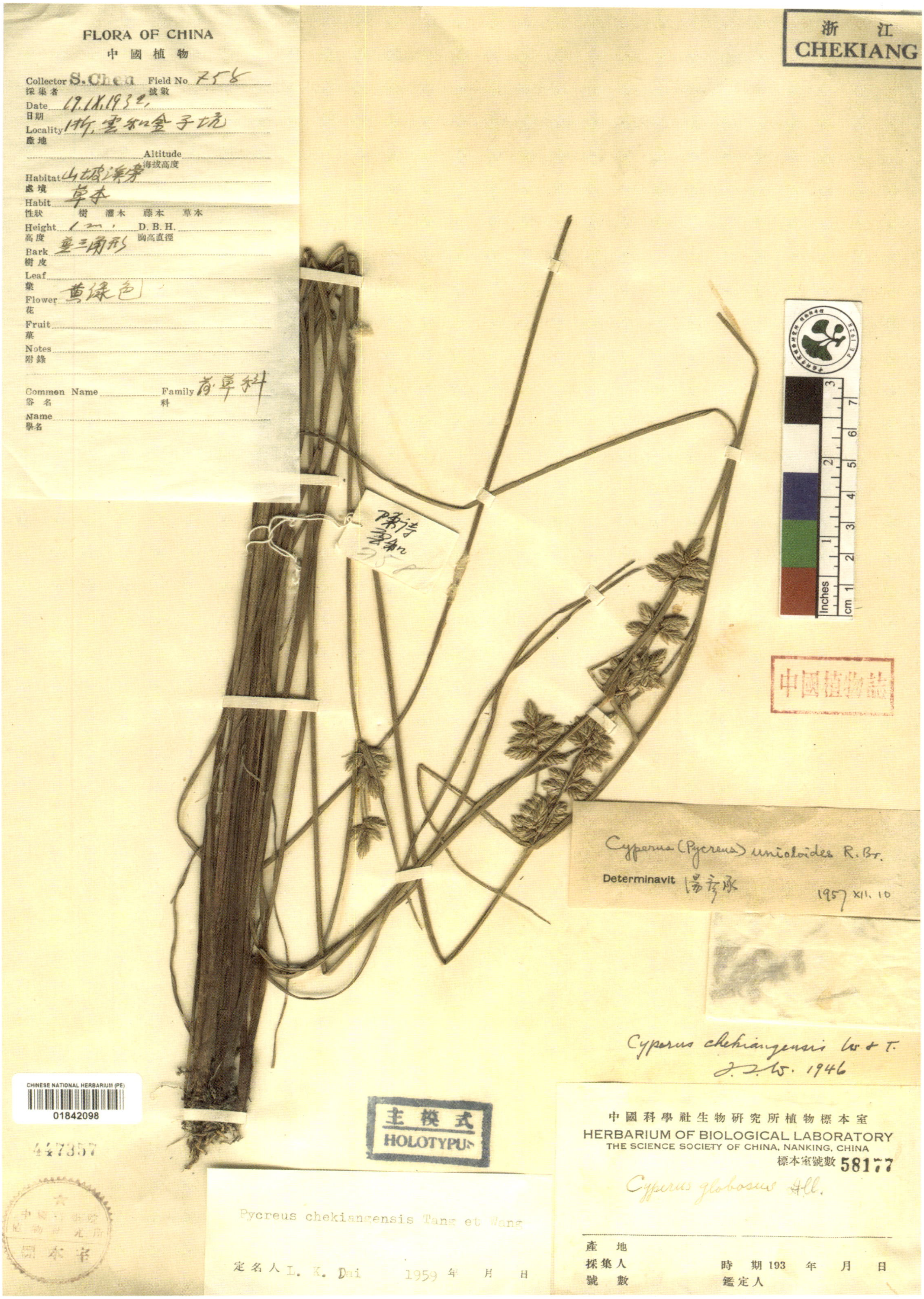

浙江扁莎 ***Pycreus chekiangensis*** Tang & F. T. Wang, Fl. Reip. Pop. Sin. 11: 165, 232, pl. 58: 5-6. 1961. **Holotype:** China. Zhejiang: Yunhe, 1932-09-19, S. Chen 758.

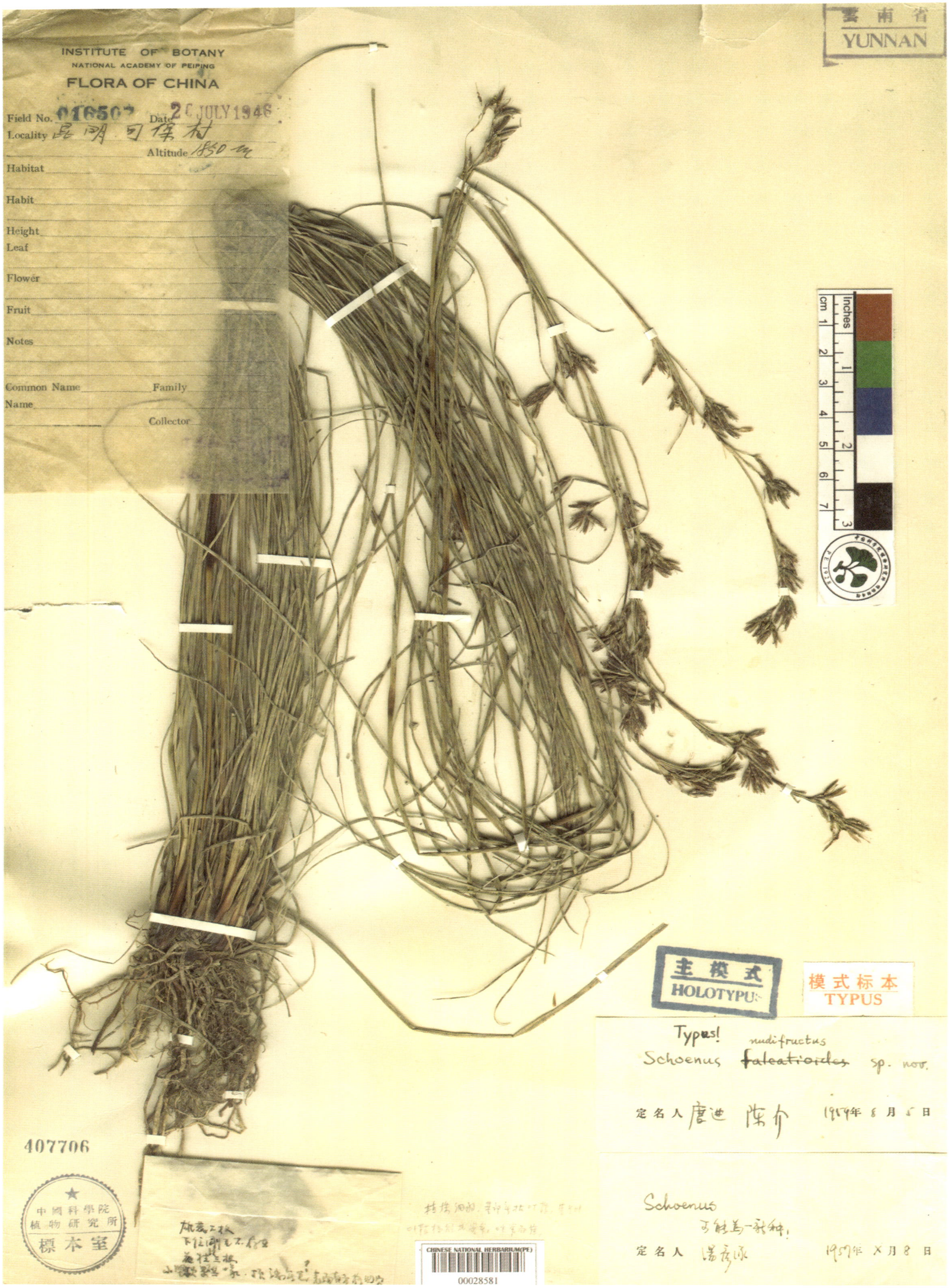

无刚毛赤箭莎 ***Schoenus nudifructus*** C. Chen, Fl. Reip. Pop. Sin. 11: 117, 230. 1961. **Holotype:** China. Yunnan: Kunming, alt. 1850 m, 1946-07-26, T. N. Liou 16507.

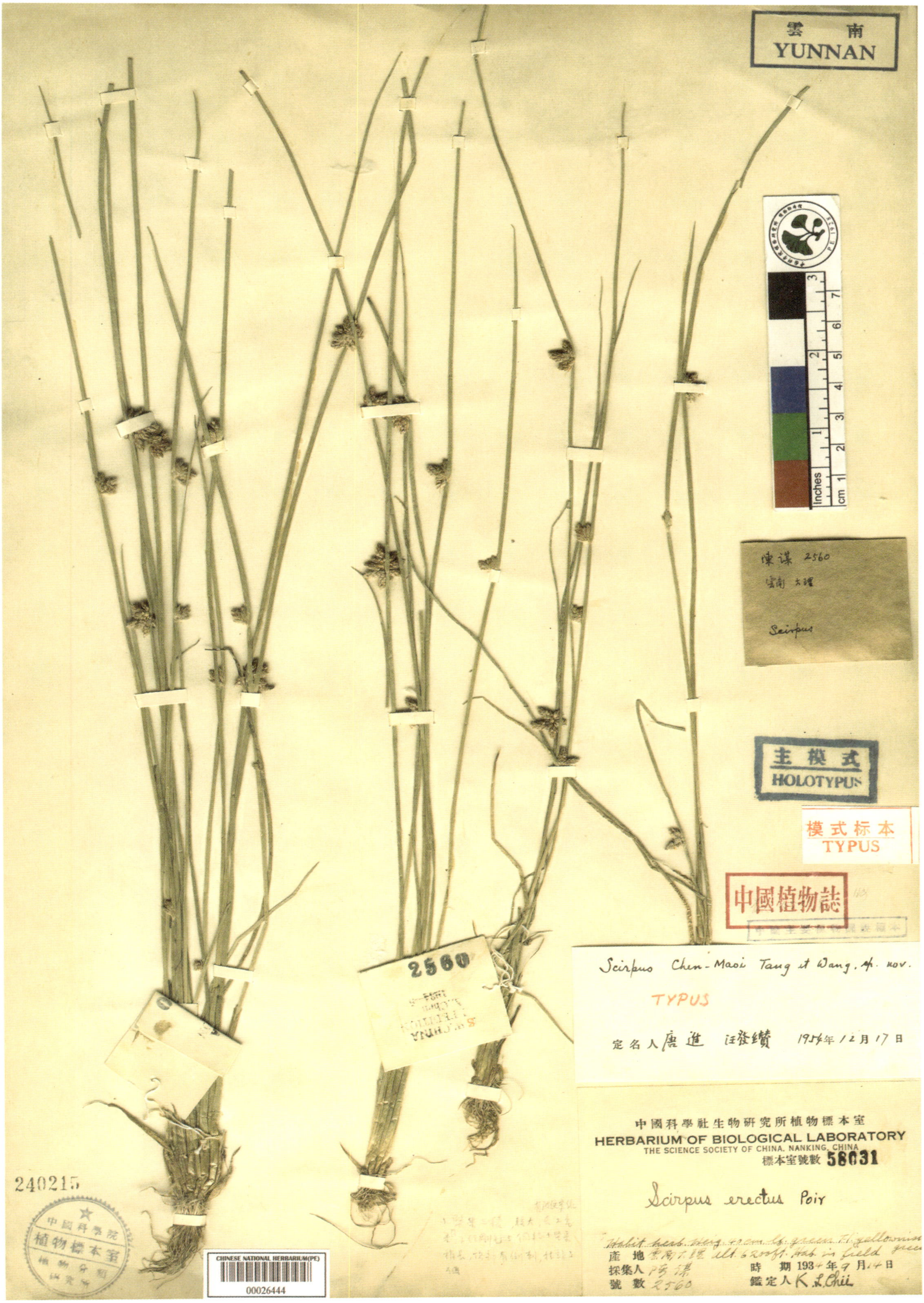

陈谋藨草 ***Scirpus chenmouii*** Tang & F. T. Wang, Fl. Reip. Pop. Sin. 11: 26, 223, pl. 12: 1-4. 1961. **Holotype:** China. Yunnan: Dali, alt. 2100 m, 1934-09-14, M. Chen 2560.

曲氏藨草 ***Scirpus chuanus*** Tang & F. T. Wang, Fl. Reip. Pop. Sin. 11: 22, 222, pl. 10, 1-5. 1961. **Holotype:** China. Jiangsu: Nanjing, 1931-08-17, K. L. Chu 341.

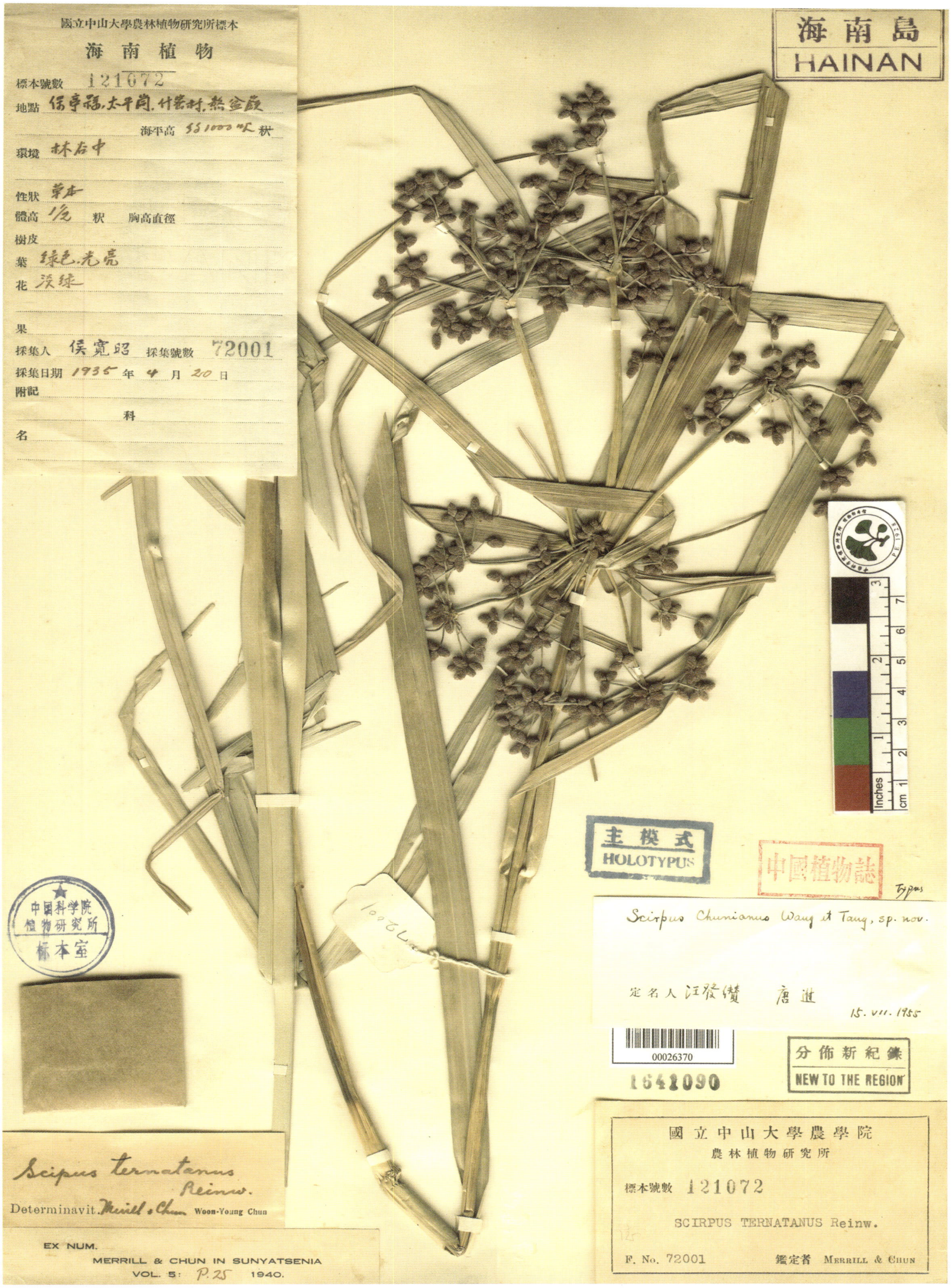

陈氏藨草 ***Scirpus chunianus*** Tang & F. T. Wang, Fl. Reip. Pop. Sin. 11: 13, 221, pl. 4: 5-8. 1961. **Holotype:** China. Hainan: Baoting, alt. 330 m, 1935-04-20, F. C. How 72001.

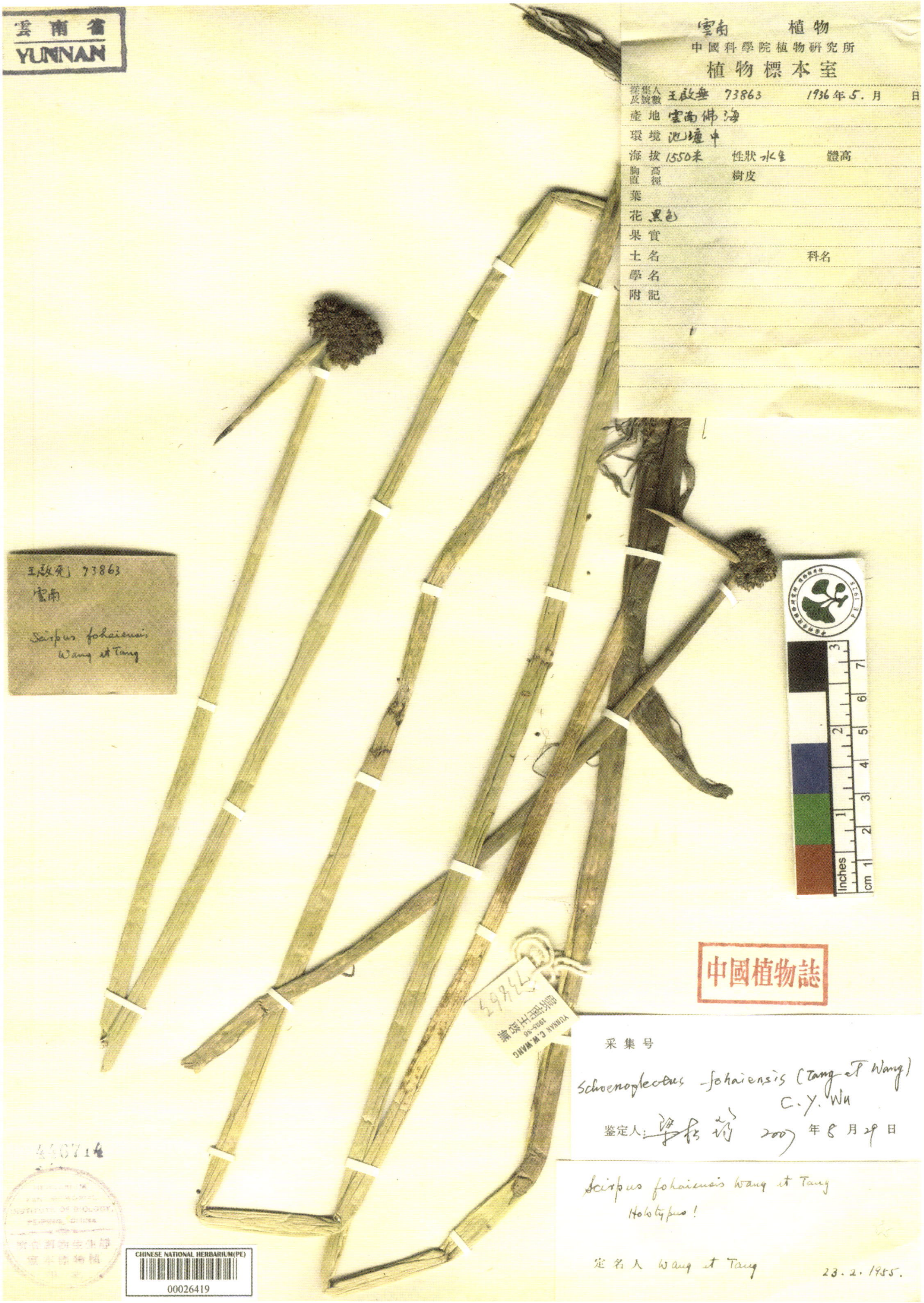

佛海藨草 ***Scirpus fohaiensis*** Tang & F. T. Wang, Fl. Reip. Pop. Sin. 11: 23, 222, pl. 10: 6-10. 1961. **Holotype:** China. Yunnan: Fohai (=Menghai), alt. 1550 m, 1936-05-??, C. W. Wang 73863.

羊胡子藨草 ***Scirpus heleocharidioides*** F. T. Wang & Tang in Acta Phototax. Sin. 1(2): 181. 1951. **Lectotype** (designated by Q. Lin & Z. R. Yang in Bull. Bot. Res., Harbin 30(2): 131. 2010.): China. Beijing: Fangshan, 1935-07-06, Y. Liu 10449.

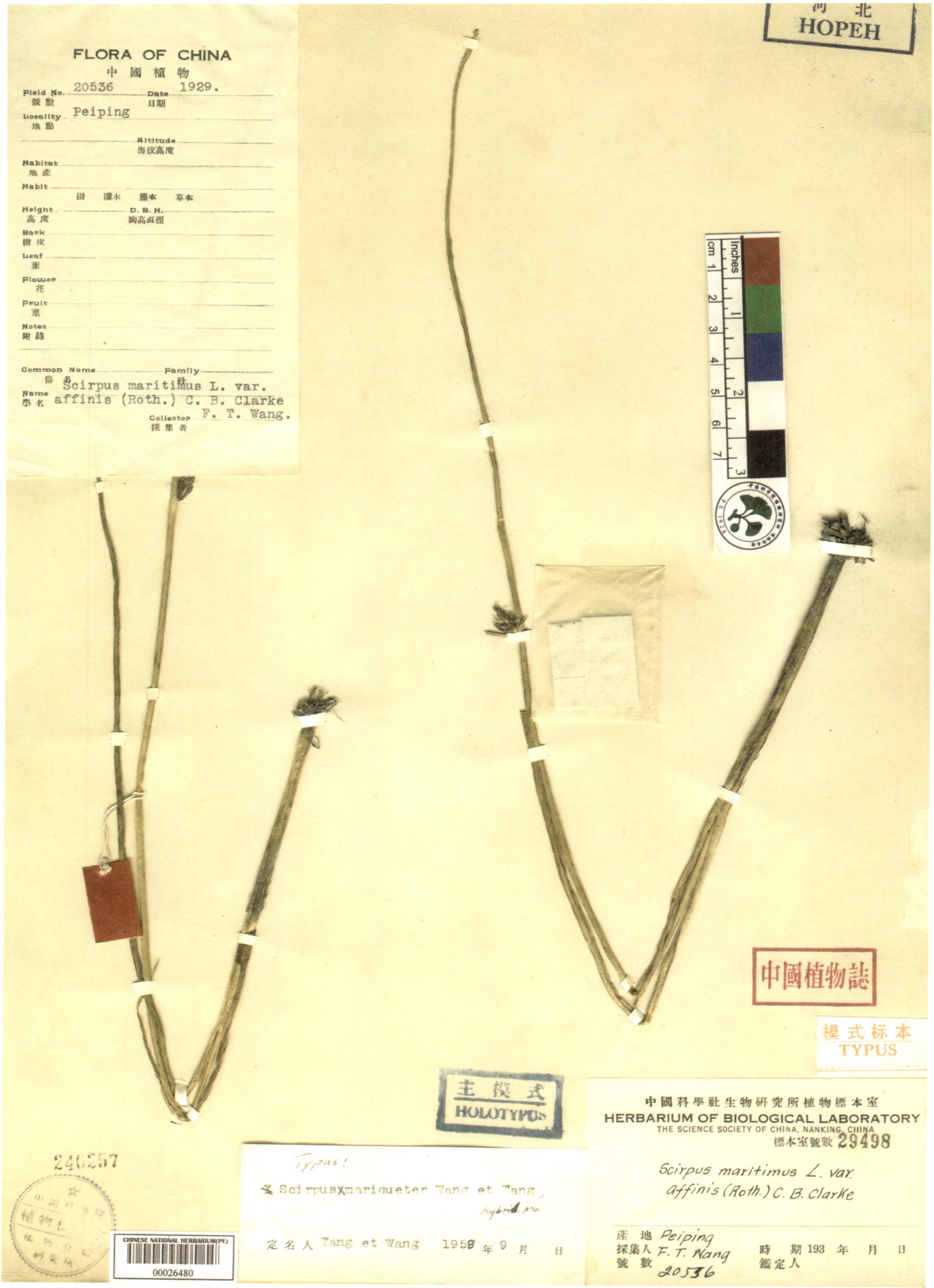

海三稜藨草 ***Scirpus* × *mariqueter*** Tang & F. T. Wang, Fl. Reip. Pop. Sin. 11: 33, 223, pl. 8: 6-10. 1961. **Holotype:** China: Beijing: Precise locality not known, 1929-??-??, F. T. Wang 20536.

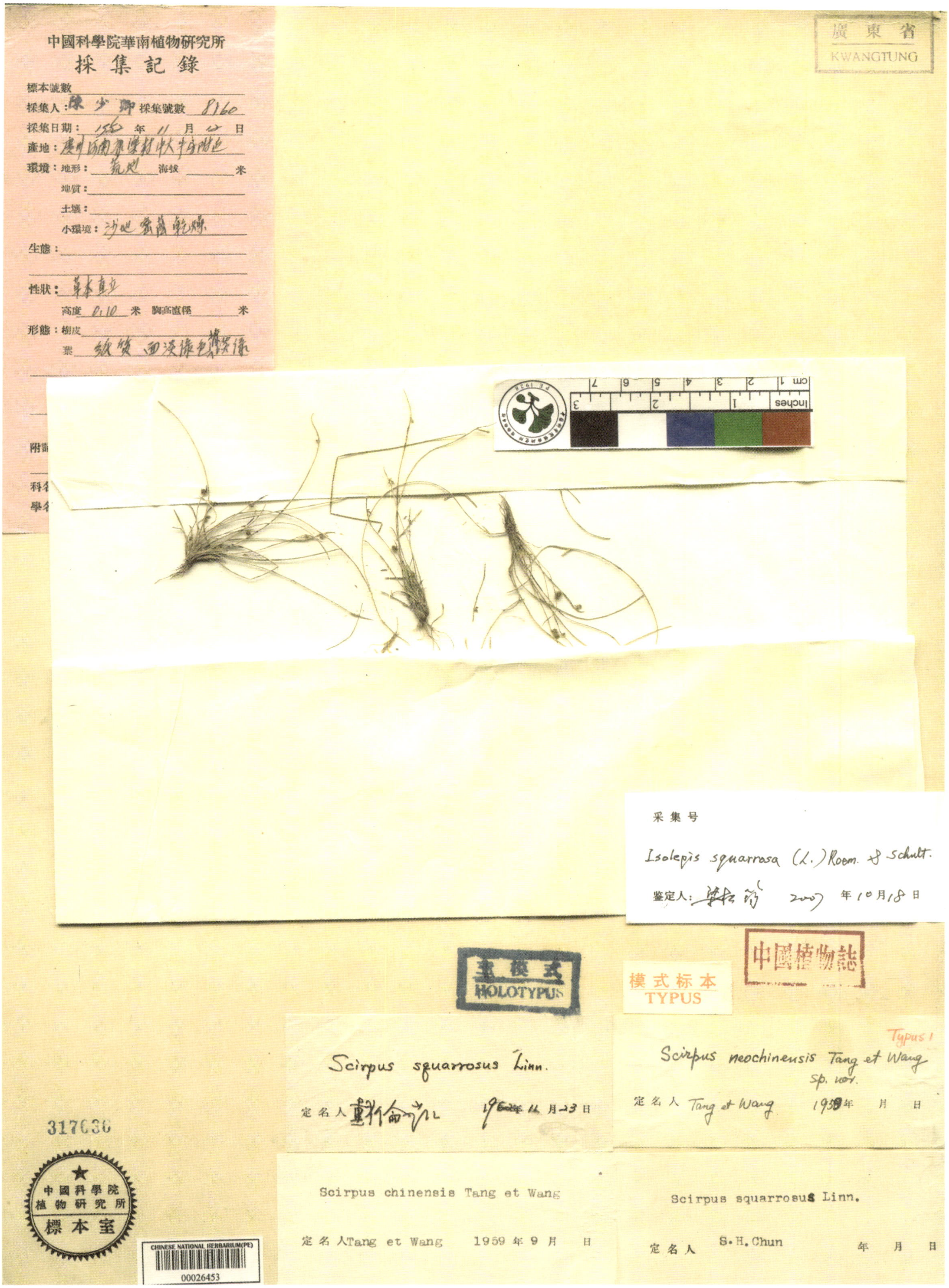

新华藨草 ***Scirpus neochinensis*** Tang & F. T. Wang, Fl. Reip. Pop. Sin. 11: 28, 223, pl. 13: 1-4. 1961. **Holotype:** China. Guangdong: Guangzhou, 1952-11-12, S. H. Chun 8160.

高山藨草 ***Scirpus paniculato-corymbosus*** Kükenth. in Acta Horti Gothob. 5: 35. 1929. **Isotype:** China. Sichuan: Tahsiangling, alt. 2000~2800 m, 1922-05-28, H. Smith 2092.

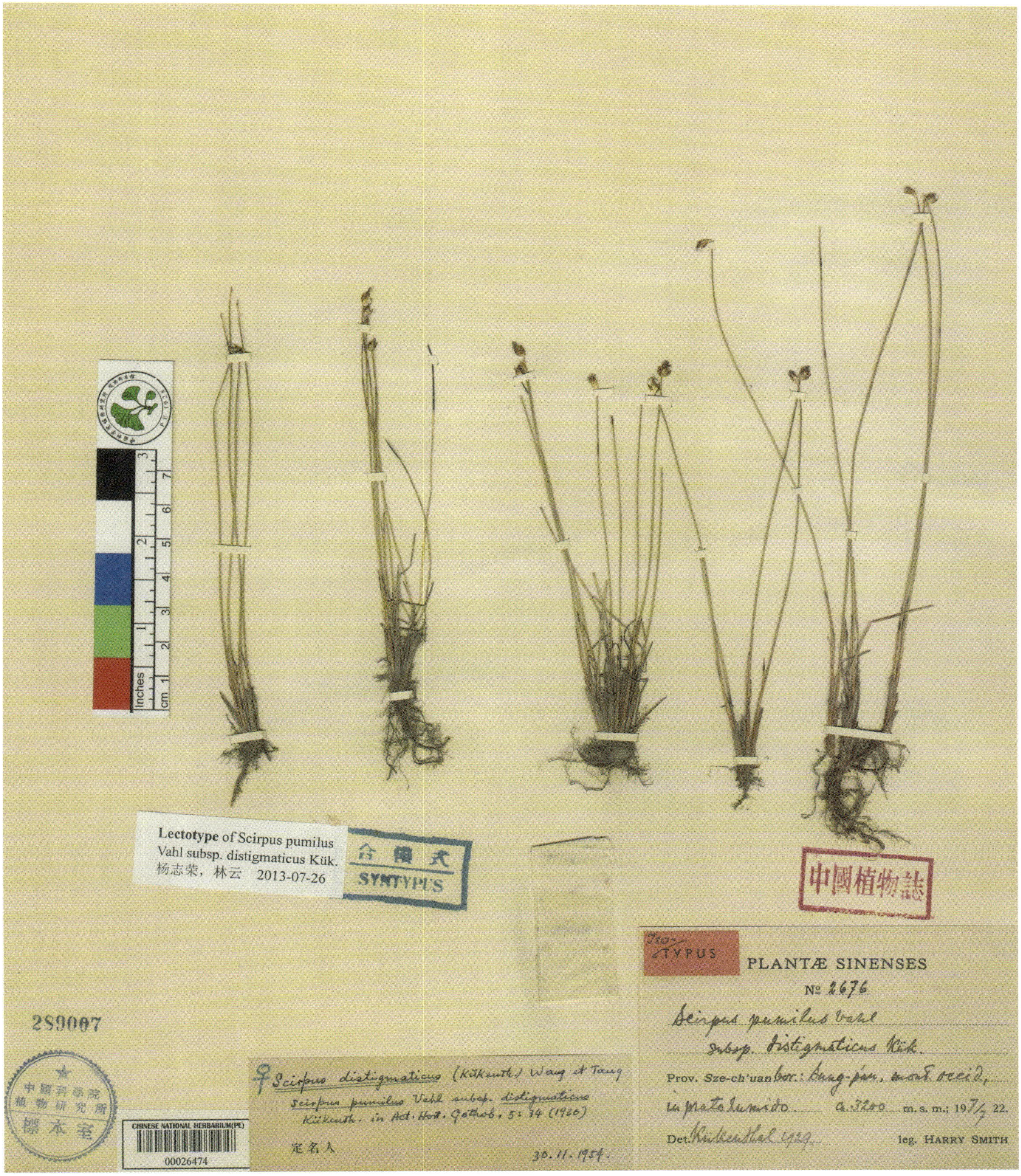

双柱头针蔺 ***Scirpus pumilus*** Vahl subsp. ***distigmaticus*** Kükenth. in Acta Horti Gothob. 5(3): 34. 1929. **Lectotype** (designated by Y. Lin & al. in Acta Bot. Bor.-Occ. Sin. 34: 412. 2014.): China. Sichuan: Songpan, alt. 3200 m, 1922-07-07, H. Smith 2676.

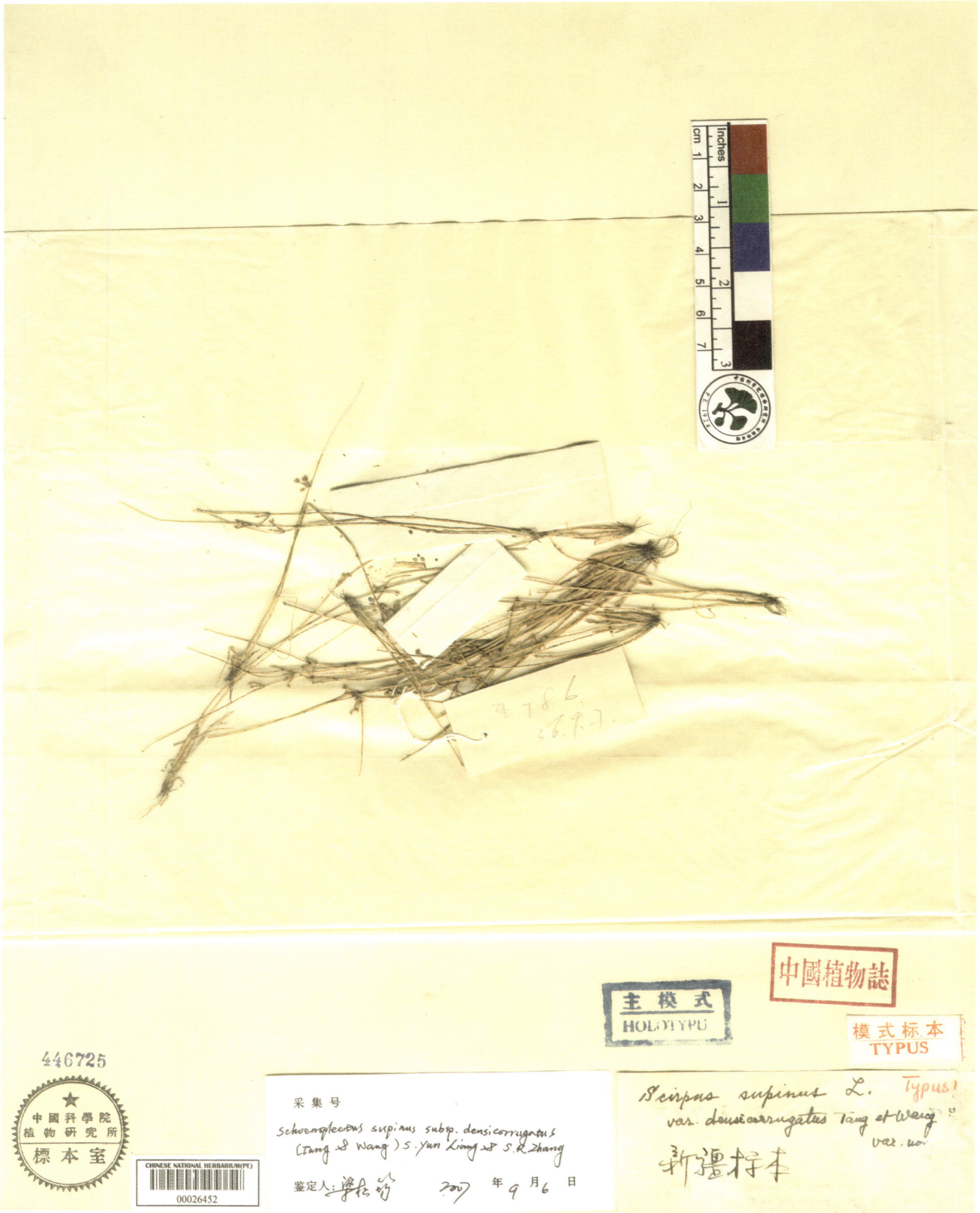

多皱纹果仰卧秆藨草 ***Scirpus supinus*** Linn. var. ***densicorrugatus*** Tang & F. T. Wang, Fl. Reip. Pop. Sin. 11: 28, 223. 1961.
Holotype: China. Xinjiang: Precise locality not known, 1956-09-07, R. C. Ching 786.

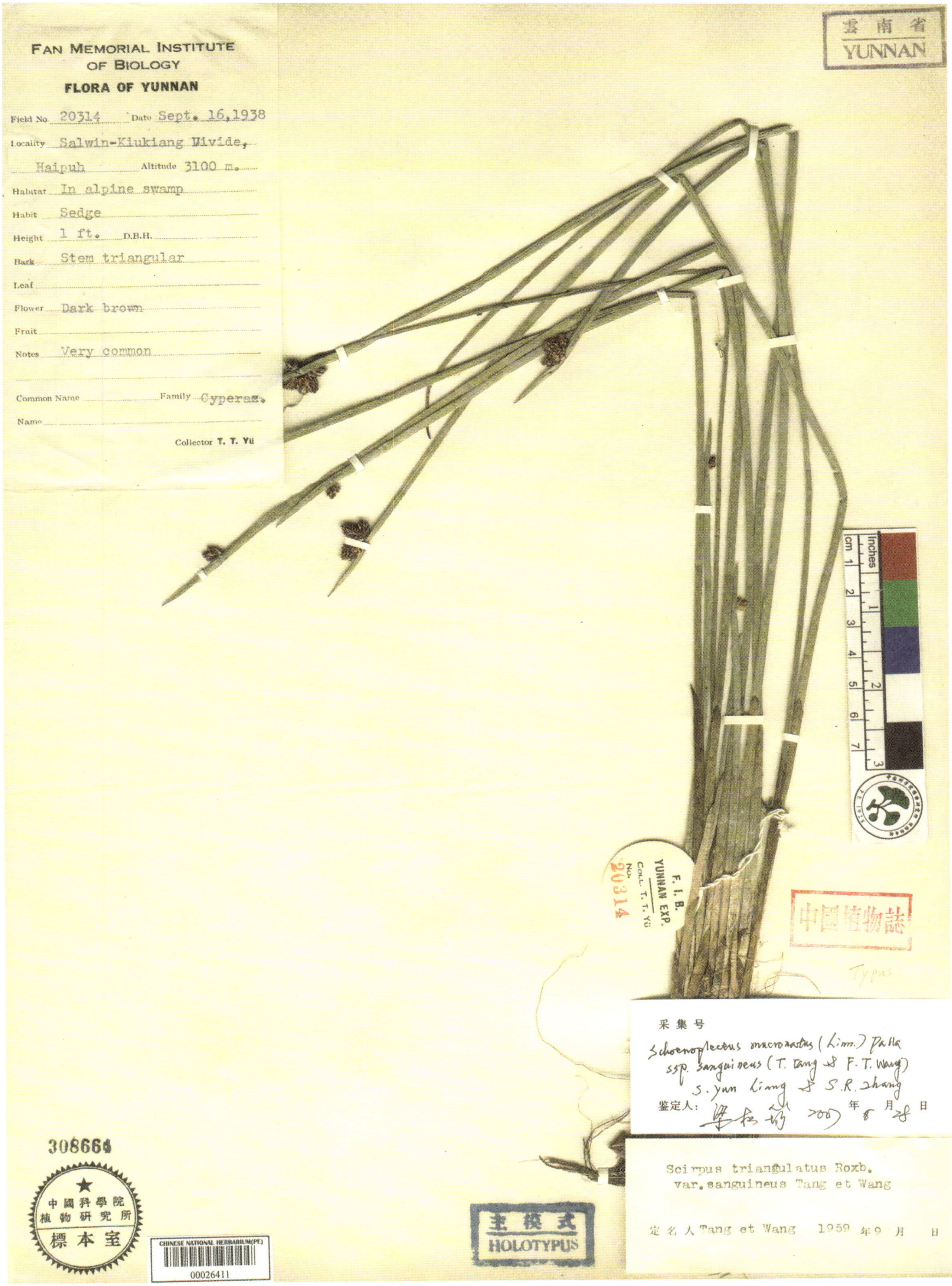

红鳞水毛花 ***Scirpus triangulates*** Roxb. var. ***sanguineus*** Tang & F. T. Wang, Fl. Reip. Pop. Sin. 11: 22, 222. 1961. **Holotype:** China. Yunnan: Salwin-kiukiang Divide, alt. 3100 m, 1938-09-16, T. T. Yu 20314.

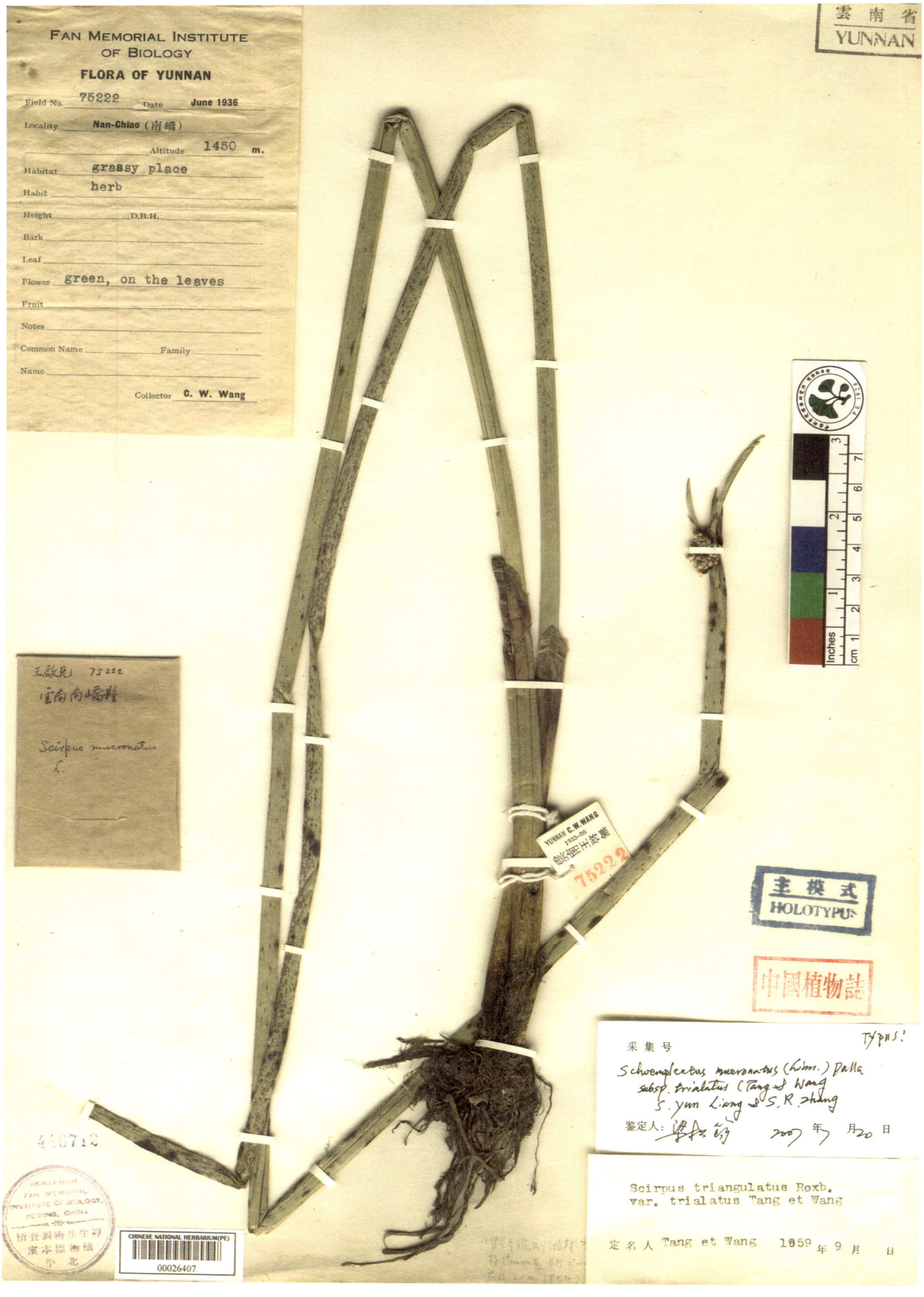

三翅水毛草 ***Scirpus triangulatus*** Roxb. var. ***trialatus*** Tang & F. T. Wang, Fl. Reip. Pop. Sin. 11: 22, 221, pl. 10: 18. 1961. **Holotype:** China. Yunnan: Menghai, alt. 1450 m, 1936-06-??, C. W. Wang 75222.

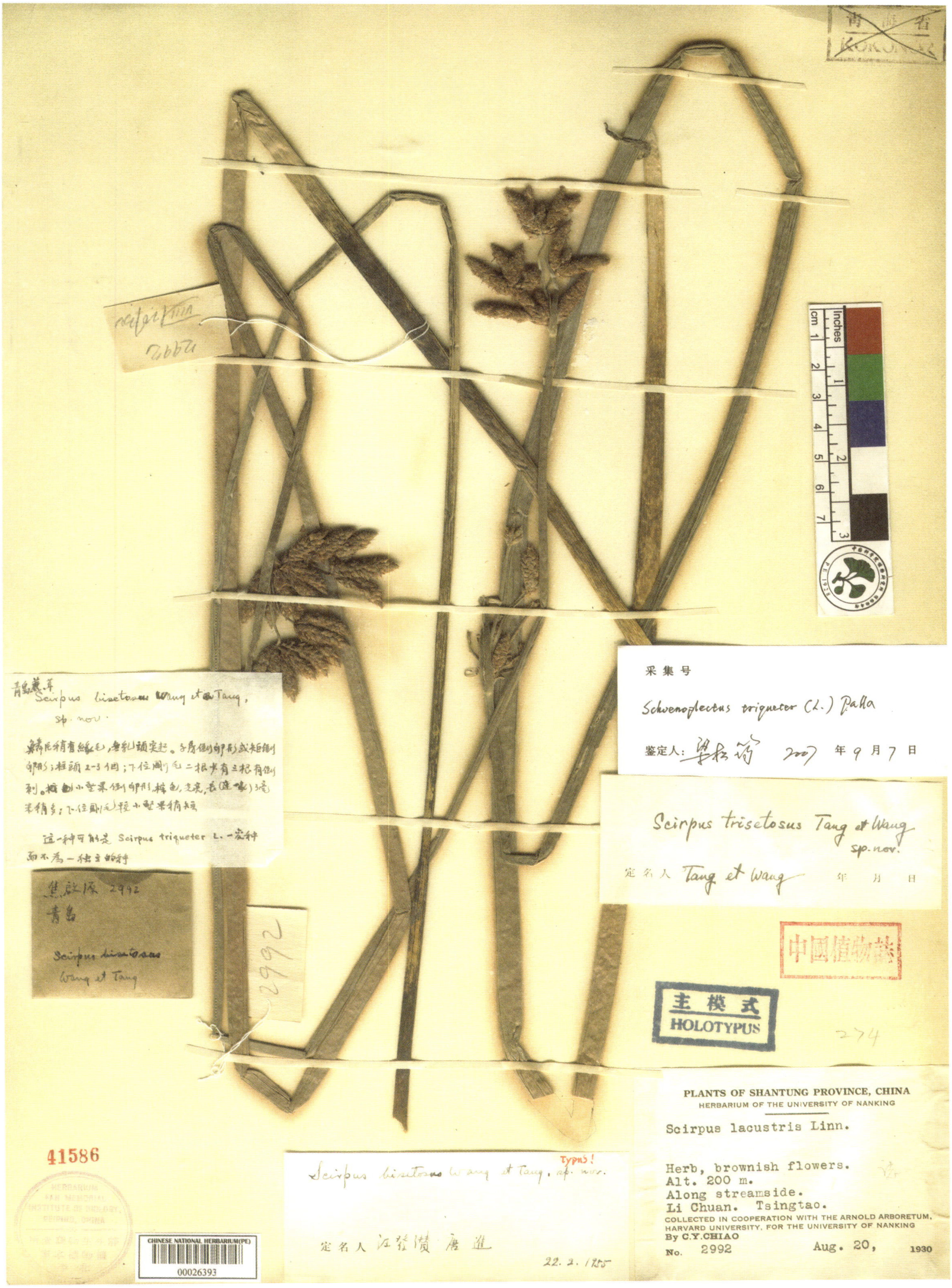

青岛藨草 ***Scirpus trisetosus*** Tang & F. T. Wang, Fl. Reip. Pop. Sin. 11: 18, 221, pl. 8: 1-5. 1961. **Holotype:** China. Shandong: Qingdao, alt. 200 m, 1930-08-20, C. Y. Chian 2992.

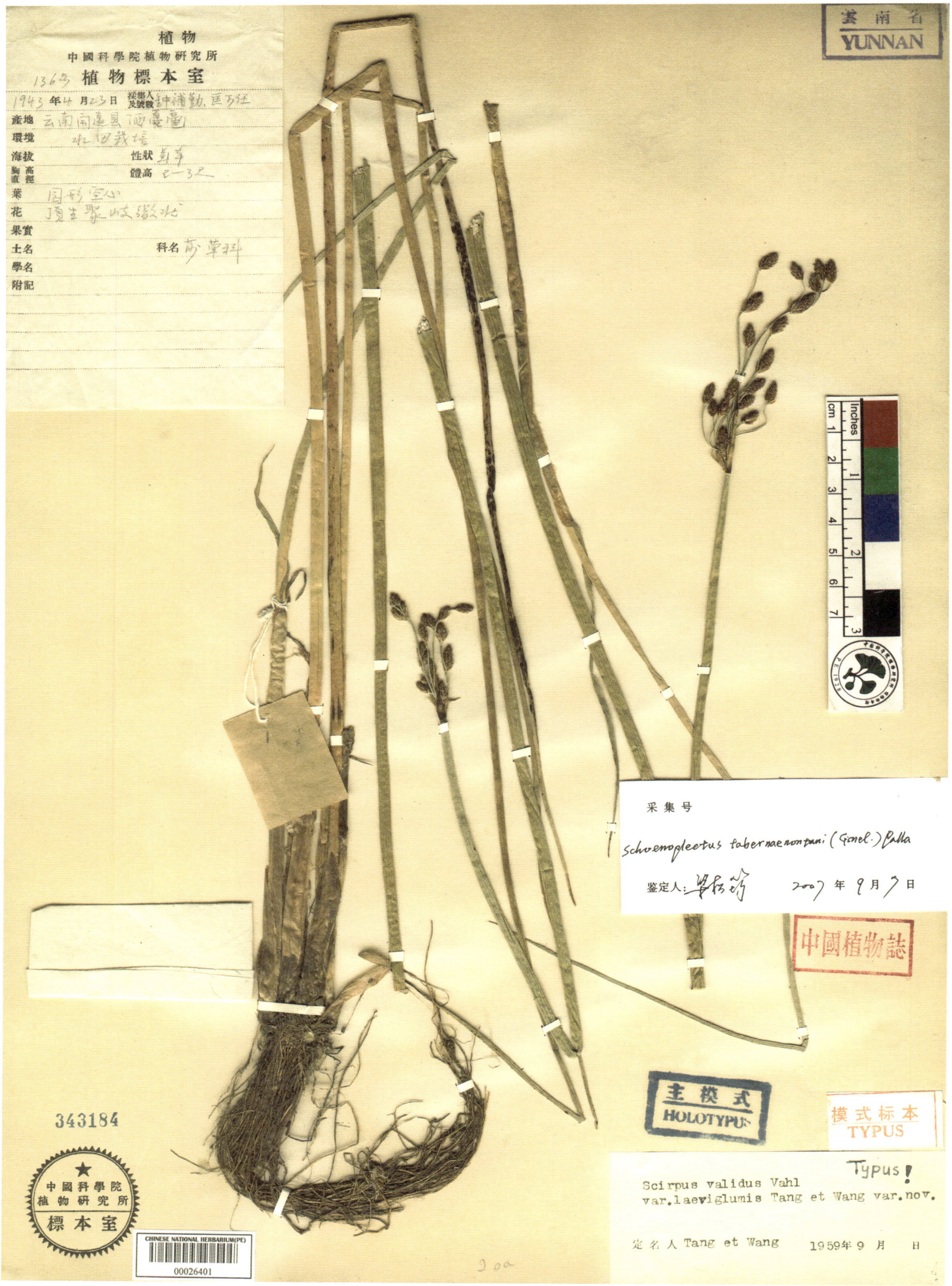

南水葱 ***Scirpus validus*** Vahl var. ***laeviglumis*** Tang & F. T. Wang, Fl. Reip. Pop. Sin. 11: 20, 221, pl. 9: 6-7. 1961. **Holotype:** China. Yunnan: Kaiyuan, 1943-04-23, P. C. Tsoong & K. R. Kuang 136.

广东珍珠茅 ***Scleria kwangtungensis*** Chun & How in Acta Phytotax. Sin. 7(1): 88, pl. 27. 1958. **Isotype:** China. Hainan: Changjiang, 1933-12-06, H. Y. Liang 66356.

光果珍珠茅 ***Scleria laeviformis*** Tang & f. T. Wang, Fl. Reip. Pop. Sin. 11: 135, 215, pl. 69: 1-3. 1961. **Holotype:** China. Yunnan: Jinping, 1956-04-19, Sino-Russia Yunnan Exped. 108.

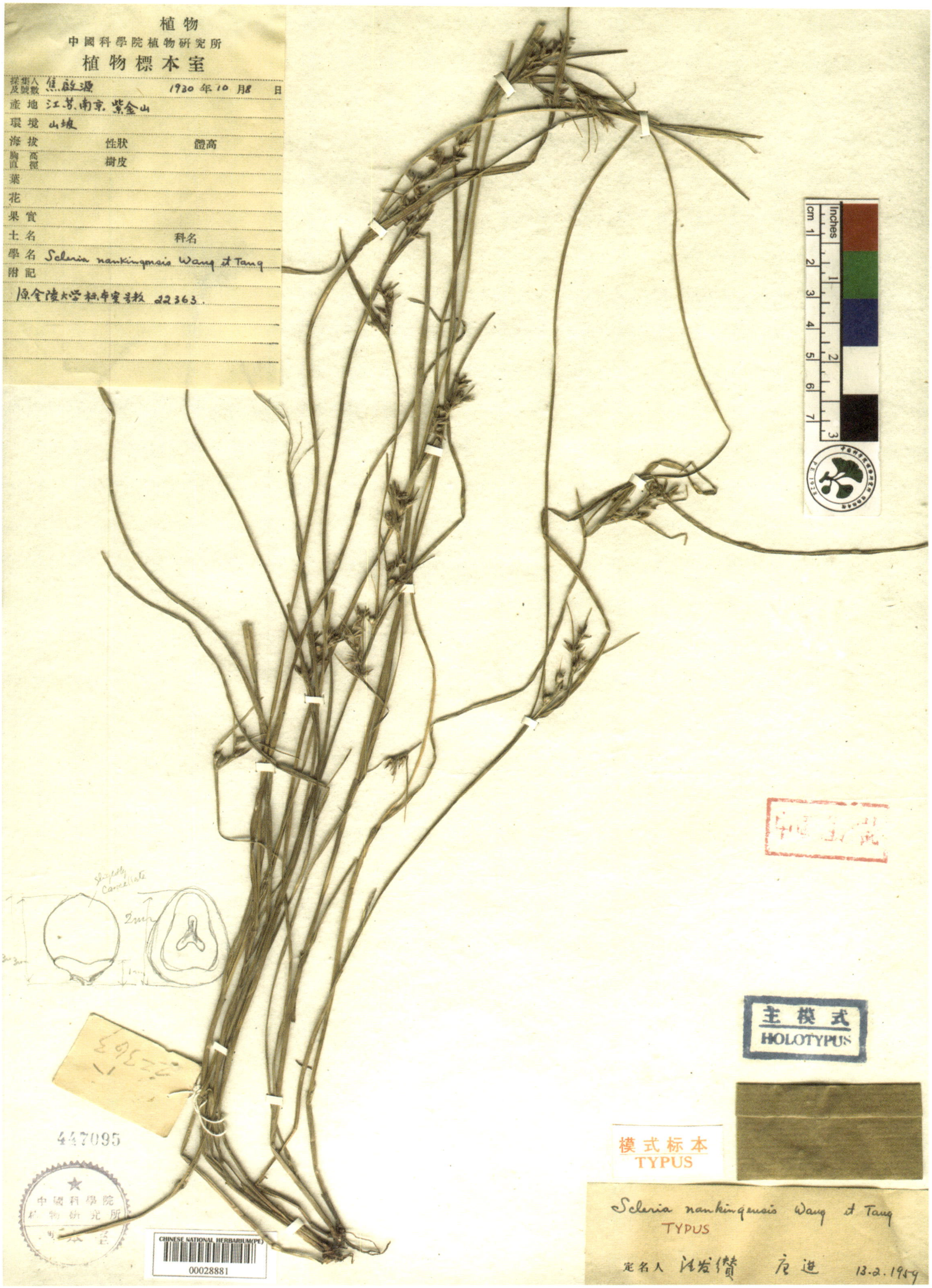

南京珍珠茅 ***Scleria nankingensis*** Tang & F. T. Wang, Fl. Reip. Pop. Sin. 11: 207, 234. 1961. **Holotype:** China. Jiangsu: Nanjing, 1930-10-08, C. Y. Chiao 22363.

Arecaceae

棕榈科

zonglüke

墨脱省藤 ***Calamus feanus*** Becc. var. ***medogensis*** S. J. Pei & S. Y. Chen in Acta Phytotax. Sin. 27(2): 137, f. 1: 10-12. 1989.
Holotype: China. Xizang: Mêdog, alt. 1850 m, 1974-09-12, Qinghai-Xizang Exped. 4725.

淡紫色省藤 ***Calamus vinosus*** Becc. in Leafl. Philipp. Bot. 8: 3061. 1919. **Isotype:** Philippines. Agusan: Cabadbaran, alt. 1600 m, 1912-10-??, A. D. E. Elmer 14158.

乌达内塔黄藤 ***Daemonorops urdanetana*** Becc. in Leafl. Philipp. Bot. 8: 3038. 1919. **Isotype:** Philippines. Agusan: Cabadbaran, alt. 1830 m, 1912-10-??, A. D. E. Elmer 14201.

Araceae

天南星科

tiannanxingke

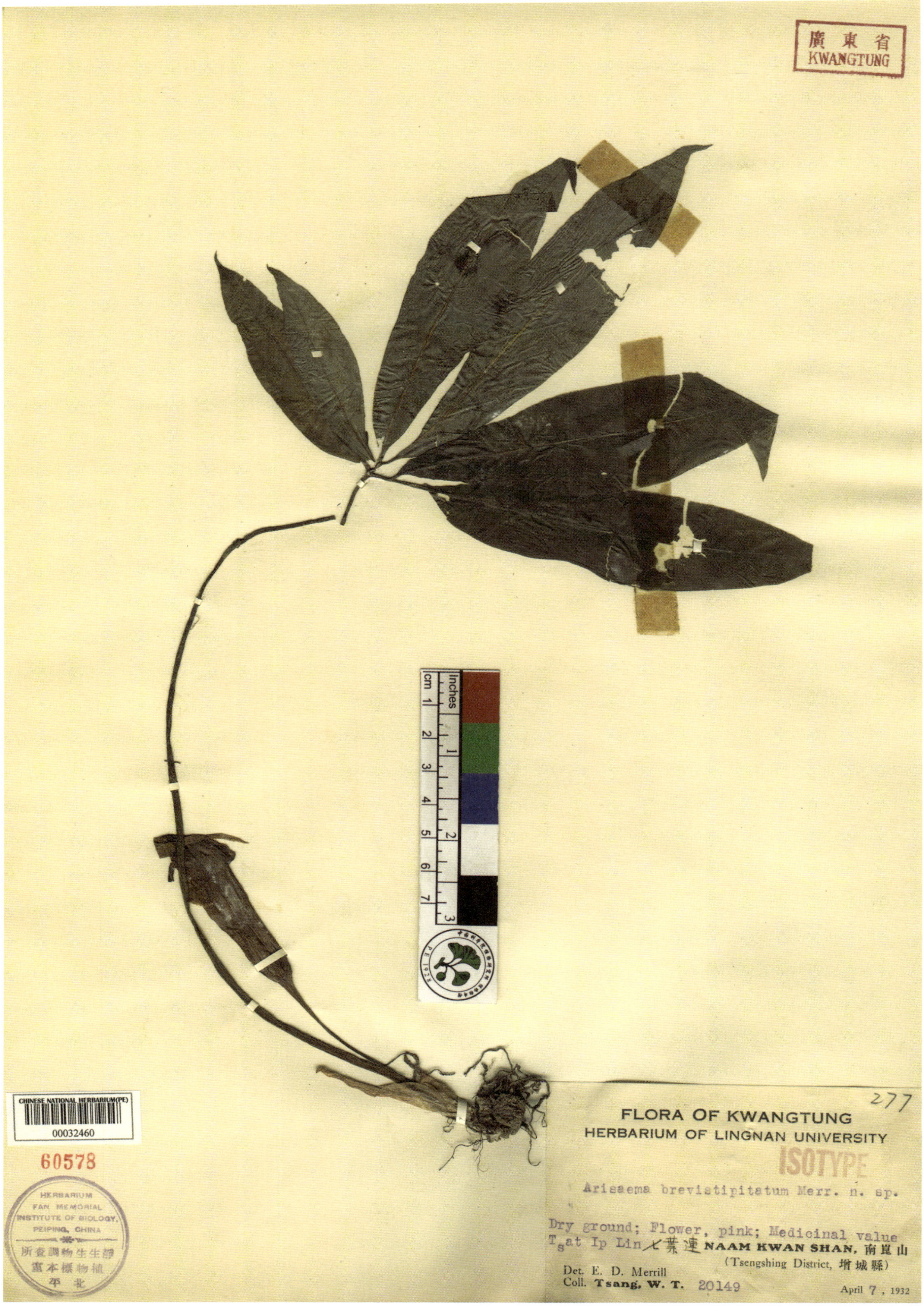

短苞南星 ***Arisaema brevistipitatum*** Merr. in Lingnan Sci. Journ. 13(1): 19. 1934. **Isotype:** China. Guangdong: Zengcheng, 1932-04-07, W. T. Tsang 20149.

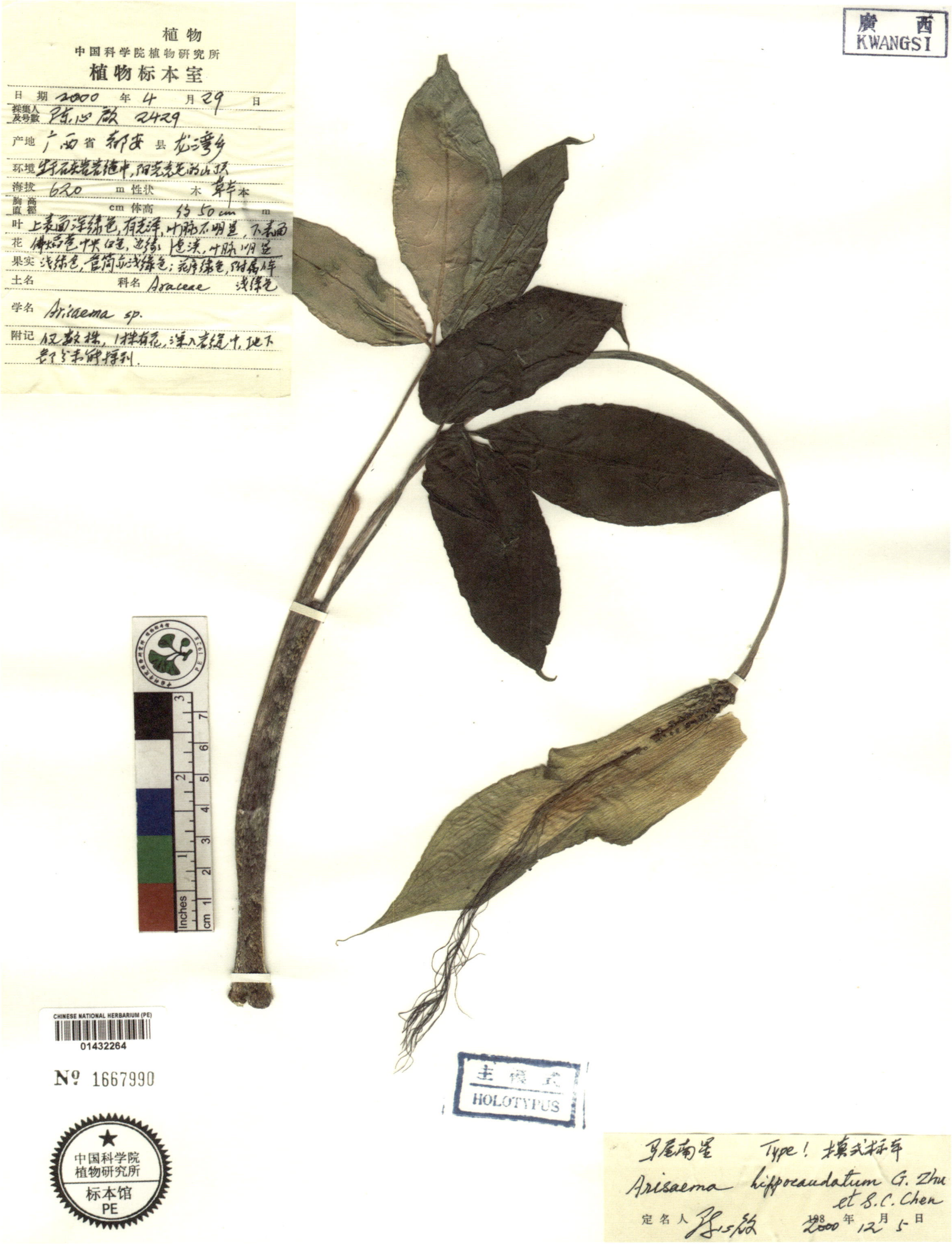

马尾南星 ***Arisaema hippocaudatum*** S. C. Chen & H. Li in Acta Bot. Yunnan. 24(5): 607, f. 1: 1-4. 2002. **Holotype:** China. Guangxi: Duan, alt. 620 m, 2000-04-29, S. C. Chen 2429.

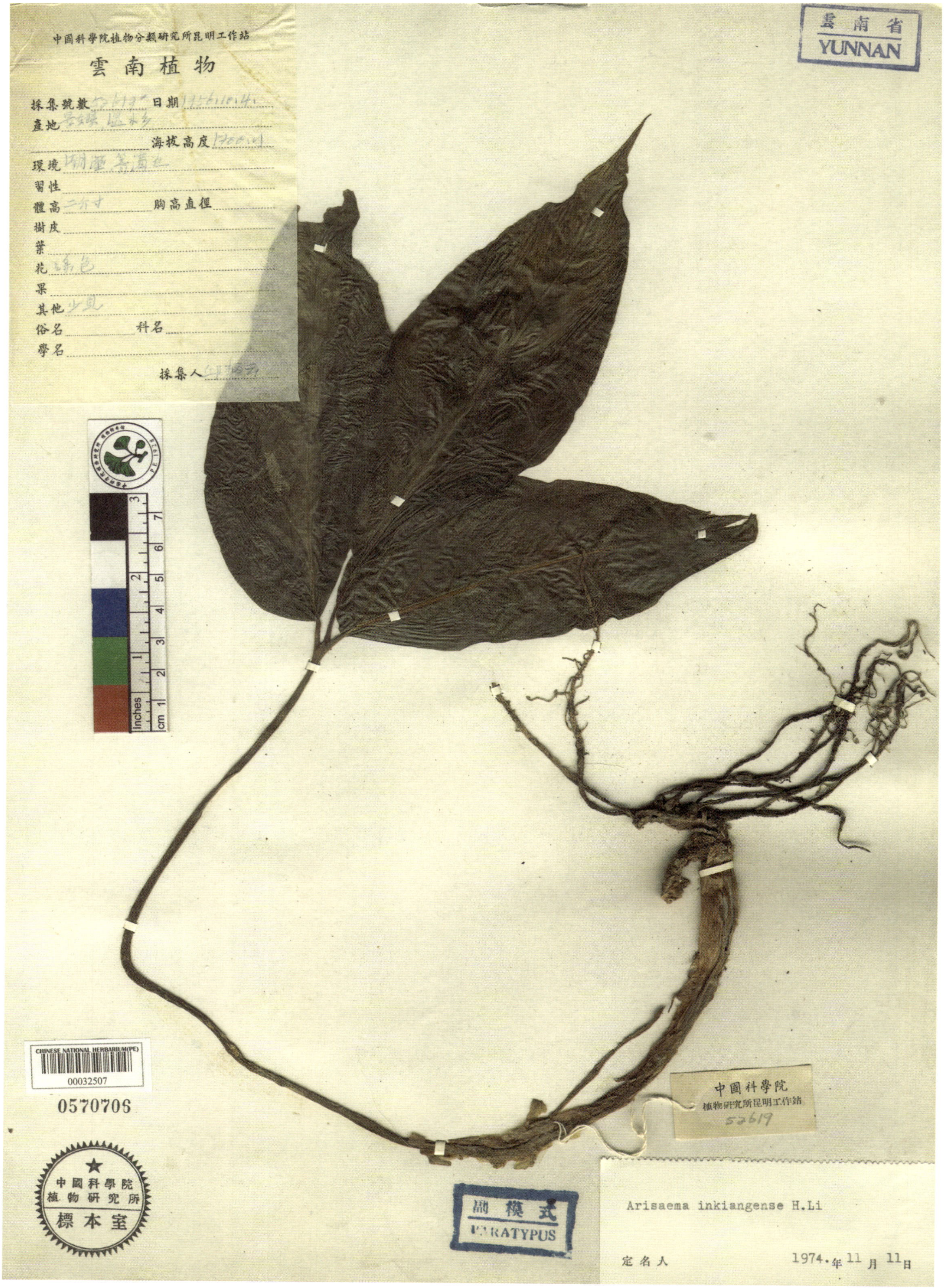

三匹箭 ***Arisaema inkiangense*** H. Li in Acta Phytotax. Sin. 15(2): 106, pl. 9: 1(6-12). 1977. **Paratype:** China. Yunnan: Jingdong, alt. 1700 m, 1956-10-04, B. Y. Qiu 52619.

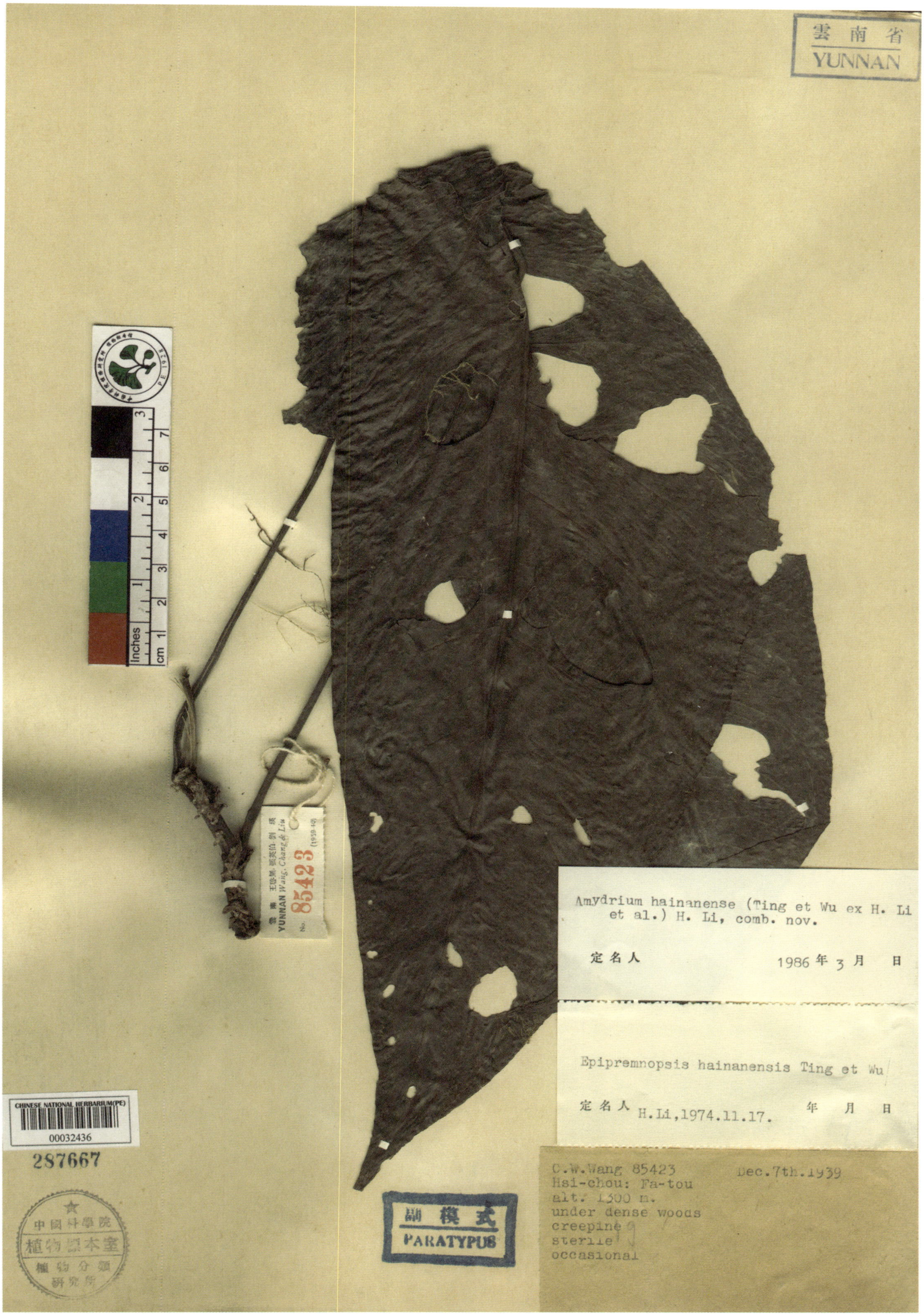

穿心藤 ***Epipremnopsis hainanensis*** K. C. Ting & T. L. Wu ex H. Li & al. in Acta Phytotax. Sin. 15(2): 102, pl. 5: 1. 1977.
Paratype: China. Yunnan: Xichou, alt. 1300 m, 1939-12-07, C. W. Wang 85423.

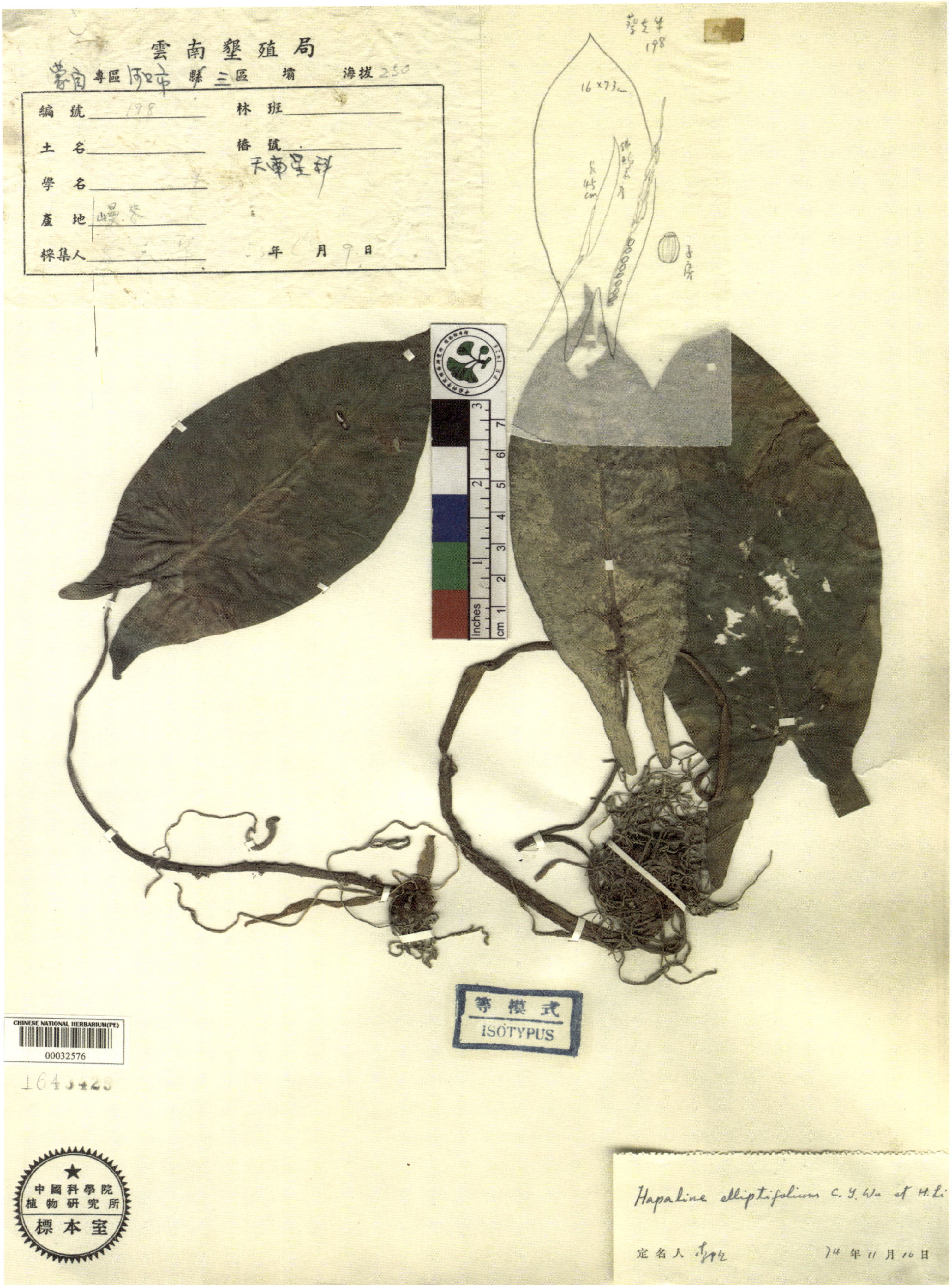

细柄芋 ***Hapaline ellipticifolium*** C. Y. Wu & H. Li in Acta Phytotax. Sin. 15(2): 104, pl. 7: 1. 1977. **Isotype:** China. Yunnan: Hekou, alt. 250 m, 1953-04-09, K. H. Cai 198.

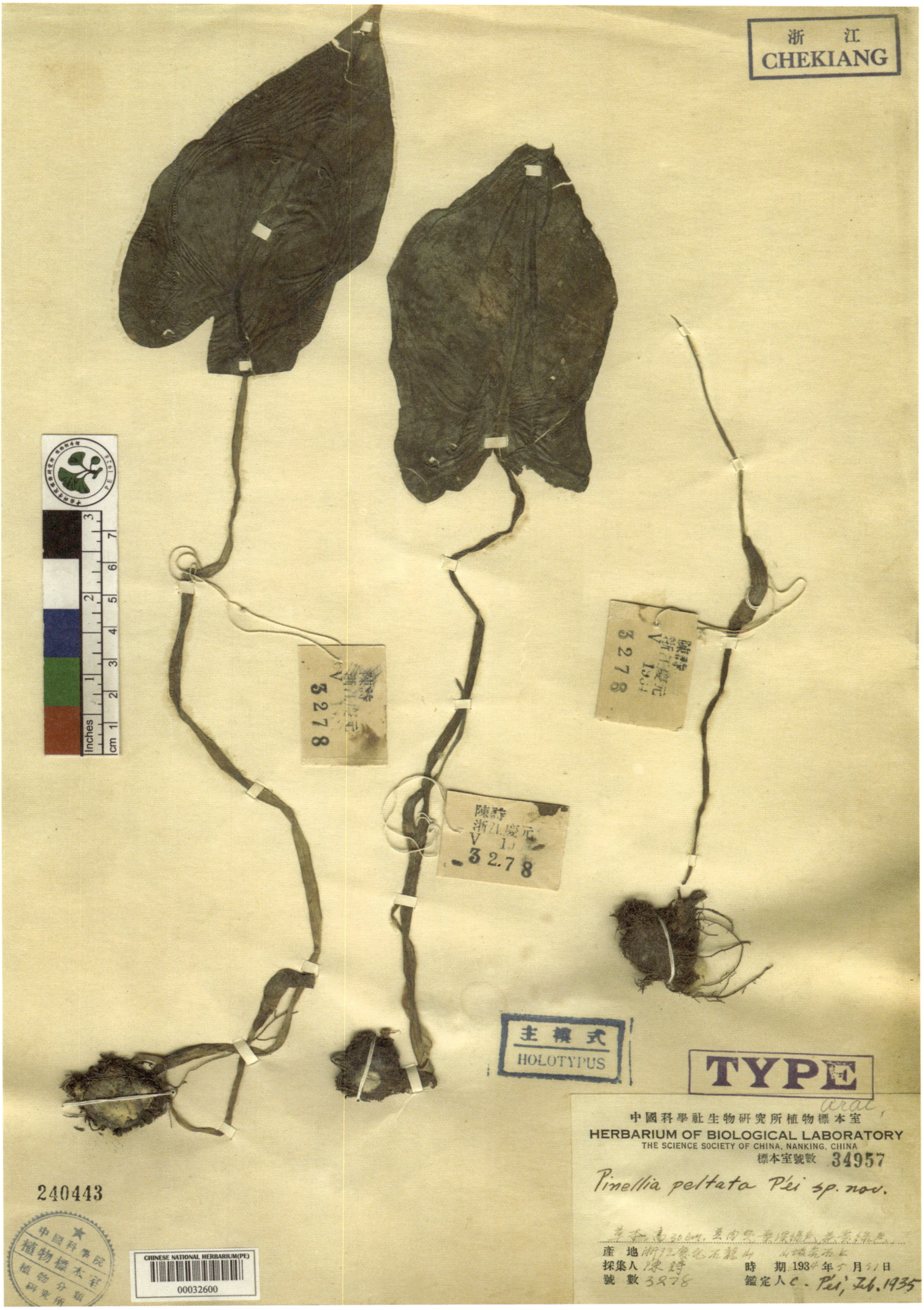

盾叶半夏 ***Pinellia peltata*** C. Pei in Contr. Biol. Lab. Sci. Soc. China 10(1): 1, f. 1. 1935. **Holotype:** China. Zhejiang: Qingyuan, 1934-05-31, S. Chen 3278.

长柄石柑 ***Pothos chinensis*** (Raf.) Merr. var. ***lotienensis*** C. Y. Wu & H. Li in Acta Phytotax. Sin. 15(2): 101. 1977. **Isotype:** China. Guizhou: Luodian, alt. 520 m, 1959-04-08, S Guizhou Exped. 731.

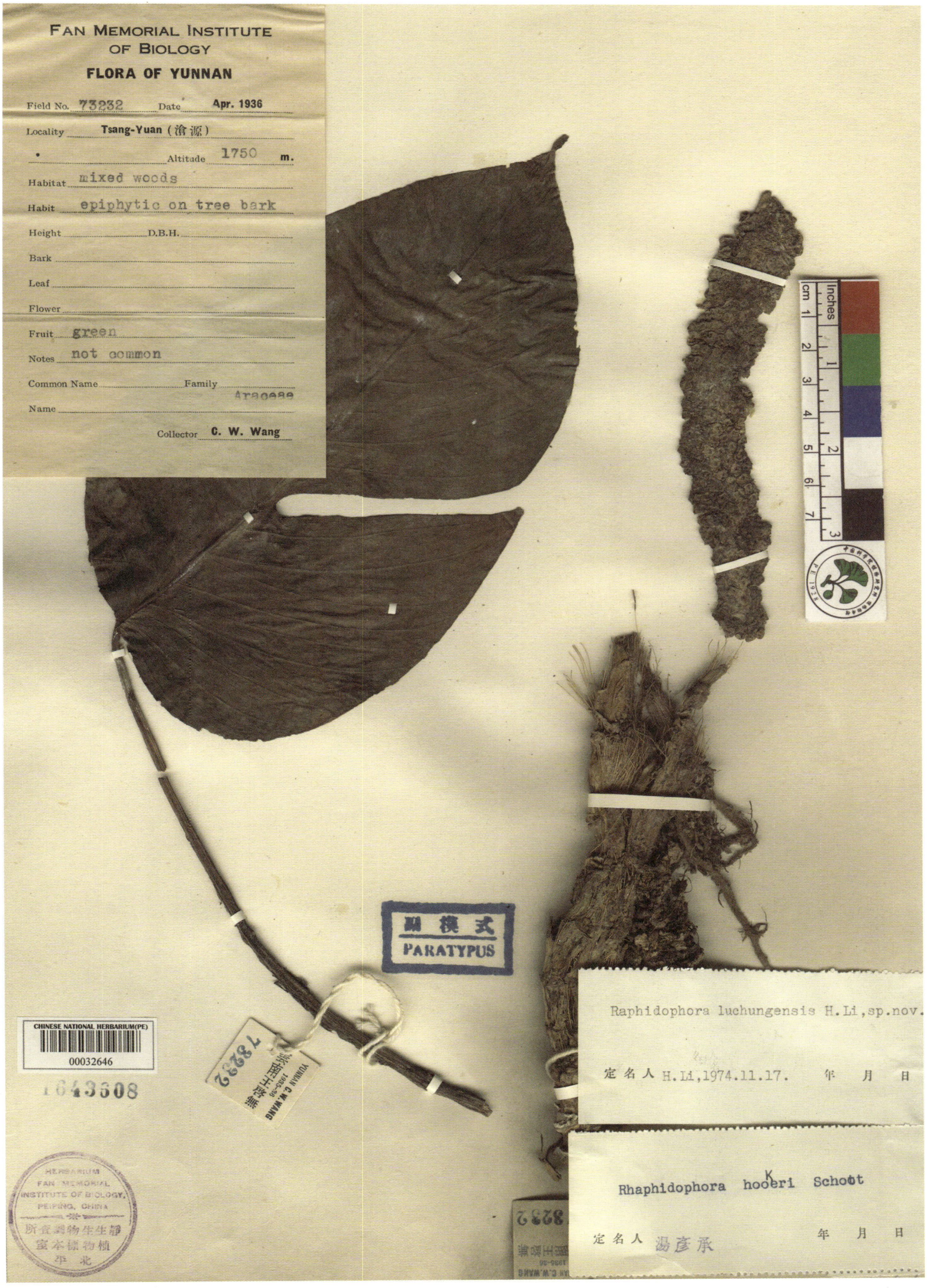

绿春崖角藤 ***Rhaphidophora luchunensis*** H. Li in Acta Phytotax. Sin. 15(2): 103, pl. 5: 2(1-5), 6: 1-3. 1977. **Paratype:** China. Yunnan: Cangyuan, alt. 1750 m, 1936-04-??, C. W. Wang 73232.

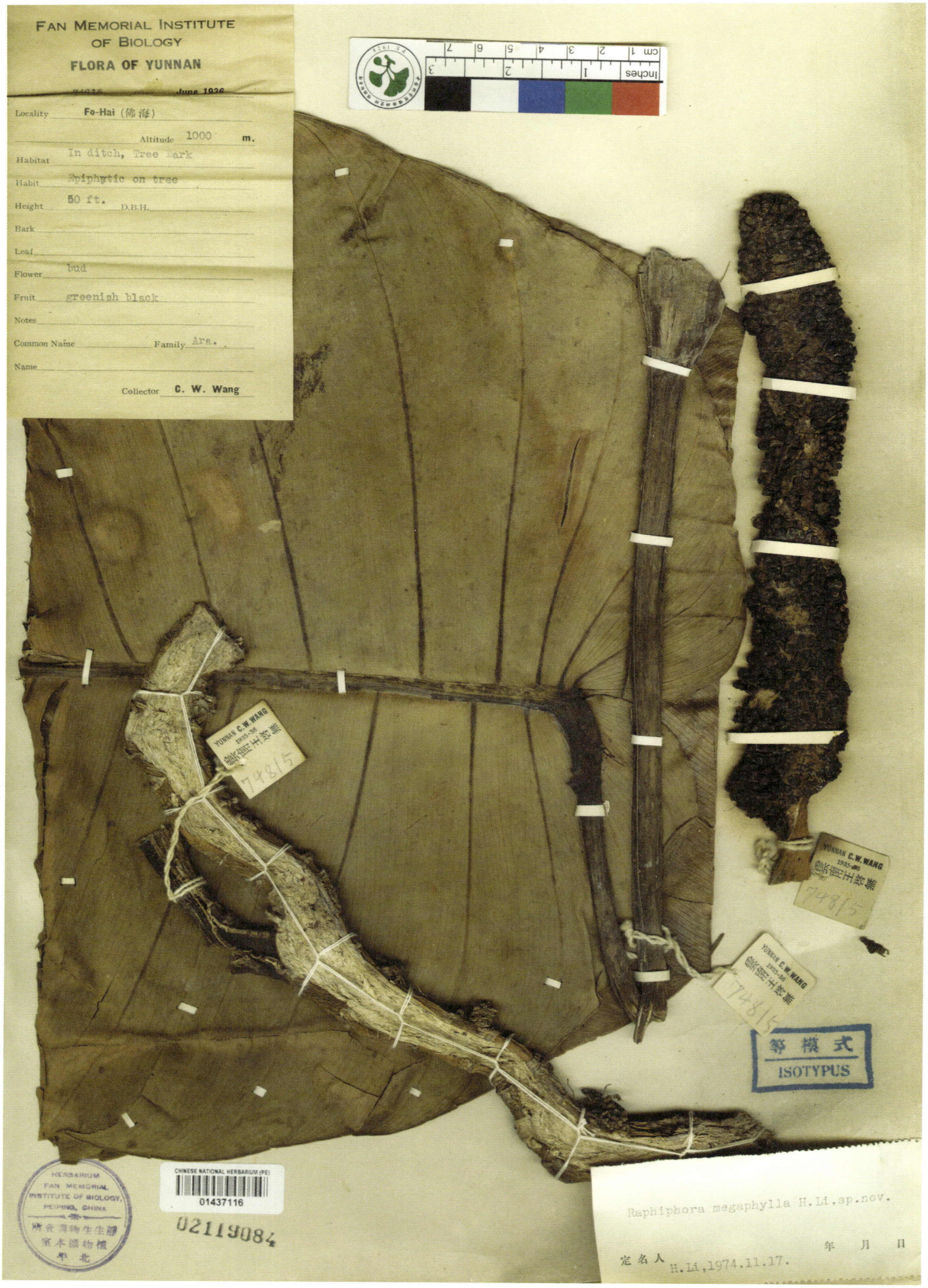

大叶崖角藤 ***Rhaphidophora megaphylla*** H. Li in Acta Phytotax. Sin. 15(2): 102, pl. 5: 2(6-10), 6: 4. 1977. **Isotype:** China. Yunnan: Fohai (=Menghai), alt. 1000 m, 1936-06-??, C. W. Wang 74815.

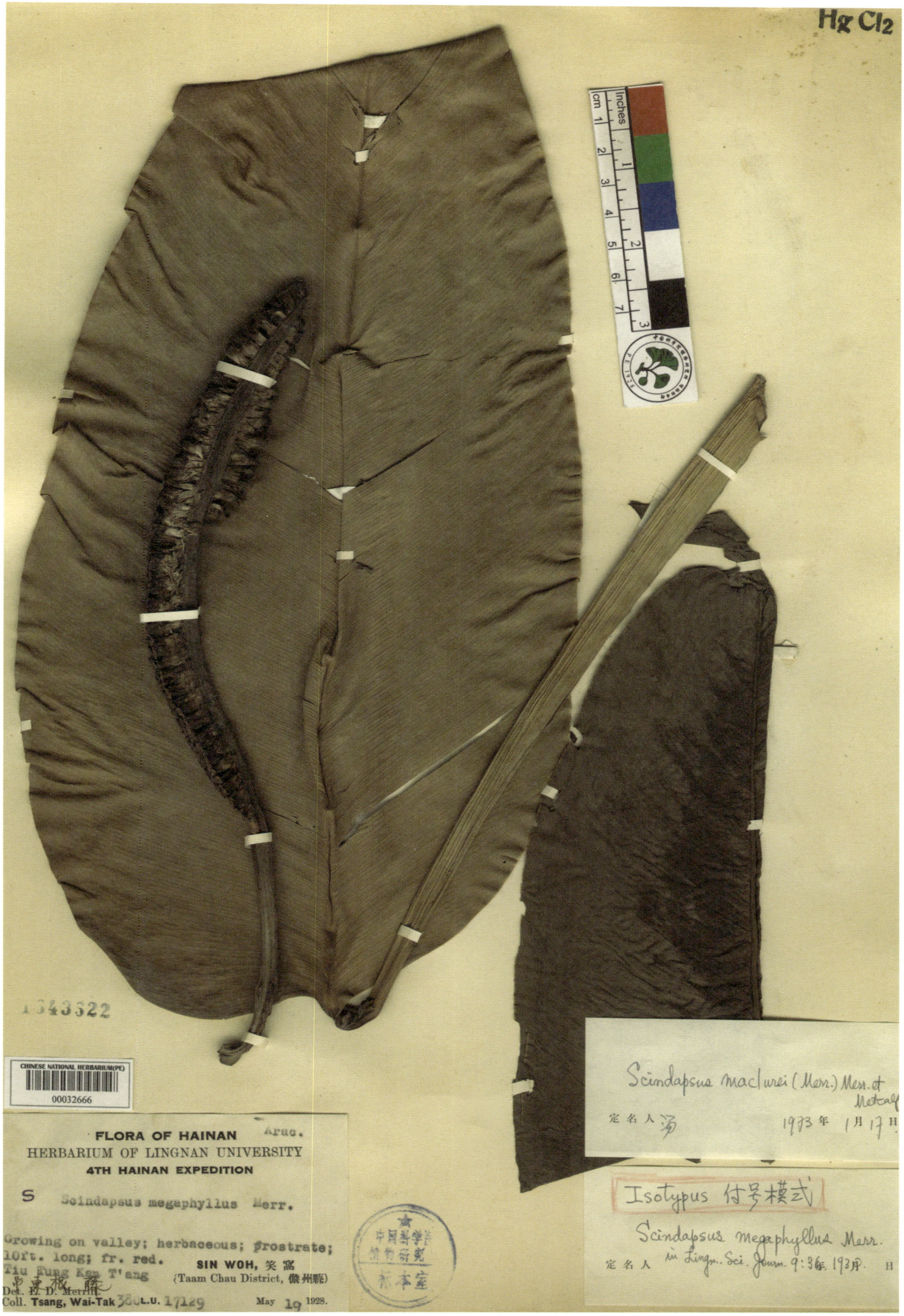

海南藤芋 ***Scindapsus megaphyllus*** Merr. in Lingnan Sci. Journ. 9(1-2): 36. 1930. **Isotype:** China. Hainan: Dan Xian, 1928-05-19, W. T. Tsang 17129.

西南犁头尖 ***Typhonium omeiense*** H. Li in Acta Phytotax. Sin. 15(2): 105, pl. 7: 2(6), 8: 4. 1977. **Isotype:** China. Sichuan: Muchuan, 1959-05-08, Sichuan Econ. Pl. Exped. 2138.

三叶犁头尖 ***Typhonium trifoliatum*** F. T. Wang & Lo ex H. Li & al. in Acta Phytotax. Sin. 15(2): 105, pl. 7: 2(4-5). 1977.
Holotype: China. Hebei: Precise locality not known, 1929-??-??, T. Tang 1224.

Centrolepidaceae

刺鳞草科

cilincaoke

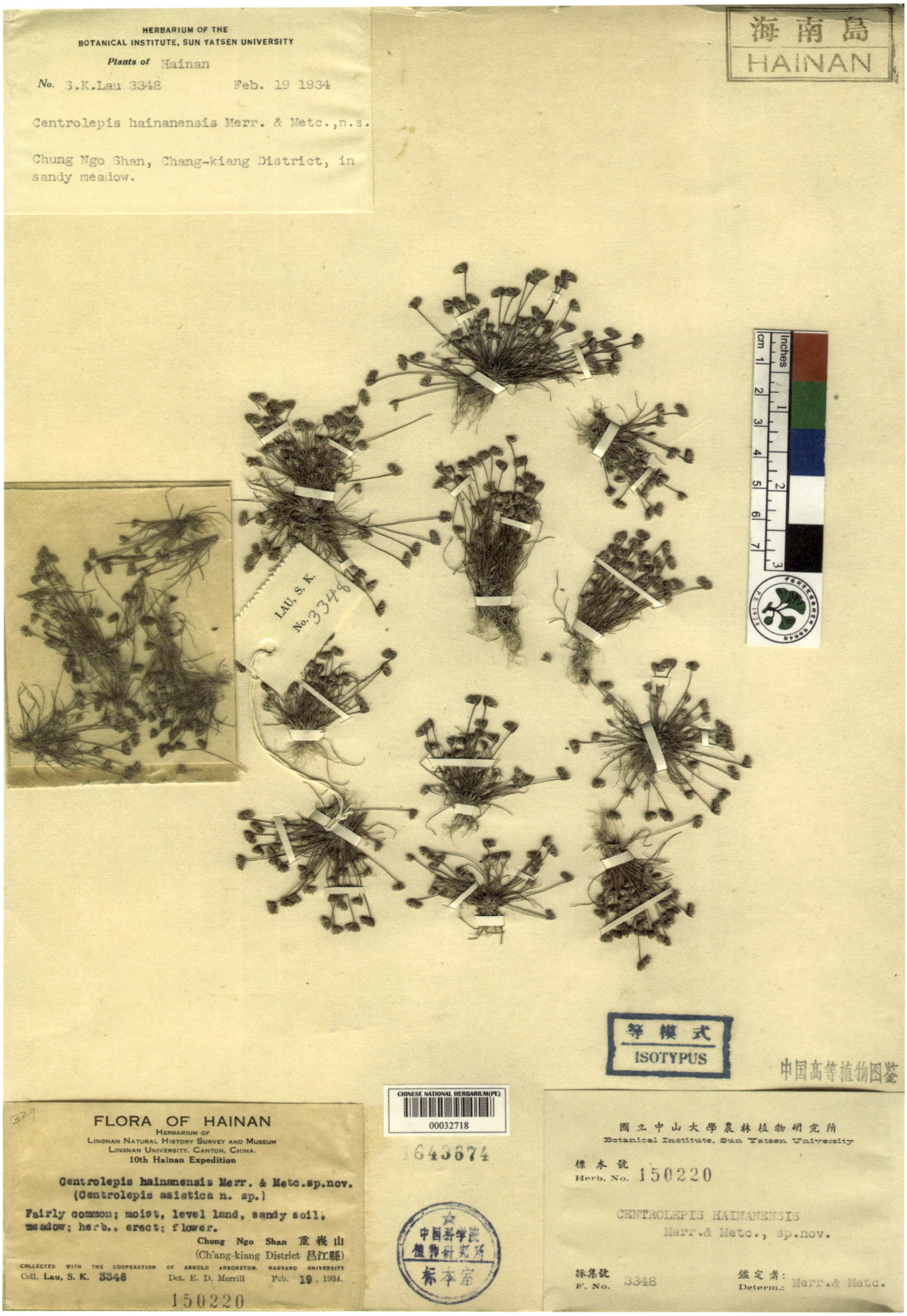

海南刺鳞草 ***Centrolepis hainanensis*** Merr. & F. P. Metc. in Lingnan Sci. Journ. 16(2): 185, f. 3. 1937. **Isotype:** China. Hainan: Changjiang, 1934-02-19, S. K. Lau 3348.

Eriocaulaceae

谷精草科

gujingcaoke

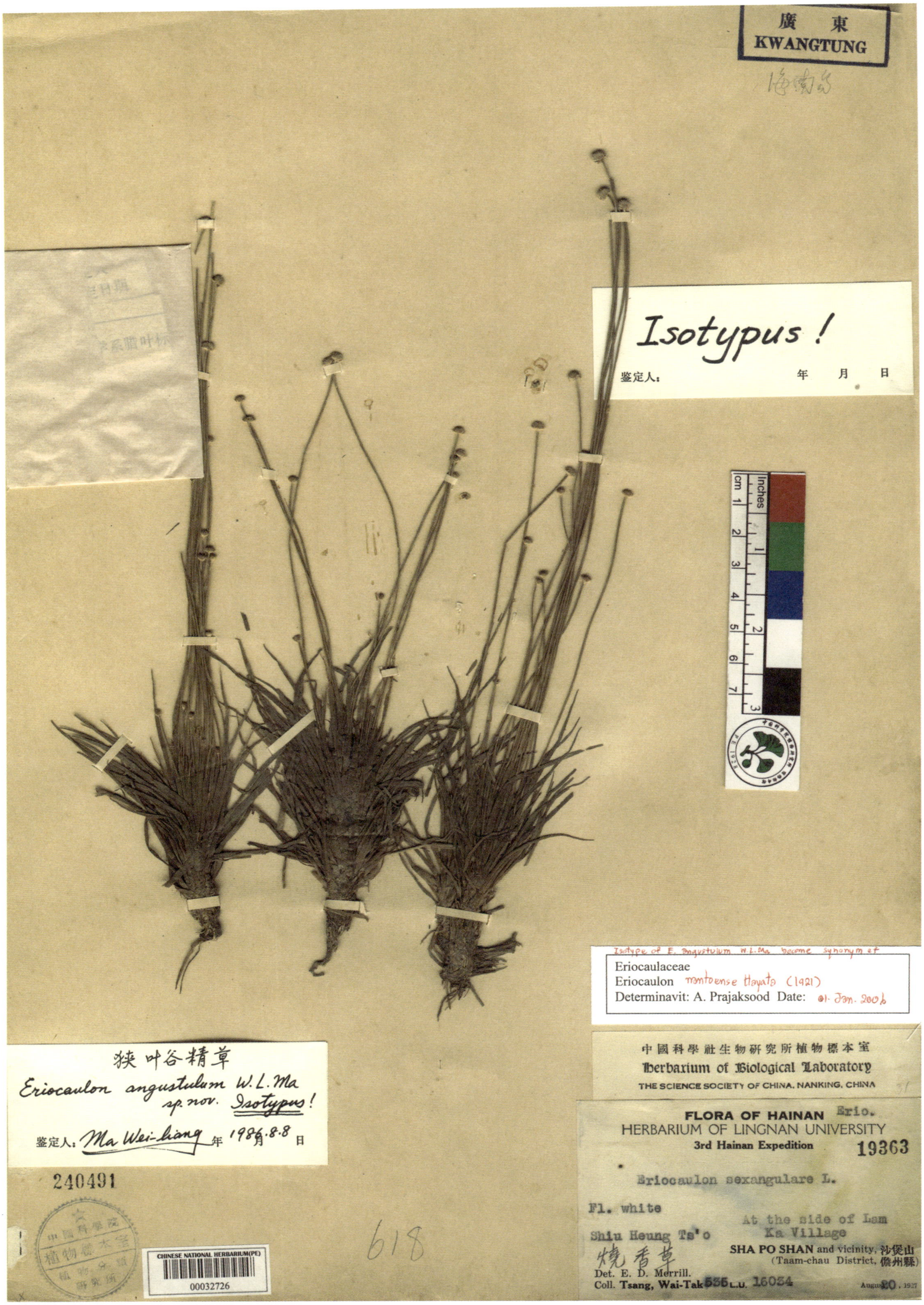

狭叶谷精草 ***Eriocaulon angustulum*** W. L. Ma in Acta Phytotax. Sin. 29(4): 295, f. 2: 1-15. 1991. **Isotype:** China. Hainan: Dan Xian, 1927-08-20, W. T. Tsang 535.

裂瓣谷精草 ***Eriocaulon bilobatum*** W. L. Ma in Acta Phytotax. Sin. 29(4): 301, f. 3: 9-20. 1991. **Holotype:** China. Yunnan: Binchuan, Jizushan, 1946-12-24, T. N. Liou 22187.

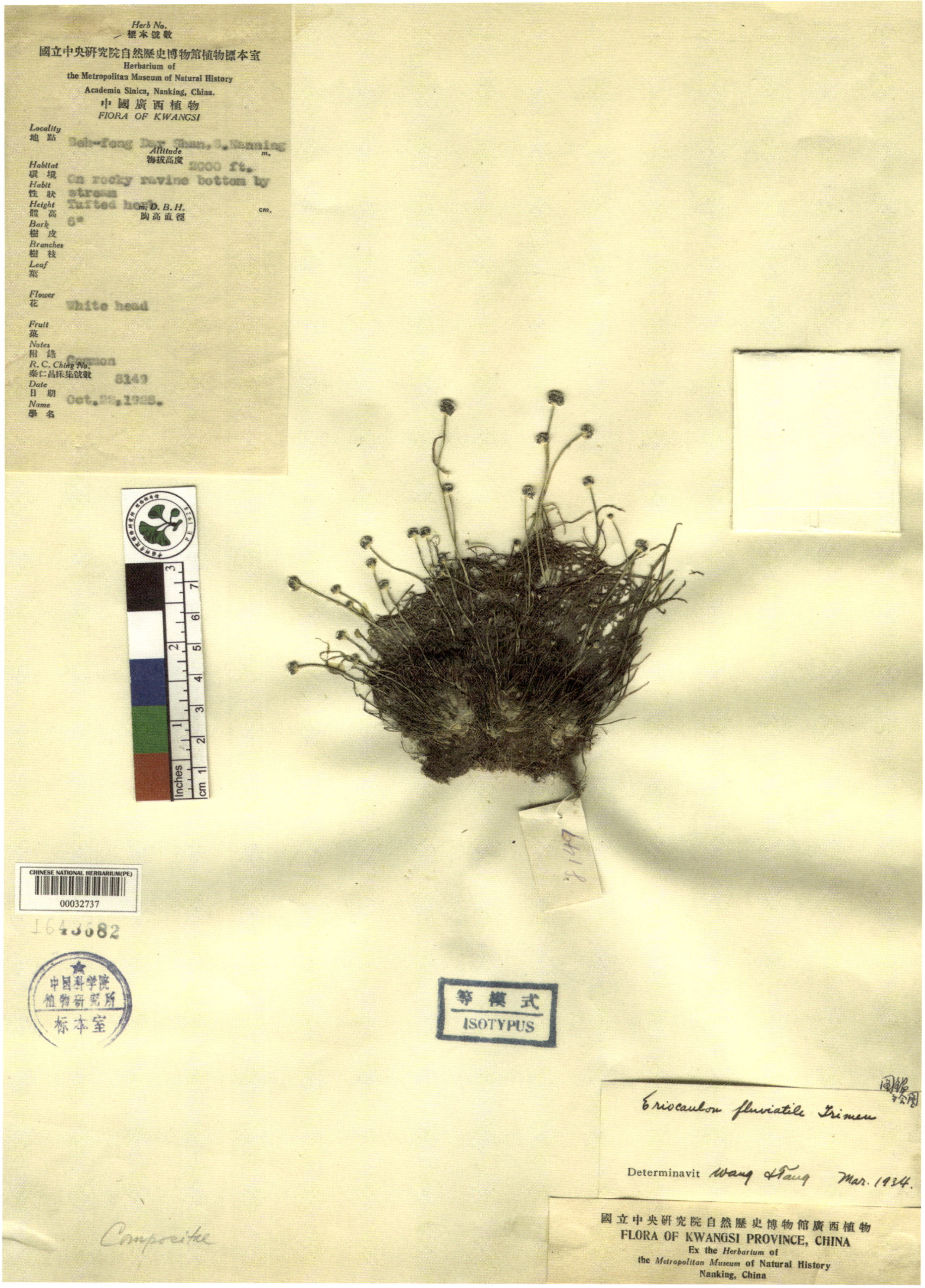

线叶谷精草 ***Eriocaulon filifolium*** Hand.-Mazz. in Sinensia 7(5): 619. 1936. **Isotype:** China. Guangxi: Nanning, alt. 700 m, 1926-10-22, R. C. Ching 8149.

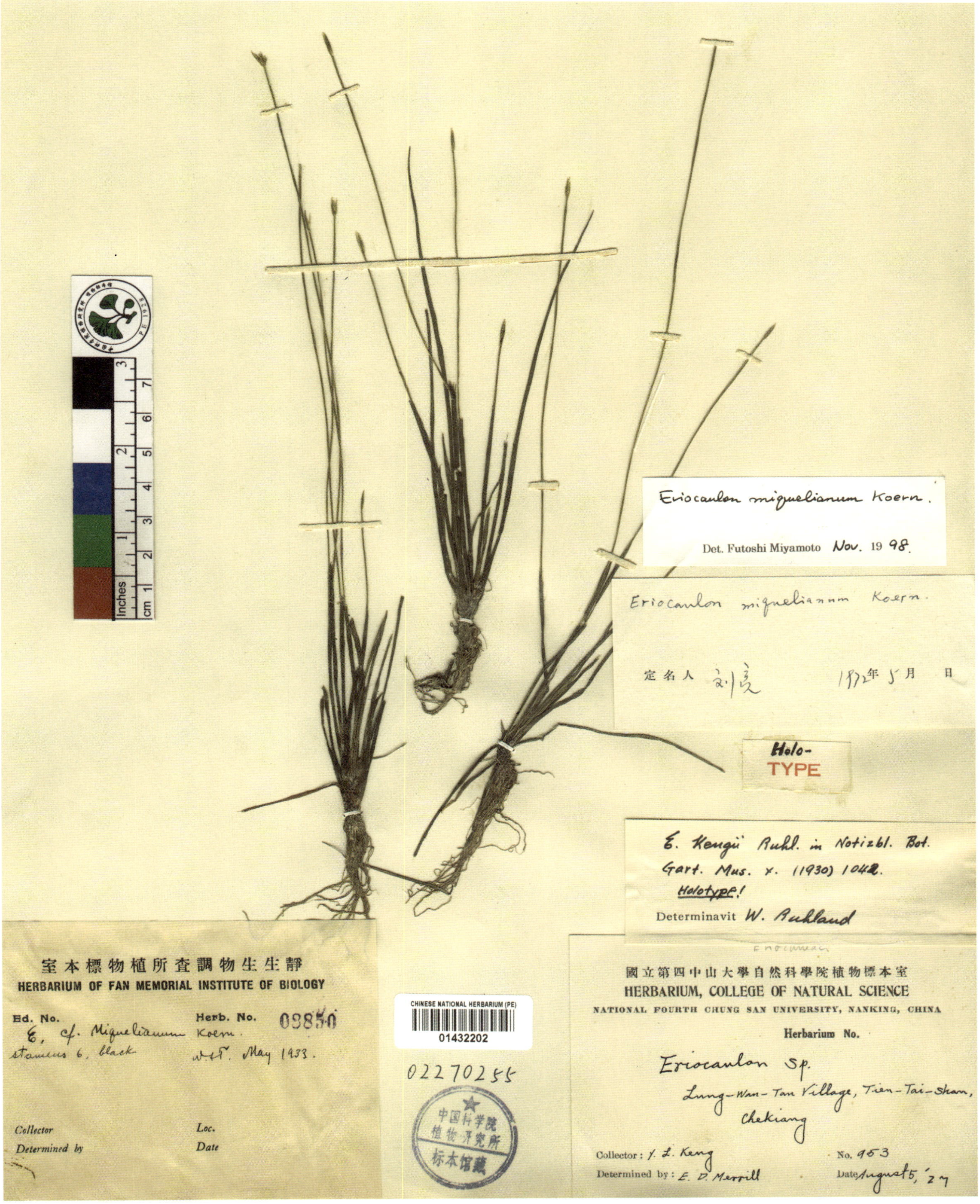

以礼谷精草 ***Eriocaulon kengii*** Ruhl. in Notizbl. Bot. Gart. Berlin 10: 1042. 1930. **Holotype:** China. Chekiang (=Zhejiang): Tiantaishan, 1927-08-05, Y. L. Keng 953.

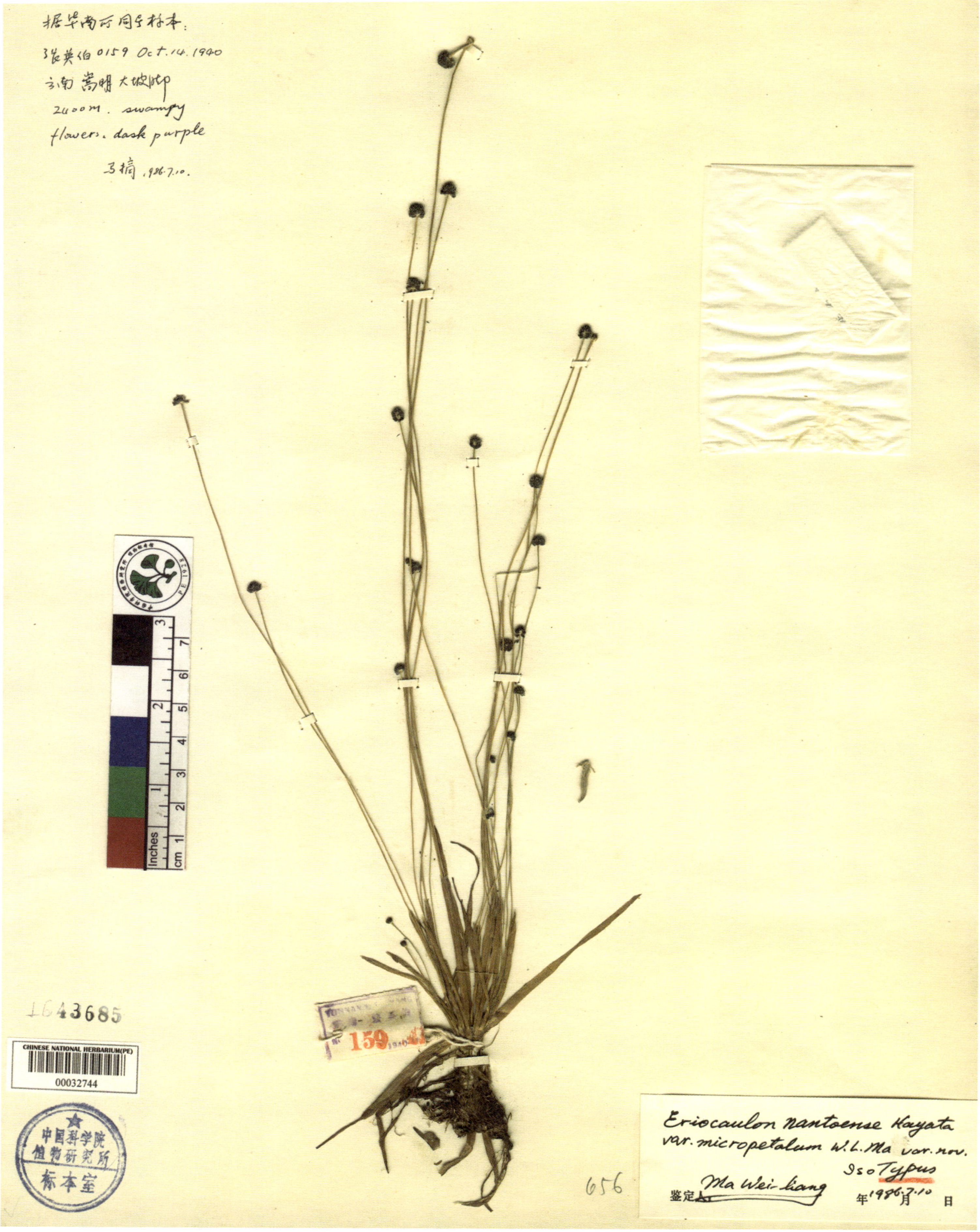

小瓣谷精草 ***Eriocaulon nantoense*** Hayata var. ***micropetalum*** W. L. Ma in Acta Phytotax. Sin. 29(4): 293. 1991. **Isotype:** China. Yunnan: Songming, 1940-10-14, Y. B. Zhang 159.

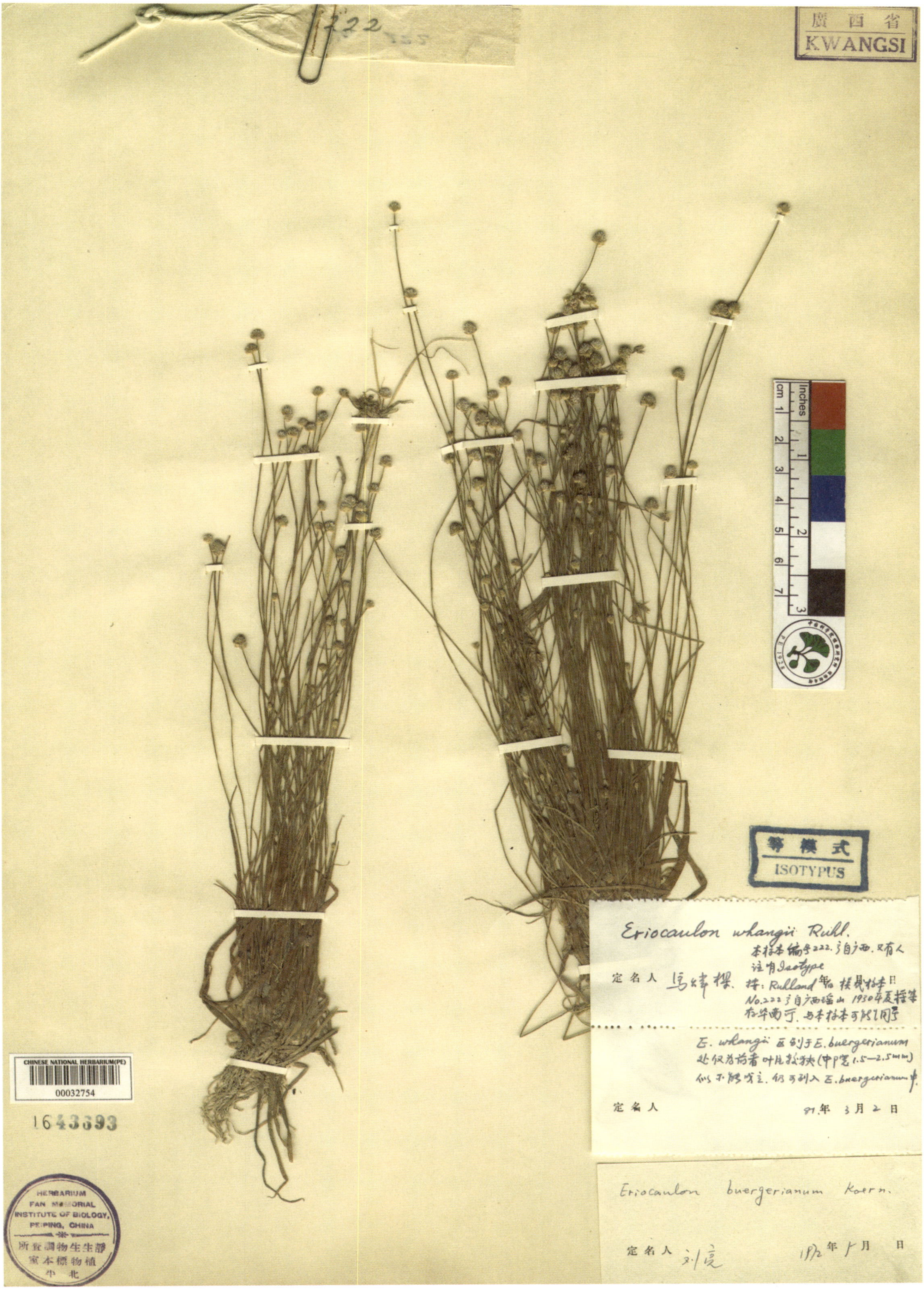

黄氏谷精草 ***Eriocaulon whangii*** Rhul. in Notizbl. Bot. Gart. Berlin 10: 1040. 1930. **Isotype:** China. Guangxi: Yaoshan, 1928-??-??, S. S. Sin & K. K. Whang 222.

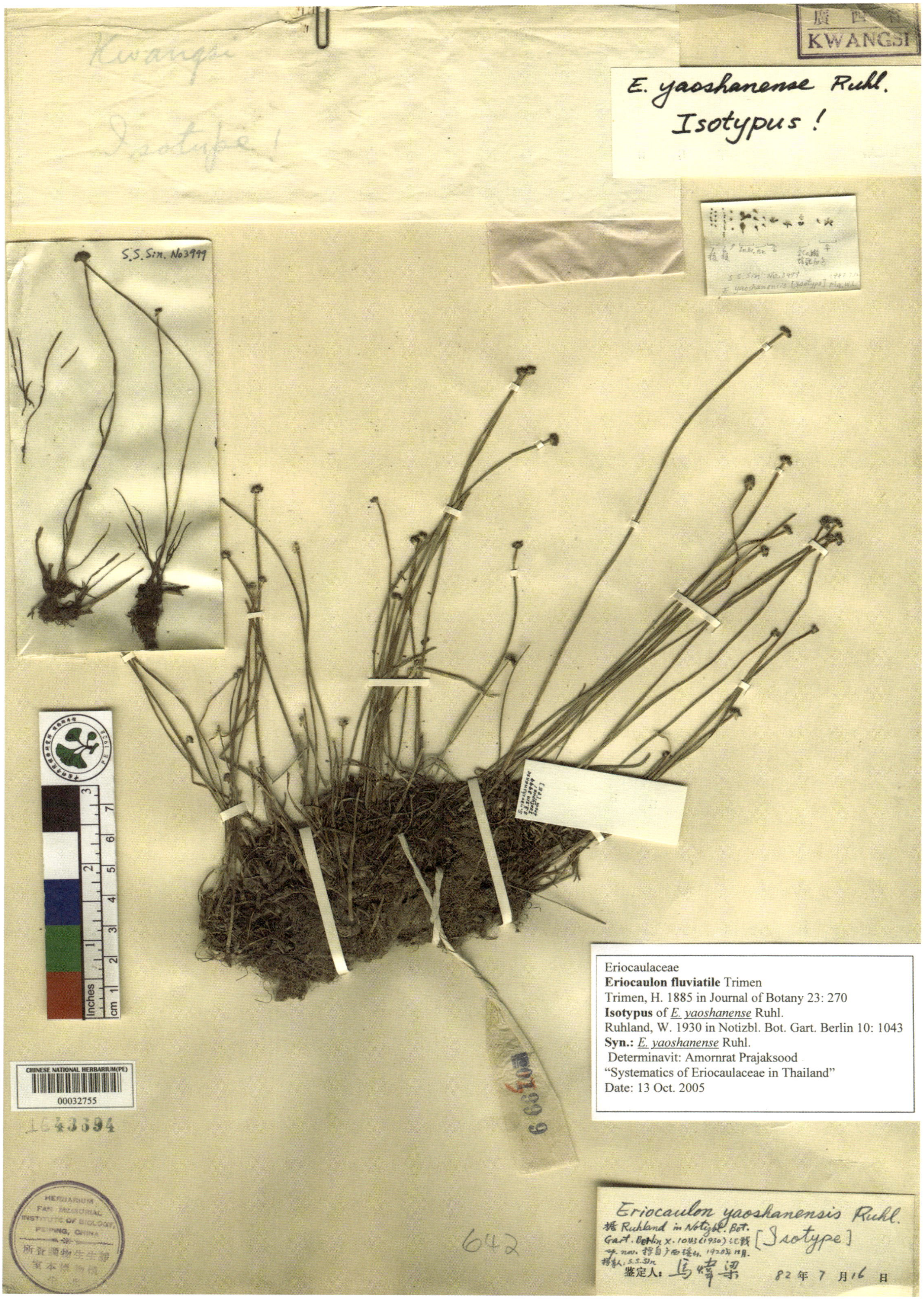

瑶山谷精草 ***Eriocaulon yaoshanense*** Ruhl. in Notizbl. Bot. Gart. Berlin 10: 1043. 1930. **Isotype:** China. Guangxi: Yaoshan, 1928-12-??, S. S. Sin 1999.

Commelinaceae

鸭跖草科

yazhicaoke

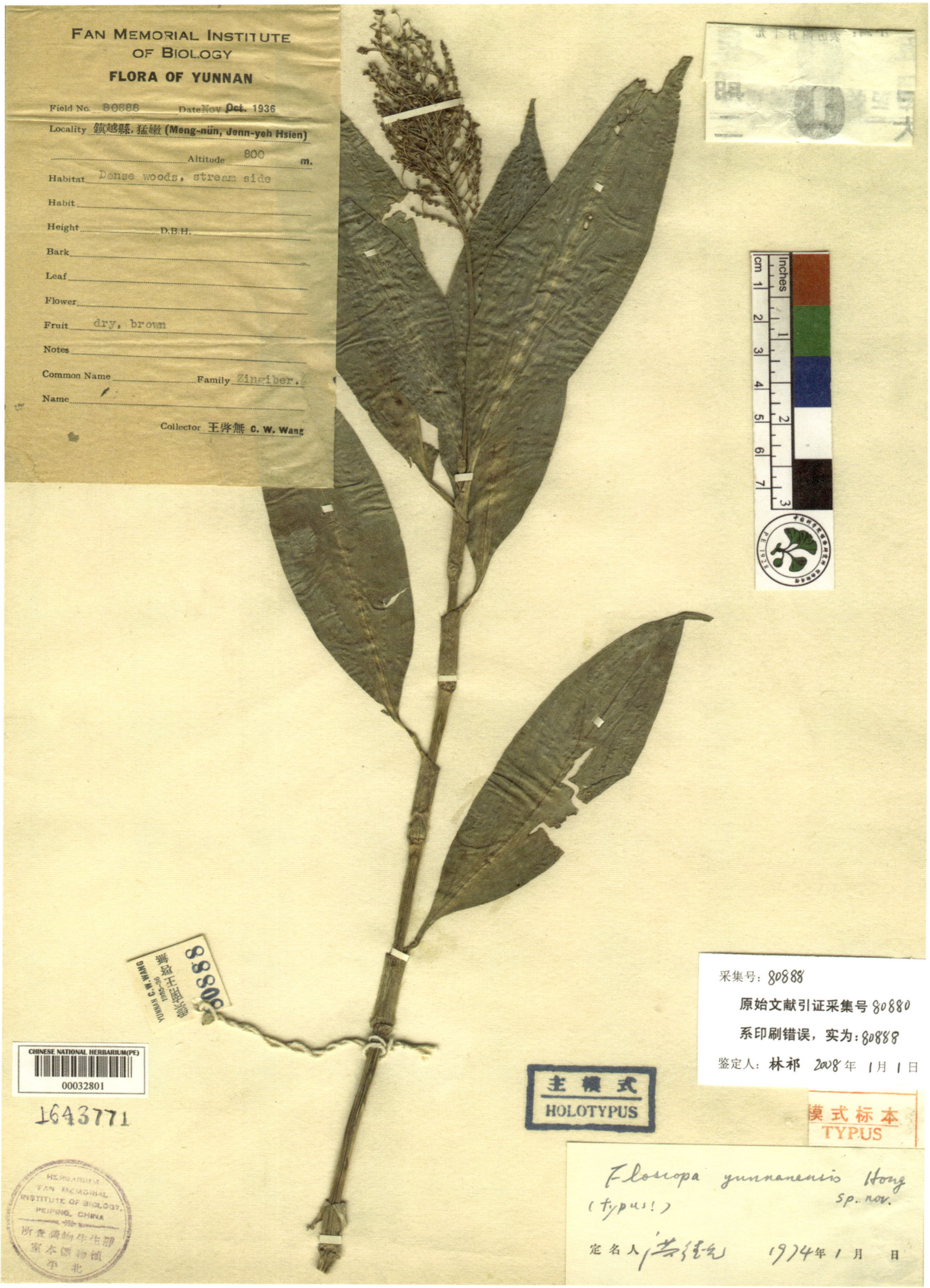

云南聚花草 ***Floscopa yunnanensis*** D. Y. Hong in Acta Phytotax. Sin. 12(4): 464, pl. 91: 1. 1974. **Holotype:** China. Yunnan: Jenn-yeh (=Mengla), alt. 800 m, 1936-10-??, C. W. Wang 80888.

大果水竹叶 ***Murdannia macrocarpa*** D. Y. Hong in Acta Phytotax. Sin. 12(4): 471, pl. 91. 1974. **Holotype:** China. Yunnan: Zhenkang, alt. 1600 m, 1938-08-19, T. T. Yu 17398.

云南水竹叶 ***Murdannia yunnanensis*** D. Y. Hong in Acta Phytotax. Sin. 12(4): 469, pl. 93: 2. 1974. **Holotype:** China. Yunnan: Cheli (=Jinghong), alt. 800 m, 1936-10-??, C. W. Wang 79880.

川杜若 ***Pollia omeiensis*** D. Y. Hong in Acta Phytotax. Sin. 12(4): 466, pl. 90. 1974. **Holotype:** China. Sichuan: Emei, Emeishan, alt. 850 m, 1957-07-14, G. H. Yang 55910.

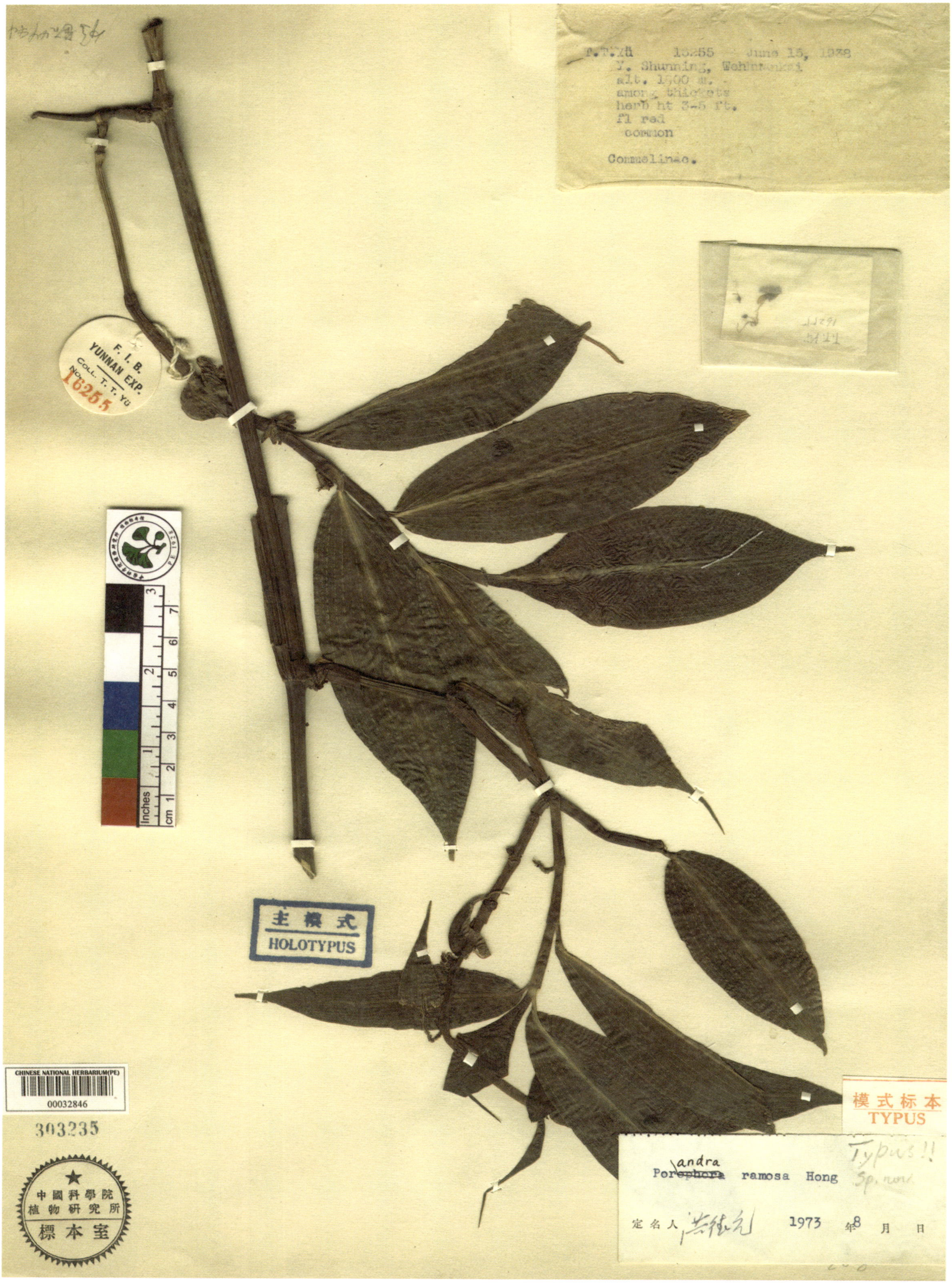

孔药花 ***Porandra ramosa*** D. Y. Hong in Acta Phytotax. Sin. 12(4): 462, pl. 89: 1-5. 1974. **Holotype:** China. Yunnan: Fengqing, alt. 1800 m, 1938-06-15, T. T. Yu 16255.

攀援孔药花 ***Porandra scandens*** D. Y. Hong in Acta Phtotax. Sin. 12(4): 462, pl. 89: 6-8. 1974. **Holotype:** China. Yunnan: Cheli (=Jinghong), alt. 1000 m, 1936-08-??, C. W. Wang 77909.

三瓣果 ***Tricarpelema chinense*** D. Y. Hong in Acta Phytotax. Sin. 12(4): 475, pl. 92. 1974. **Holotype:** China. Sichuan: Leshan, alt. 1500 m, 1931-08-25, T. Tang 23593.

Juncaceae

灯心草科

dengxincaoke

丽江灯心草 ***Juncus fimbristyloides*** Noltie in Edinburgh Journ. Bot. 55(1): 39, f. 1B, 2E-J. 1998. **Isotype:** China. Yunnan: Lijiang, alt. 3040 m, 1990-10-11, CLD-90 1122.

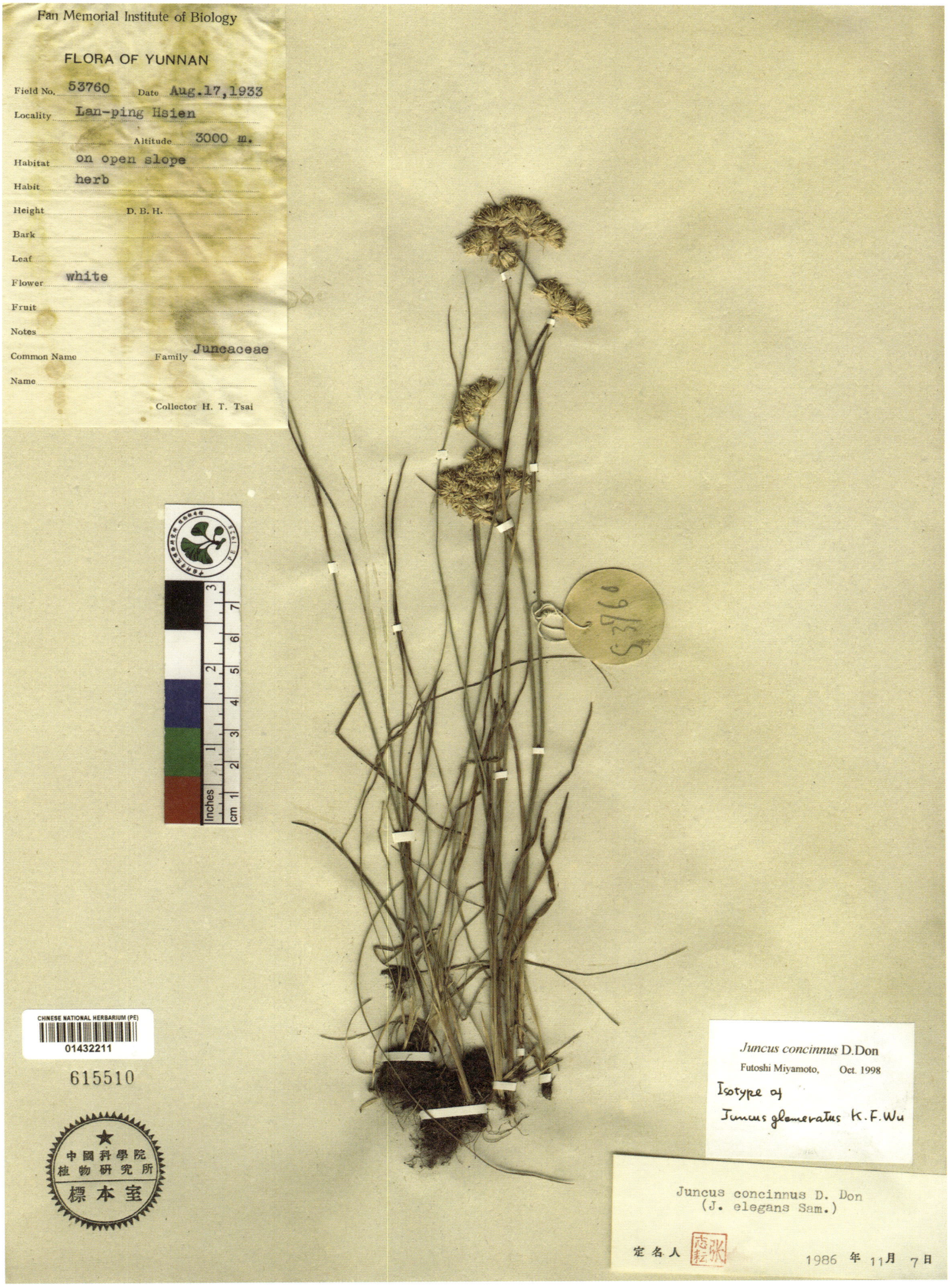

密花灯心草 ***Juncus glomeratus*** K. F. Wu in Acta Phytotax. Sin. 32(5): 462, f. 6: 1-5. 1994. **Isotype:** China. Yunnan: Lanping, alt. 3000 m, 1933-08-17, H. T. Tsai 53760.

西南灯心草 ***Juncus inflexus*** Linn. subsp. ***austro-occidentalis*** K. F. Wu, Fl. Reip. Pop. Sin. 13(3): 261. 1997. **Isotype:** China. Sichuan: Shimian, 1955-??-??, C. C. Hsieh 42386.

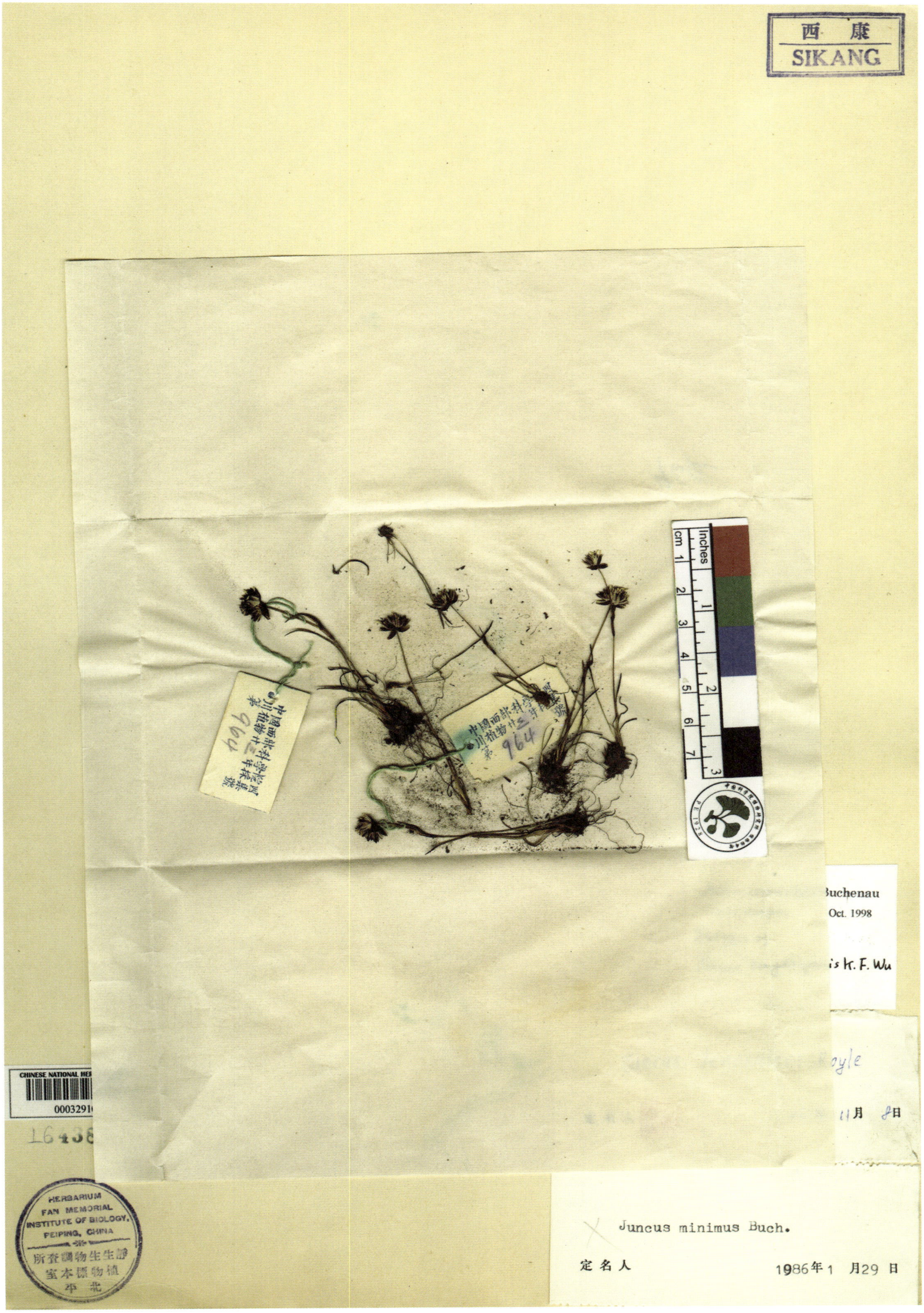

康定灯心草 ***Juncus kangdingensis*** K. F. Wu in Acta Phytotax. Sin. 32(5): 460, f. 4: 9-10. 1994. **Isotype:** China. Sichuan: Kangding, alt. 3400 m, 1934-??-??, Z. S. Liu 964.

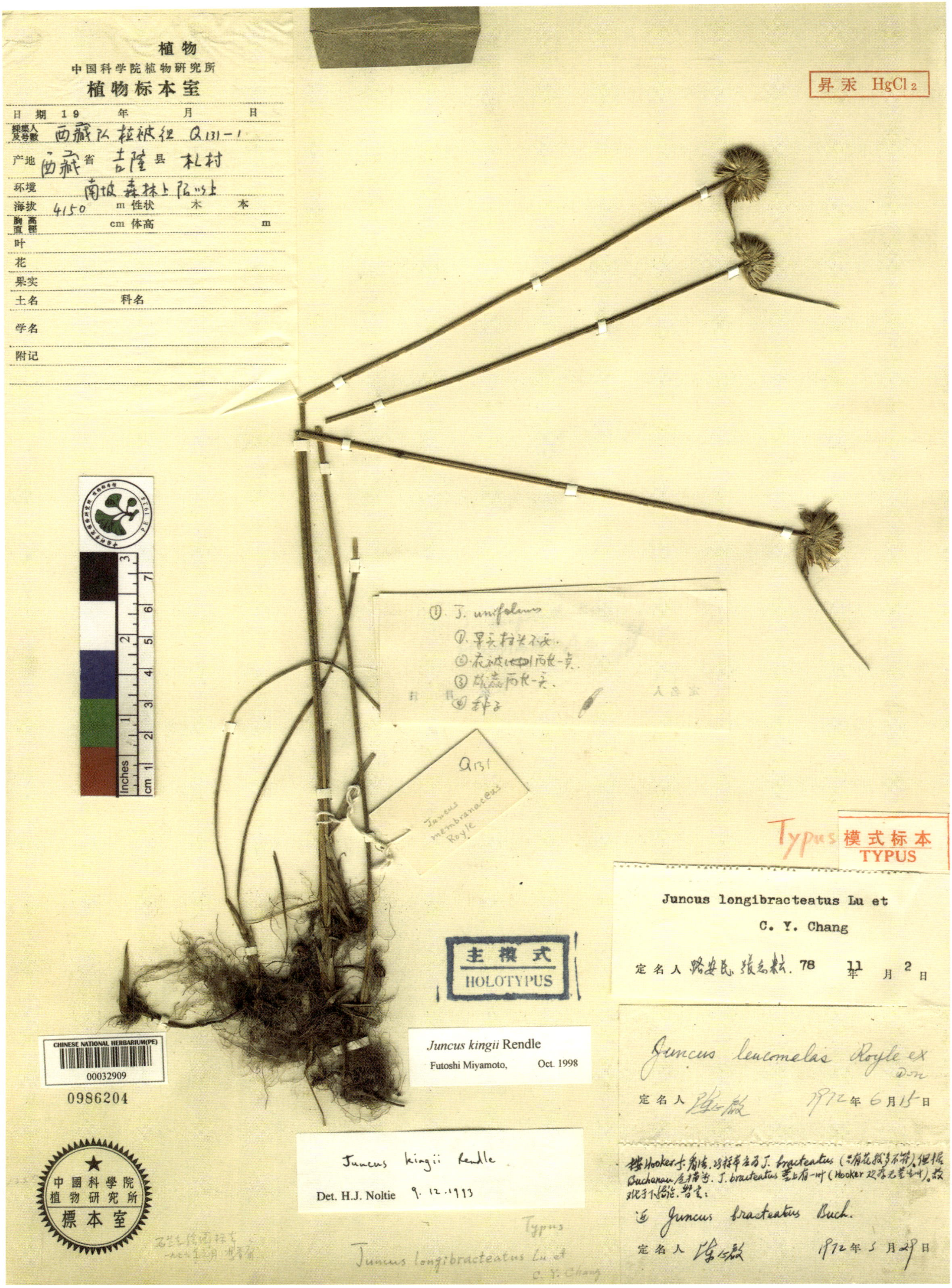

吉隆灯心草 ***Juncus longibracteatus*** A. M. Lu & Z. Y. Zhang in Acta Phytotax. Sin. 17(3): 126, f. 1: 5-8. 1979. **Holotype:** China. Xizang: Gyirong, alt. 4150 m, Qinghai-Xizang Exped. Q 131-1.

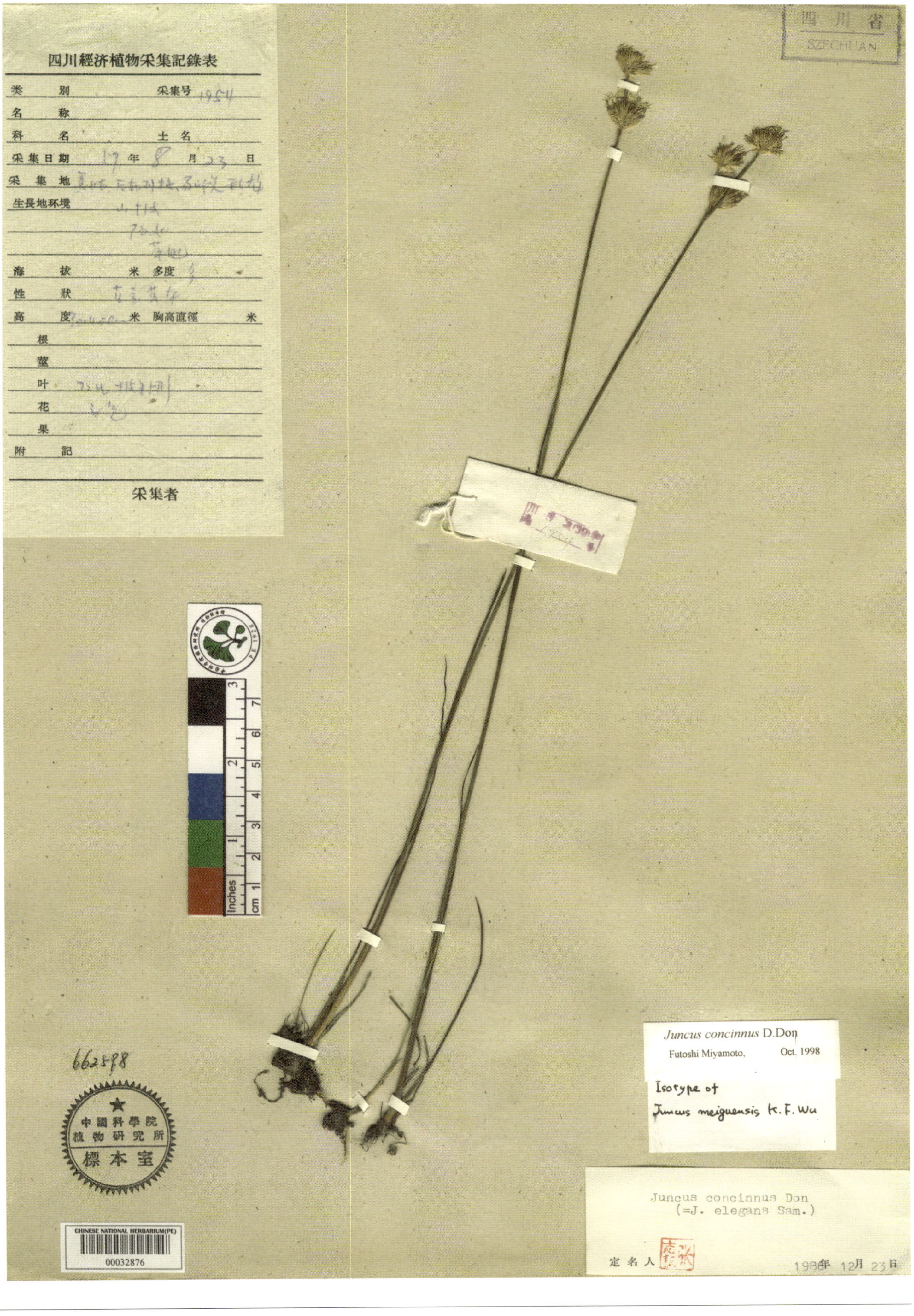

美姑灯心草 ***Juncus meiguensis*** K. F. Wu in Acta Phytotax. Sin. 32(5): 463, f. 6: 6-7. 1994. **Isotype:** China. Sichuan: Meigu, 1959-08-23, Sichuan Econ. Pl. Exped. 1954.

米拉山灯心草 ***Juncus milashanensis*** A. M. Lu & Z. Y. Zhang in Acta Phytotax. Sin. 17(3): 127, f. 2: 6-10. 1979. **Holoype:** China. Xizang: Milashan, alt. 5200 m, 1965-09-16, Y. T. Chang & K. Y. Lang 2734.

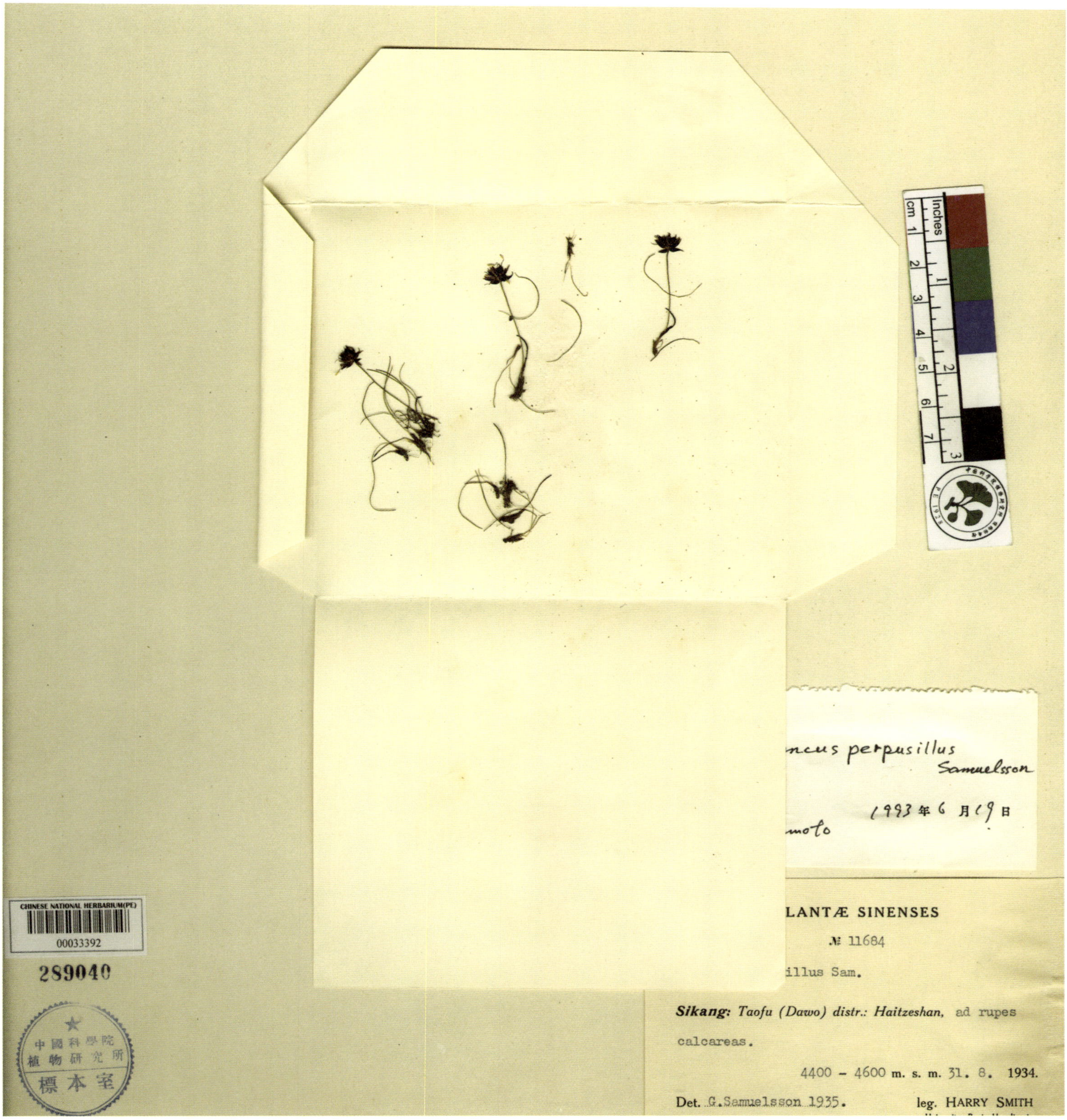

短茎灯心草 ***Juncus perpusillus*** G. Sam. in Hand.-Mazz, Symb. Sin. 7: 1237. 1936. **Isotype:** China. Xizang: Haitze-schan, alt. 4400~4600 m, 1934-08-31, H. Smith 11684.

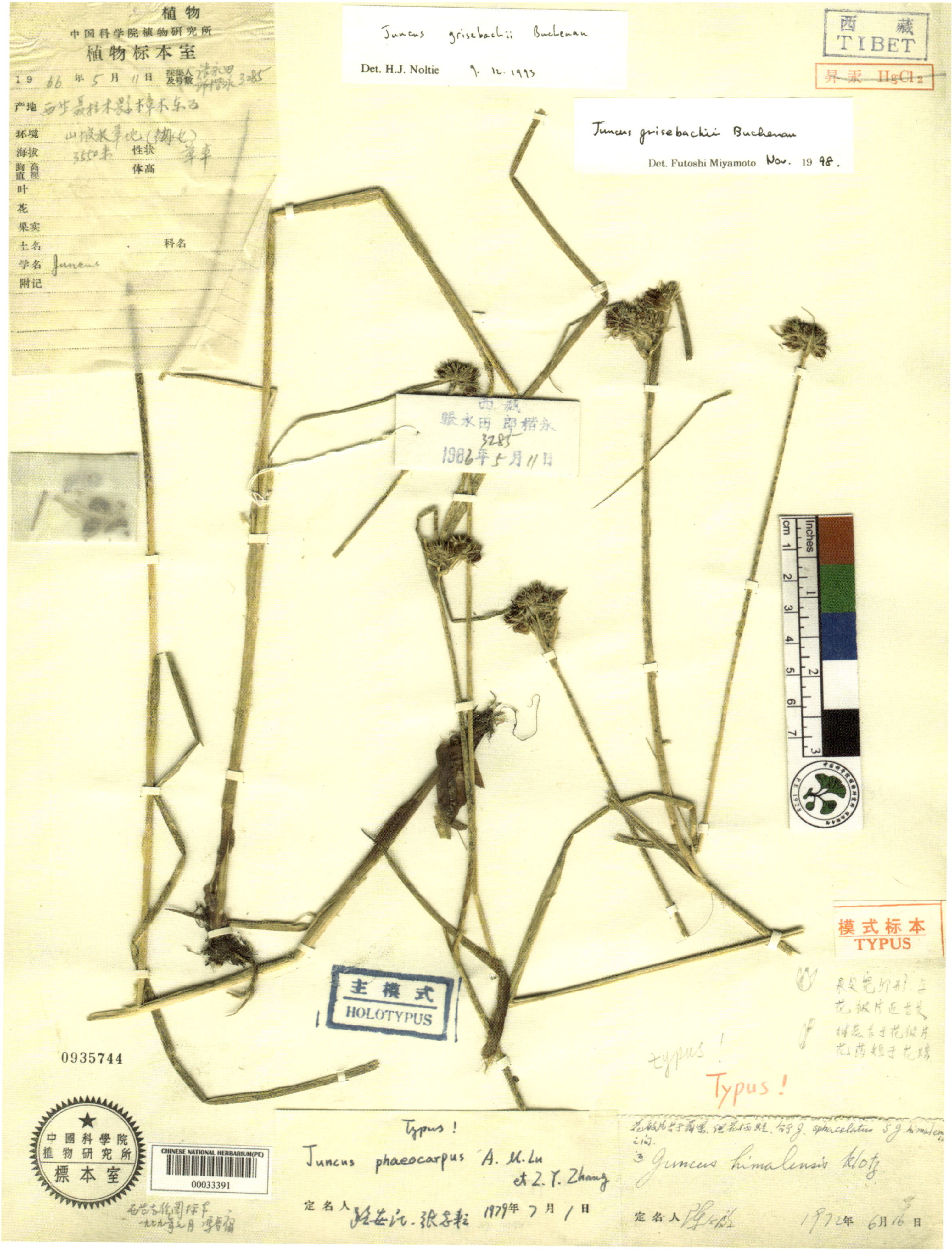

短果灯心草 ***Juncus phaeocarpus*** A. M. Lu & Z. Y. Zhang in Acta Phytotax. Sin. 17(3): 126, f. 2: 1-5. 1979. **Holotype:** China. Xizang: Nyalam, alt. 3550 m, 1966-05-11, Y. T. Chang & K. Y. Lang 3285.

圆柱叶灯心草 ***Juncus prismatocarpus*** R. Br. subsp. ***teretifolius*** K. F. Wu in Acta Phytotax. Sin. 32(5): 466. 1994. **Isotype:** China. Guangdong: Longmen, 1932-05-31, W. T. Tsang 20636.

德钦灯心草 ***Juncus sikkimensis*** Hook. f. var. ***helvolus*** K. F. Wu in Acta Phytotax. Sin. 32(5): 461. 1994. **Holotype:** China. Yunnan: Dêqên, alt. 3800~4000 m, 1940-08-13, K. M. Feng 6594.

锡金灯心草 ***Juncus tobdenii*** Noltie in Edinburgh Journ. Bot. 55(1): 42, f. 1A, 2A-D. 1998. **Isotype:** India. Sikkim, Yakche, N of Lachung, alt. 3105 m, 1996-07-14, D. G.. Long & H. J. Noltie 209.

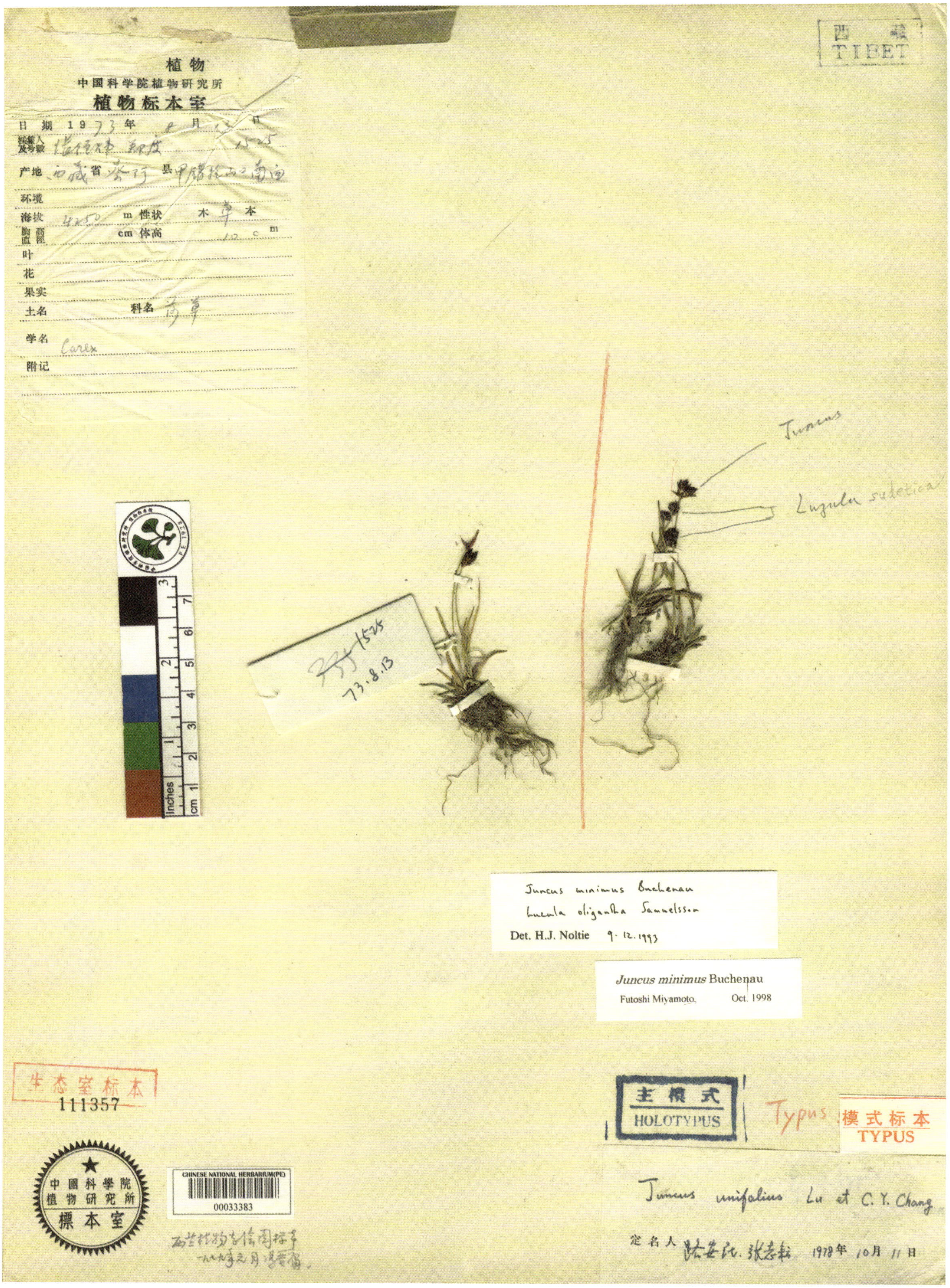

单叶灯心草 ***Juncus unifolius*** A. M. Lu & Z. Y. Zhang in Acta Phytotax. Sin. 17(3): 125, f. 1: 1-4. 1979. **Holotype:** China. Xizang: Zayü, alt. 4250 m, 1973-08-23, G. W. Chang & D. Cheng 1525.

波密地杨梅 ***Luzula bomiensis*** K. F. Wu in Journ. E. China Norm. Univ., Nat. Sci. 1992(3): 100, pl. 4: 1-5. 1992. **Holotype:** China. Xizang: Bomi, alt. 4000 m, 1965-08-23, J. S. Ying & D. Y. Hong 650991.

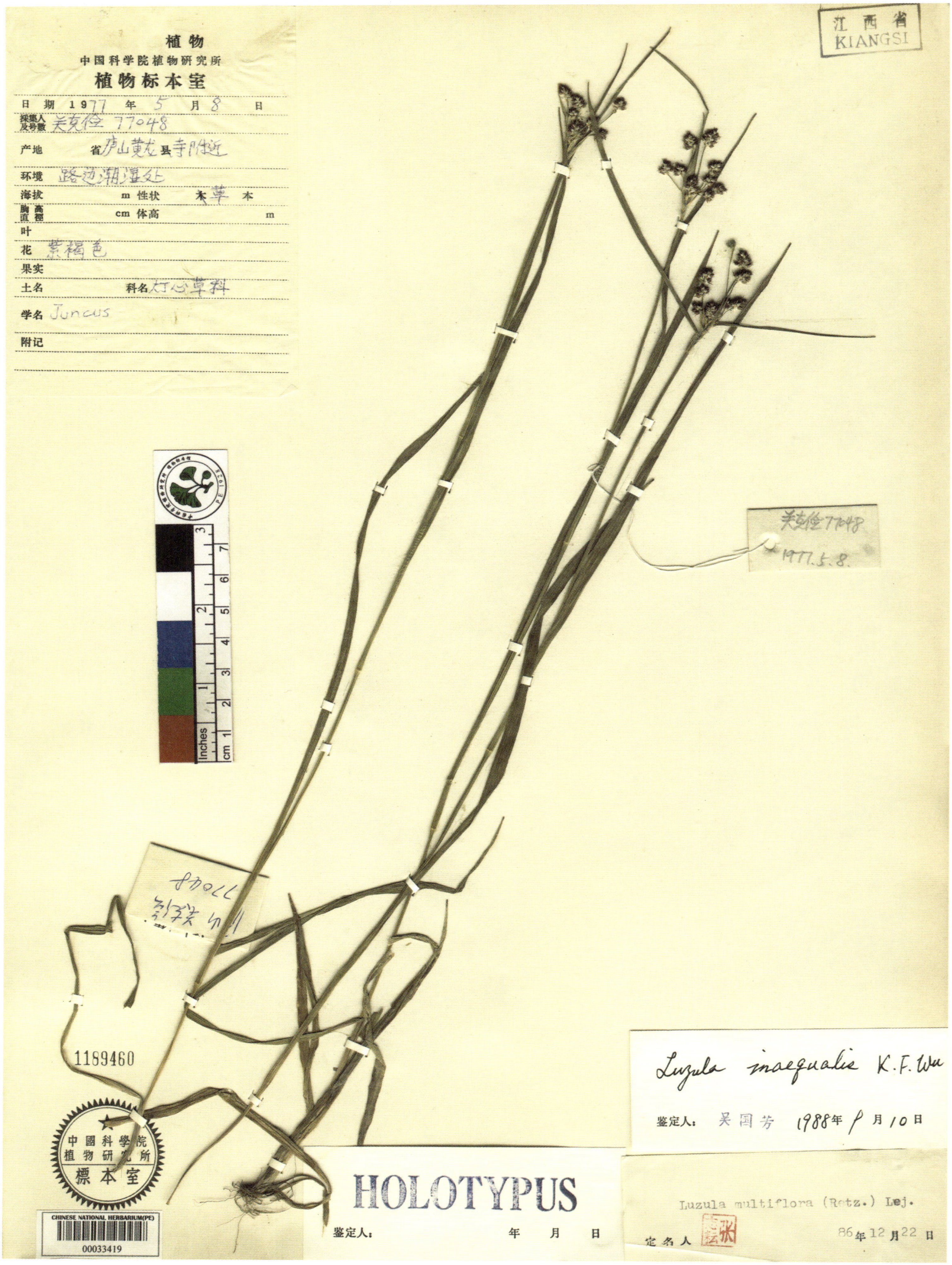

异被地杨梅 ***Luzula inaequalis*** K. F. Wu in Journ. E. China Norm. Univ., Nat. Sci. 1992(3): 97, pl. 3: 1-8. 1992. **Holotype:** China. Jiangxi: Lushan, 1977-05-08, K. C. Kuan 77048.

四川地杨梅 ***Luzula sichuanensis*** K. F. Wu in Journ. E. China Norm. Univ., Nat. Sci. 1992(3): 92, pl. 1. 1992. **Holotype:** China. Sichuan: Maerkang, alt. 4000 m. 1922-09-13, H. Smith 4385.

Liliaceae

百合科

baiheke

头花粉条儿菜 ***Aletris capitata*** F. T. Wang & Tang, Fl. Reip. Pop. Sin. 15: 172, 254, pl. 58: 1-3. 1978. **Holotype:** China. Sichuan: Baoxing, 1936-08-15, K. L. Chu 3604.

灰鞘粉条儿菜 ***Aletris cinerascens*** F. T. Wang & Tang, Fl. Reip. Pop. Sin. 15: 174, 254, pl. 58: 4-6. 1978. **Holotype:** China. Guangxi: Xingan, 1953-06-??, Guangxi Exped. 4248.

高粉条儿菜 ***Aletris elata*** F. T. Wang & Tang in Bull. Fan Mem. Inst. Biol., New Ser. 1(1): 108. 1943. **Holotype:** China. Sichuan: Precise locality not known, 1933-??-??, T. T. Yu 2123.

纤细粉条儿菜 ***Aletris gracilipes*** F. T. Wang & Tang in Bull. Fan Mem. Inst. Biol., New Ser. 1(1): 107. 1943. **Holotype:** China. Sichuan: Precise locality not known, 1930-??-??, F. T. Wang 20674.

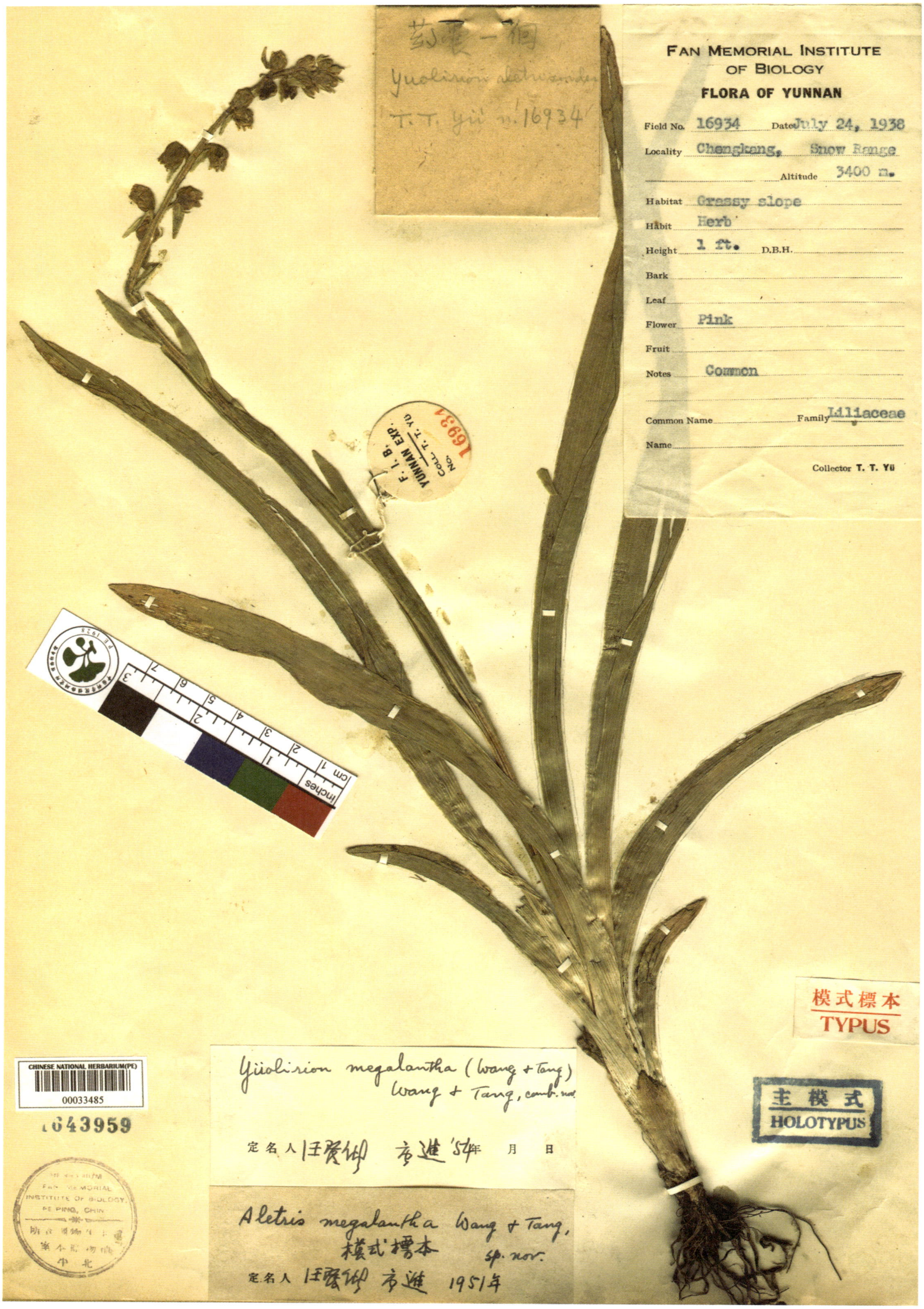

大花粉条儿菜 ***Aletris megalantha*** F. T. Wang & Tang in Acta Phytotax. Sin. 1(1): 119. 1951. **Holotype:** China. Yunnan: Chengkang, alt. 3400 m, 1938-07-24, T. T. Yu 16934.

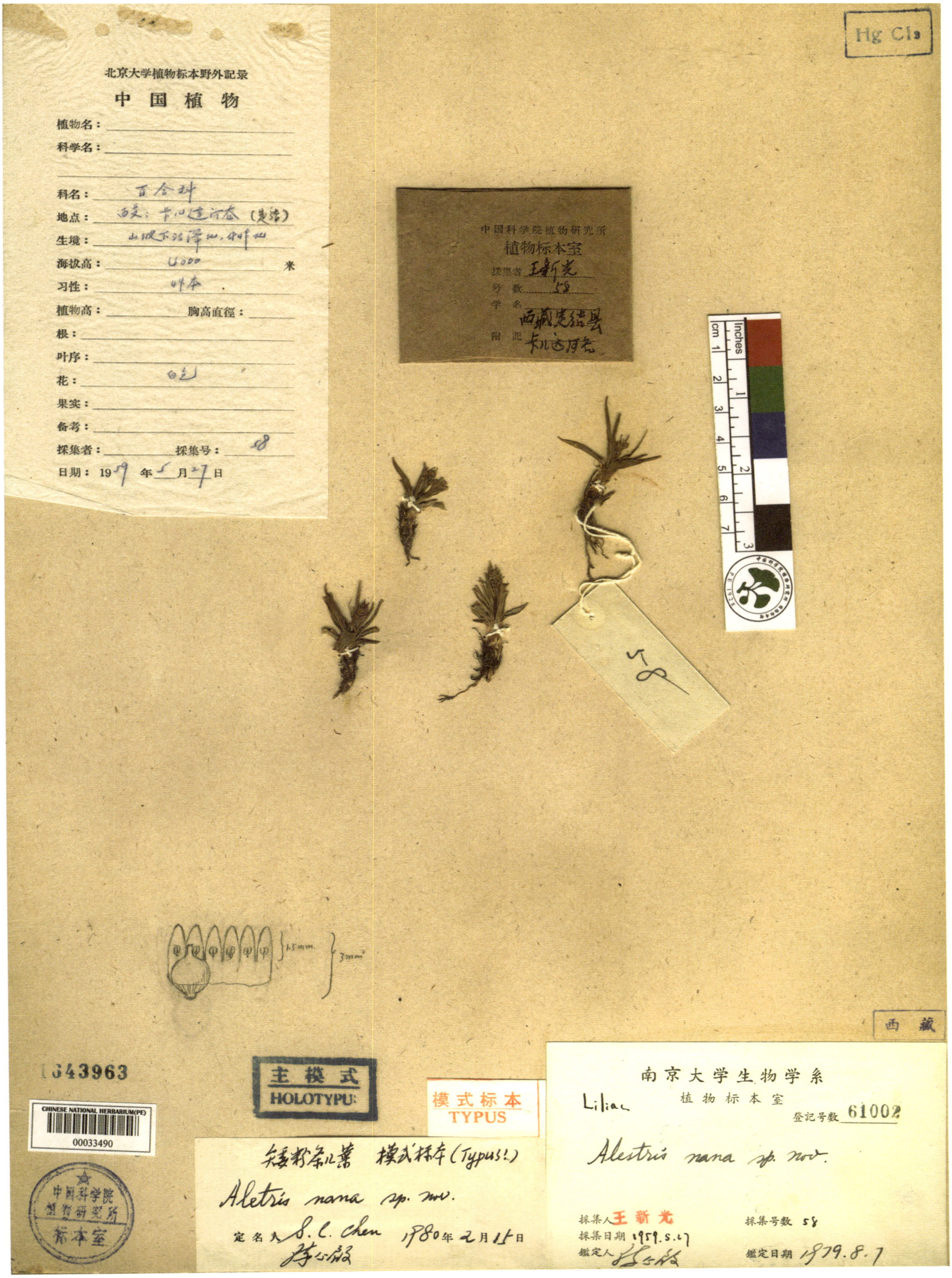

矮粉条儿菜 ***Aletris nana*** S. C. Chen in Acta Phytotax. Sin. 19(4): 503, f. 1: 6-8. 1981. **Holotype:** China. Xizang: Dinggye, alt. 4000 m, 1959-05-27, X. G. Wang 58.

长柄粉条儿菜 ***Aletris pedicellata*** F. T. Wang & Tang in Bull. Fan Mem. Inst. Biol., New Ser. 1(1): 109. 1943. **Holotype:** China: Sichuan: Precise locality not known, 1934-??-??, T. T. Yu 2982.

城口粉条儿菜 ***Aletris spicata*** Franch. var. ***fargesii*** Franch. in Journ. Bot. 10(12): 200. 1896. **Isotype:** China. Chongqing: Chengkou, R. P. Farges 437.

狭瓣粉条儿菜 ***Aletris stenoloba*** Franch. in Journ. Bot. 10(12): 203. 1896. **Isotype:** China. Yunnan: precise locality not known, 1882-05-??, Delavay 2282.

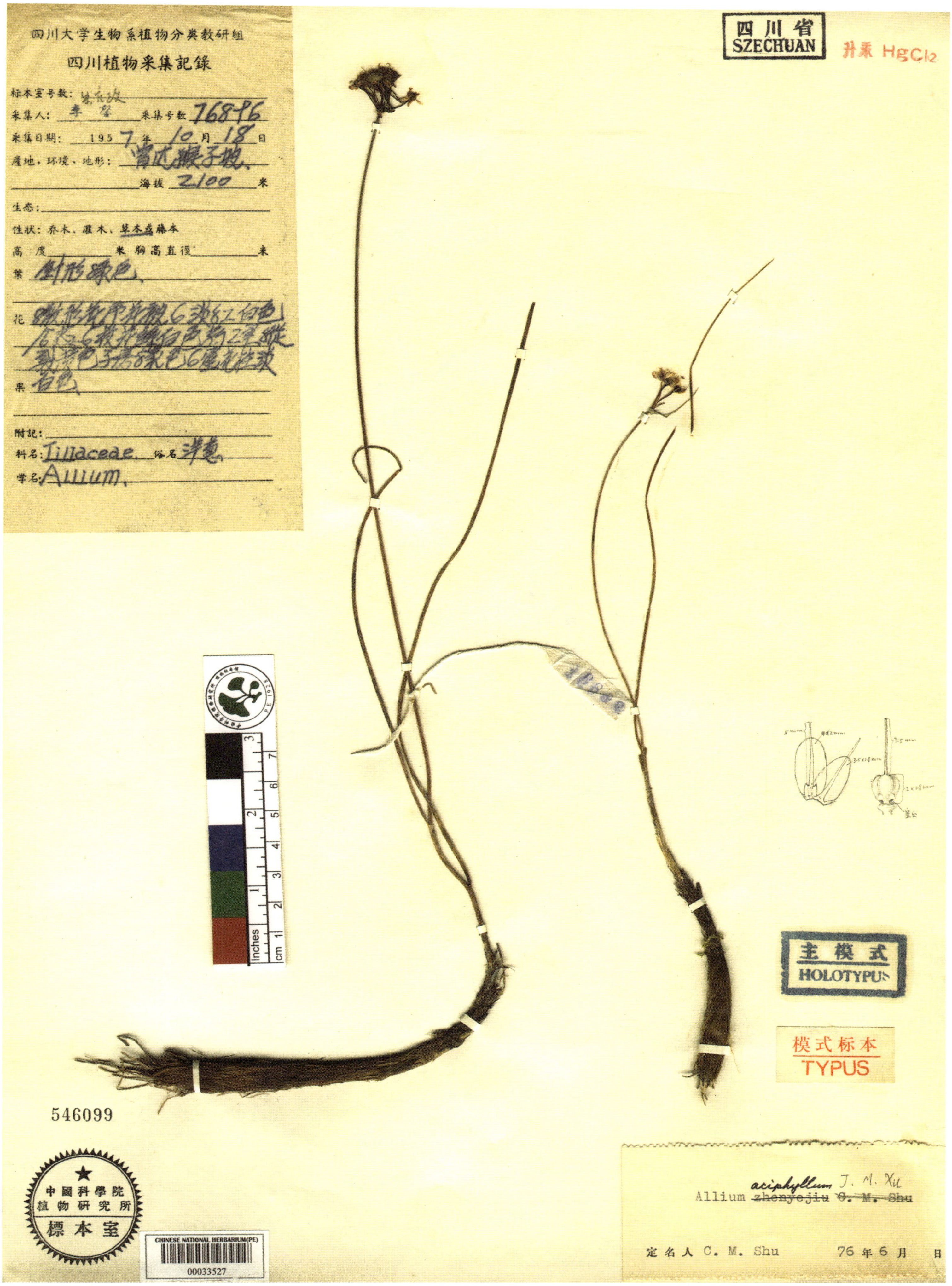

针叶韭 ***Allium aciphyllum*** J. M. Xu, Fl. Reip. Pop. Sin. 14: 230, 284, f. 55. 1980. **Holotype:** China. Sichuan: Jinchuan, alt. 2100 m, 1957-10-18, G. C. Chu & H. Li 76896.

广元蒜 ***Alliun albostellerianum*** F. T. Wang & Tang in Bull. Fan Mem. Inst. Biol., Bot. 7(6): 293. 1937. **Holotype:** China. Sichuan: Kwangyuan (=Guangyuan), alt. 1750 m, 1930-09-25, F. T. Wang 22589.

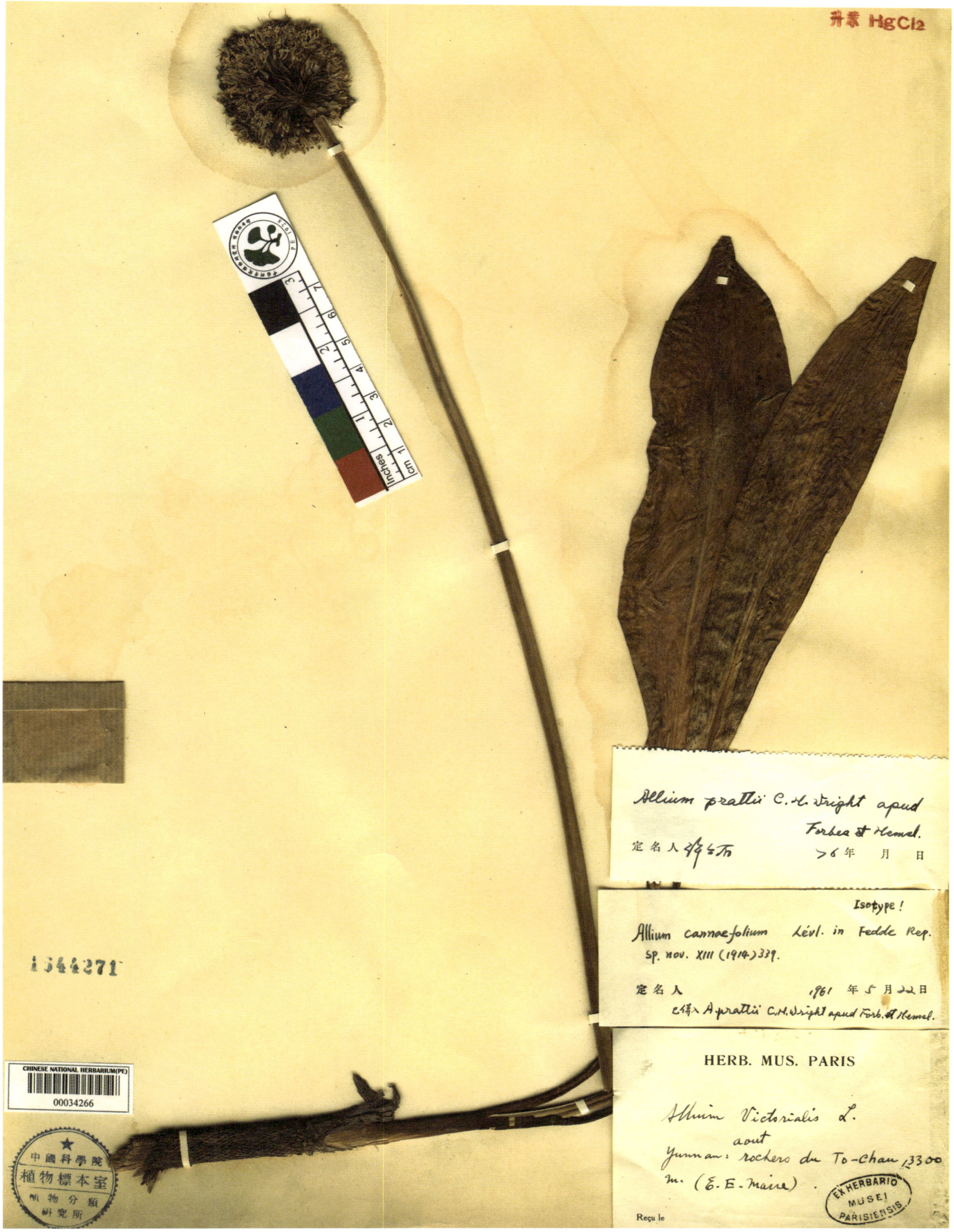

美人蕉叶韭 ***Allium cannaefolium*** Lévl. in Fedde, Rep. Sp. Nov. Reg. Veg. 13: 339. 1914. **Isotype:** China. Yunnan: Rochers du Io-Chan, alt. 3300 m, 1913-??-??, E. E. Maire s. n.

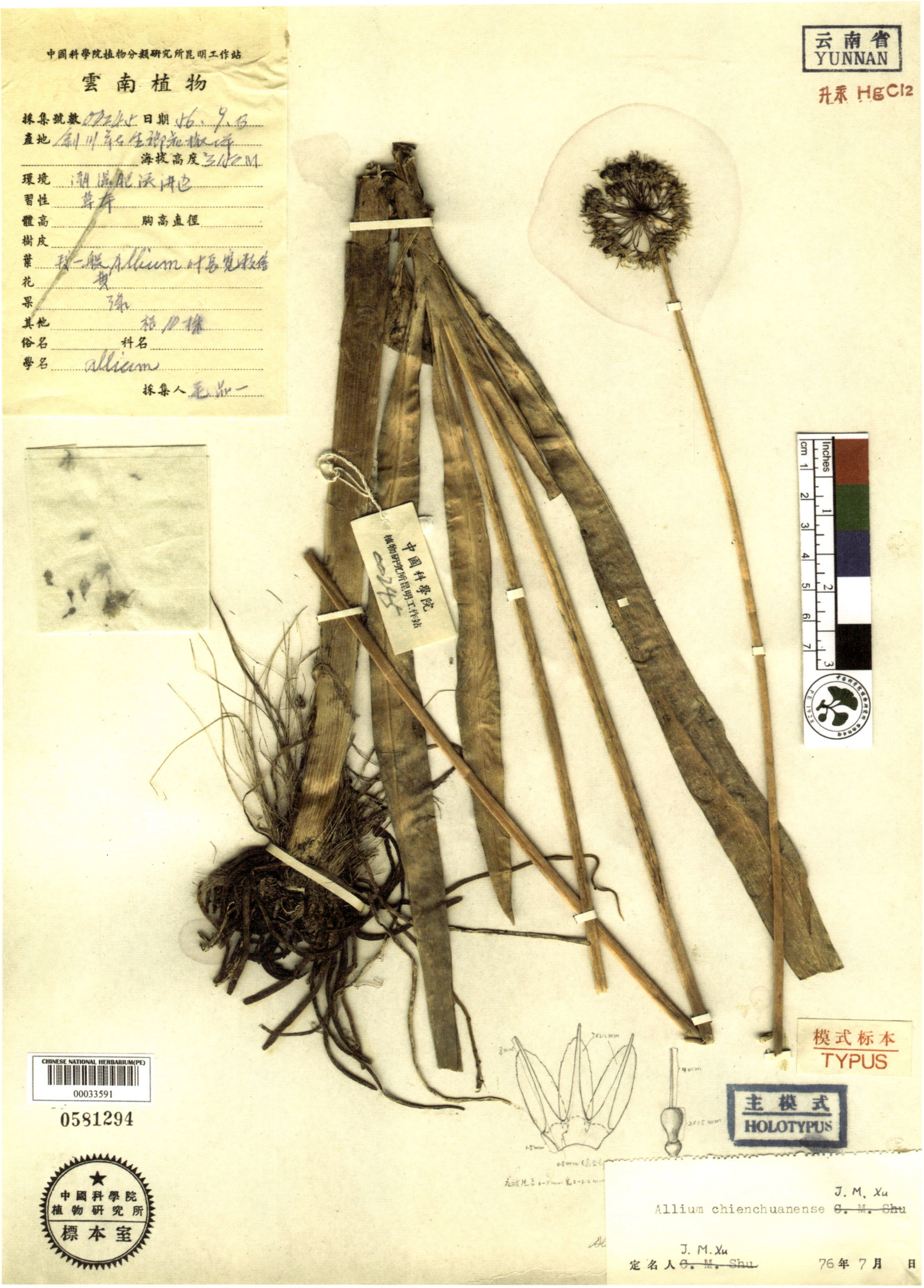

剑川韭 ***Allium chienchuanense*** J. M. Xu, Fl. Reip. Pop. Sin. 14: 211, 284, f. 35. 1980. **Holotype:** China. Yunnan: Jianchuan, alt. 3150 m, 1956-09-08, P. I. Mao 245.

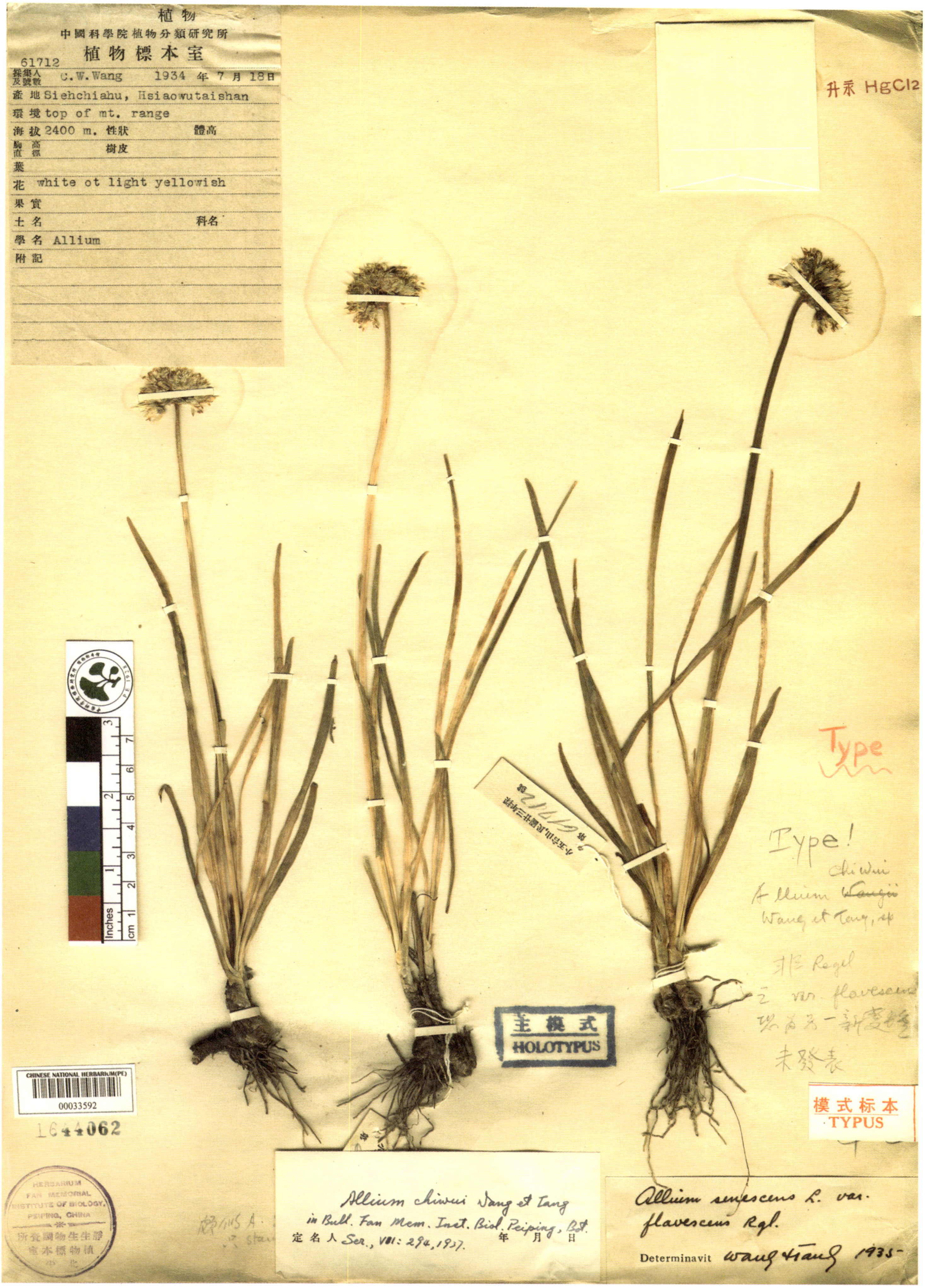

冀韭 ***Allium chiwui*** F. T. Wang & Tang in Bull. Fan Mem. Inst. Biol., Bot. 7(6): 294. 1937. **Holotype:** China. Hebei: Xiaowutaishan, alt. 2400 m, 1934-07-18, C. W. Wang 61712.

心叶韭 ***Allium cordifolium*** J. M. Xu, Fl. Reip. Pop. Sin. 14: 205, 284, f. 27. 1980. **Holotype:** China. Sichuan: Xiaojin, alt. 3500 m, 1958-07-25, H. S. Chang & Y. S. Ren 6210.

天蓝韭 ***Allium cyaneum*** Regel in Acta Hoit. Petiop. 3(2): 174. 1875. **Isotype:** China. Gansu: Precise locality not known, 1872-08-10, Przewalski s. n.

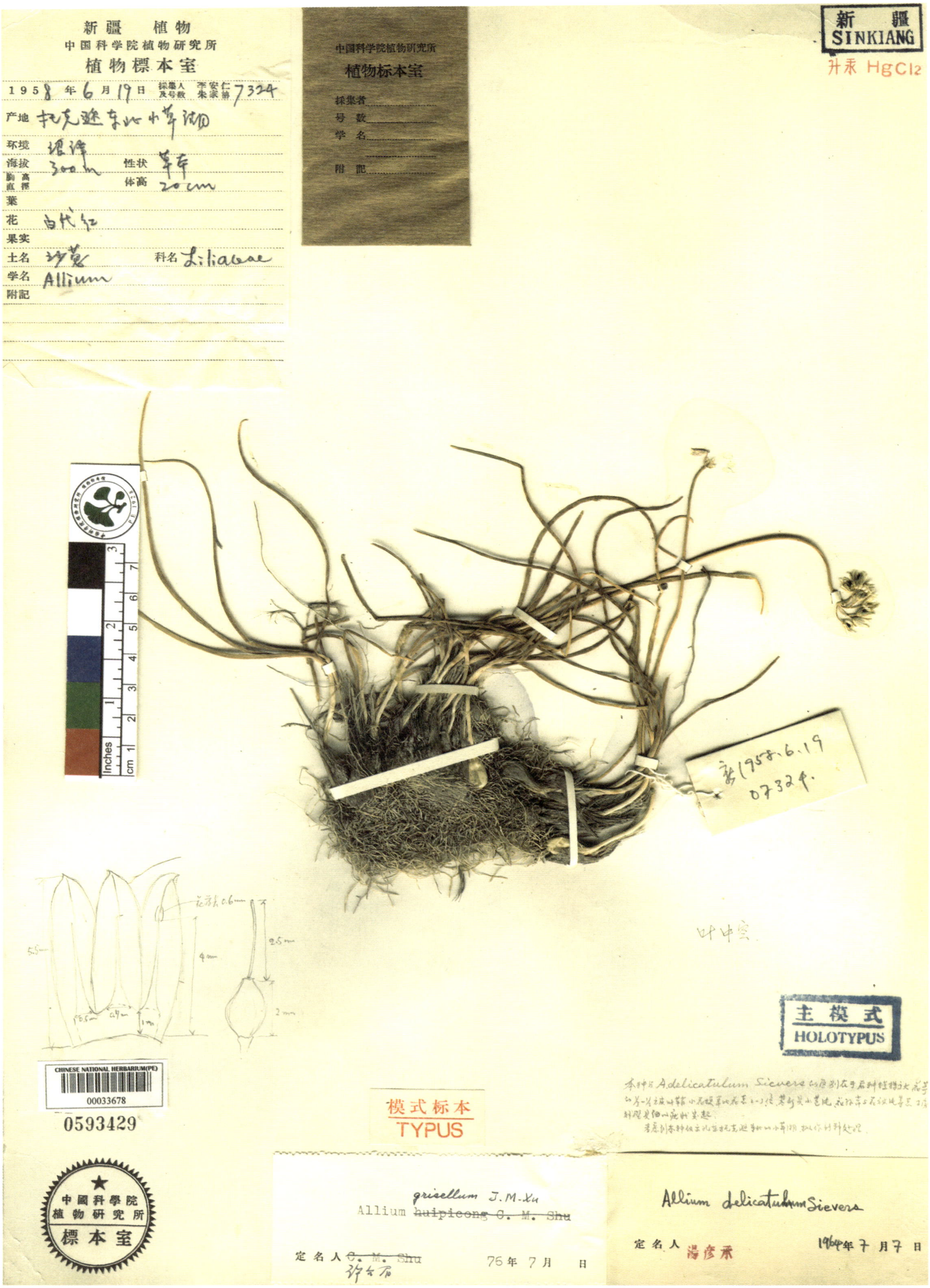

灰皮葱 ***Allium grisellum*** J. M. Xu, Fl. Reip. Pop. Sin. 14: 259, 286, f. 83. 1980. **Holotype:** China. Xinjiang: Toksum, alt. 300 m, 1958-06-19, A. J. Li & C. L. Chu 7324.

疏花韭 ***Allium henryi*** C. H. Wright in Kew Bull. 119. 1895. **Isotype:** China. Hubei: Hsingshan (=Xingshan), 1889-12-26, A. Henry 6924.

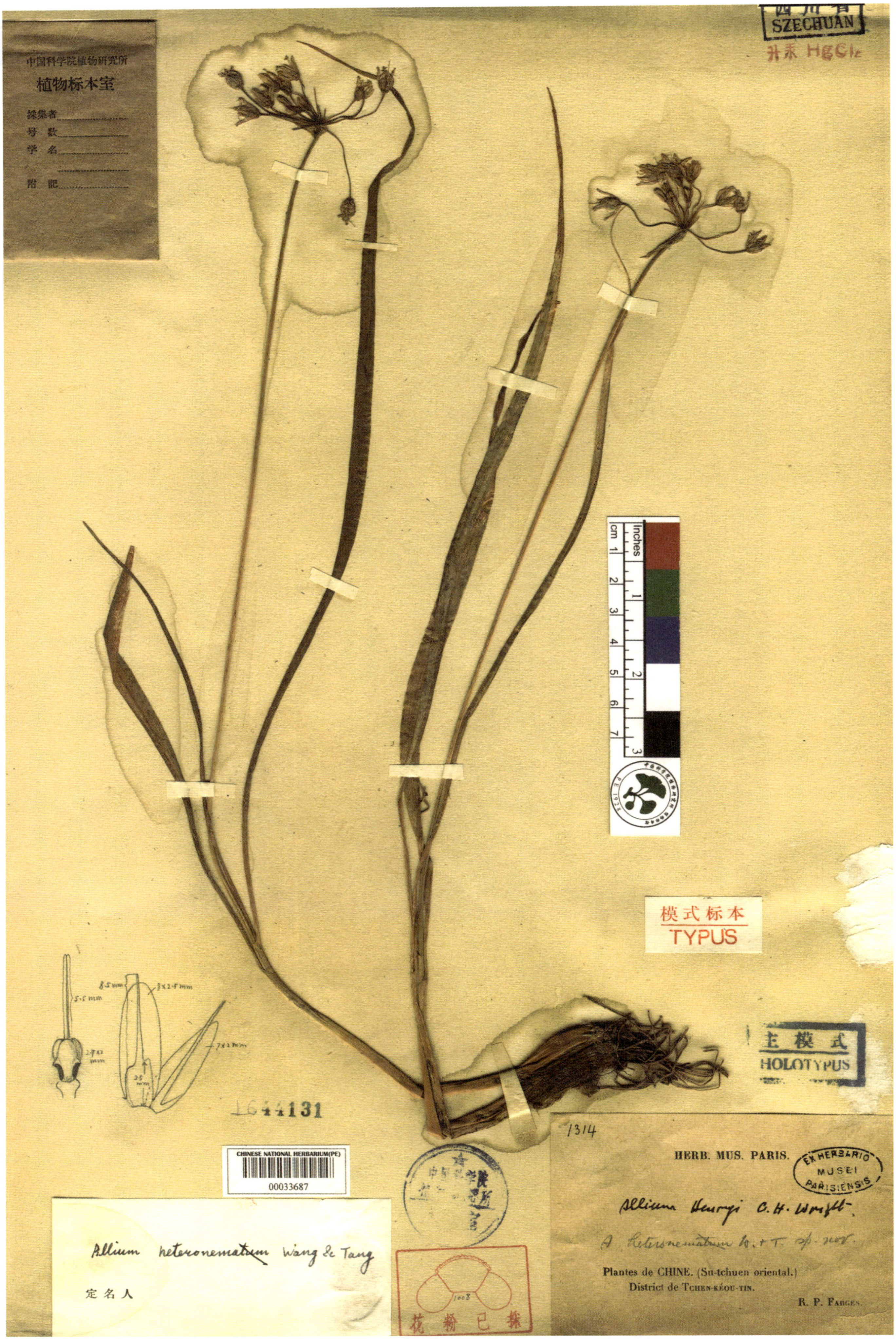

异 梗 韭 ***Allium heteronema*** F. T. Wang & Tang, Fl. Reip. Pop. Sin. 14: 231, 285, f. 57. 1980. **Holotype:** China. Chongqing: Chengkou: R. P. Farges 1314.

三柱韭 ***Allium humile*** Kunth var. ***trifurcatum*** F. T. Wang & Tang, Fl. Reip. Pop. Sin. 14: 208, 284. 1980. **Holotype:** China. Yunnan: Chongtien (=Zhongdian), alt. 3400 m, 1937-06-26, T. T. Yu 11792.

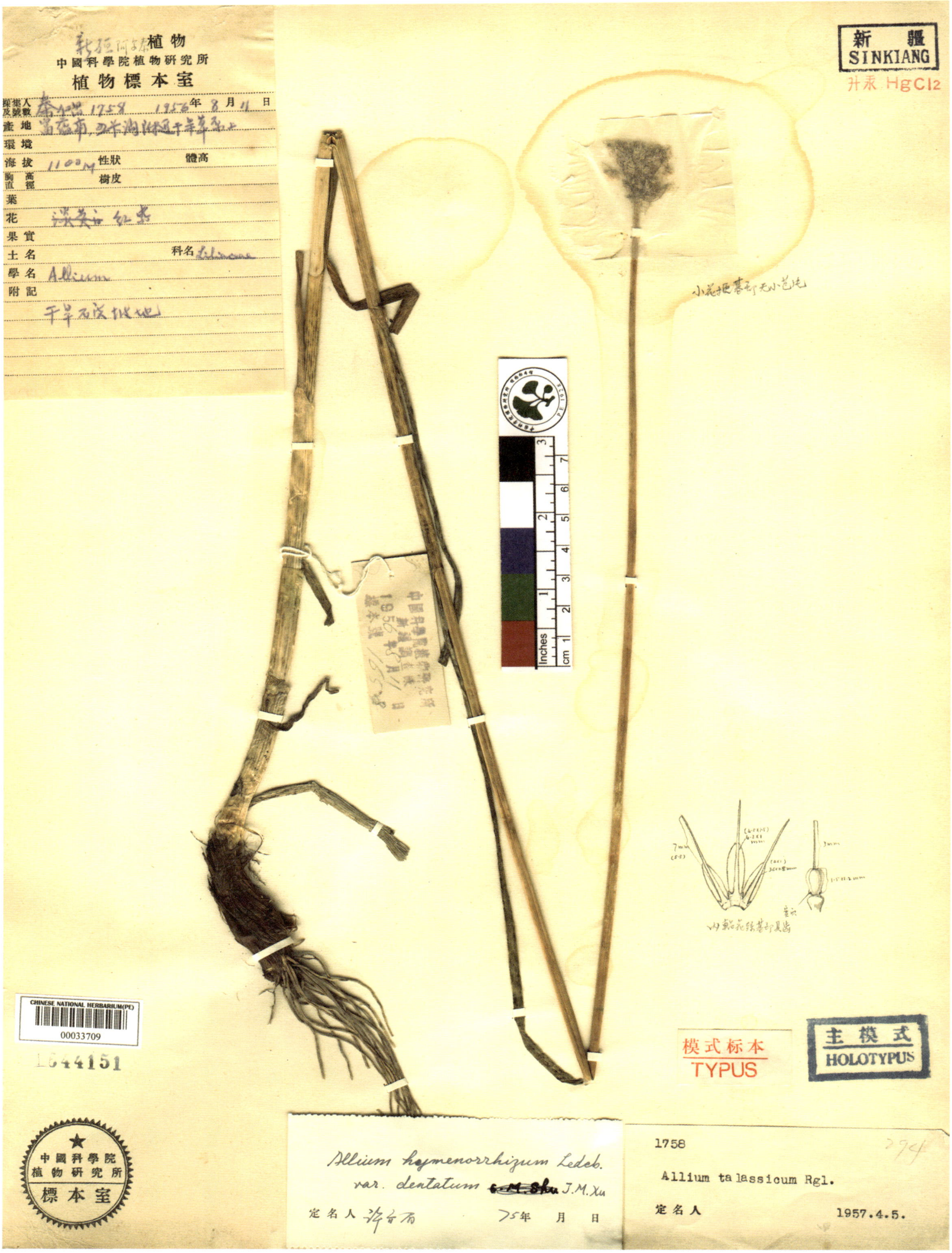

旱生韭 ***Allium hymenorrhizum*** Ledeb. var. ***dentatum*** J. M. Xu, Fl. Reip. Pop. Sin. 14: 245, 285, f. 68. 1980. **Holotype:** China. Xinjiang: Fuwun, alt. 1100 m, 1956-08-11, R. C. Ching 1758.

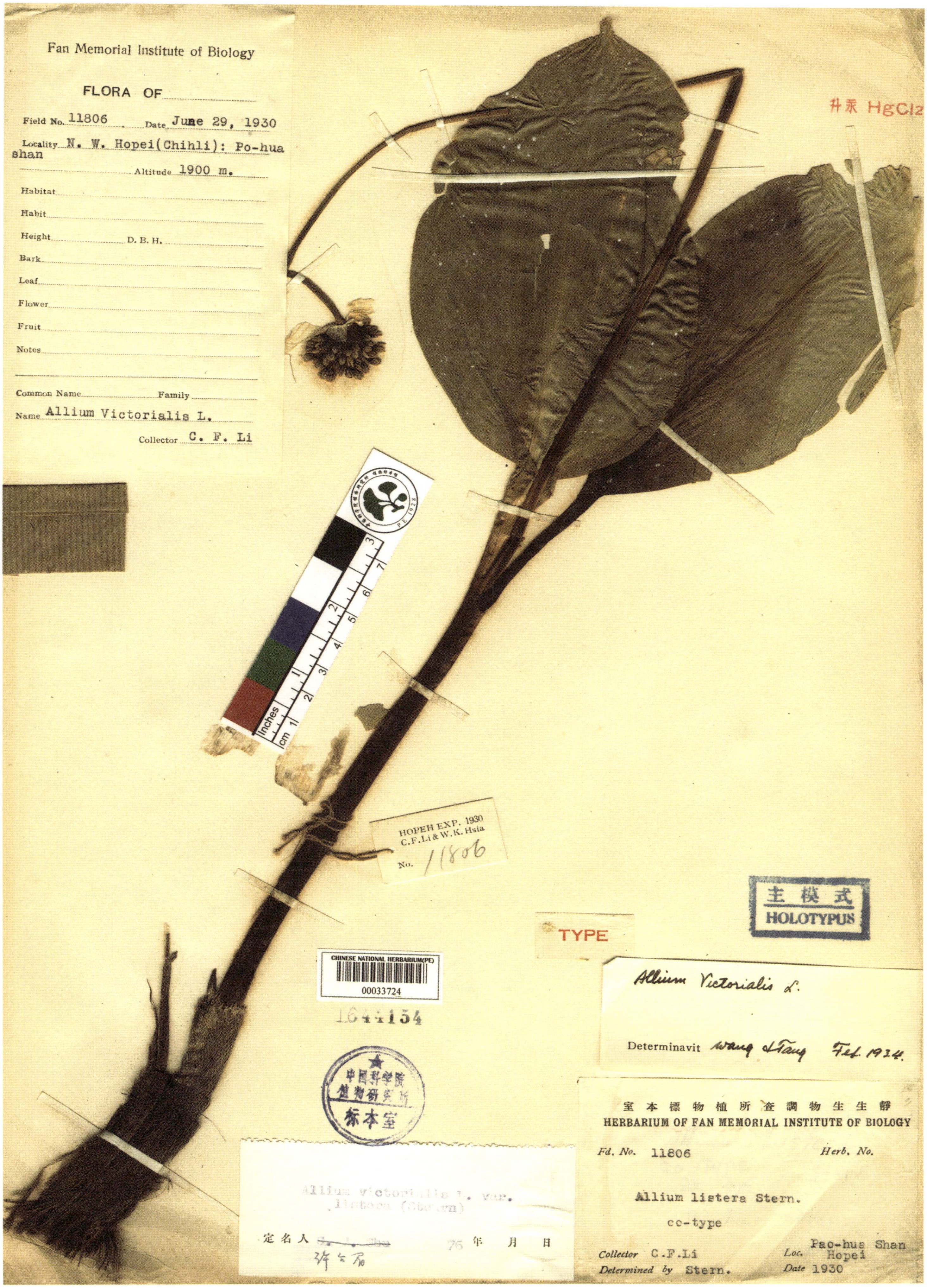

对叶韭 ***Allium listera*** Stern in Bull. Fan Mem. Inst. Biol., Bot. 5(6): 326. 1934. **Holotype:** China. Hebei: Baohuashan, alt. 1900 m, 1930-06-29, C. F. Li 11806.

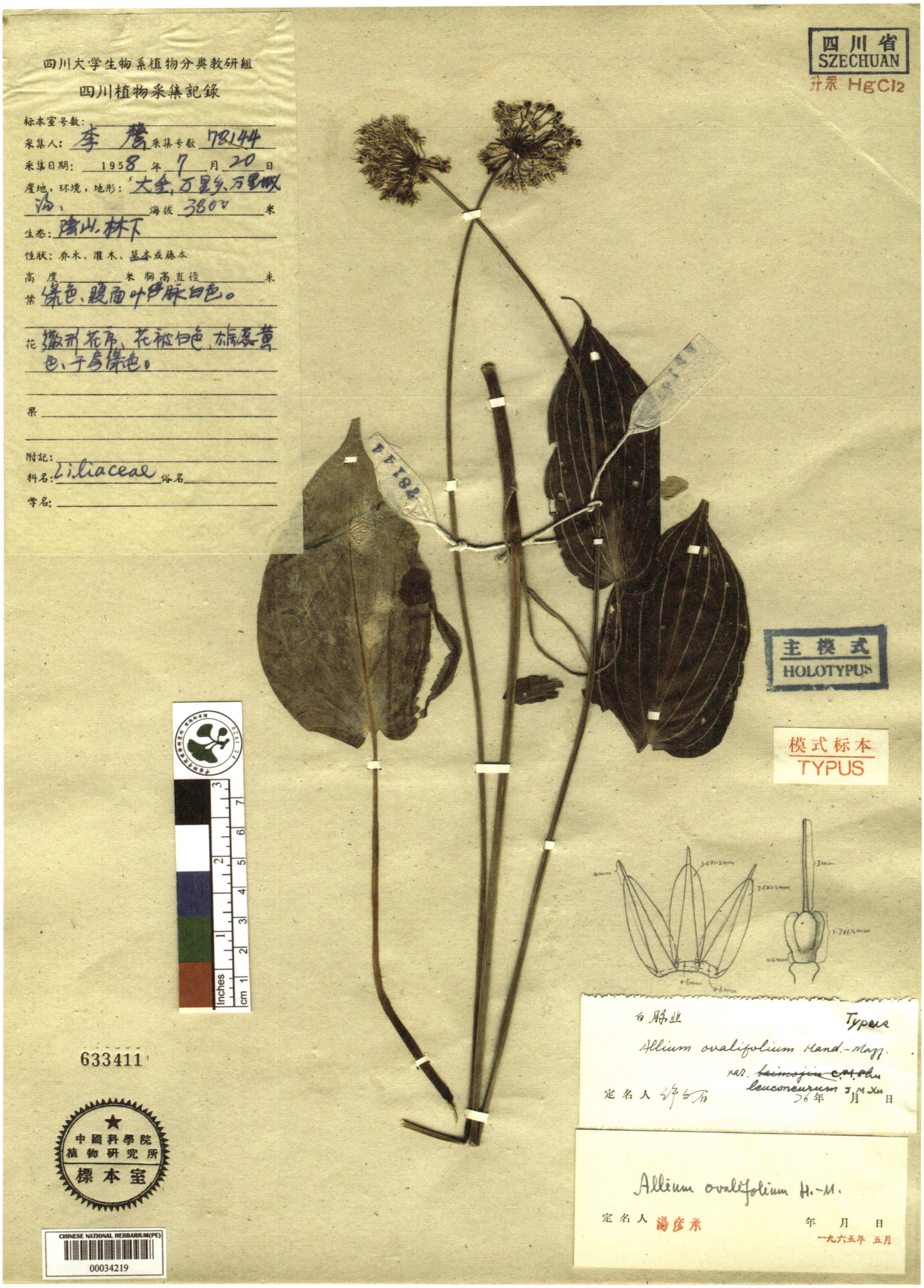

卵叶韭 ***Allium ovalifolium*** Hand.-Mazz. var. ***leuconeurum*** J. M. Xu, Fl. Reip. Pop. Sin. 14: 204, 283. 1980. **Holotype:** China. Sichuan: Dajin, alt. 3800 m, 1958-07-20, H. Li 78144.

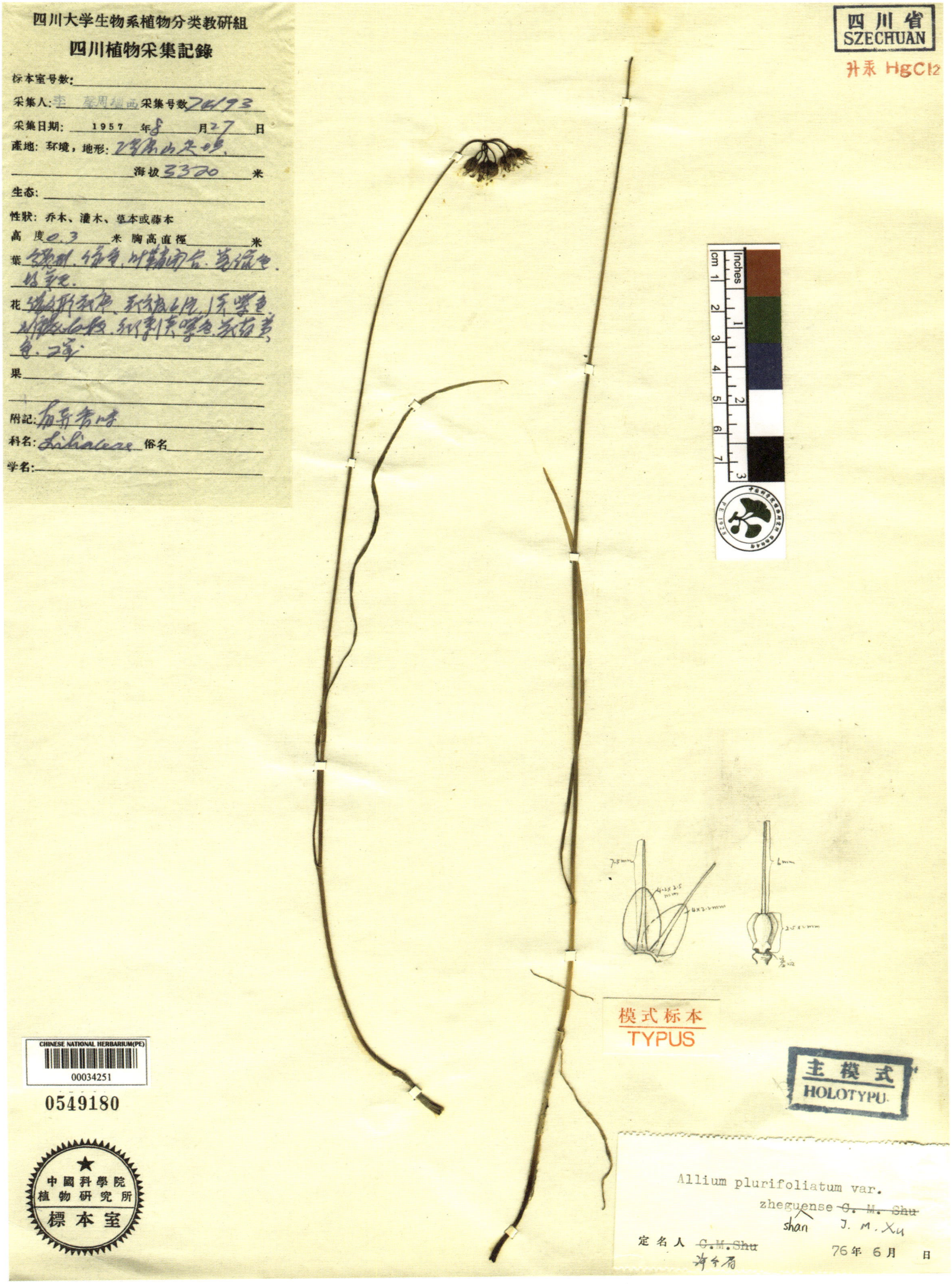

多叶韭 ***Allium plurifoliatum*** Rendle var. ***zhegushanensis*** J. M. Xu, Fl. Reip. Pop. Sin. 14: 233, 285. 1980. **Holotype:** China. Sichuan: Li Xian, alt. 3320 m, 1957-08-27, H. Li & T. H. Chou 74193.

宽叶太白韭 ***Allium prattii*** C. H. Wright var. ***latifoliatum*** F. T. Wang & Tang in Bull. Fan Mem. Inst. Biol., Bot. 7(6): 297. 1937.
Holotype: China. Sichuan: Yutong, alt. 3280 m, 1933-09-??, D. H. Du 4443.

喙果韭 ***Allium rhynchogynum*** Diels in Notes Roy. Bot. Gard. Edinburgh 25: 302. 1912. **Isotype:** China. Yunnan: Chu Tien, alt. 2000~2330 m, 1904-??-??, G. Forrest. 267.

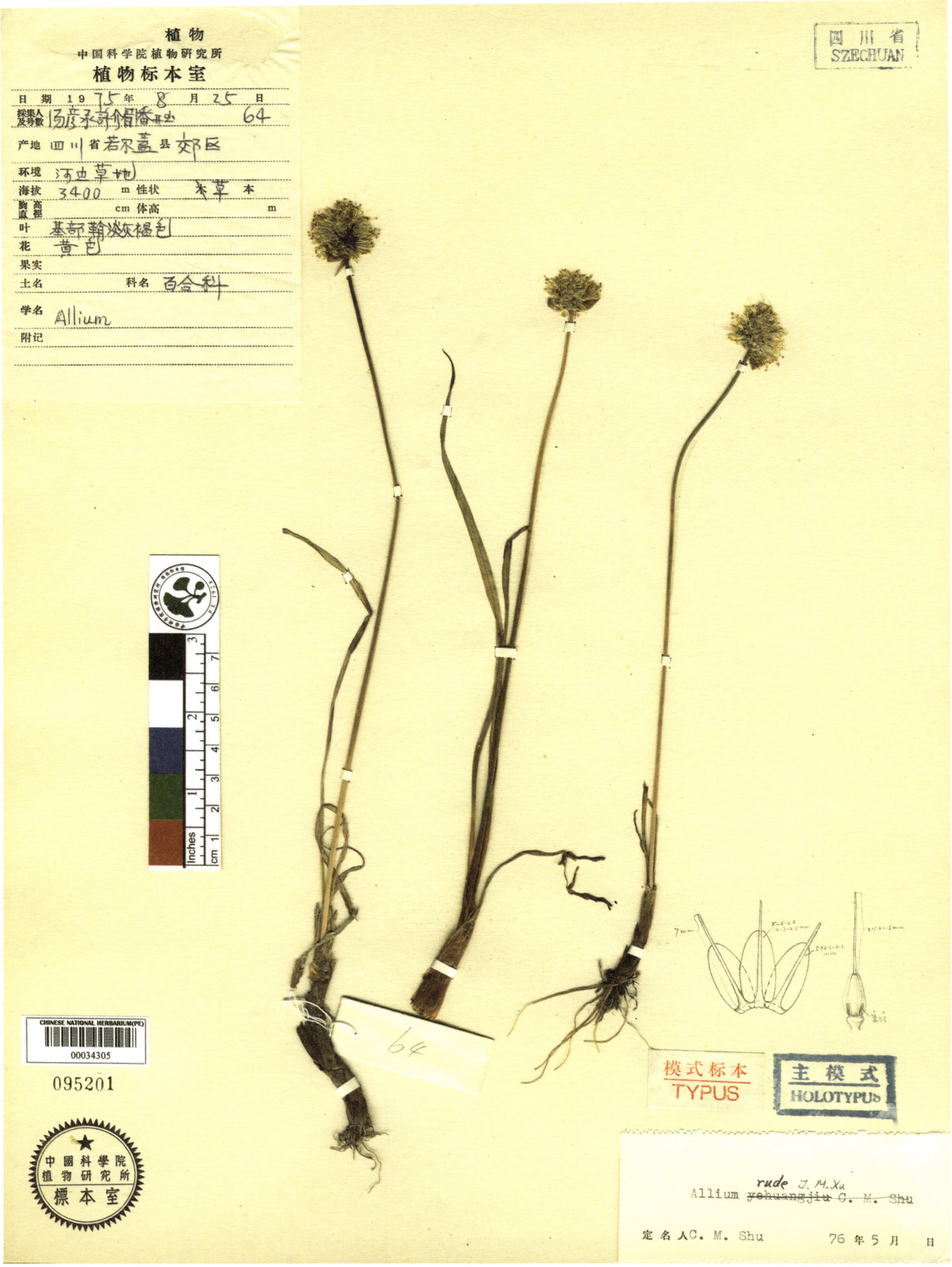

野黄韭 ***Allium rude*** J. M. Xu, Fl. Reip. Pop. Sin. 14: 249, 286, pl. 50: 6. 1980. **Holotype:** China. Sichuan: Zoige, alt. 3400 m, 1975-08-25, Y. C. Tang & al. 64.

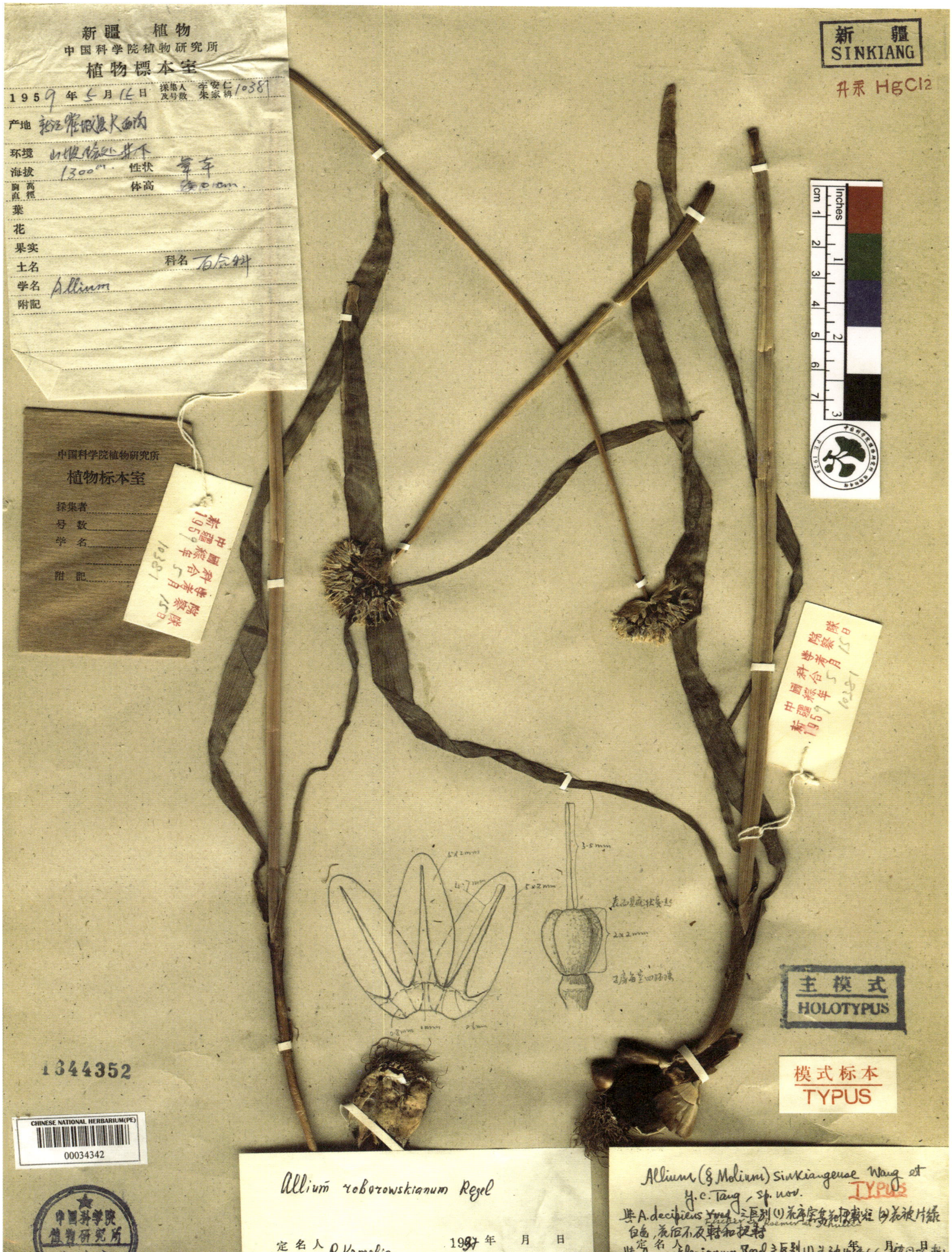

新疆蒜 ***Allium sinkiangense*** F. T. Wang & Y. C. Tang, Fl. Reip. Pop. Sin. 14: 269, 286, f. 97. 1980. **Holotype:** China. Xinjiang: Huocheng, alt. 1300 m, 1959-05-15, A. J. Li & C. L. Chu 10381.

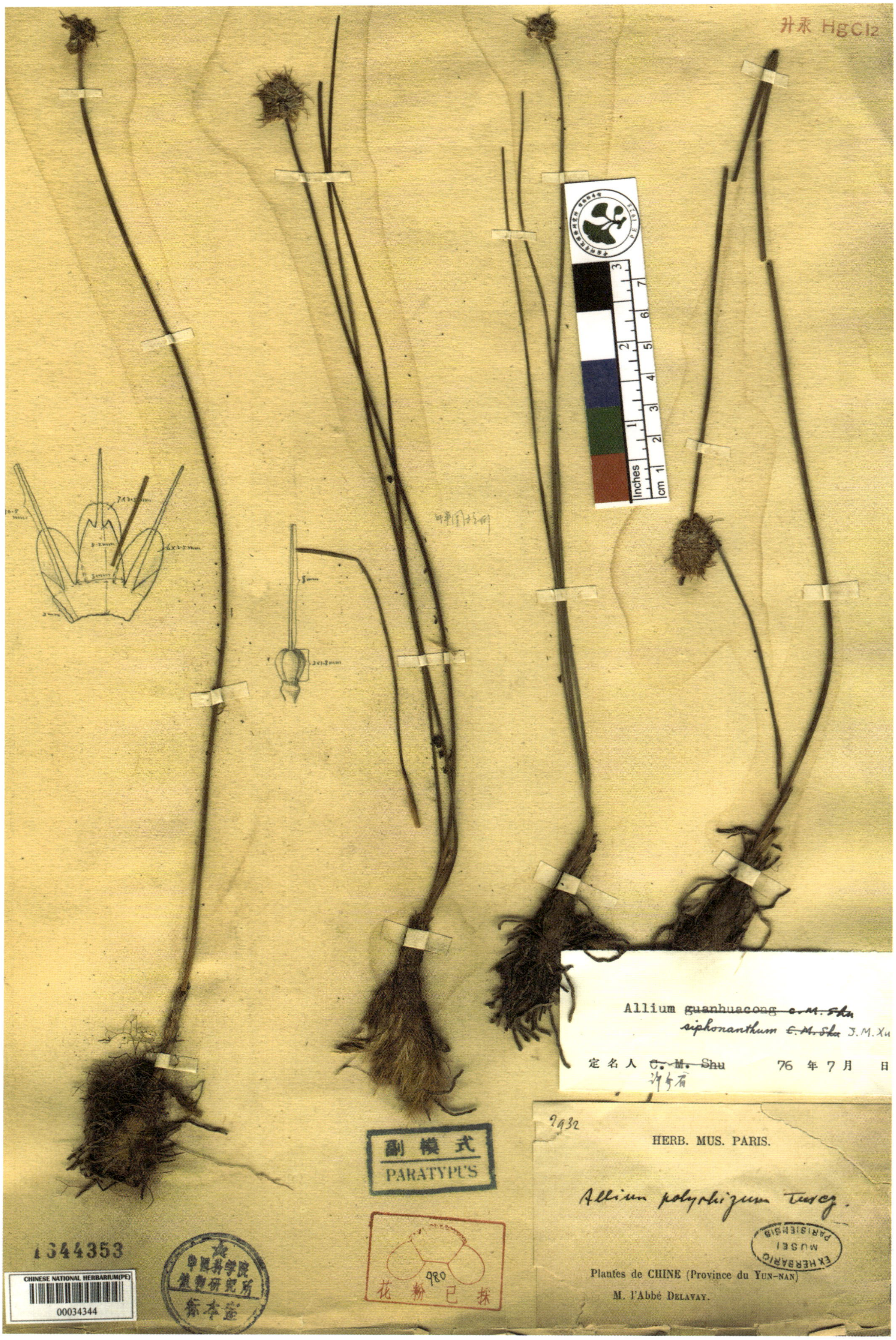

管花葱 ***Allium siphonanthum*** J. M. Xu, Fl. Reip. Pop. Sin. 14: 217, 284, f. 41. 1980. **Paratype:** China. Yunnan: Precise locality not known, Delavay 3932.

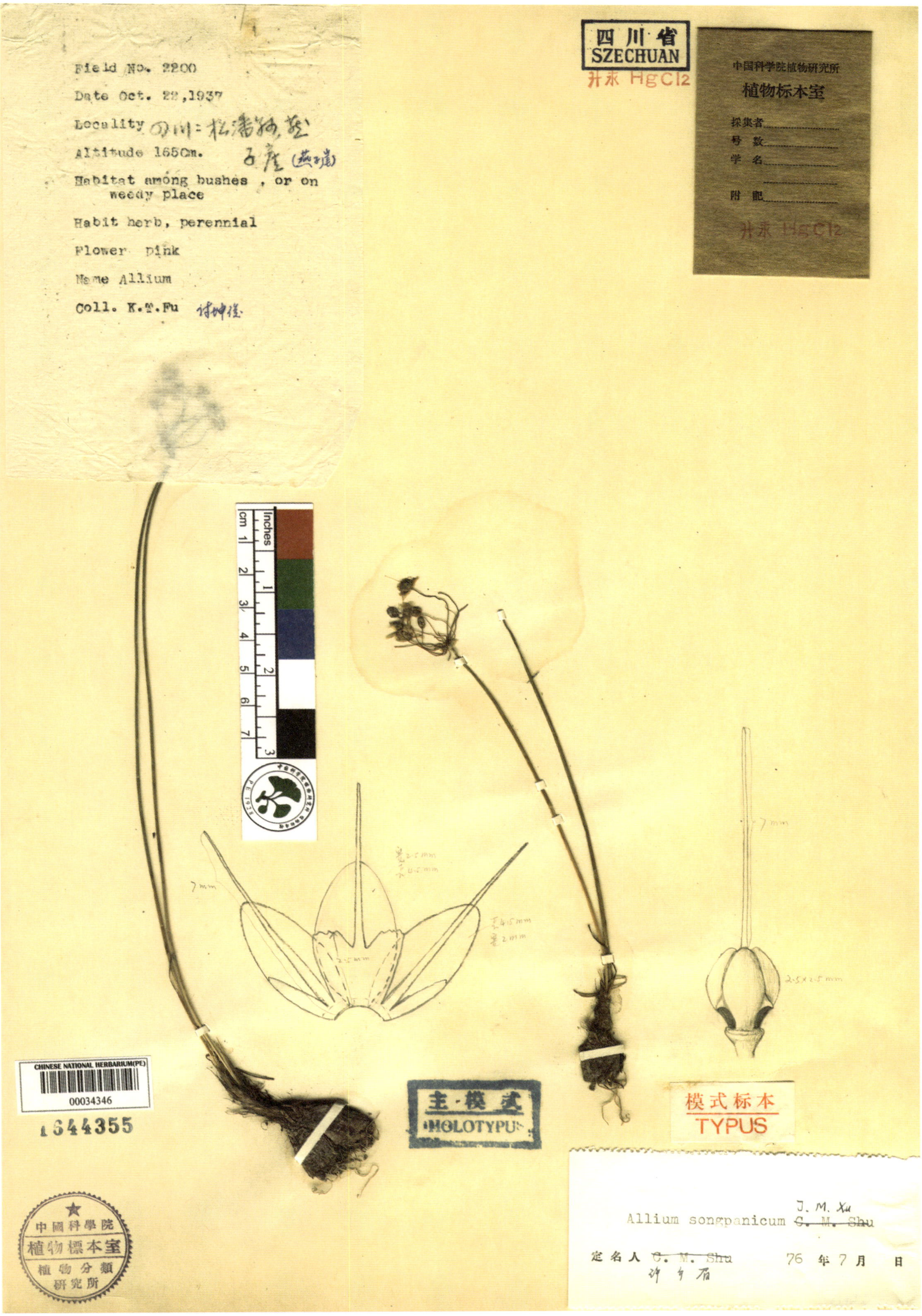

松潘韭 ***Allium songpanicum*** J. M. Xu, Fl. Reip. Pop. Sin. 14: 261, 286, f. 86. 1980. **Holotype:** China. Sichuan: Songpan, 1937-10-22, K. T. Fu 2200.

川韭 ***Alliun szechuanicum*** F. T. Wang & Tang in Bull Fab Mem. Inst. Biol., Bot. 7(6): 296. 1937. **Holotype:** China. Sichuan: Sungpan (=Songpan), alt. 2800 m, 1933-09-??, T. T. Yu 2611.

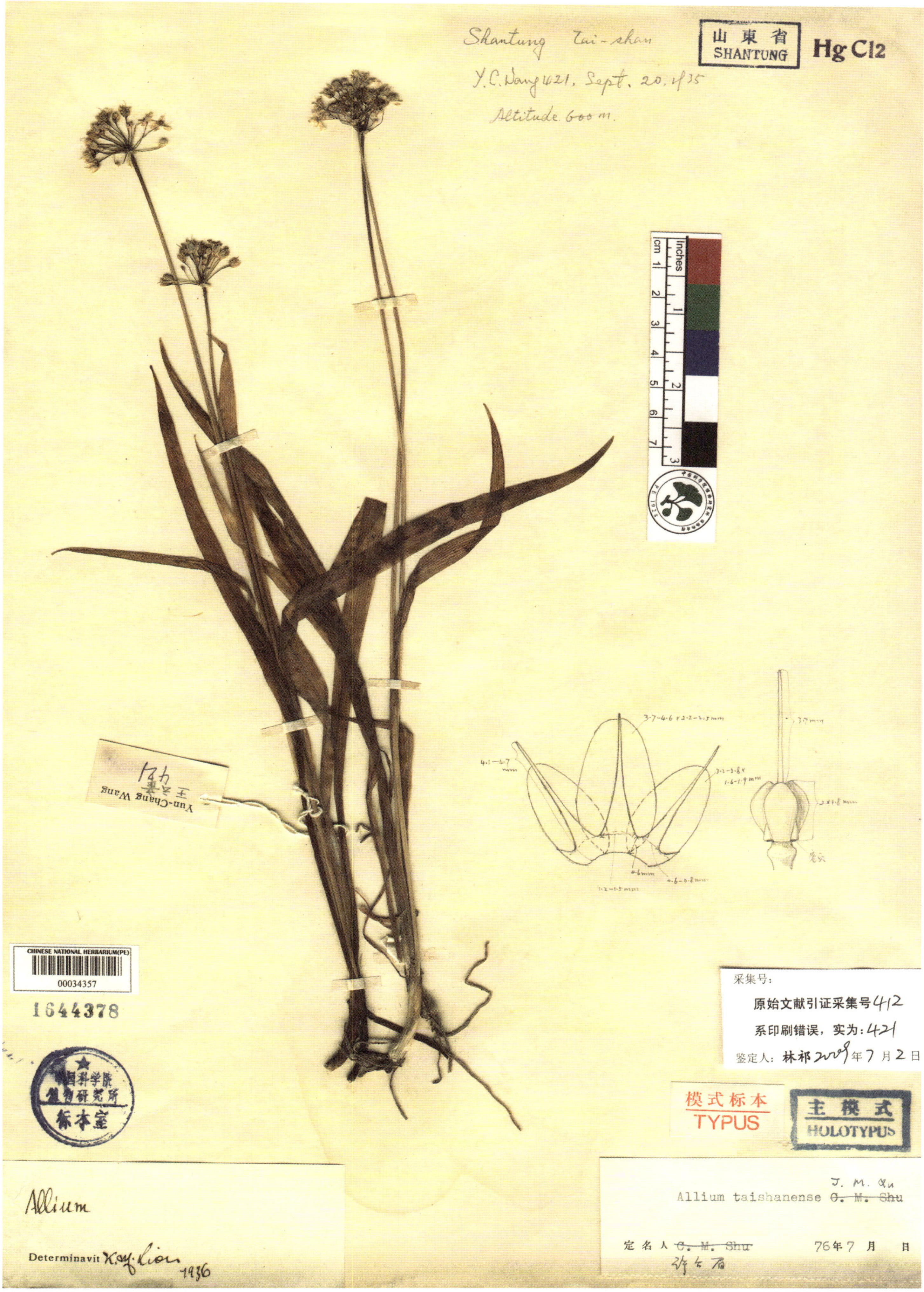

泰山韭 ***Allium taishanense*** J. X. Xu, Fl. Reip. Pop. Sin. 14: 240, 285, f. 64. 1980. **Holotype:** China. Shandong: Taishan, alt. 600 m, 1935-09-20, Y. C. Wang 421.

唐古韭 ***Allium tanguticum*** Regel in Acta Hort. Petrop. 10: 317. 1887. **Isotype:** China. Gansu: Datong He, 1880-07-27, N. M. Przewalski 752.

观光韭 ***Allium tsoongii*** F. T. Wang & Tang in Bull. Fan Mem. Inst. Biol., Bot. 7(6): 292. 1937. **Holotype:** China. Zhejiang: Precise locality not known, 1921-09-14, K. K. Tsoong 3732.

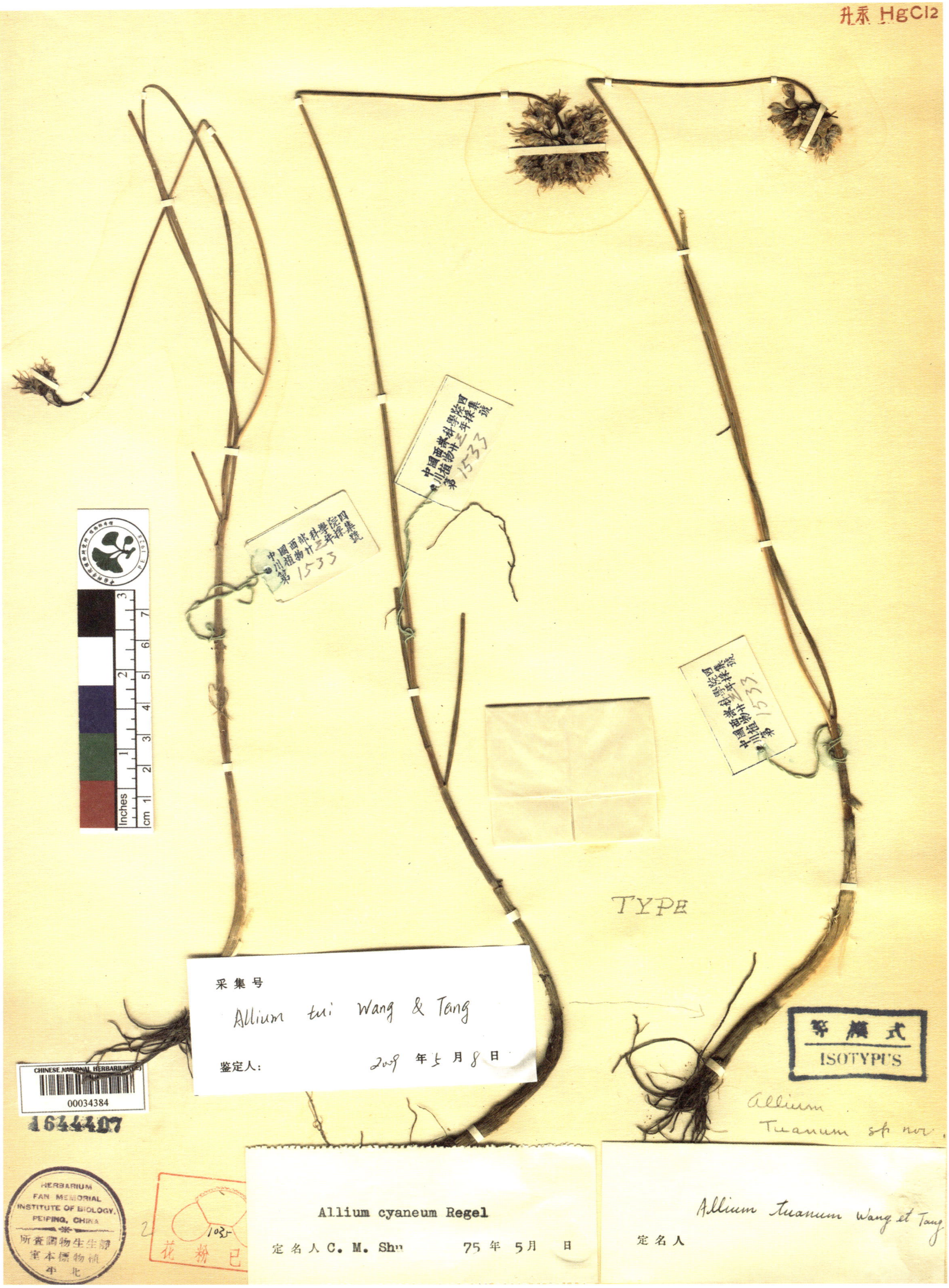

大华韭 ***Allium tui*** F. T. Wang & Tang in Bull. Fan Mem. Inst. Biol., Bot. 7(6): 295. 1937. **Isotype:** China. Sichuan: Huili, alt. 2850 m, 1934-09-??, T. H. Tu 1533.

雅致韭 ***Allium venustum*** C. H. Wright in Journ. Linn. Soc. 36: 126. 1903. **Isotype:** China. Sichuan: Tatsienleu, 1893-??-??, J. A. Soulie 609.

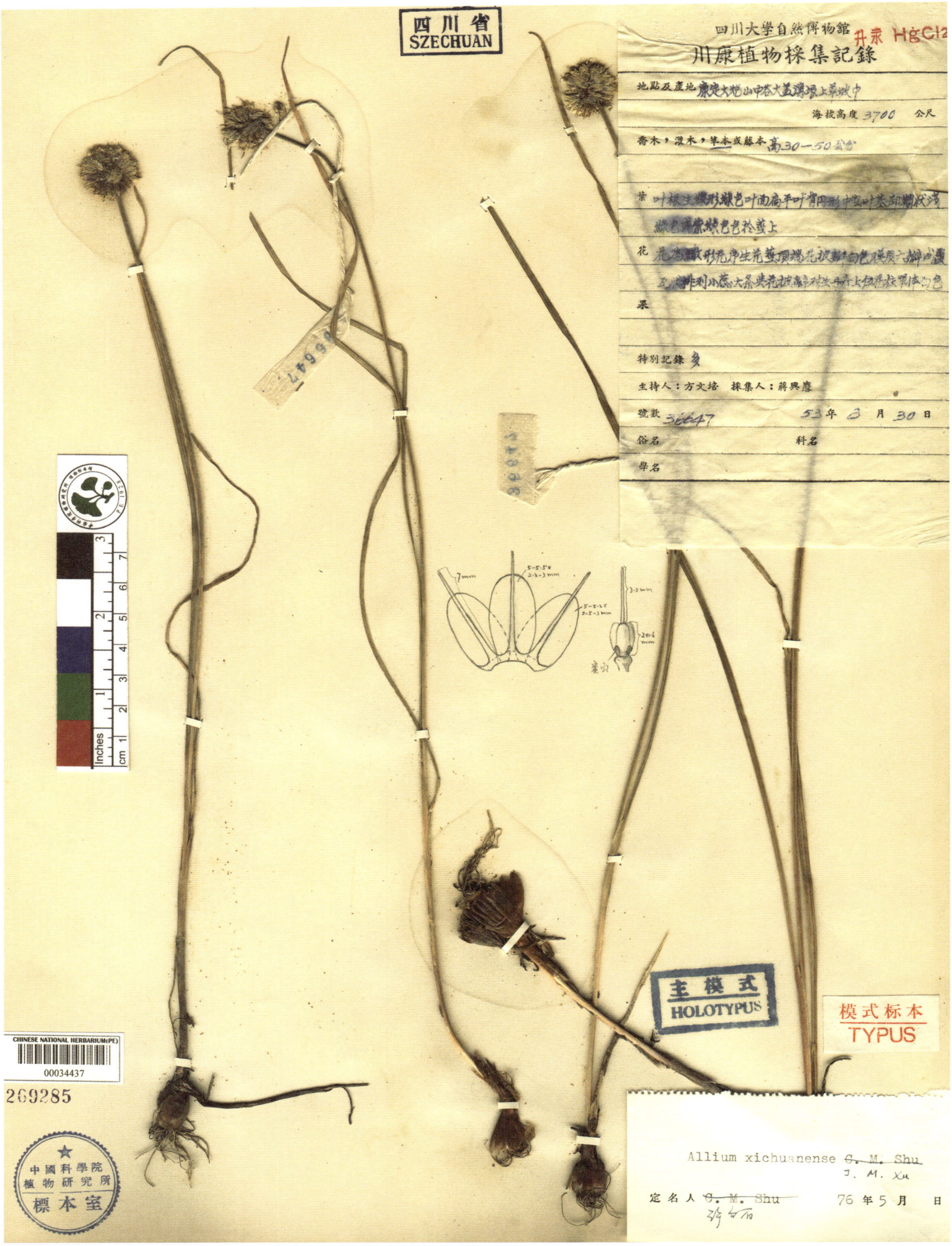

西川韭 ***Allium xichuanense*** J. M. Xu, Fl. Reip. Pop. Sin. 14: 249, 285, pl. 50: 1-5. 1980. **Holotype:** China. Sichuan: Kangding, alt. 3700 m, 1953-08-30, H. L. Tsiang 36647.

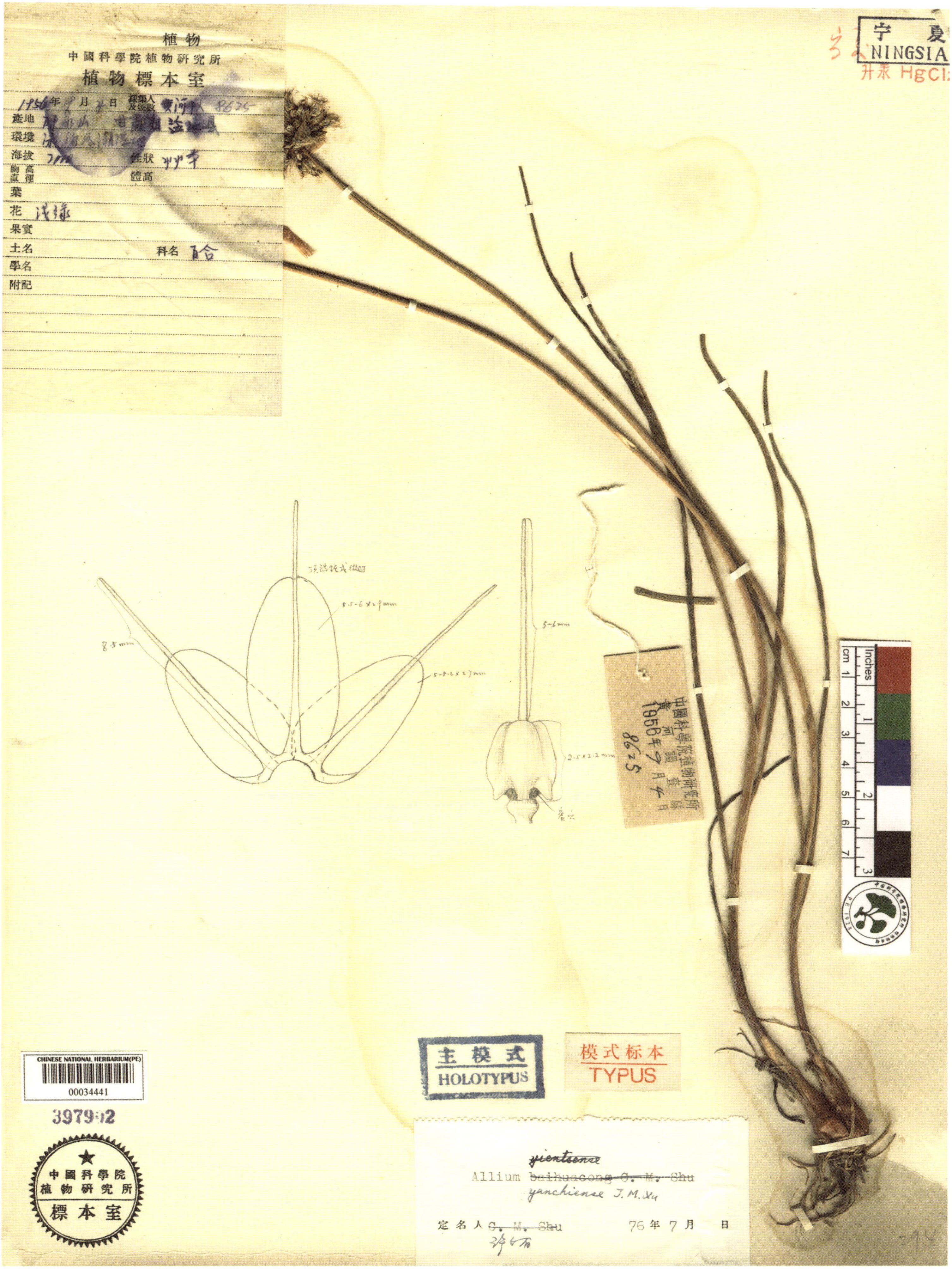

白花葱 ***Allium yanchiense*** J. M. Xu, Fl. Reip. Pop. Sin. 14: 260, 286, f. 85. 1980. **Holotype:** China. Ningxia: Yanchi, alt. 2000 m, 1956-09-04, Huanghe Exped. 8625.

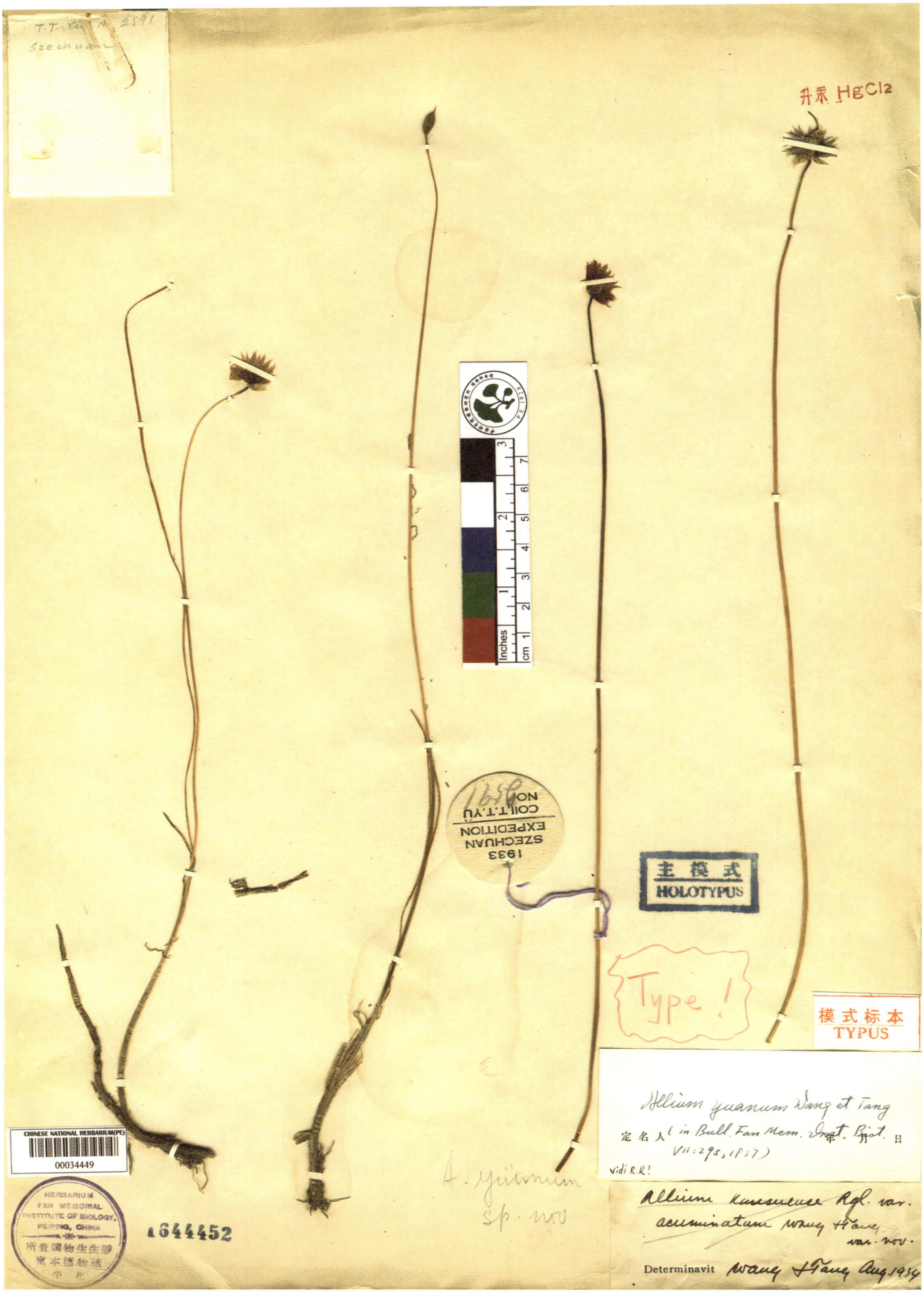

齿被韭 ***Allium yuanum*** F. T. Wang & Tang in Bull. Fan Mem Inst. Biol., Bot. 7(6): 295. 1937. **Holotype:** China. Sichuan: Sungpan (=Songpan), alt. 3180 m, 1933-09-??, T. T. Yu 2591.

山文竹 ***Asparagus acicularis*** F. T. Wang & S. C. Chen in Acta Phytotax. Sin. 16(1): 93, pl. 3: 2. 1978. **Holotype:** China. Guangxi: Wuming, 1958-11-20, Inst. Materia Medica, Acad. Med. Sci. 842.

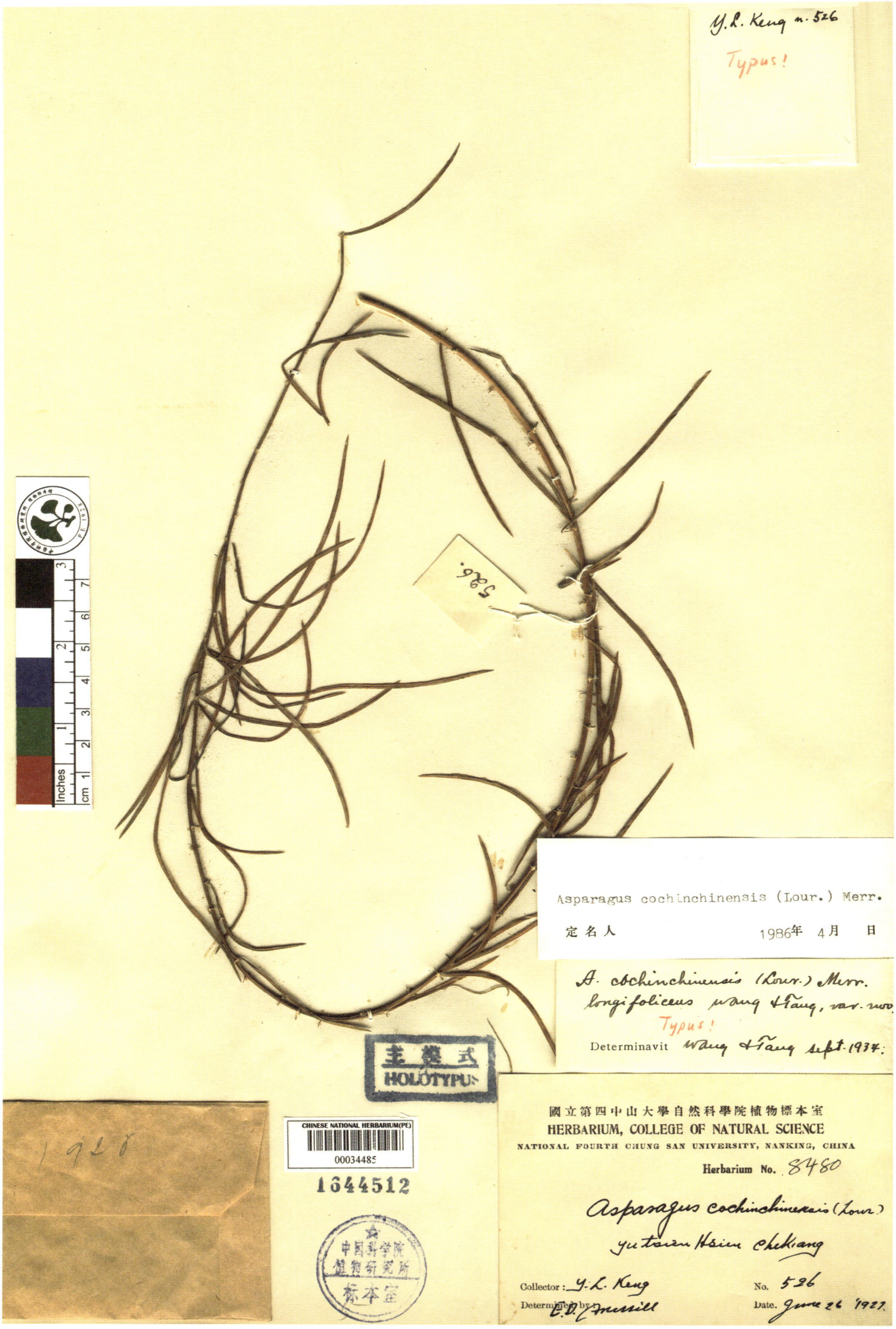

长叶天门冬 ***Asparagus cochinchinensis*** (Lour.) Merr. var. ***longifoliatus*** F. T. Wang & Tang in Bull. Fan Mem. Inst. Biol., Bot. 7(6): 291. 1937. **Holotype:** China. Zhejiang: Yutsien (= Quzhou), 1927-06-26, Y. L. Keng 526.

大叶羊齿天门冬 ***Asparagus filicinus*** Ham. var. ***megaphyllus*** F. T. Wang & Tang in Bull. Fan Mem. Inst. Biol., Bot. 7(6): 290. 1937. **Lectotype** (designated by Q. Lin & Z. R. Yang in Bull. Bot. Res., Harbin 30: 131. 2010.): China. Yunnan: Precise locality not known, 1934-??-??, H. T. Tsai 51101.

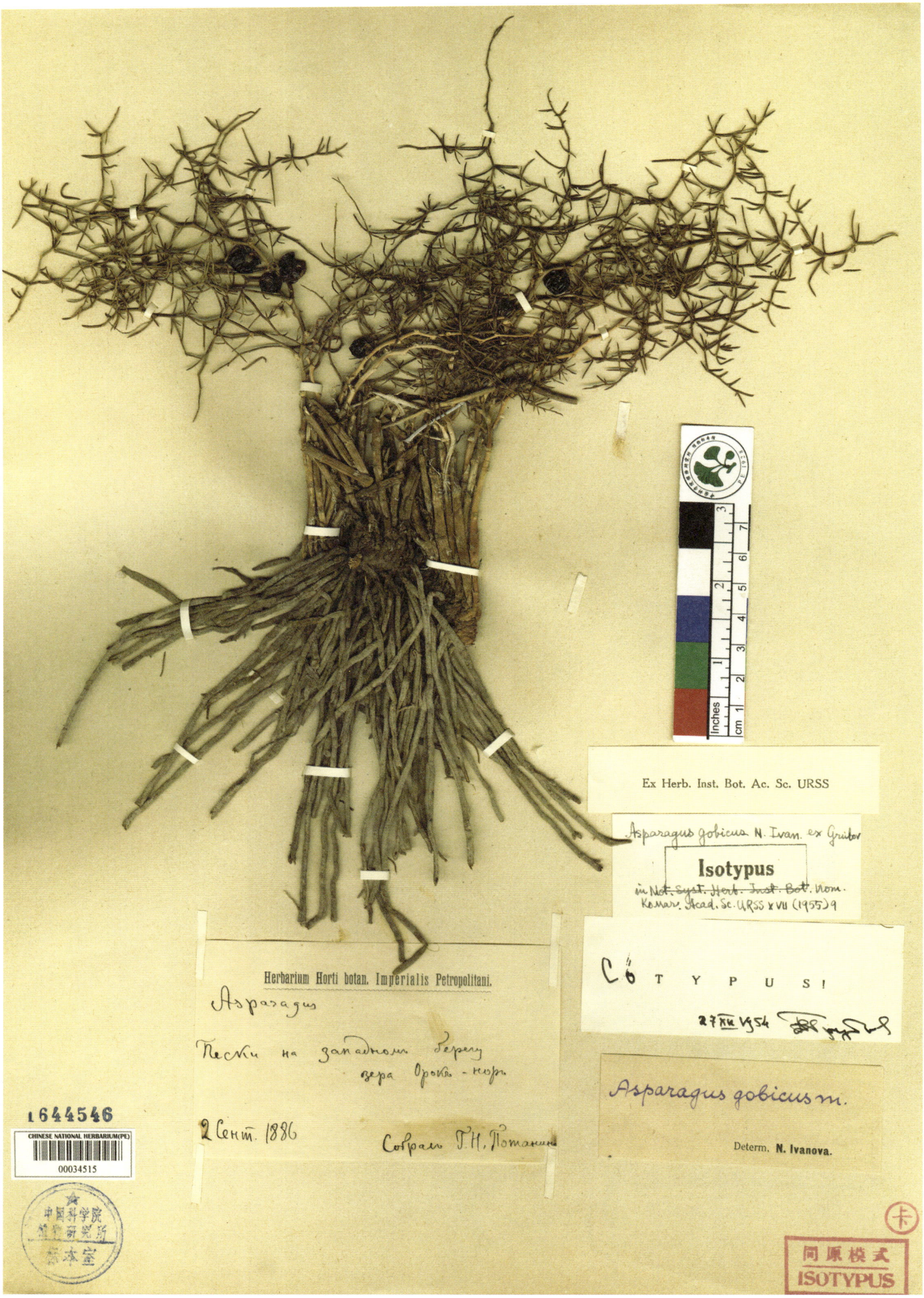

蒙古天门冬 ***Asparagus gobicus*** Ivan. ex Grubov in Not. Syst. Herb. Inst. Bot. Nom. Kom. Acad. Sci. URSS. 17: 9. 1955. **Isotype:** Mongolia. Lacus Orok-nur, 1886-09-02, G. Potanin s. n.

甘肃天门冬 ***Asparagus kansuensis*** F. T. Wang & Tang ex S. C. Chen in Acta Phytotax. Sin. 16(1): 94. 1978. **Holotype:** China. Gansu: Wen Xian, alt. 910 m, 1930-06-15, K. S. Hao 416.

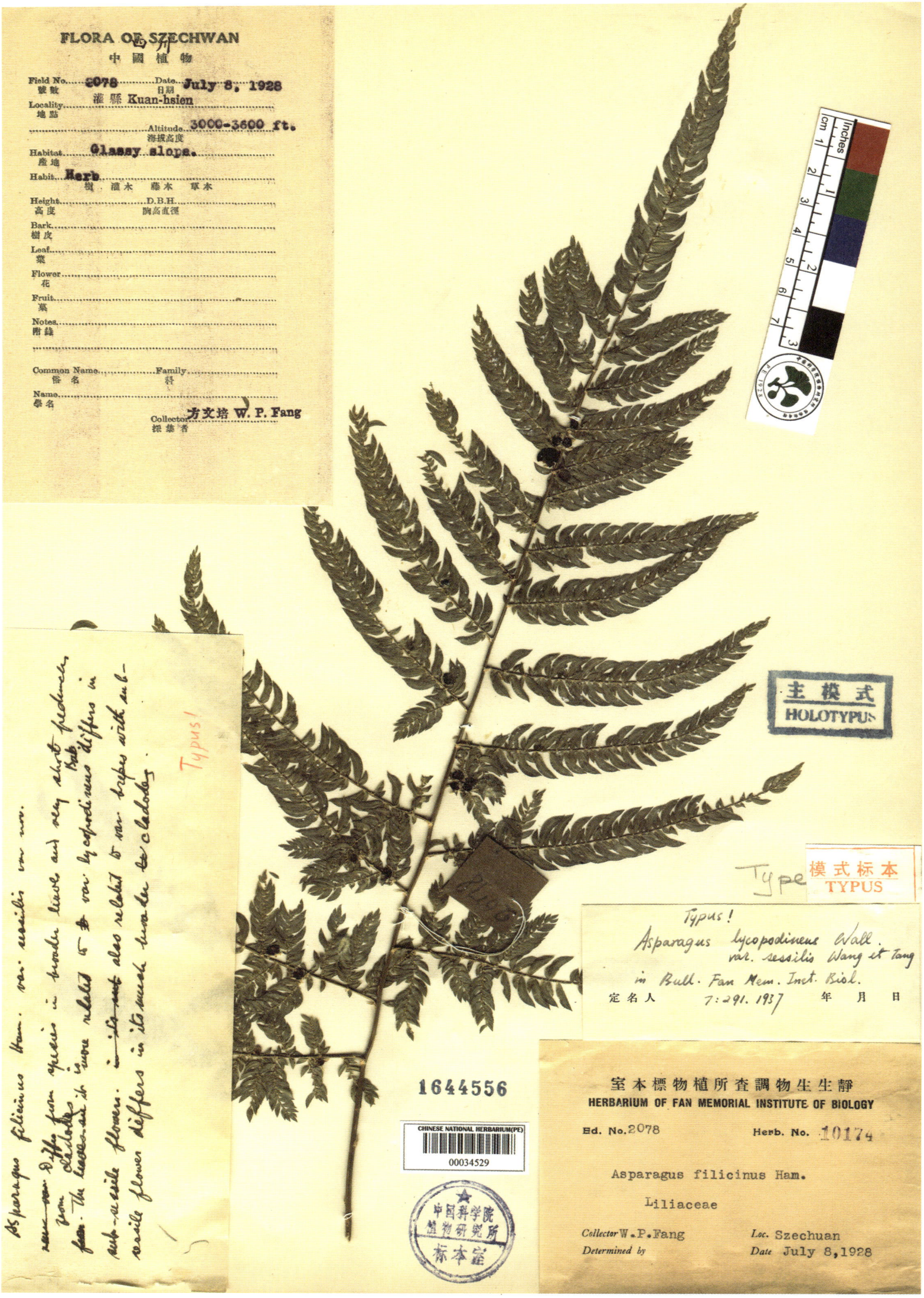

无梗天门冬 ***Asparagus lycopodineus*** Wall. var. ***sessilis*** F. T. Wang & Tang in Bull. Fan Mem. Inst. Biol., Bot. 7(6): 291. 1937.

Holotype: China. Sichuan: Kuan Hsien (=Dujiangyan), alt. 1000~1200 m, 1928-07-08, W. P. Fang 2078.

细枝天门冬 ***Asparagus meioclados*** Lévl. var. ***trichoclados*** F. T. Wang & Tang in Bull. Fan Mem. Inst. Biol., Bot. 7(6): 290. 1937. **Holotype:** China. Yunnan: Chengyuan (=Mengla), alt. 1150 m, 1933-11-01, Y. Tsiang 12500.

西南天门冬 ***Asparagus munitus*** F. T. Wang & S. C. Chen in Acta Phytotax. Sin. 16(1): 91, pl. 2: 2. 1978. **Holotype:** China. Yunnan: Yunning (=Yongning), alt. 2400 m, 1937-05-01, T. T. Yu 5212.

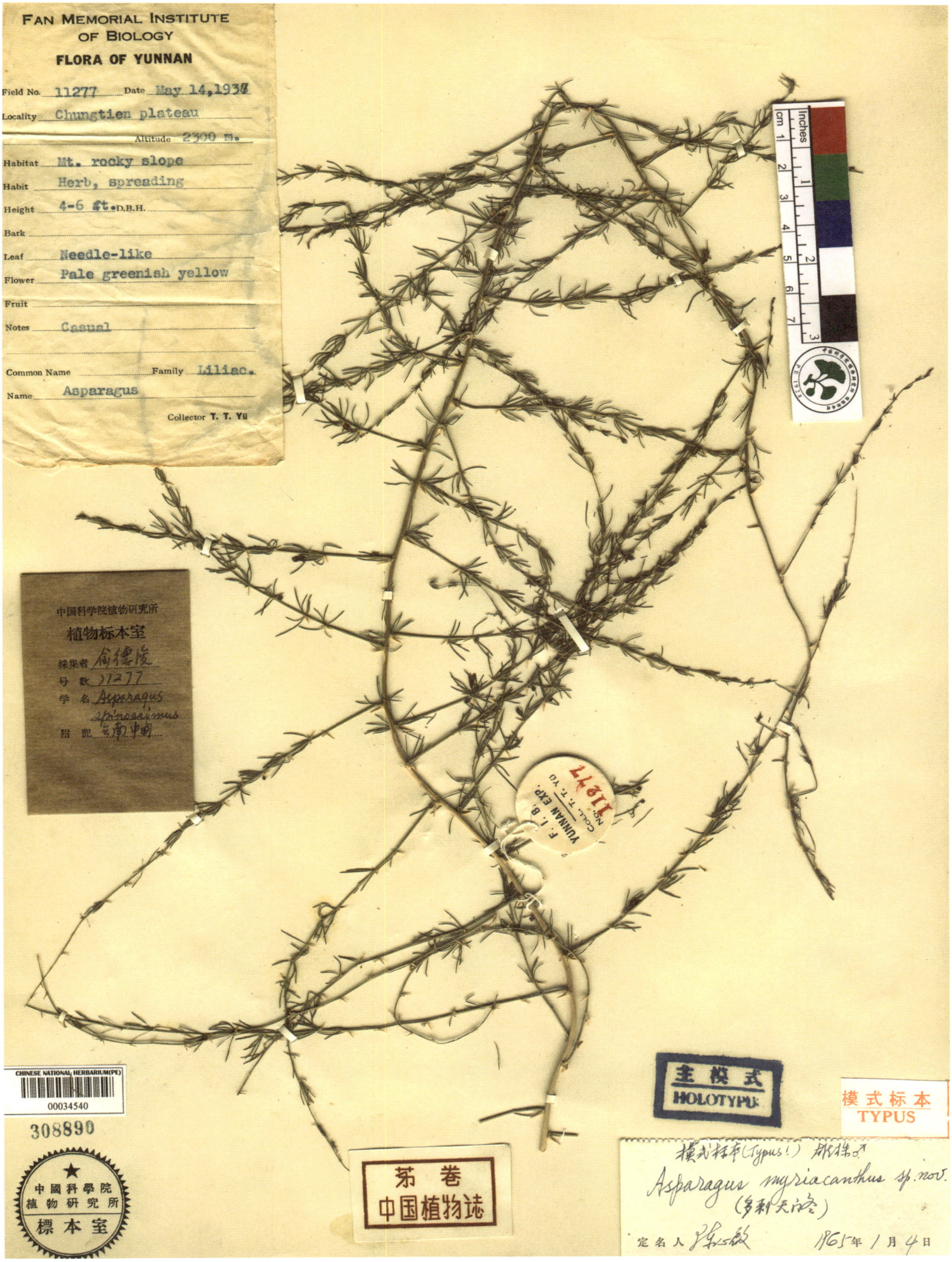

多刺天门冬 ***Asparagus myriacanthus*** F. T. Wang & S. C. Chen in Acta Phytotax. Sin. 16(1): 92, pl. 2: 3. 1978. **Holotype:** China. Yunnan: Weixi, alt. 2300 m, 1937-05-14, T. T. Yu 11277.

四川天门冬 ***Asparagus sichuanicus*** S. C. Chen & D. Q. Liu in Acta Phytotax. Sin. 22(5): 418, f. 1: 2. 1984. **Holotype:** China. Sichuan: Wenchuan, 1952-04-23, Z. He 12270.

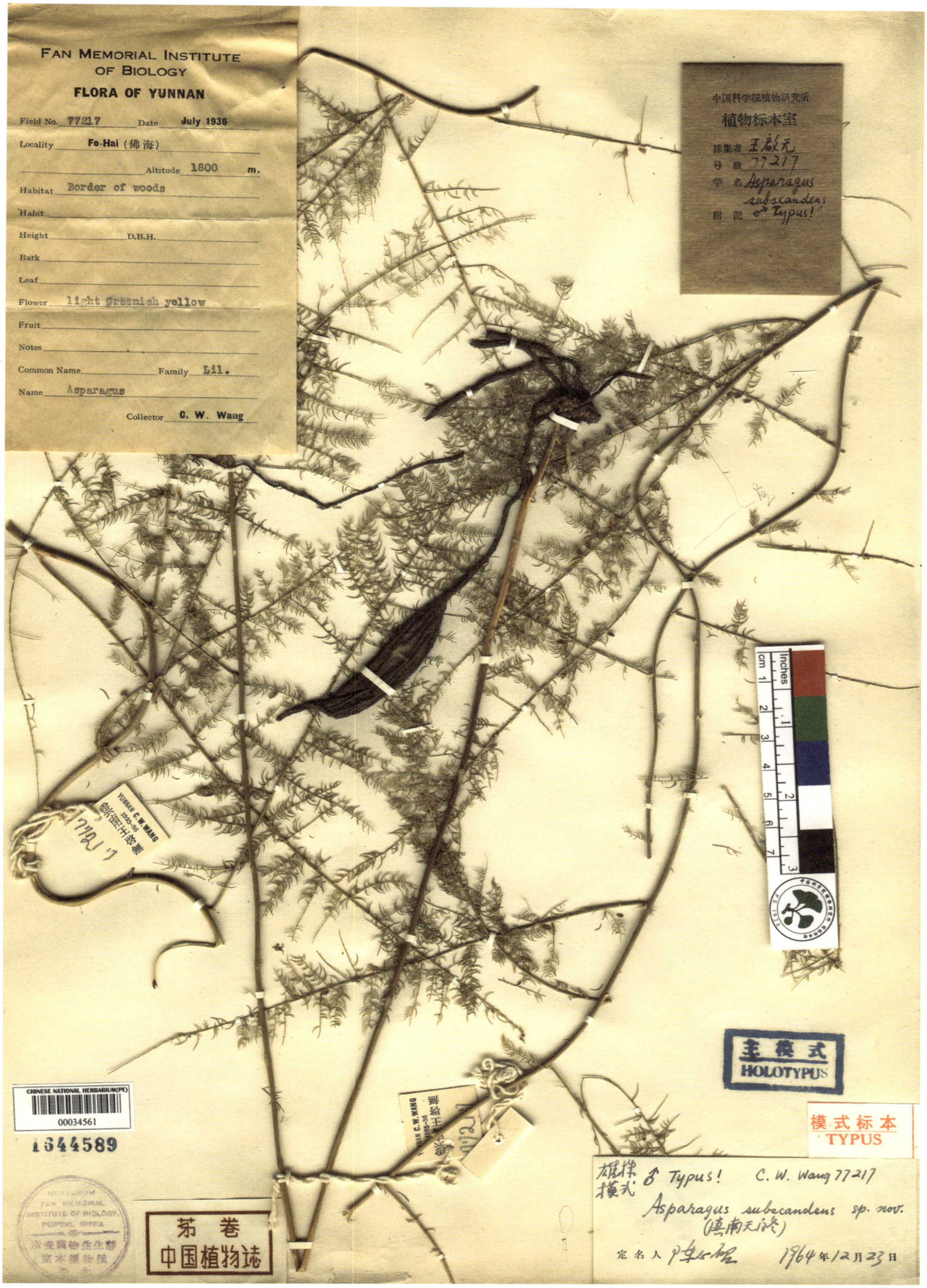

滇南天门冬 ***Asparagus subscandens*** F. T. Wang & S. C. Chen in Acta Phytotax. Sin. 16(1): 92, pl. 2: 4. 1978. **Holotype:** China. Yunnan: Fohai (=Menghai), alt. 1800 m, 1936-07-??, C. W. Wang 77217.

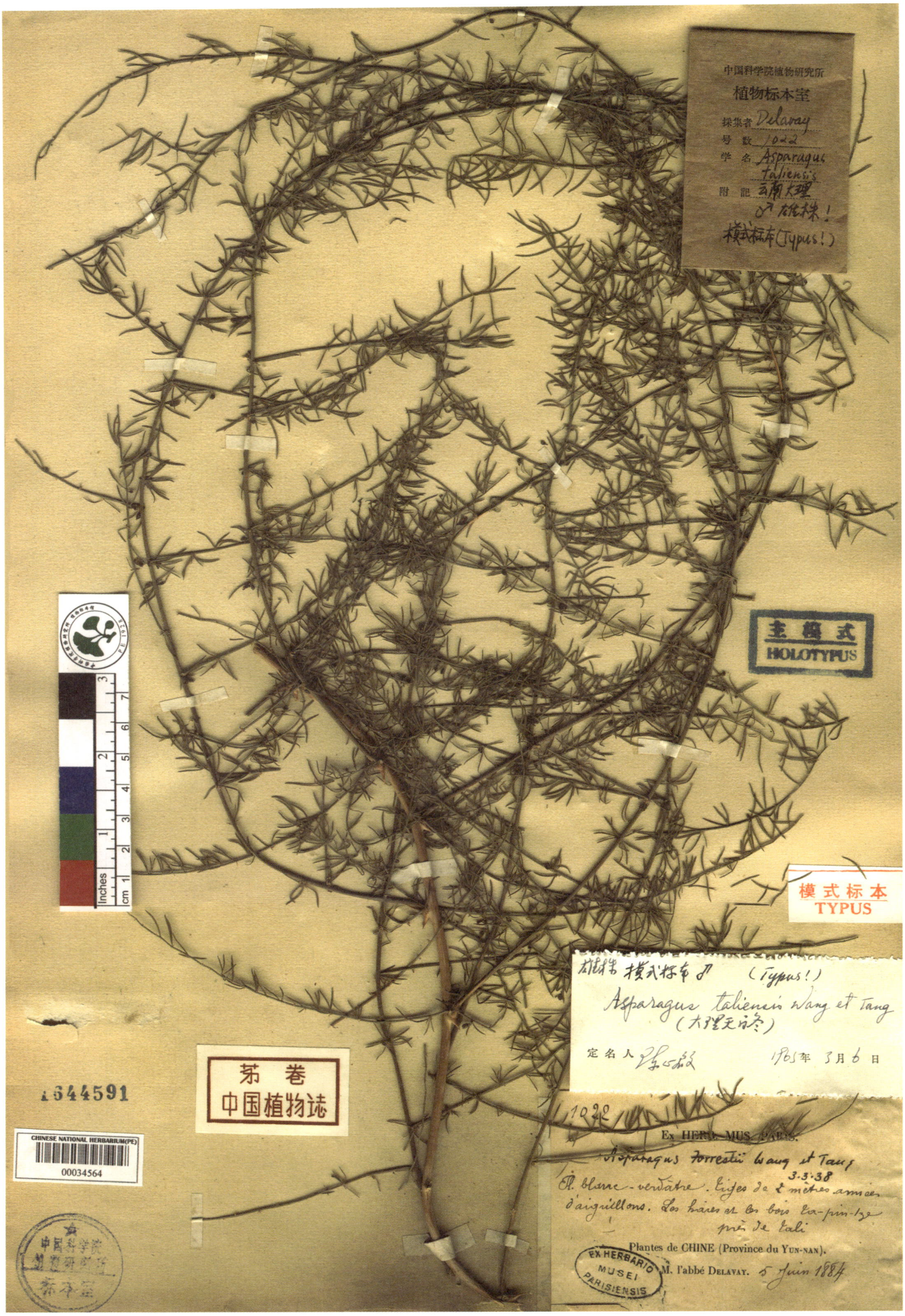

大理天门冬 ***Asparagus taliensis*** F. T. Wang & Tang ex. S. C. Chen in Acta Phytotax. Sin. 16(1): 91, pl. 2: 1. 1978. **Holotype:** China. Yunnan: Dali, 1884-06-05, Delavay 1022.

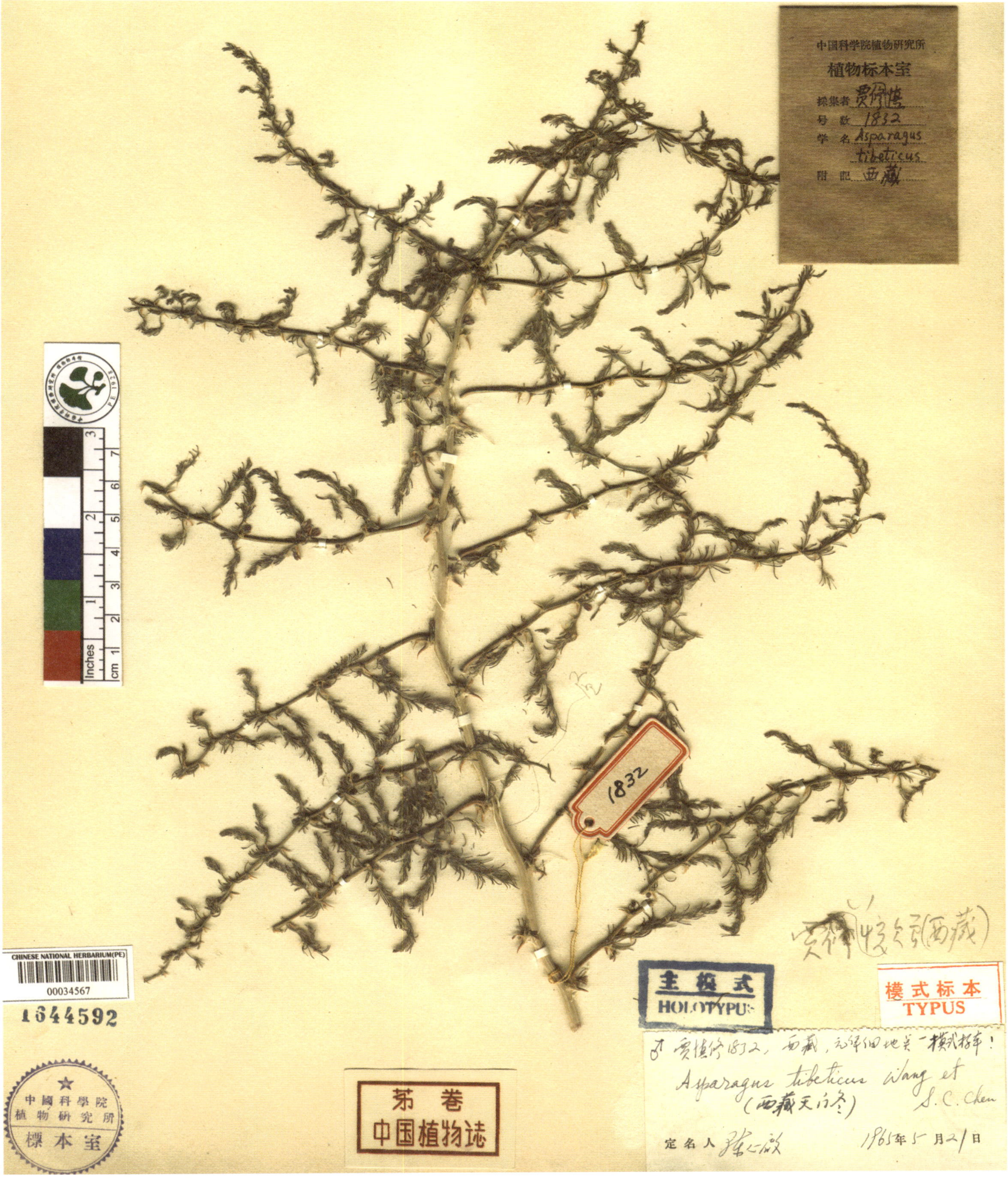

西藏天门冬 ***Asparagus tibeticus*** F. T. Wang & S. C. Chen in Acta Phytotax. Sin. 16(1): 93, pl. 3: 1. 1978. **Holotype:** China. Xizang: Precise locality not known, S. X. Jia 1832.

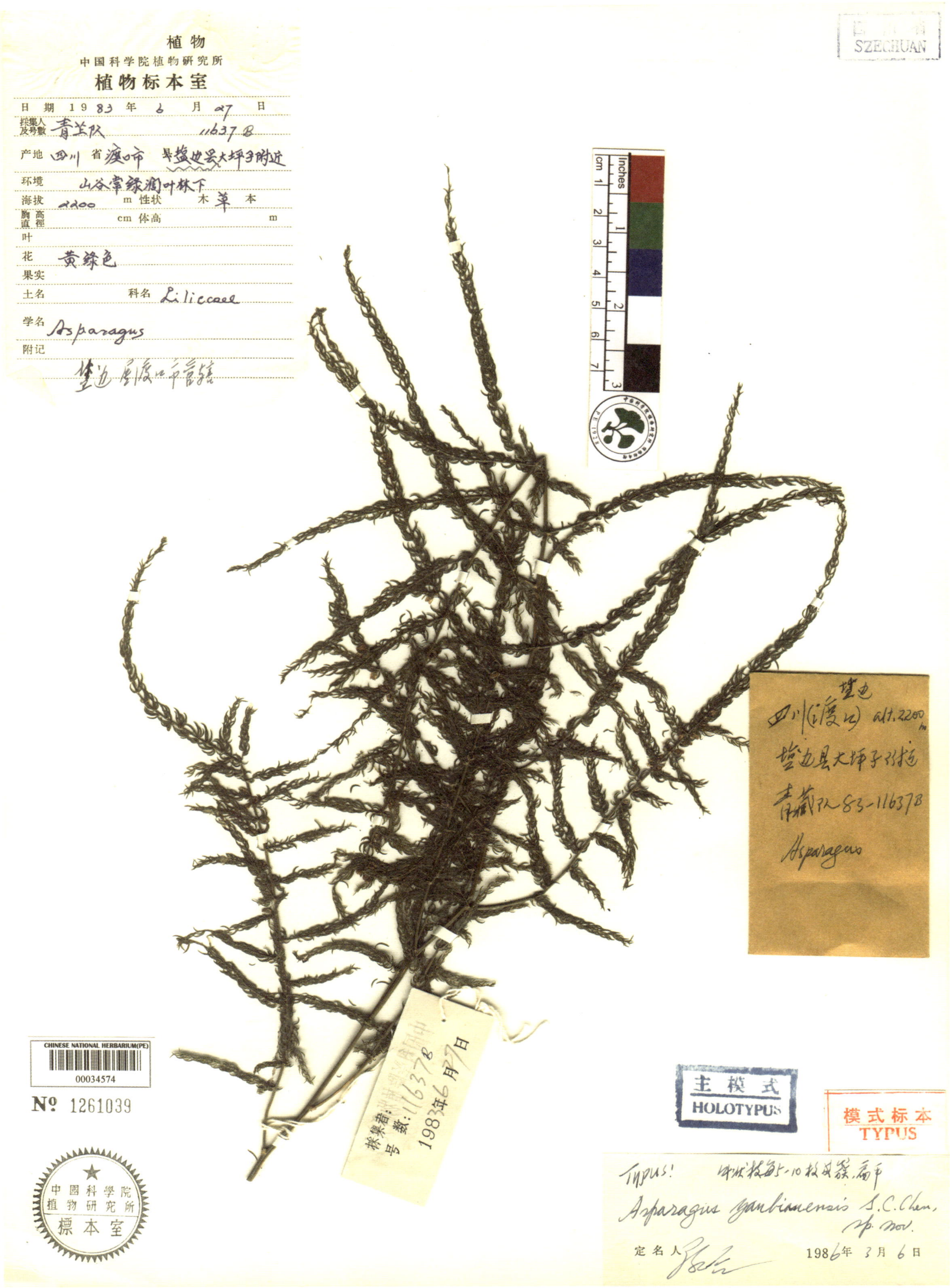

盐边天门冬 ***Asparagus yanbianensis*** S. C. Chen in Acta Phytotax. Sin. 26(2): 139, pl. 1: 1. 1988. **Holotype:** China. Sichuan: Yanbian, alt. 2200 m, 1983-06-27, Qinghai-Xizang Exped. 11637B.

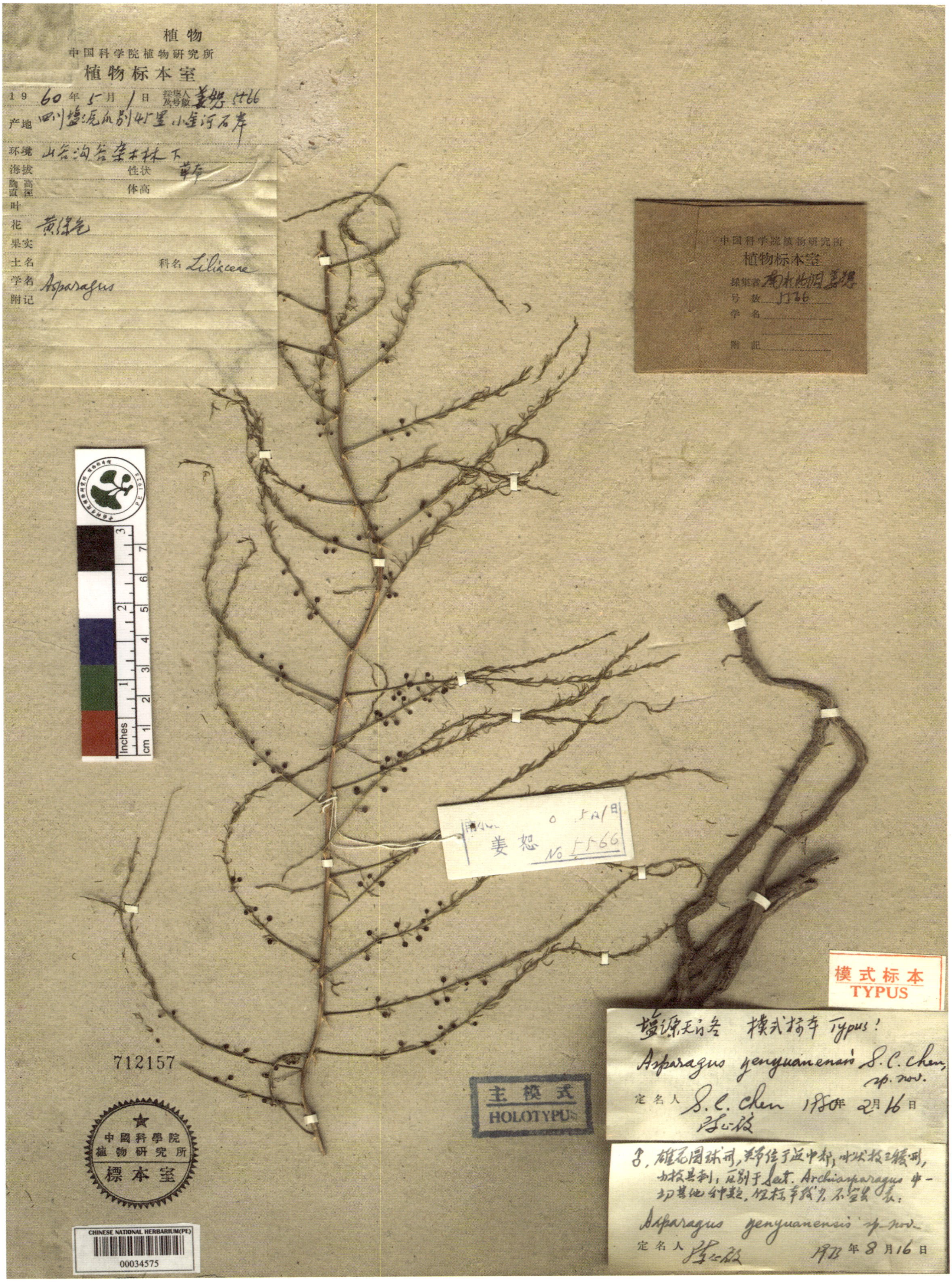

盐源天门冬 ***Asparagus yanyuanensis*** S. C. Chen in Acta Phytotax. Sin. 19(4): 501, f. 2: 3-5. 1981. **Holotype:** China. Sichuan: Yanyuan, 1960-05-01, S. Jiang 5566.

丛生蜘蛛抱蛋 ***Aspidistra caespitosa*** C. Pei in Contr. Biol. Lab. Sci. Soc. China, Bot. Ser. 12(2): 101. 1939. **Lectotype** (designated by Y. Lin & al. in Acta Bot. Bor.-Occ. Sin. 34: 412. 2014.): China. Chongqing: Pehpei (=Beibei), Jinyunshan, 1939-04-06, C. Pei 7073.

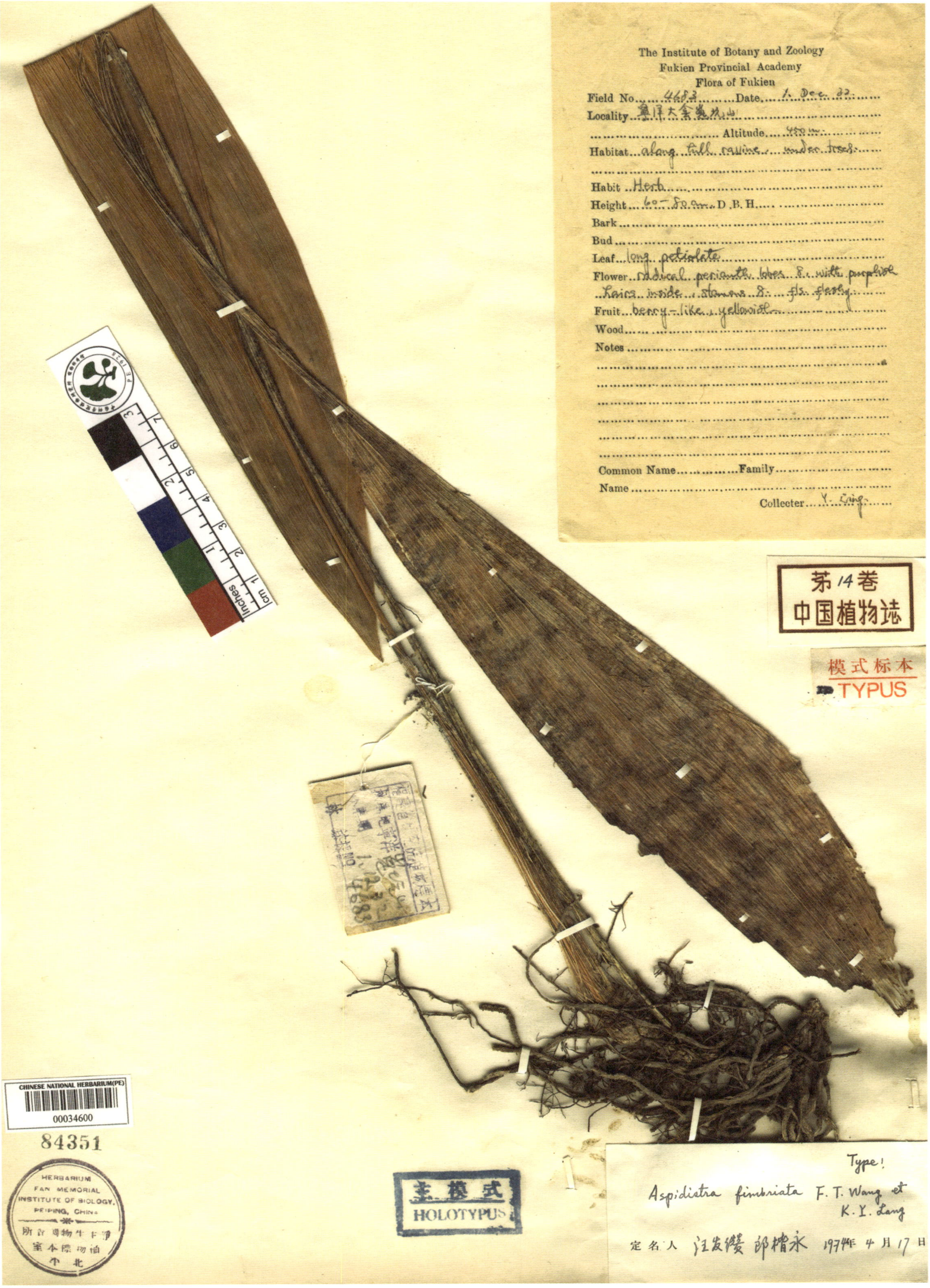

流苏蜘蛛抱蛋 ***Aspidistra fimbriata*** F. T. Wang & K. Y. Lang in Acta Phytotax. Sin. 16(1): 76, f. 1. 1978. **Holotype:** China. Fujian: Ningyang, alt. 450 m, 1932-12-01, Y. Ling 4683.

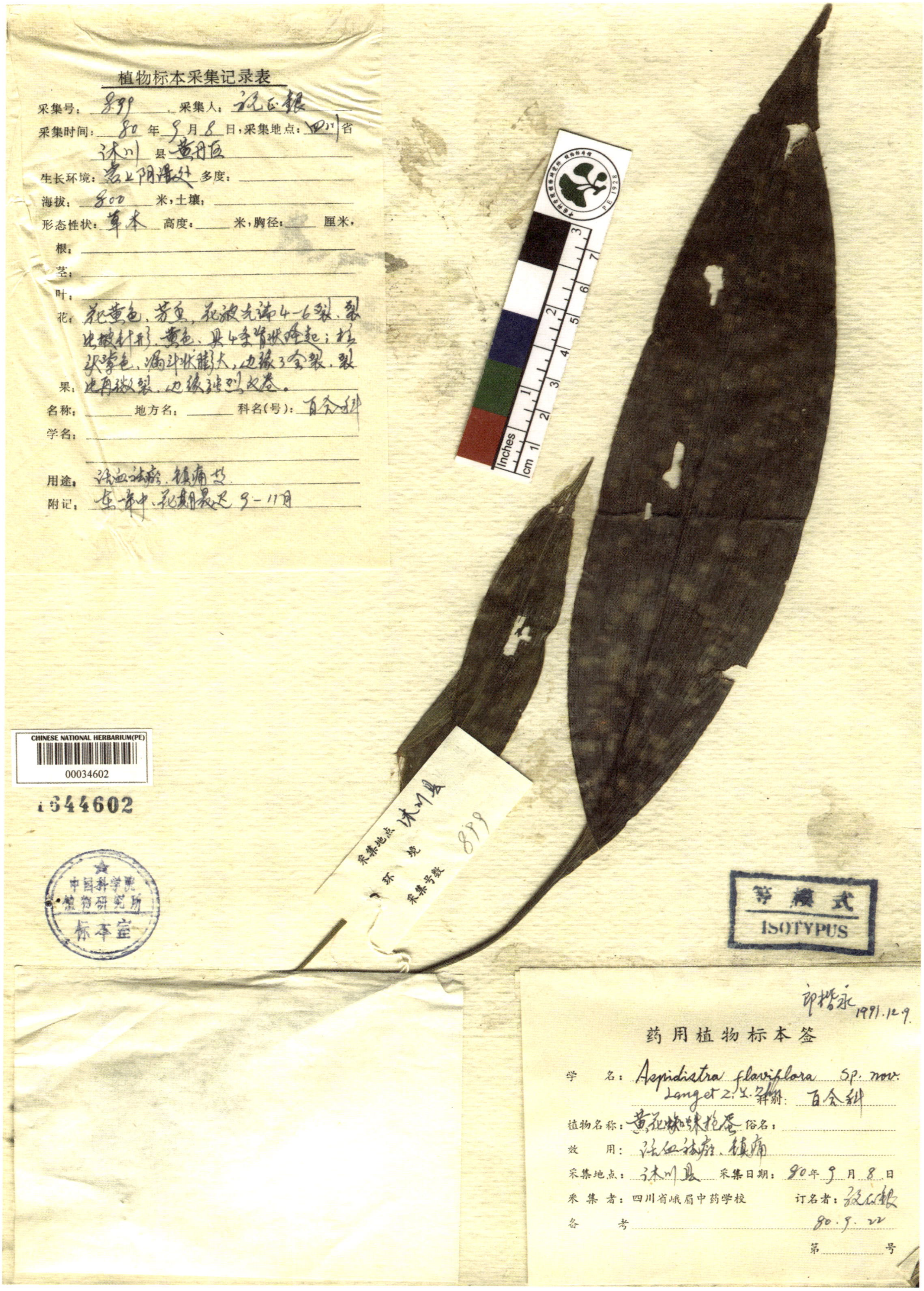

黄花蜘蛛抱蛋 ***Aspidistra flaviflora*** K. Y. Lang & Z. Y. Zhu in Acta Phytotax. Sin. 20(4): 485, f. 1: 1-5. 1982. **Isotype:** China. Sichuan: Muchuan, alt. 800 m, 1980-09-08, Z. Y. Zhu 899.

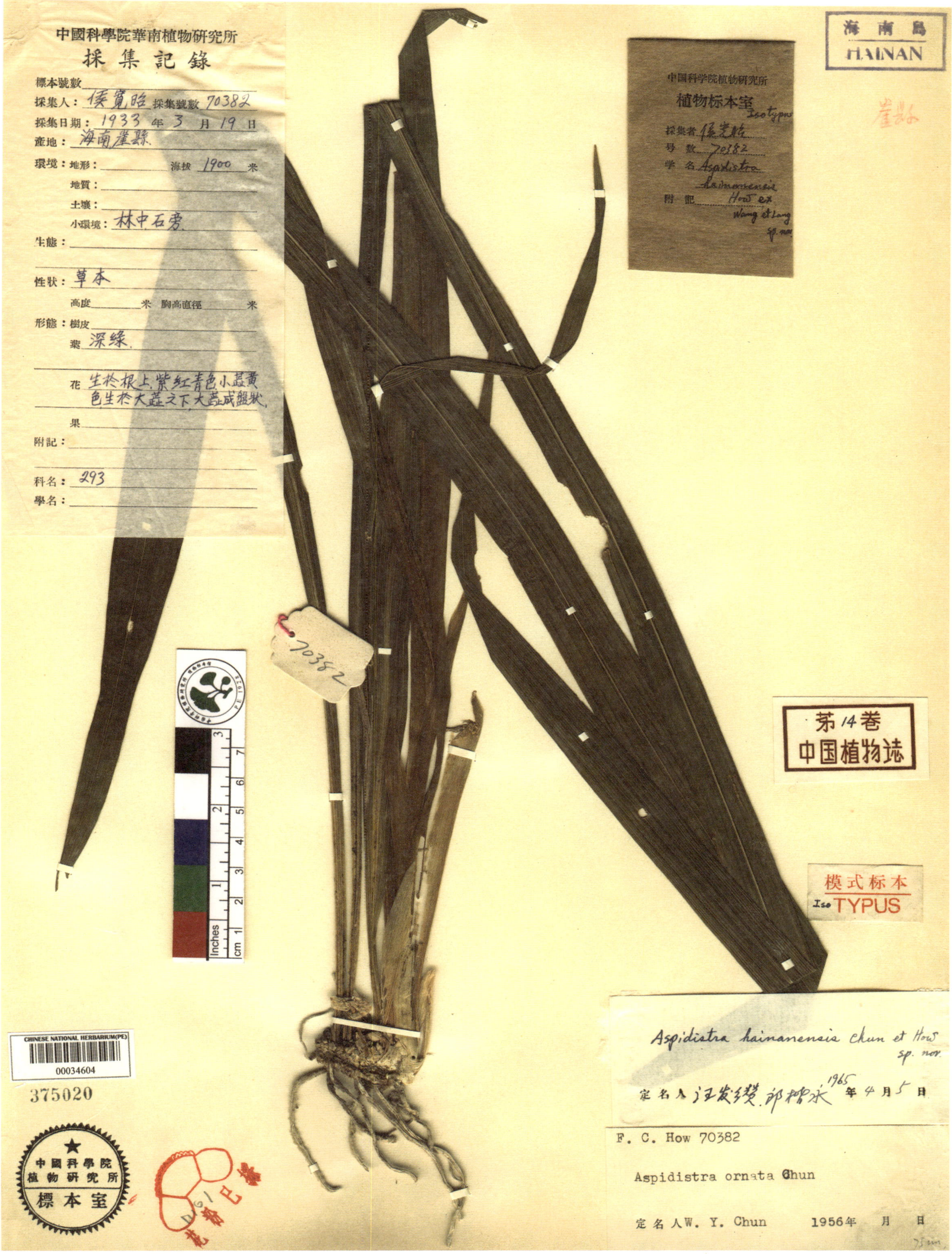

海南蜘蛛抱蛋 ***Aspidistra hainanensis*** Chun & How, Fl. Hainan 4: 114, 533. 1977. **Isotype:** China. Hainan: Ya Xian, alt. 1900 m, 1933-03-19, F. C. How 70382.

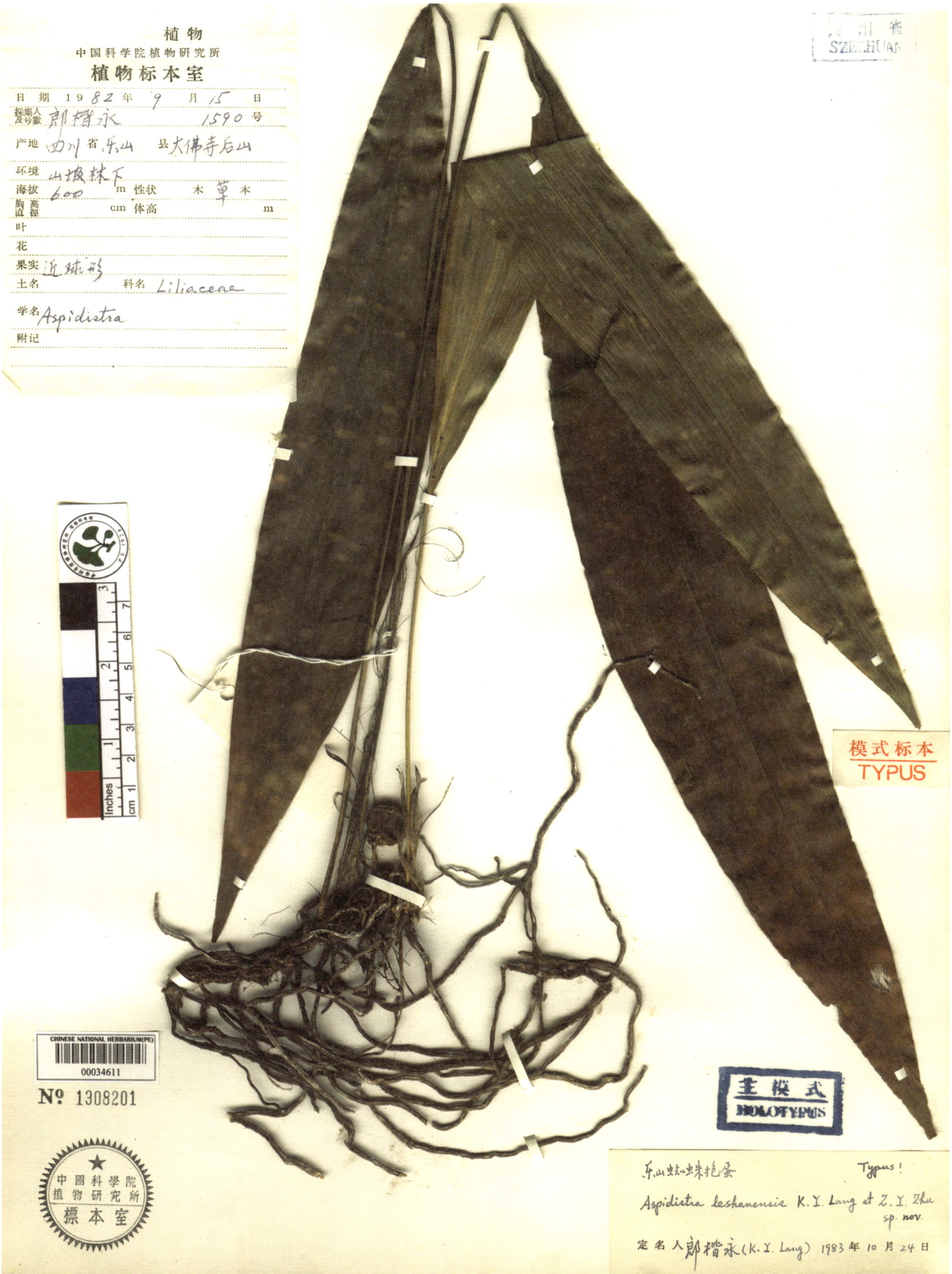

乐山蜘蛛抱蛋 ***Aspidistra leshanensis*** K. Y. Lang & Z. Y. Zhu in Acta Bot. Yunnan. 6(4): 385, f. 1: 1-4. 1984. **Holotype:** China. Sichuan: Leshan, alt. 600 m, 1982-09-15, K. Y. Lang 1590.

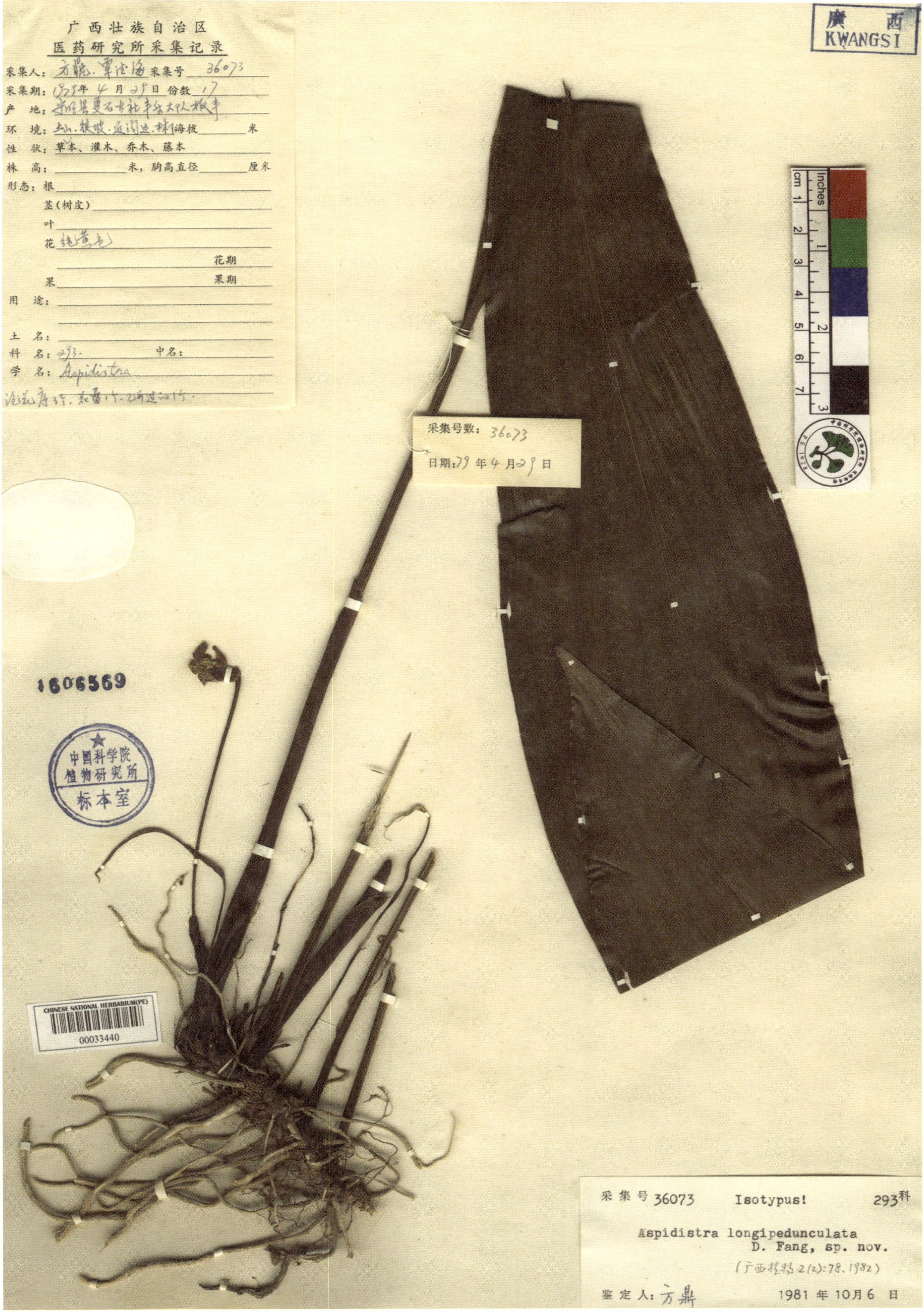

长梗蜘蛛抱蛋 ***Aspidistra longipedunculata*** D. Fang in Guihaia 2(2): 78, f. 2. 1982. **Isotype:** China. Guangxi: Ningming, 1979-04-29, D. Fang & D. H. Qin 36073.

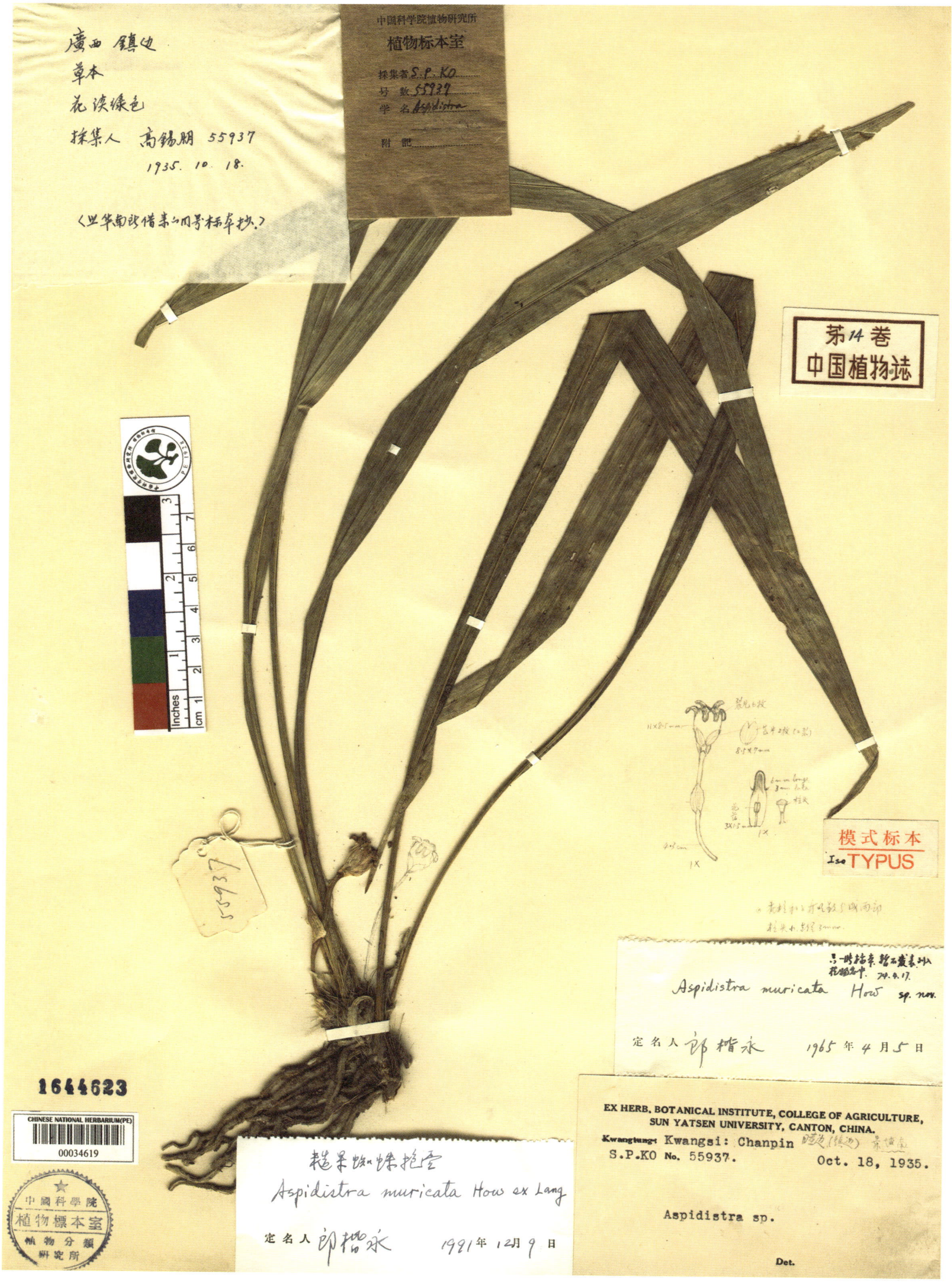

糙果蜘蛛抱蛋 ***Aspidistra muricata*** How ex K. Y. Lang in Acta Phytotax. Sin. 19(3): 383, f. 1: 1-4. 1981. **Isotype:** China. Guangxi: Chenpien (=Napo), 1935-10-18, S. P. Ko 55937.

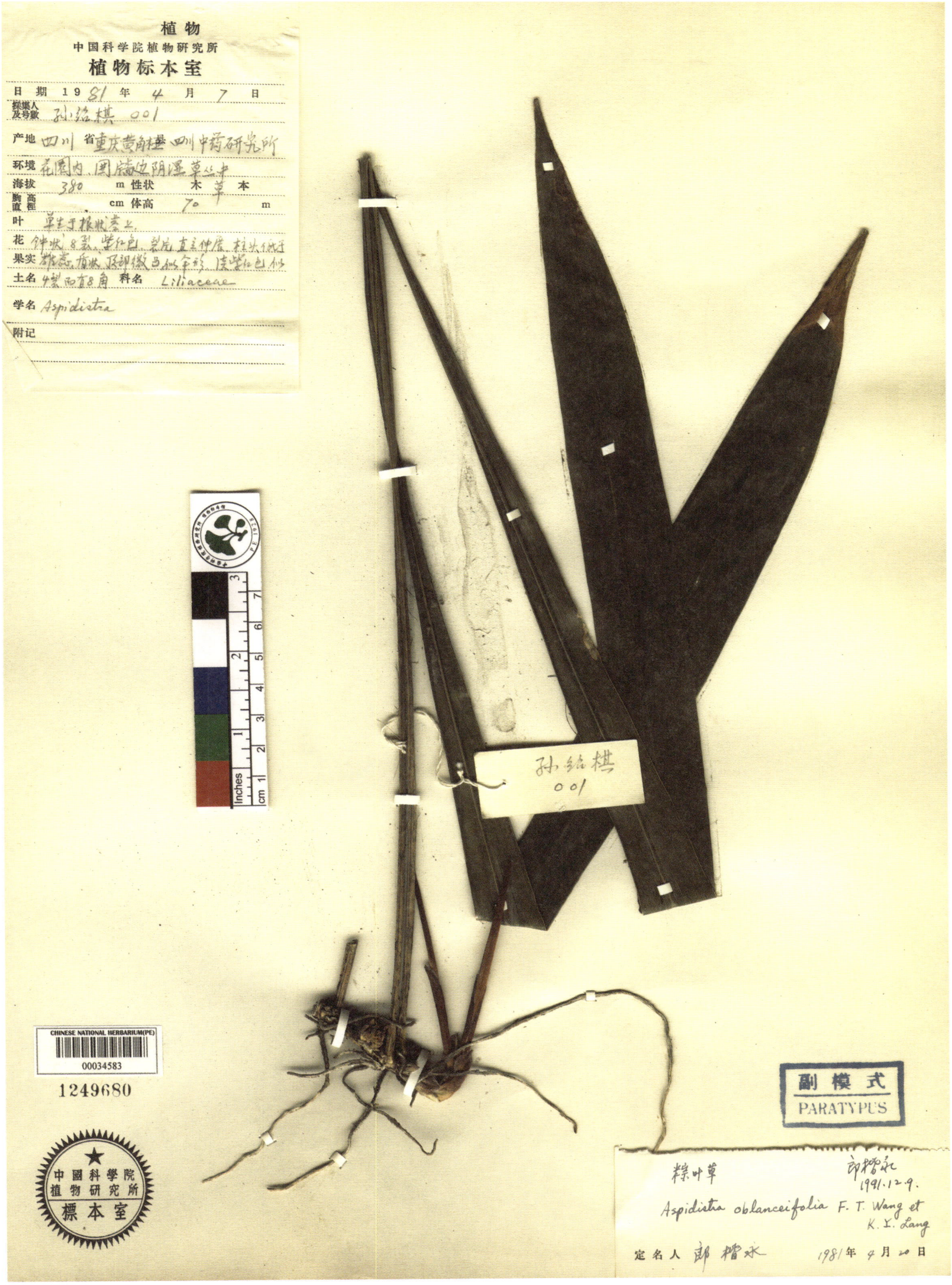

粽叶草 ***Aspidistra oblanceifolia*** F. T. Wang & K. Y. Lang in Acta Phytotax. Sin. 20(4): 487, f. 2: 1-4. 1982. **Paratype:** China. Chongqing, alt. 380 m, 1981-04-07, S. Q. Sun 1.

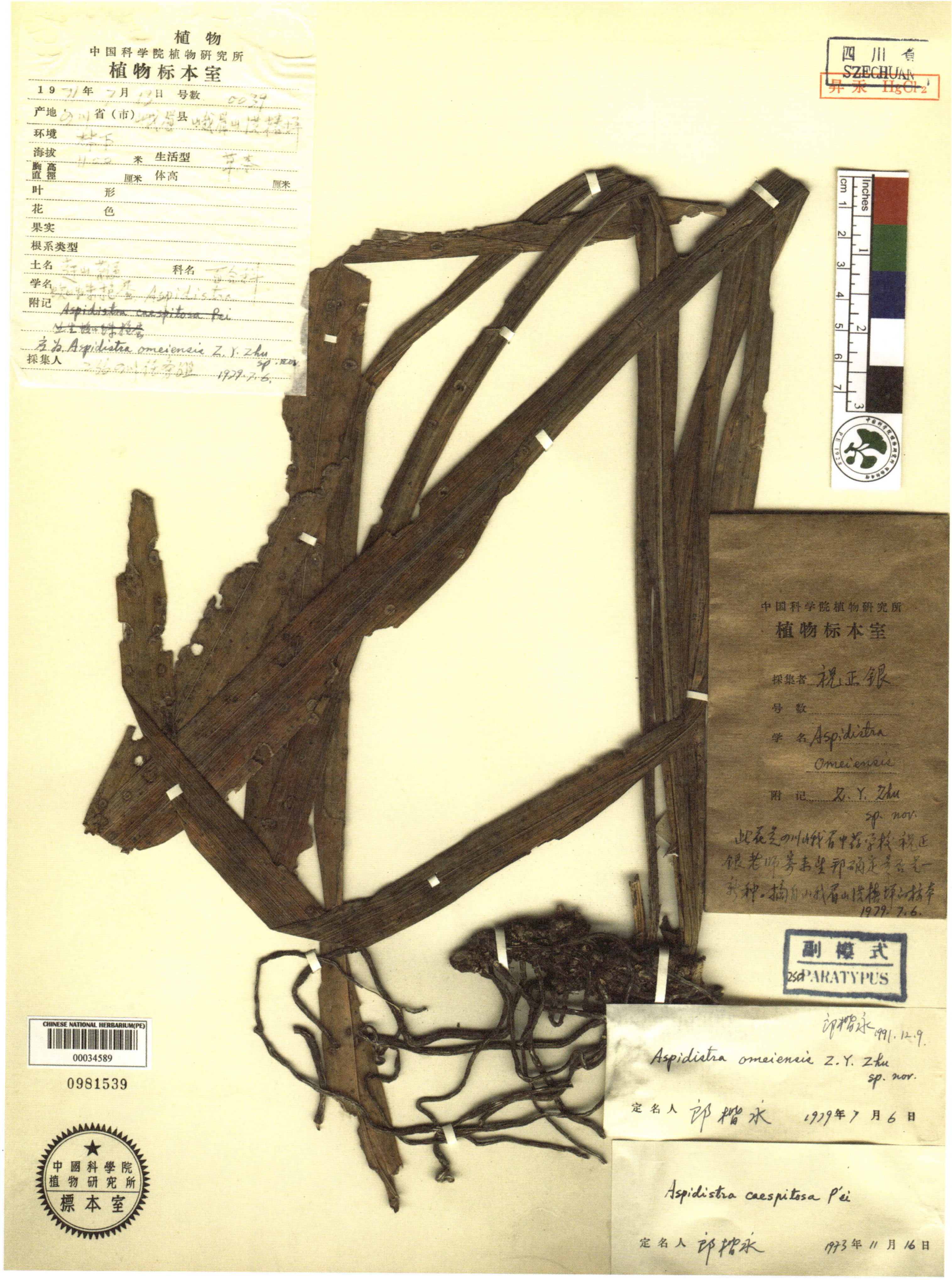

峨眉蜘蛛抱蛋 ***Aspidistra omeiensis*** Z. Y. Zhu & J. L. Zhang in Acta Phytotax. Sin. 19(3): 386, f. 1. 1981. **Isoparatype:** China. Sichuan: Emei, Emeishan, alt. 1100 m, 1971-07-13, Z.Y. Zhu & al. 39.

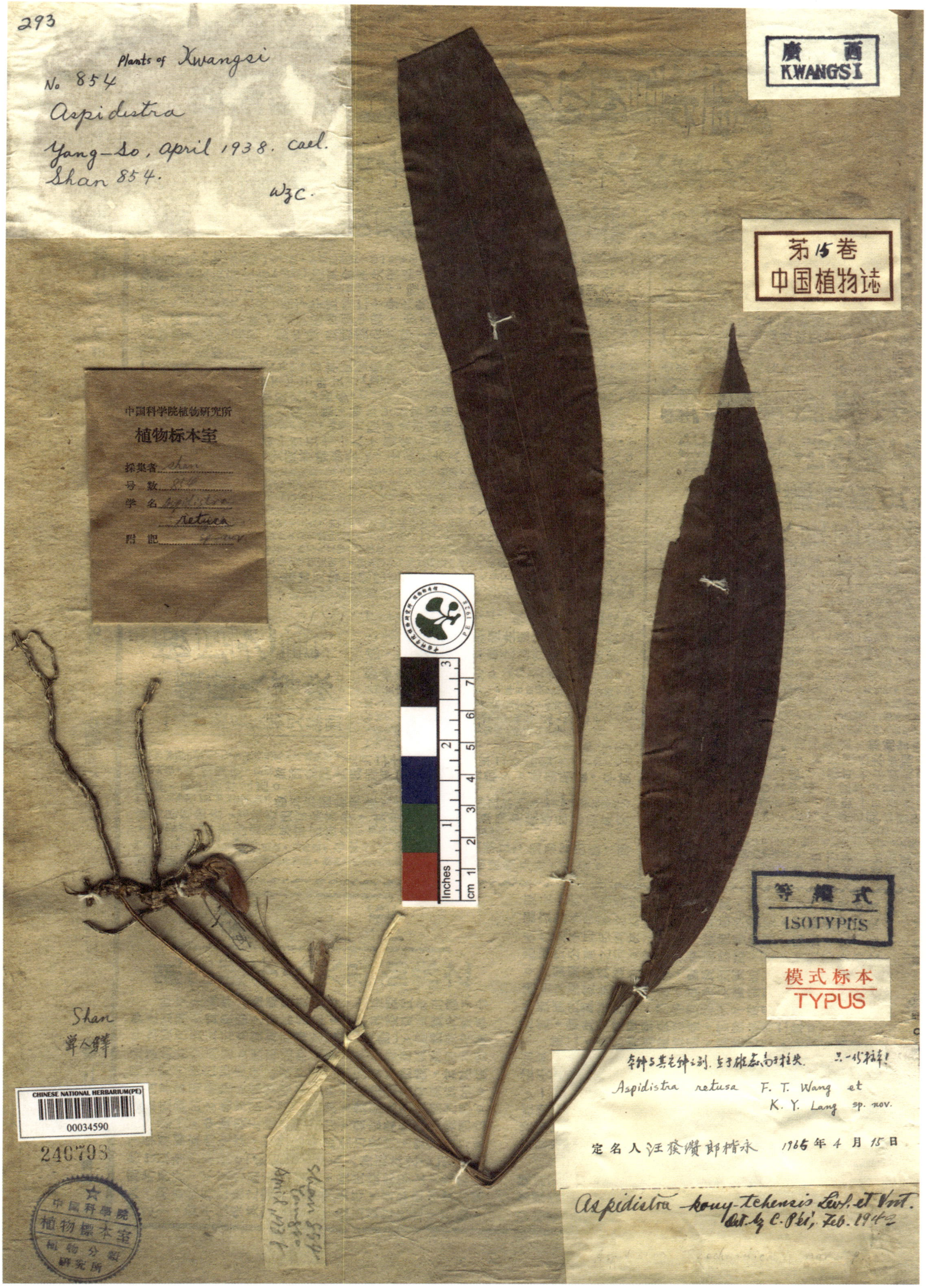

广西蜘蛛抱蛋 ***Aspidistra retusa*** K. Y. Lang & S. Z. Huang in Acta Phytotax. Sin. 19(3): 379, f. 1: 8-10. 1981. **Isotype:** China. Guangxi: Guilin, Yangshuo, 1938-04-??, R. H. Shan 854.

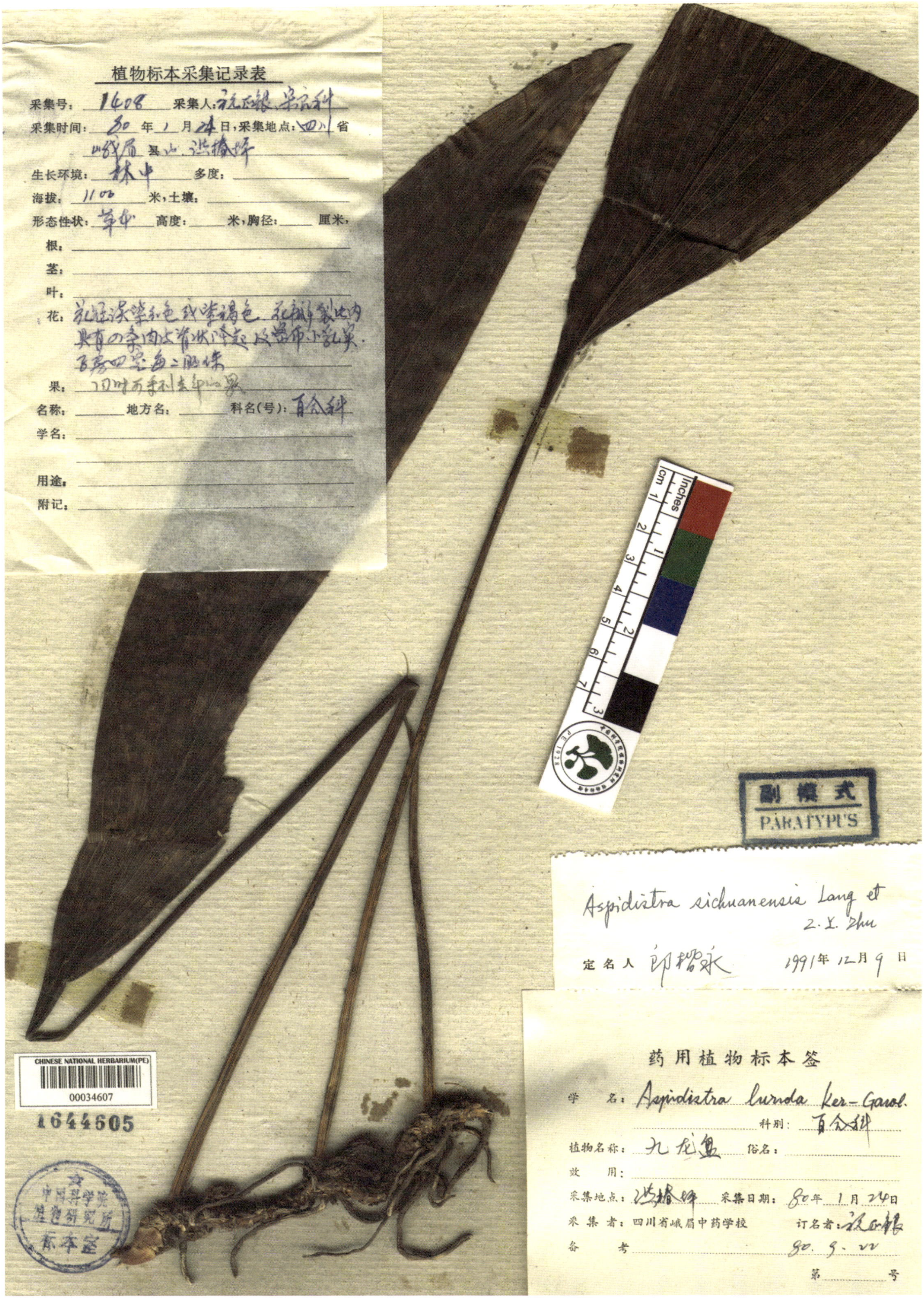

四川蜘蛛抱蛋 ***Aspidistra sichuanensis*** K. Y. Lang & Z. Y. Zhu in Acta Bot. Yunnan. 6(4): 387, f. 1: 5-6. 1984. **Paratype:** China. Sichuan: Emei, Emeishan, alt. 1100 m, 1980-01-24, Z. Y. Zhu 1408.

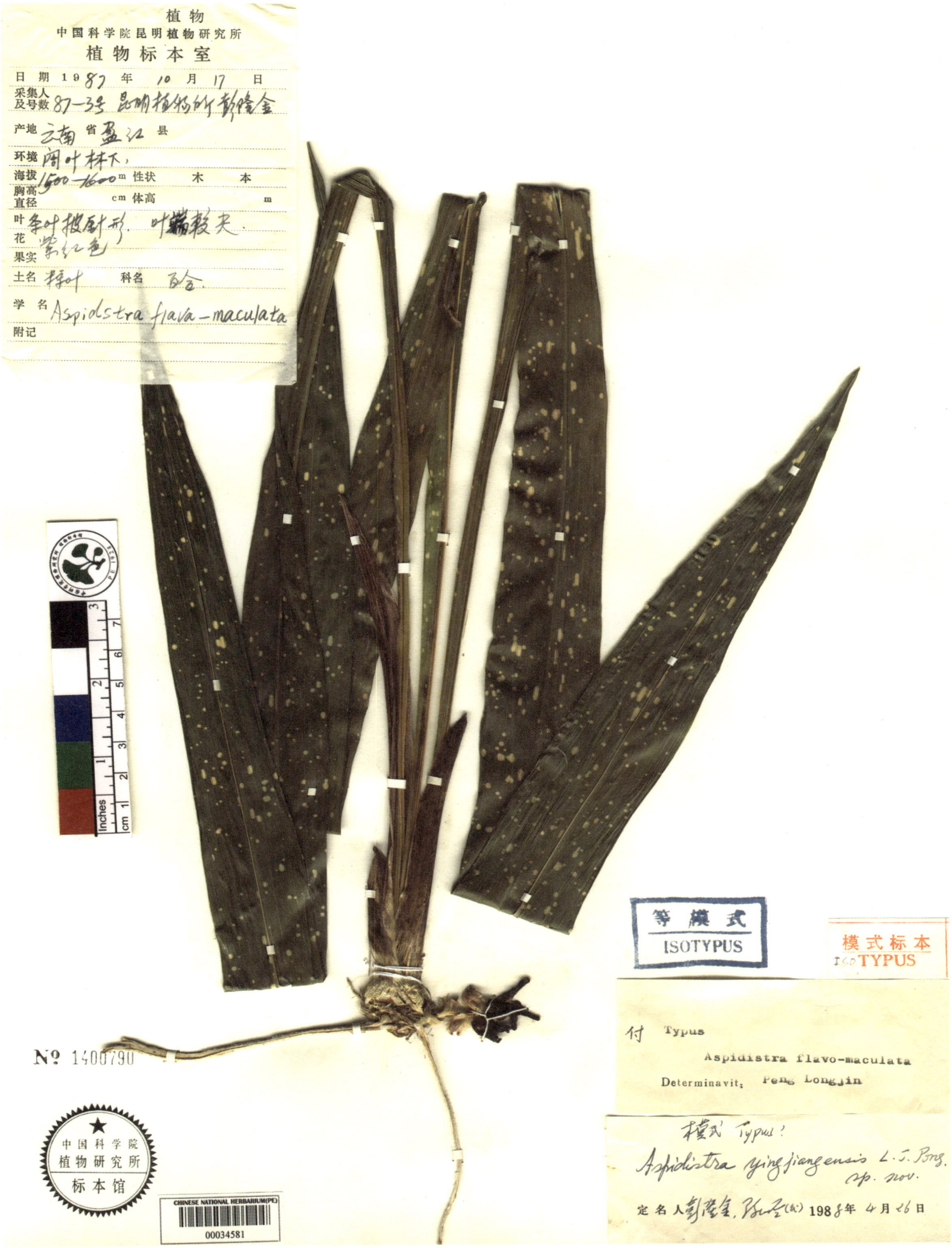

盈江蜘蛛抱蛋 ***Aspidistra yingjiangensis*** L. J. Peng in Acta Bot. Yunnan. 11(2): 173, f. 1. 1989. **Isotype:** China. Yunnan: Yingjiang, alt. 1500~1600 m, 1987-10-17, L. J. Peng 87-3.

粽粑叶 ***Aspidistra zongbayi*** K. Y. Lang & Z. Y. Zhu in Acta Phytotax. Sin. 20(4): 486, f. 1: 6-10. 1982. **Isotype:** China. Sichuan: Emei, Emeishan, alt. 1200 m, 1980-01-27, Z. Y. Zhu 1419.

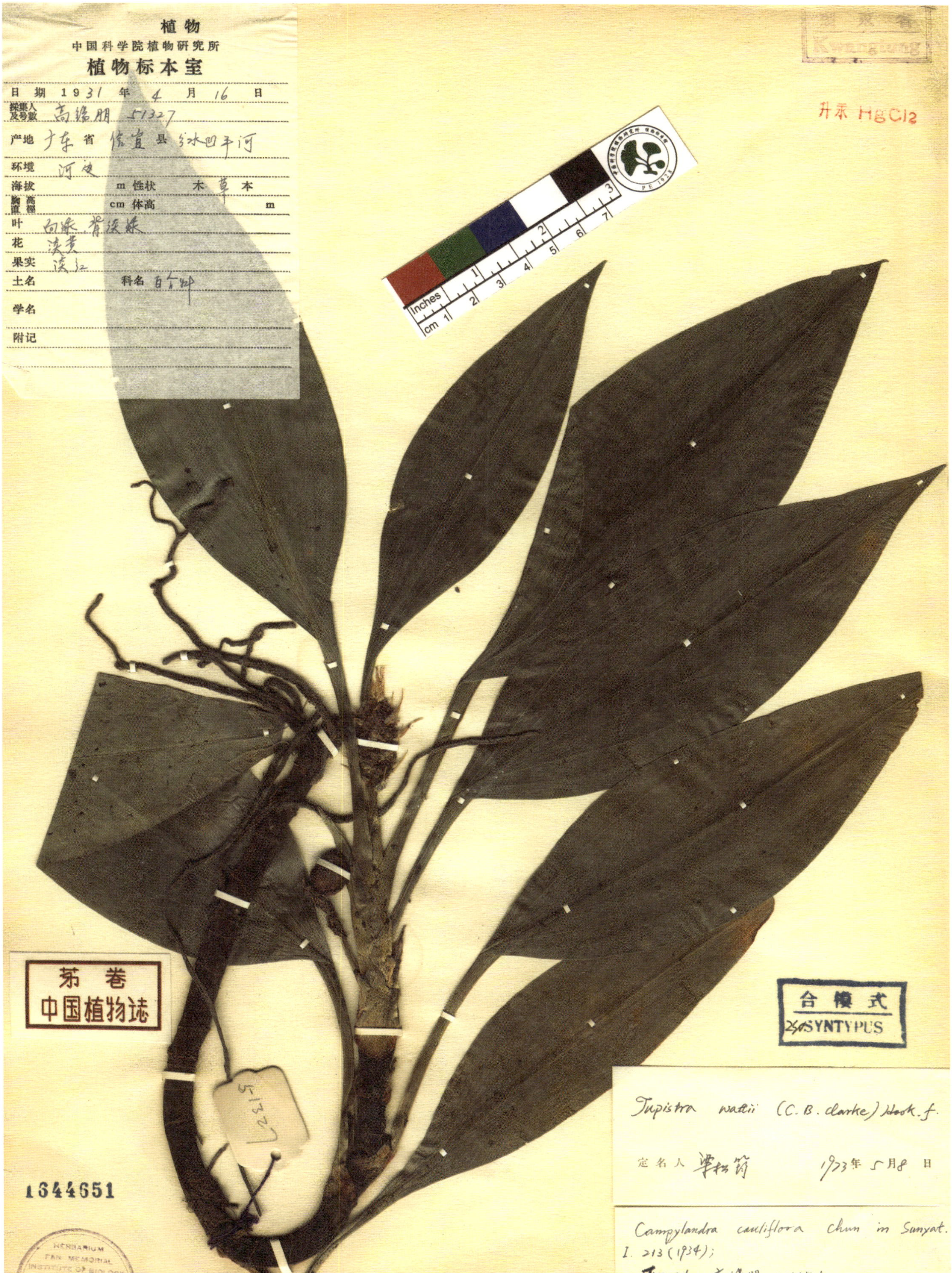

茎生花开口箭 ***Campylandra cauliflora*** Chun in Sunyatsenia 1(4): 213. 1934. **Isosyntype:** China. Guangdong: Xinyi, 1931-04-16, S. P. Ko 51327.

长苞开口箭 ***Campylandra longibracteata*** F. T. Wang & Tang in Contr. Inst. Bot. Nat. Acad. Peiping 6(1): 17. 1949. **Lectotype** (designated by Y. Lin & al. in Acta Bot. Bor.-Occ. Sin. 34: 412. 2014.): China. Yunnan: Malipo, alt. 1000 m, 1940-01-03, C. W. Wang 86154.

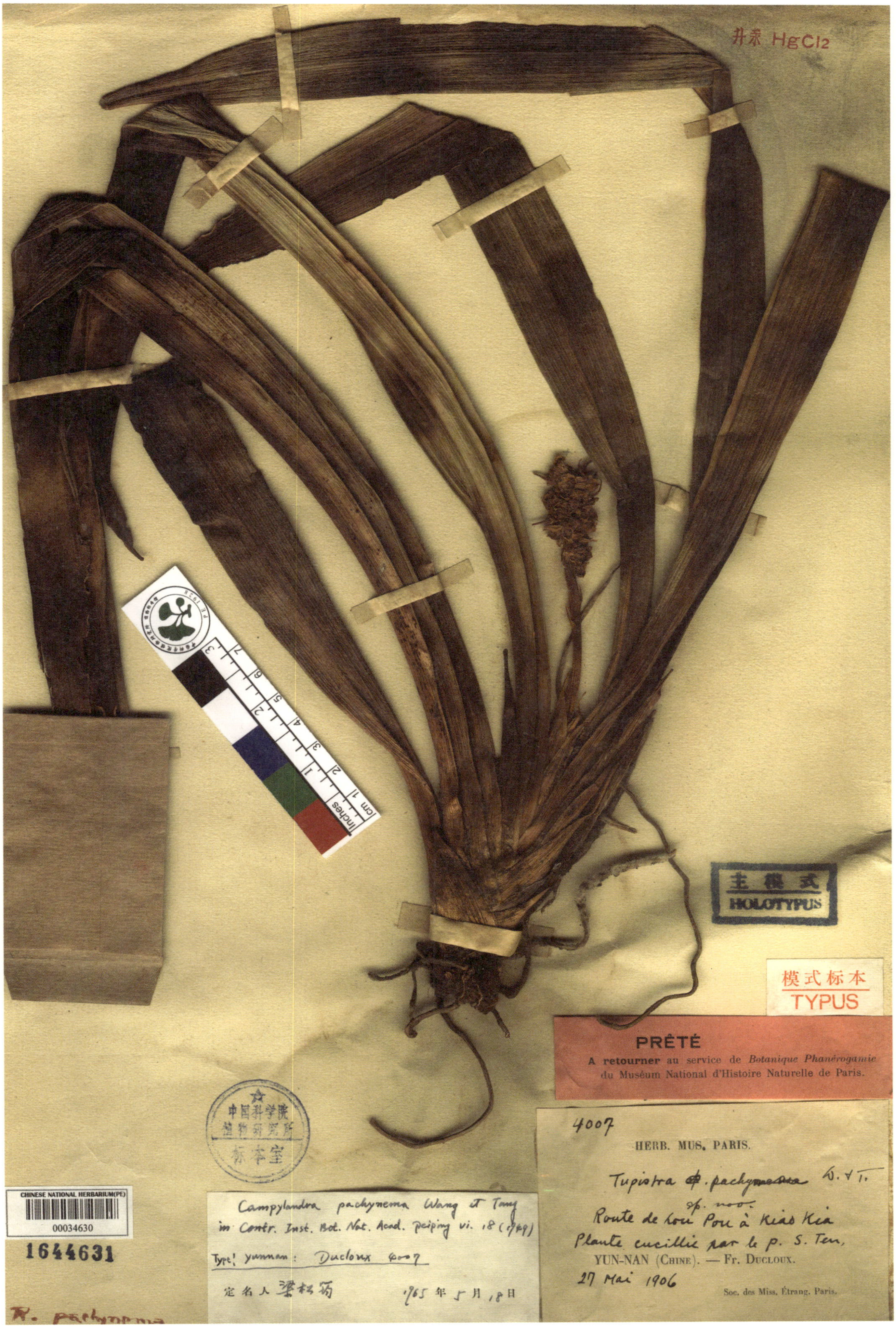

粗花丝开口箭 ***Campylandra pachynema*** F. T. Wang & Tang in Contr. Inst. Bot. Nat. Acad. Peiping 6(1): 18. 1949. **Holotype:** China. Yunnan: Loupou to Kiaokia, 1906-05-27, P. S. Ten 4007.

中国白丝草 ***Chionographis chinensis*** K. Krause in Notizbl. Bot. Gart. Mus. Berlin-Dahlem 10(98): 807. 1929. **Lectotype** (designated by Y. Lin & al. in Acta Bot. Bor.-Occ. Sin. 34: 413. 2014.): China. Guangxi: Yaoshan, 1929-??-??, S. S. Sin 8962.

广西白丝草 ***Chionographis merrilliana*** H. Hara in Journ. Japan. Bot. 43(9): 264, pl. 17: c, d. 1968. **Isoparatype:** China. Guangxi: Precise locality not known, 1936-05-08, C. Wang 39123.

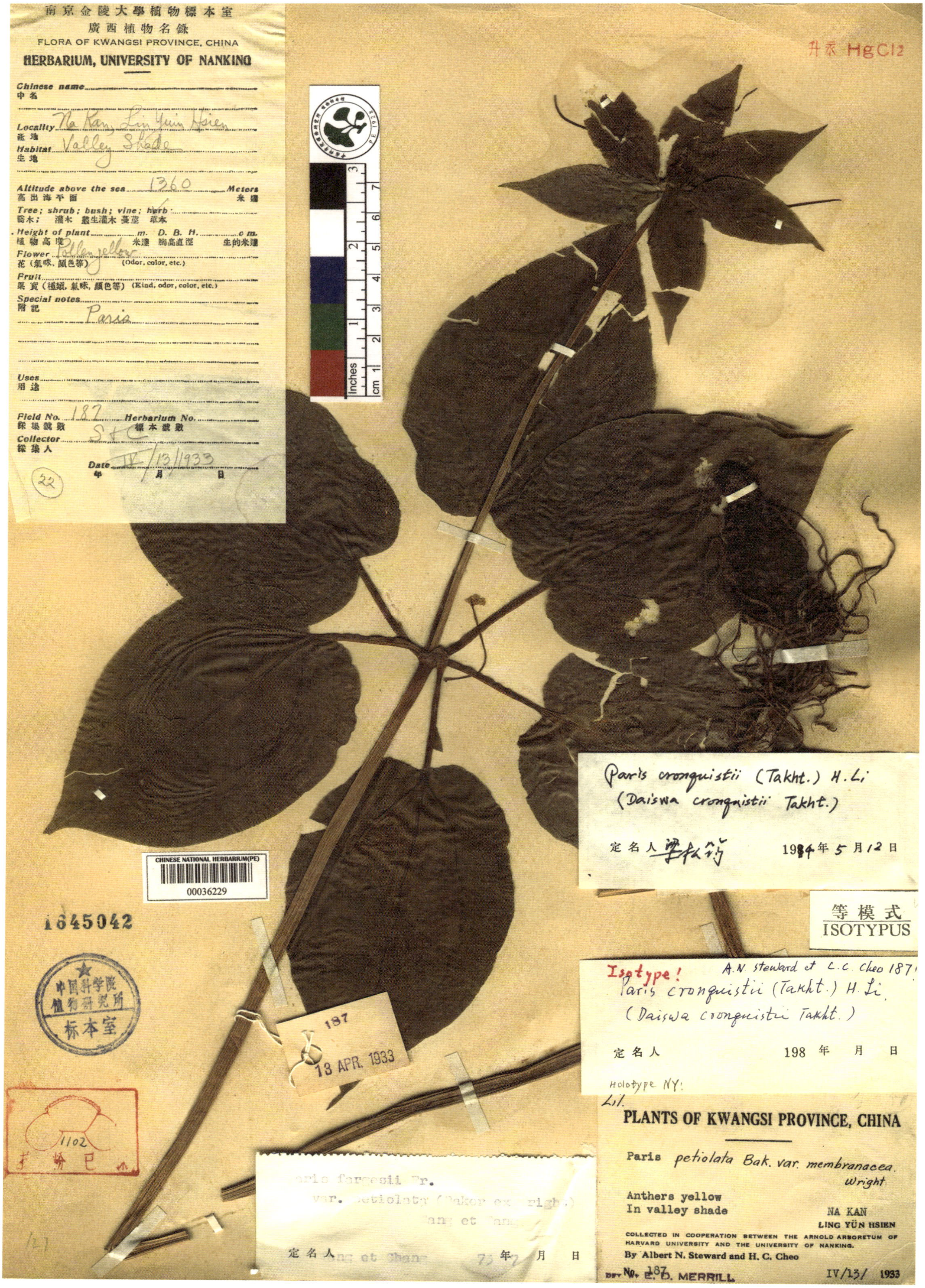

凌云重楼 ***Daiswa cronquistii*** Takht. in Brittonia 35(3): 262, f. 4. 1983. **Isotype:** China. Guangxi: Lingyun, alt. 1360 m, 1933-04-13, A. N. Steward & C. C. Cheo 187.

长柱重楼 ***Daiswa forrestii*** Takht. in Brittonia 35(3): 268, f. 5. 1983. **Isoparatype:** China. Yunnan: Weixi, alt. 2300 m, 1935-06-??, C. W. Wang 63641.

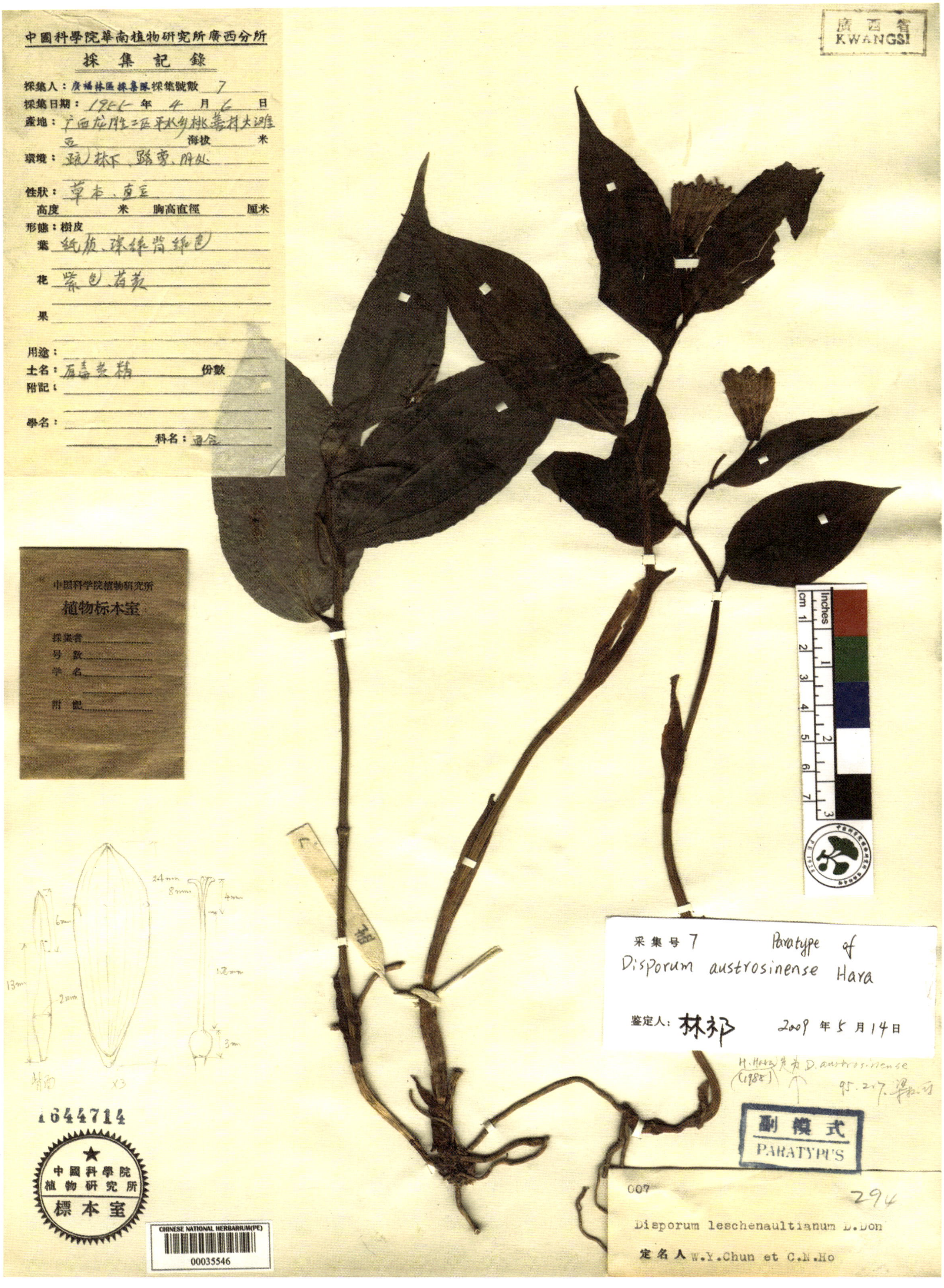

华南万寿竹 ***Disporum austrosinense*** H. Hara in Himalayan Plants 1: 192, f. 13e-h, 15. 1988. **Paratype:** China. Guangxi: Longsheng, 1955-04-06, Guangfu For. Exped. 7.

大花万寿竹 ***Disporum megalanthum*** F. T. Wang & Tang, Fl. Reip. Pop. Sin. 15: 45, 250, pl. 16: 1. 1978. **Holotype:** China. Hubei: Xingshan, alt. 1600 m, 1957-05-27,Y. Liu 498.

南川鹭鸶草 ***Diuranthera inarticulata*** F. T. Wang & K. Y. Lang, Fl. Reip. Pop. Sin. 14: 46, 282, pl. 11: 1-5. 1980. **Isotype:** China. Chongqing: Nanchuan, 1958-07-??, C. Y. Chu 1181.

安徽贝母 ***Fritillaria anhuiensis*** S. C. Chen & S. F. Yin in Acta Phytotax. Sin. 21(1): 100, f. 1: 1. 1983. **Isotype:** China. Anhui: Huoqiu, alt. 200~300 m, 1981-04-10, S. F. Yin & T. X. Ling 135.

白花贝母 ***Fritillaria anhuiensis*** S. C. Chen & S. F. Yin var. ***albiflora*** S. C. Chen & S. F. Yin in Acta Bot. Yunnan. 7(3): 307, f. 1, 5. 1985. **Isotype:** China. Anhui: Jinzhai, alt. 500 m, 1984-04-12, S. F. Yin 8434.

粗茎贝母 ***Fritillaria crassicaulis*** S. C. Chen in Acta Phytotax. Sin. 15(2): 36, f. 2: 1-5. 1977. **Holotype:** China. Yunnan: Zhongdian, alt. 3000 m, T. T. Yu 11319.

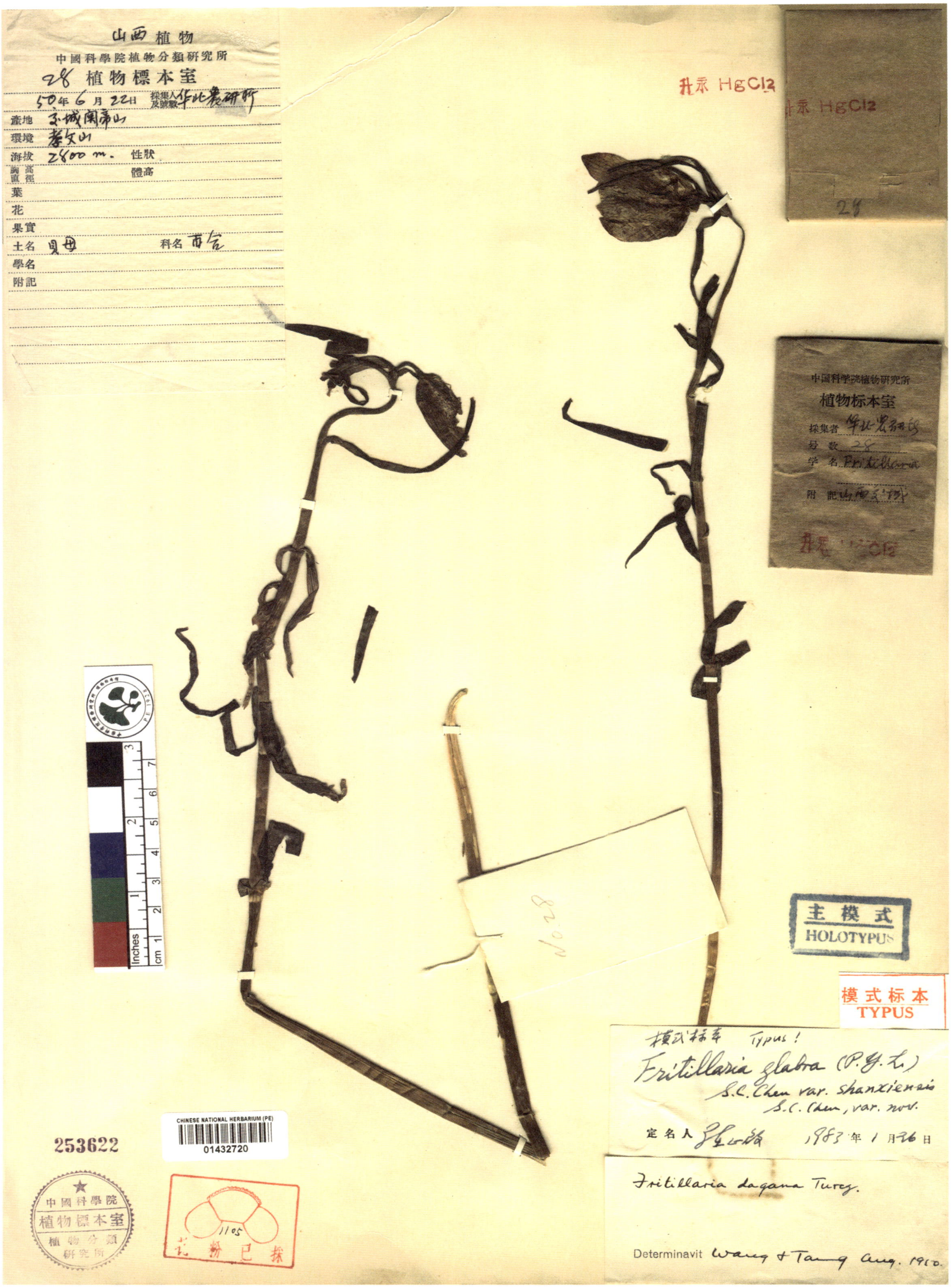

山西贝母 ***Fritillaria glabra*** (P. Y. Li) S. C. Chen var. ***shanxiensis*** S. C. Chen in Acta Bot. Yunnan. 5(4): 373. 1983. **Holotype:** China. Shanxi: Jiaochen, alt. 2800 m, 1950-06-22, Inst. Agric. Sinae Boreal. 28.

重瓣额敏贝母 ***Fritillaria meleagroides*** Patrin ex Schultes var. ***plena*** X. Z. Duan & X. J. Zheng in Acta Phytotax. Sin. 27(4): 309. 1989. **Holotype:** China. Xinjiang: Emin, alt. 910 m, 1985-05-06, X. Z. Duan 259.

铜陵黄花贝母 ***Fritillaria monantha*** Migo var. ***tonglingensis*** S. C. Chen & S. F. Yin in Acta Phytotax. Sin. 21(1): 101, f. 1: 2. 1983. **Isotype:** China. Anhui: Tongling, alt. 400 m, 1981-03-30, S. F. Yin & T. X. Ling 79.

宁国贝母 ***Fritillaria ningguoensis*** S. C. Chen & S. F. Yin in Acta Bot. Yunnan 7(3): 306, f. 1: 1-4. 1985. **Isotype:** China. Anhui: Ningguo, alt. 1000 m, 1984-04-12, M. M. Fang 840405.

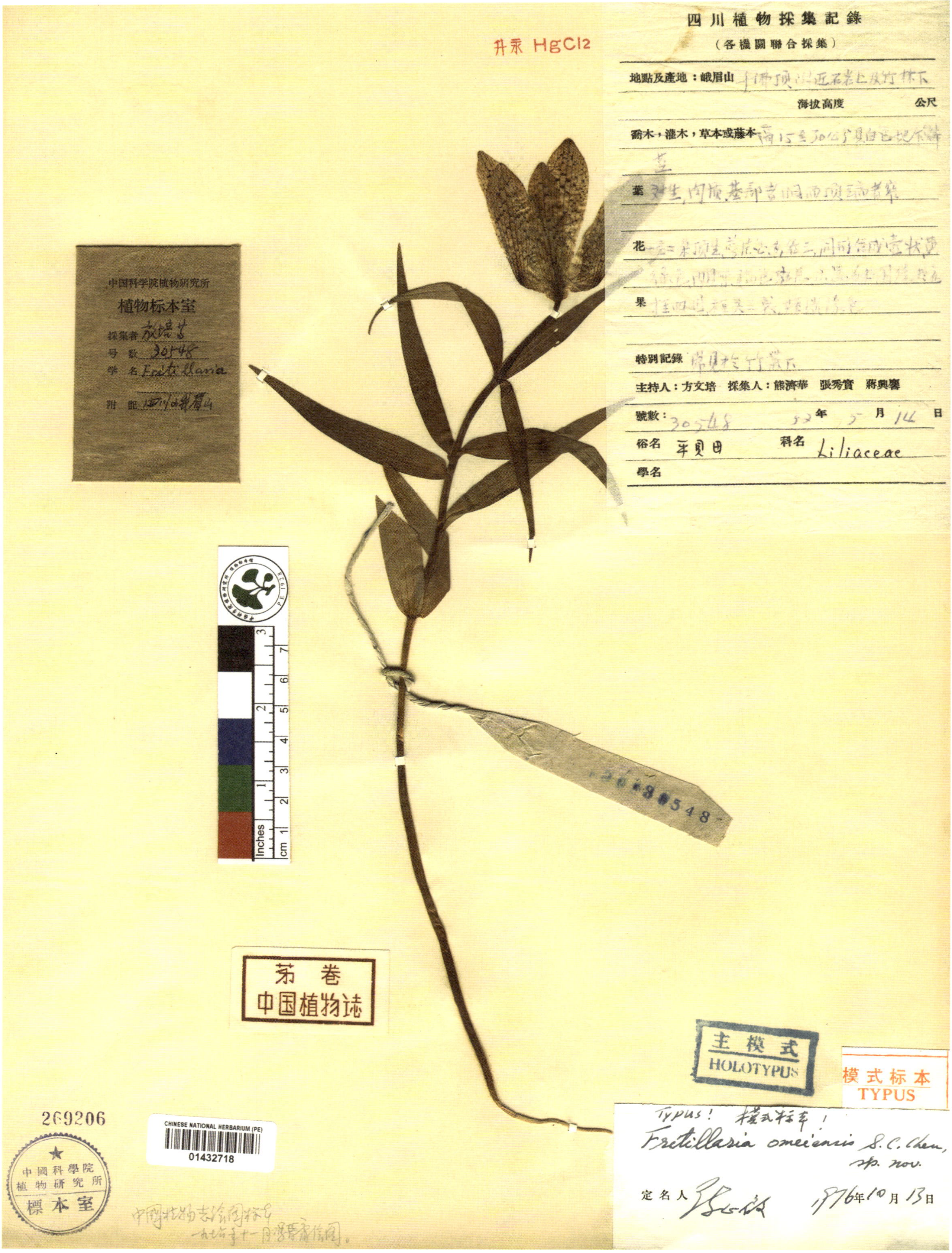

峨眉贝母 ***Fritillaria omeiensis*** S. C. Chen in Acta Phytotax. Sin. 15(2): 39, f. 3: 1. 1977. **Holotype:** China. Sichuan: Emei, Emeishan, 1952-05-14, J. H. Xiong & al. 30548.

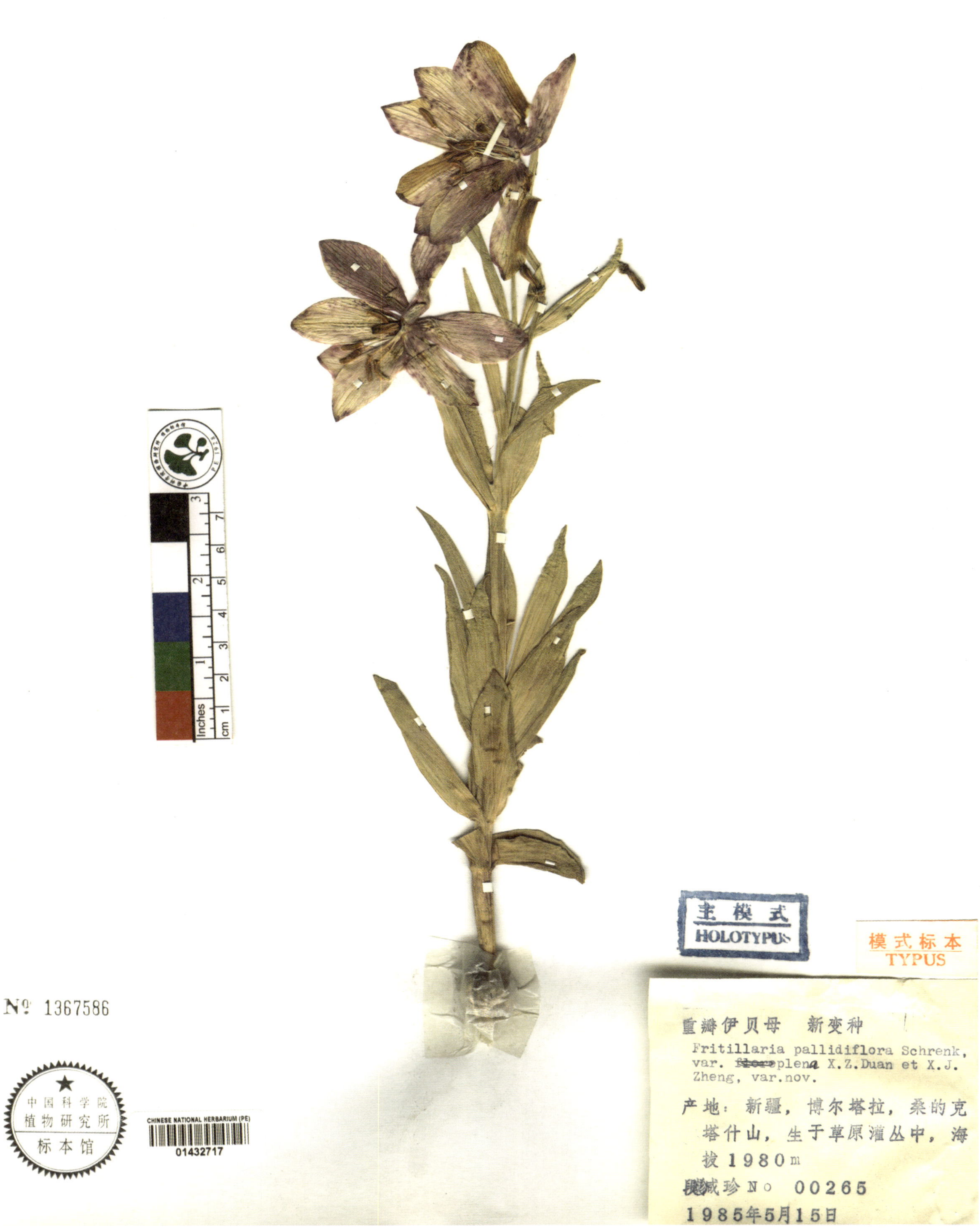

重瓣伊贝母 ***Fritillaria pallidiflora*** Schrenk var. ***plena*** X. Z. Duan & X. J. Zheng in Acta Phytotax. Sin. 27(4): 306. 1989. **Holotype:** China. Xinjiang: Bortala, alt. 1980 m, 1985-05-15, X. Z. Duan 265.

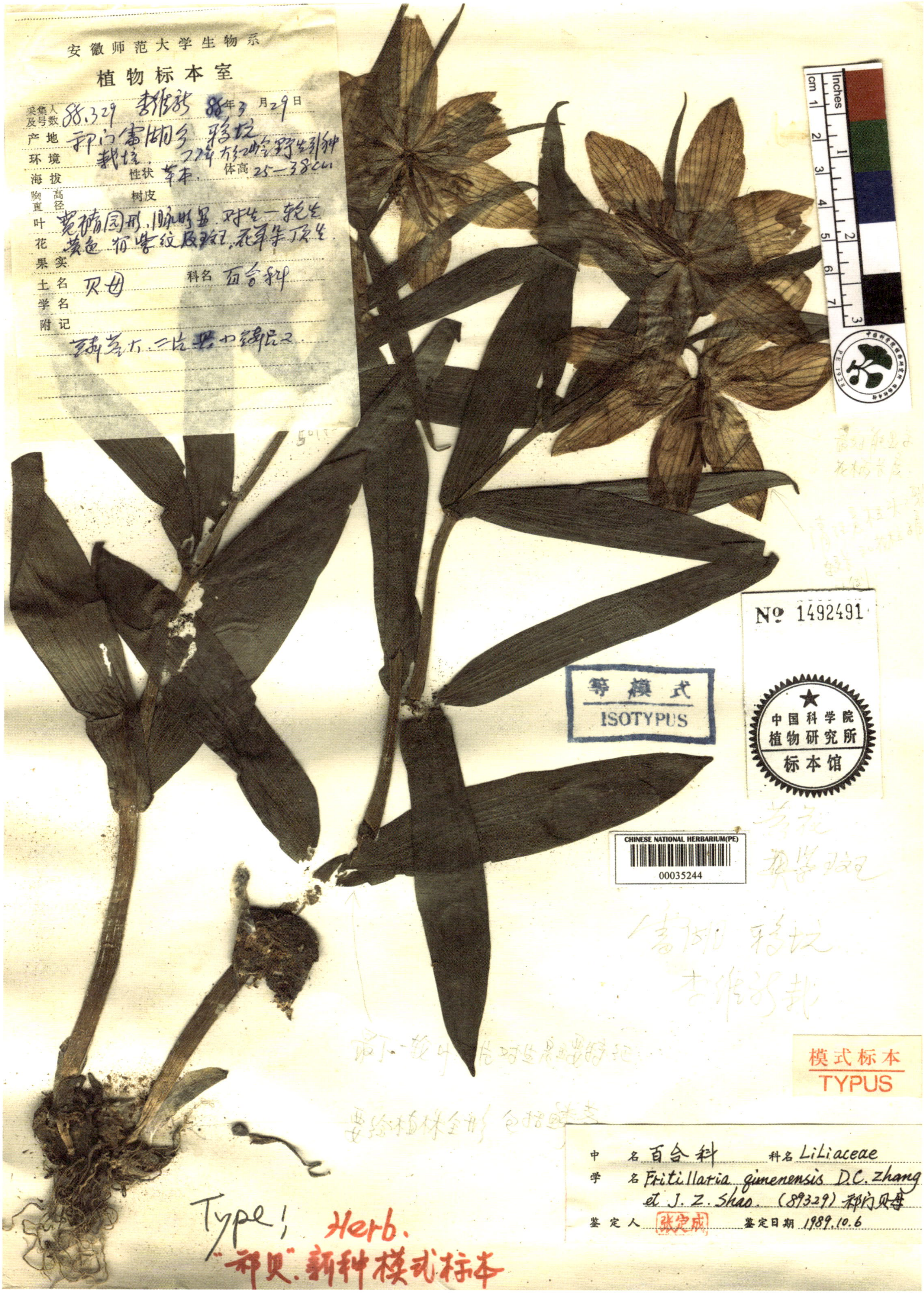

祁门贝母 ***Fritillaria qimenensis*** D. G. Zhang & J. Z. Shao in Acta Phytotax. Sin. 29(5): 474, f. 1. 1991. **Isotype:** China. Anhui: Qimen, alt. 850 m, 1988-03-29, W. X. Li 88329.

中华贝母 ***Fritillaria sinica*** S. C. Chen in Acta Phytotax. Sin. 19(4): 500, f. 1: 1-5. 1981. **Holotype:** China. Sichuan: Tianquan, alt. 3400 m, 1953-05-27, H. L. Tsiang & J. H. Xiong 34237.

白花托里贝母 ***Fritillaria tortifolia*** X. Z. Duan & X. J. Zheng var. ***albiflora*** X. Z. Duan & X. J. Zheng in Acta Phytotax. Sin. 27(4): 307. 1989. **Holotype:** China. Xinjiang: Yumin, alt. 1450 m, 1986-05-14, X. Z. Duan 289.

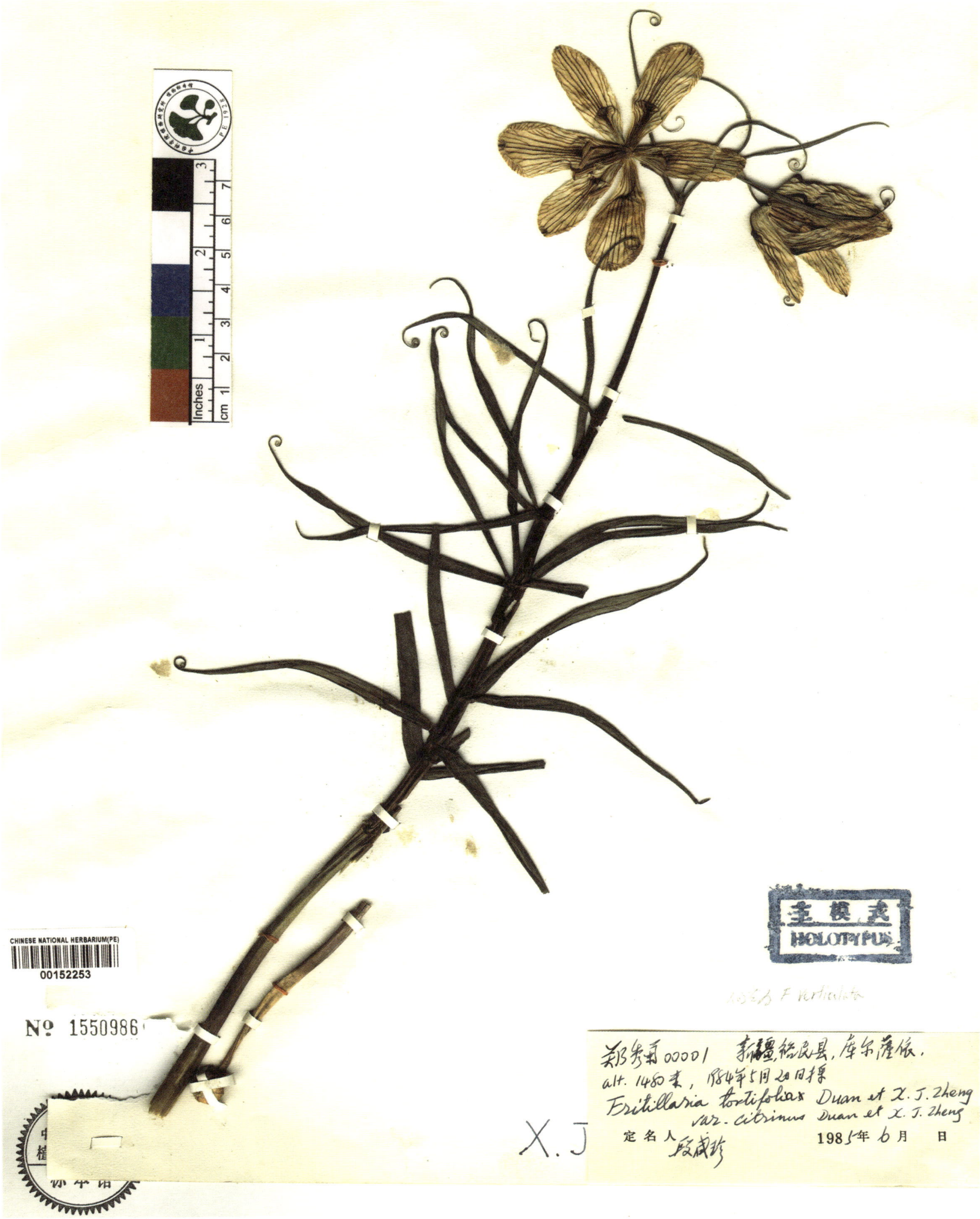

托里贝母黄花变种 ***Fritillaria tortifolia*** X. Z. Duan & X. J. Zheng var. ***citrina*** X. Z. Duan & X. J. Zheng in Acta Phytotax. Sin. 25(1): 60. 1987. **Holotype:** China. Xinjiang: Yumin, alt. 1480 m, 1984-05-20, X. J. Zheng 1.

小花托里贝母 ***Foritillaria tortifolia*** X. Z. Duan & X. J. Zheng var. ***parviflora*** X. Z. Duan & X. J. Zheng in Acta Phytotax. Sin. 27(4): 307. 1989. **Holotype:** China. Xinjiang: Yumin, alt. 1520 m, 1986-05-01, X. J. Zheng 10.

重瓣托里贝母 ***Fritillaria tortifolia*** X. Z. Duan & X. J. Zheng var. ***plena*** X. Z. Duan & X. J. Zheng in Acta Phytotax. Sin. 27(4): 307. 1989. **Holotype:** China. Xinjiang: Yumin, alt. 1480 m, 1985-05-05, X. J. Zheng 5.

暗紫贝母 ***Fritillaria unibracteata*** Hsiao & K. C. Hsia in Acta Phytotax. Sin. 15(2): 39, f. 4: 2-3. 1977. **Paratype:** China. Sichuan: Songpan, 1928-08-08, W. P. Fang 4034.

重瓣新疆贝母 ***Fritillaria walujewii*** Regel var. ***plena*** X. Z. Duan & X. J. Zheng in Acta Phytotax. Sin. 27(4): 306. 1989. **Holotype:** China. Xinjiang: Shawan, alt. 1640 m, 1986-05-07, X. Z. Duan 277.

Index to Scientific Names / 拉丁学名索引

A

Index to Chinese Names / 中名索引

5 画

6 画

9 画

10 画

Authors and Addresses / 作者及工作单位地址

BAN Qin, CHEN Shurong, DU Yufen, LIN Qi, MA Xintang, SUN Qian, TIAN Xiya, WANG Zhongtao, YANG Zhirong
China National Herbarium (PE), Institute of Botany, Chinese Academy of Sciences, Xiangshan, Beijing 100093, Beijing.

LI Qianyun, WU Tingting
Department of Biology, Taiyuan Normal University, Taiyuan 030031, Shanxi.

LIN Yun
Hunan Medication Vestibule School, Changsha 410208, Hunan.

SHI Qingchun
College of Life Sciences, Hebei Normal University, Shijiazhuang 050016, Hebei

班　勤　陈淑荣　杜玉芬　林　祁　马欣堂　孙　茜　田希娅　王忠涛　杨志荣
邮政编码：100093，北京市，香山，中国科学院植物研究所 国家植物标本馆（PE）

李倩云　吴婷婷
邮政编码：030031，山西省，太原市，太原师范学院 生物系

林　云
邮政编码：410208，湖南省，长沙市，湖南省医药技工学校

石青春
邮政编码：050016，河北省，石家庄市，河北师范大学 生命科学学院